Hotspots of Subterranean Biodiversity—2nd Volume

Hotspots of Subterranean Biodiversity—2nd Volume

Editors

Louis Deharveng
David C. Culver
Tanja Pipan

Basel • Beijing • Wuhan • Barcelona • Belgrade • Novi Sad • Cluj • Manchester

Editors

Louis Deharveng
Museum National d'Histoire
Naturelle
Paris
France

David C. Culver
American University
Washington, DC
USA

Tanja Pipan
Karst Research Institute ZRC
SAZU
Postojna
Slovenia

Editorial Office
MDPI
St. Alban-Anlage 66
4052 Basel, Switzerland

This is a reprint of articles from the Special Issue published online in the open access journal *Diversity* (ISSN 1424-2818) (available at: https://www.mdpi.com/journal/diversity/special_issues/Subterranean_Biodiversity_II).

For citation purposes, cite each article independently as indicated on the article page online and as indicated below:

Lastname, A.A.; Lastname, B.B. Article Title. *Journal Name* **Year**, *Volume Number*, Page Range.

ISBN 978-3-7258-1033-8 (Hbk)
ISBN 978-3-7258-1034-5 (PDF)
doi.org/10.3390/books978-3-7258-1034-5

Cover image courtesy of Didier Rigal

Contents

Editorial

Hotspots of Subterranean Biodiversity Redux

Louis Deharveng [1], Tanja Pipan [2], Anne Bedos [1] and David C. Culver [3,*]

[1] Institute de Systématique, Évolution, Biodiversité (ISYEB), UMR7205, CNRS, Muséum National D'Histoire Naturelle, Sorbonne Université, EPHE, 45 Rue Buffon, 75005 Paris, France
[2] Research Center of the Slovenian Academy of Sciences and Arts, Karst Research Institute, Novi Trg 2, SI-1000 Ljubljana, Slovenia
[3] Department of Environmental Science, American University, 4400 Massachusetts Ave. NW, Washington, DC 20016, USA
* Correspondence: dculver@american.edu; Tel.: +1-703-508-9509

Citation: Deharveng, L.; Pipan, T.; Bedos, A.; Culver, D.C. Hotspots of Subterranean Biodiversity Redux. *Diversity* **2022**, *14*, 794. https://doi.org/10.3390/d14100794

Received: 16 September 2022
Accepted: 22 September 2022
Published: 24 September 2022

Publisher's Note: MDPI stays neutral with regard to jurisdictional claims in published maps and institutional affiliations.

For most plants and animals the broad outlines of global species richness are well known, and often in some detail. The same cannot be said of subterranean communities in general and cave communities in particular. A set of challenges face any attempt to describe the biodiversity of cave communities. First, cave habitats are often difficult to access, and the fauna is numerically rare. Second, the ranges of cave species are typically small [1,2], requiring extensive sampling to capture most of the species richness in a region. Third, the cave fauna, which shows highly convergent morphology at the gross morphological level (i.e., the loss of eyes and pigment, and the elongation of appendages [3]), is often considered as a unit comprising many individual clades. Some individual clades are diverse enough for patterns to emerge [4], but this is rarely the case. Fourth, many species remain undescribed and undiscovered [5].

Despite these challenges, there has been speculation concerning the pattern of subterranean biodiversity since at least the 1960s. Early ideas about cave colonization relied heavily on a climate-forcing model, where species were driven into caves by the climate changes instigated by the Pleistocene glaciations [6,7]. Thus, the highest diversity would be expected in those areas where climate effects were strongest, and opportunities for colonization greatest, i.e., near glacial margins in northern temperate zones. With their discovery of obligate cave-dwelling invertebrates in the tropics, Mitchell [8] and Howarth [9] raised objections to both the Pleistocene model of cave colonization and the scarcity of cave-limited species in the tropics. It is curious that obligate cave-dwelling fish were known from tropical Africa and Mexico well before that [10,11], but had little impact on the early discussions of species diversity.

These early studies raised doubts about the temperate richness model, but actual data were slow in coming; contrastingly, the amount of data now accumulated is significant. In 2009, Gibert and Culver [12] reported more than 3500 known species of stygobionts (aquatic species limited to subterranean habitats, typically caves). Of these, 2000 were from Europe—surely an overrepresentation due to much more thorough collection and description. While more data are accumulating outside Europe and the United States [13], the preponderance of data remains European (and, to a lesser extent, from the U.S.A). For example, Zagmajster et al. [14] report on the diversity patterns of 1570 stygobiotic European species, and Christman et al. [15] report on the diversity patterns of 750 stygobiotic and troglobiotic U.S. species based on nearly 10,000 records. For Europe, the continental pattern for both stygobionts and troglobionts is one of a ridge of high diversity at 45° N [14,16], along the spine of the Pyrenees and through the Dinaric karst of Italy, Slovenia, Croatia, Serbia, Bosnia and Hercegovina, and Montenegro. The pattern in the U.S. is less resolved, but there is an overall hotspot of troglobiotic (terrestrial) species richness in northeast Alabama and adjoining parts of Tennessee [16,17]. Explanations for these regional patterns are complex and highly scale dependent [16,18].

Information on the patterns of global subterranean biodiversity derive almost exclusively from data on individual "hotspot" caves [19]. Outside of Europe and the United States, no extensive regional collection of the scores of caves needed for species accumulation curves [20] has been conducted. Well-sampled tropical caves are listed by Deharveng and Bedos [13]. Given the high β diversity compared to α-diversity [19], inferences from the data about a few very diverse caves would seem unlikely to be informative. However, in their classic paper on subterranean biodiversity, Gibert and Deharveng [1] point out that regional diversity is a good predictor of local diversity and vice versa. This was buttressed by later findings that species accumulation curves infrequently crossed, and thus, the regional pattern could be captured by a relatively small number of samples [16,21]. Culver and Sket [22] took this idea to its logical extreme and considered only caves and karst wells with the highest species richness, originally finding 20 sites with 20 or more species specialized for subterranean life. While the coverage of large numbers of caves in a relatively small area was (and is) limited to Europe and the United States, they reasoned that at least a few outstanding caves, which were extensively sampled, were known from most large karst areas. The publication of their paper stimulated both the further sampling of high diversity caves throughout the world, and the compilation of species lists for high diversity caves. Due to this activity, the criterion for the inclusion of a cave in the hotspot list in 2021 was set at 25, with a total of 23 sites meeting this criterion [19,22]. Of these 23 caves, six had over 50 stygobionts and troglobionts [23–27].

An explanation of the observed hotspot patterns, especially for terrestrial hotspots, has proved elusive. For stygobionts, Culver and Sket [22] note that they tended to be from the Dinaric karst, were chemoautotrophic or anchialine, or were connected to ground (phreatic) water. Culver et al. [18] also noted that the stygobiotic hotspots tended to be in chemoautotrophic or in the Dinaric karst. Terrestrial hotspots are more dispersed and occur in both temperate and tropical zones. Part of the problem may be data limitations. There are regions of high species richness but without any one cave being rich, as is apparently the case for a cave region in Brazil [28]. Second, there may be unsampled cave hotspots, ones which will clarify the causes of the patterns.

In a Special Issue of *Diversity*, which was published in 2021 [29], 13 of 23 hotspot caves and their fauna were described in detail, and one other was described elsewhere [30]. Due to the positive response to calls for papers about hotspot caves and the fact that nine hotspot caves require updating, we have added a second Special Issue devoted to this topic. As in the first Special Issue, there will be a species list for each hotspot cave—information that is often unpublished for these caves. This is especially important given the controversy around the ecological status of cave species. Deharveng and Bedos [31] pointed out that considerable confusion exists in the literature about the terms troglobiont—which should be used only for species not found outside of caves, irrespective of their morphology—and troglomorph [32,33], i.e., species with reduced eyes and pigment and elongated appendages. The two are not identical, a problem that arises not only with guanobionts, but also with all species without troglomorphic features that are found in caves [34]. The Special Issue will also provide a physical setting for the caves and groundwater habitats, including their hydrogeological and environmental context, their use by humans, the nature of the karst in which they are situated, and the knowledge on nearby cave biodiversity.

Author Contributions: Conceptualization, L.D.; methodology, L.D., T.P., A.B. and D.C.C.; data curation, L.D., A.B.; writing—original draft preparation, D.C.C.; writing—review and editing, L.D., T.P., A.B.; visualization—L.D., T.P., A.B. and D.C.C. All authors have read and agreed to the published version of the manuscript.

Funding: This research received no external funding.

Conflicts of Interest: The authors declare no conflict of interest.

References

1. Gibert, J.; Deharveng, L. Subterranean ecosystems: A truncated functional diversity. *Bioscience* **2002**, *52*, 473–481. [CrossRef]

2. Lamoreaux, J. Stygobites are more wide ranging than troglobites. *J. Cave Karst Stud.* **2004**, *66*, 18–19.

3. Culver, D.C.; Pipan, T. *The Biology of Caves and Other Subterranean Habitats*, 2nd ed.; Oxford University Press: Oxford, UK, 2019.

4. Faille, A.; Ribera, I.; Deharveng, L.; Bourdeau, C.; Garnery, L.; Quéinnec, E.; Deuve, T. A molecular phylogeny shows the single origin of the Pyrenean subterranean Trechini ground beetles (Coleoptera: Carabidae). *Molec. Phylog. Evol.* **2010**, *54*, 97–106. [CrossRef]

5. Culver, D.C.; Trontelj, P.; Zagmajster, M.; Pipan, T. Paving the way for standardized and comparable subterranean biodiversity studies. *Subterr. Biol.* **2012**, *10*, 43–50. [CrossRef]

6. Jeannel, R. *Les Fossiles Vivants des Cavernes*; Gallimard: Paris, France, 1944.

7. Barr, T.C. Cave ecology and the evolution of troglobites. *Evol. Biol.* **1968**, *2*, 35–105.

8. Mitchell, R.W. A comparison of temperate and tropical cave communities. *Southwest. Nat.* **1969**, *14*, 73–88. [CrossRef]

9. Howarth, F.G. Cavernicoles in lava tubes on the island of Hawaii. *Science* **1972**, *175*, 325–326. [CrossRef]

10. Berti, R.; Messana, G. Subterranean fishes of Africa. In *Biology of Subterranean Fishes*; Trajano, E., Bichuette, M.E., Kapoor, B.G., Eds.; Science Publishers: Enfield, NH, USA, 2010; pp. 357–396.

11. Ornelas-Garcia, C.P.; Pedraza-Lara, C. Phylogeny and evolutionary history of *Astyanax mexicanus*. In *Biology and Evolution of the Mexican Cavefish*; Keene, A.C., Yoshizawa, M., McGaugh, S.E., Eds.; Elsevier: Amsterdam, The Netherlands, 2016; pp. 77–92.

12. Gibert, J.; Culver, D.C. Assessing and conserving groundwater biodiversity: An introduction. *Freshw. Biol.* **2009**, *54*, 639–648. [CrossRef]

13. Deharveng, L.; Bedos, A. Biodiversity in the tropics. In *Encyclopedia of Caves*, 3rd ed.; White, W.B., Culver, D.C., Pipan, T., Eds.; Academic Press: Waltham, MA, USA, 2019; pp. 146–162.

14. Zagmajster, M.; Eme, D.; Fišer, C.; Galassi, D.; Marmonier, P.; Stoch, F.; Cornu, J.; Malard, F. Geographic variation in range size and beta diversity of groundwater crustaceans: Insights from habitats with low thermal seasonality. *Glob. Ecol. Biogeogr.* **2014**, *23*, 1135–1145. [CrossRef]

15. Christman, M.C.; Doctor, D.H.; Niemiller, M.L.; Weary, D.J.; Young, J.A.; Zigler, K.S.; Culver, D.C. Predicting the occurrence of cave-inhabiting fauna based on features of the Earth surface environment. *PLoS ONE* **2016**, *11*, e0160408. [CrossRef]

16. Culver, D.C.; Deharveng, L.; Bedos, A.; Lewis, J.J.; Madden, M.; Reddell, J.R.; Sket, B.; Trontelj, P.; White, D. The mid-latitude biodiversity ridge in terrestrial cave fauna. *Ecography* **2006**, *29*, 120–128. [CrossRef]

17. Niemiller, M.L.; Zigler, K.S. Patterns of cave biodiversity and endemism in the Appalachians and interior plateau of Tennessee. *PLoS ONE* **2013**, *8*, e64177. [CrossRef]

18. Zagmajster, M.; Malard, F.; Eme, D.; Culver, D.C. Subterranean biodiversity patterns from global to regional scales. In *Cave Ecology*; Moldovan, O.T., Kováč, L., Halse, S., Eds.; Springer: Cham, Switzerland, 2018; pp. 195–228.

19. Culver, D.C.; Deharveng, L.; Pipan, T.; Bedos, A. An overview of subterranean biodiversity hotspots. *Diversity* **2021**, *13*, 487. [CrossRef]

20. Malard, F.; Boutin, C.; Camacho, A.I.; Ferreira, D.; Michel, G.; Sket, B.; Stoch, F. Diversity patterns of stygobiotic crustaceans across multiple spatial scales in western Europe. *Freshw. Biol.* **2009**, *54*, 756–776. [CrossRef]

21. Dole-Olivier, M.J.; Castellarini, F.; Coineau, N.; Galassi, D.M.P.; Martin, P.; Mori, N.; Valdecasas, A.; Gibert, J. Towards an optimal sampling strategy to assess groundwater biodiversity: Comparison across six regions of Europe. *Freshw. Biol.* **2009**, *54*, 777–796. [CrossRef]

22. Culver, D.C.; Sket, B. Hotspots of subterranean biodiversity in caves and wells. *J. Cave Karst Stud.* **2000**, *62*, 11–17.

23. Polak, S.; Pipan, T. The subterranean fauna of Križna jama, Slovenia. *Diversity* **2021**, *13*, 210. [CrossRef]

24. Camacho, A.I.; Puch, C. Ojo Guareña, a hotspot of subterranean biodiversity in Spain. *Diversity* **2021**, *13*, 199. [CrossRef]

25. Zagmajster, M.; Polak, S.; Fišer, C. Postojna Planina Cave System in Slovenia, a hotspot of subterranean biodiversity and a cradle of speleobiology. *Diversity* **2021**, *13*, 271. [CrossRef]

26. Iliffe, T.M.; Calderón-Gutiérrez, F. Bermuda's Walsingham Caves: A global hotspot for anchialine stygobionts. *Diversity* **2021**, *13*, 352. [CrossRef]

27. Hutchins, B.T.; Gibson, J.R.; Diaz, P.H.; Schwartz, B.F. Stygobiont diversity in the San Marcos Artesian Well and Edwards Aquifer groundwater ecosystem, Texas, USA. *Diversity* **2021**, *13*, 234. [CrossRef]

28. Trajano, E.; Gallão, J.E.; Bichuette, M.E. Spots of high diversity of troglobites in Brazil: The challenge of measuring subterranean diversity. *Biodivers. Conserv.* **2016**, *25*, 1805–1828. [CrossRef]

29. Pipan, T.; Culver, D.C.; Deharveng, L. (Eds.) *Hotspots of Subterranean Biodiversity*; MDPI: Basel, Switzerland, 2021.

30. Souza-Silva, M.; Cerqueira, R.; Pellegrini, T.G.; Ferreira, R.L. Habitat selection of cave-restricted fauna in a new hotspot of subterranean biodiversity in the tropica. *Biodiv. Cons* **2021**, *30*, 4223–4250. [CrossRef]

31. Deharveng, L.; Bedos, A. Diversity of terrestrial invertebrates in subterranean habitats. In *Cave Ecology*; Moldovan, O.T., Kováč, L., Halse, S., Eds.; Springer: Cham, Switzerland, 2018; pp. 107–172.

32. Christiansen, K.A. Proposition pour la classification des animaux cavernicoles. *Spelunca Mem.* **1962**, *2*, 76–78.

33. Christiansen, K.A. Morphological adaptation. In *Encyclopedia of Caves*, 2nd ed.; White, W.B., Culver, D.C., Eds.; Academic Press: Waltham, MA, USA, 2012; pp. 517–528.

34. Culver, D.C.; Pipan, T. *Shallow Subterranean Habitats. Ecology, Evolution, and Conservation*; Oxford University Press: Oxford, UK, 2014.

diversity

Article

The Crystal-Wonder Cave System: A New Hotspot of Subterranean Biodiversity in the Southern Cumberland Plateau of South-Central Tennessee, USA

Matthew L. Niemiller [1,*], Kirk S. Zigler [2], Amata Hinkle [1], Charles D. R. Stephen [3], Brendan Cramphorn [1], Jared Higgs [1], Nathaniel Mann [4], Brian T. Miller [5], K. Denise Kendall Niemiller [1], Kelly Smallwood [6] and Jason Hardy [6]

[1] Department of Biological Sciences, The University of Alabama in Huntsville, 301 Sparkman Dr NW, Huntsville, AL 35899, USA; amatahinkle@gmail.com (A.H.); btc0011@uah.edu (B.C.); jph0029@uah.edu (J.H.); dk0047@uah.edu (K.D.K.N.)
[2] Department of Biology, University of the South, Sewanee, TN 37383, USA; kzigler@sewanee.edu
[3] Department of Biology, Missouri State University, Springfield, MO 65897, USA; cdr.stephen@gmail.com
[4] Spencer Mountain Grotto, Spencer, TN 38585, USA; thany@blomand.net
[5] Department of Biology, Middle Tennessee State University, Murfreesboro, TN 37132, USA; brian.miller@mtsu.edu
[6] Tennessee Cave Survey, 770 Three Forks Road, South Pittsburg, TN 37380, USA; rowland7840@bellsouth.net (K.S.); wmjhardy@yahoo.com (J.H.)
* Correspondence: matthew.niemiller@uah.edu or cavemander17@gmail.com

Abstract: The Crystal-Wonder Cave System developed in the Western Escarpment of the southern Cumberland Plateau in the Interior Low Plateau karst region of south-central Tennessee, USA is a global hotspot of cave-limited biodiversity. We combined historical literature, museum accessions, and database occurrences with new observations from bio-inventory efforts conducted between 2005 and 2022 to compile an updated list of troglobiotic and stygobiotic biodiversity for the Crystal-Wonder Cave System. The list of cave-limited fauna includes 31 species (23 troglobionts and 8 stygobionts) with 28 and 18 species documented from the Crystal and Wonder caves, respectively, which represents five phyla, ten classes, nineteen orders, and twenty-six families (six arachnids, three springtails, two diplurans, three millipedes, six insects, three terrestrial snails, one flatworm, five crustaceans, and two vertebrates, respectively). The Crystal-Wonder Cave System is the type locality for six species—*Anillinus longiceps, Pseudanophthalmus humeralis, P. intermedius, Ptomaphagus hatchi, Tolus appalachius,* and *Chitrella archeri.* The carabid beetle *Anillinus longiceps* is endemic to the Crystal-Wonder Cave System. Sixteen species are of conservation concern, including twelve taxa with NatureServe conservation ranks of G1–G3. The exceptional diversity of the Crystal-Wonder Cave System has been associated with several factors, including a high dispersal potential of cave fauna associated with expansive karst exposures along the Western Escarpment of the southern Cumberland Plateau, a high surface productivity, and a favorable climate throughout the Pleistocene.

Keywords: checklist; karst; species richness; stygobiont; troglobiont

Citation: Niemiller, M.L.; Zigler, K.S.; Hinkle, A.; Stephen, C.D.R.; Cramphorn, B.; Higgs, J.; Mann, N.; Miller, B.T.; Niemiller, K.D.K.; Smallwood, K.; et al. The Crystal-Wonder Cave System: A New Hotspot of Subterranean Biodiversity in the Southern Cumberland Plateau of South-Central Tennessee, USA. *Diversity* **2023**, *15*, 801. https://doi.org/10.3390/d15070801

Academic Editors: Tanja Pipan, David C. Culver and Louis Deharveng

Received: 26 May 2023
Revised: 12 June 2023
Accepted: 19 June 2023
Published: 23 June 2023

1. Introduction

The escarpments of the southern Cumberland Plateau of Tennessee, Alabama, and Georgia (TAG), USA in the Interior Low Plateau karst region contain thousands of caves that harbor an exceptional subterranean biodiversity [1–5]. The high density of cave systems and exposed karst in this region is a contributing factor for high levels of species richness and endemism in the southern Cumberland Plateau [1,2,6,7], which has been recognized as a global regional hotspot of subterranean biodiversity [1,2,7]. Several cave systems support a significant biodiversity at the local scale, including 19 caves with >12 cave-limited (i.e., permanent inhabitants of subterranean habitats) species documented [2,8], Niemiller and Zigler, unpublished data]. In particular, two cave systems in the southern Cumberland Plateau region have been recognized as hotspots for cave-limited biodiversity [sensu 9]: Shelta Cave in Madison County, Alabama with 24 species [9], and Fern Cave in Jackson County, Alabama with 27 species, respectively [10]. Here, we add a third cave system—the Crystal-Wonder Cave System in Grundy County, Tennessee—that is the most speciose cave system with respect to the cave-limited fauna in the southern Cumberland Plateau.

1.1. Description of the Crystal-Wonder Cave System

The Crystal-Wonder Cave System in southwestern Grundy County, Tennessee, USA is located 6.4 km (4 mi) north of the former resort town Monteagle, at the base of Cedar Ridge on the north side of Layne Cove on the Western Escarpment of the Cumberland Plateau in the Upper Elk River watershed (Figure 1). The stream in the Wonder Cave (Tennessee Cave Survey no. TGD30), called the Mystic River, emerges from the main spring entrance, and then flows west for ca. 91 m (300 ft) before sinking into the insurgence entrance of the Wonder Natural Bridge Cave (TGD179) for ca. 46 m (150 ft), then resurging at the spring entrance of the Wonder Natural Bridge Cave, flowing on the surface for ca. 31 m (100 ft) before sinking into the insurgence entrance of the Crystal Cave (TGD10). In total, the Crystal-Wonder Cave System has ca. 5.74 km (18,828 ft) of passage, which includes 4474 m (14,678 ft) in the Wonder Cave, 46 m (150 ft) in the Wonder Natural Bridge Cave, and 1219 m (4000 ft) in the Crystal Cave, respectively.

The Wonder Cave is developed from the Mississippian-aged Monteagle Limestone. From the spring entrance (E1 black in Figure 1), the cave extends east for 2134 m (7000 ft). The Mystic River flows on the north side of passage; the passage averages 7.6 m (25 ft) wide and 2.4 m (8 ft) high for the first 335 m (1100 ft), respectively, and then enlarges to 13.7 m (45 ft) high with an extensive, well-decorated upper level with numerous dripstone formations. The upper level is 7.6 m (25 ft) above the Mystic River and ranges 4.5–6 m (15–20 ft) high, 9–24 m (30–80 ft) wide, and 107 m (350 ft) long, respectively with two main chambers—Statuary and Cathedral halls. At ca. 1402 m (4600 ft) from the entrance, the Mystic River flows from beneath a breakdown pile, marking the entry climb into the Pyramid Room, which is a dome chamber 37 m (120 ft) high and 61 m (200 ft) in diameter, respectively. Two passages continue from this room. One heads north for ca. 1500 m (5000 ft) in a dry upper-level avenue 6 m (20 ft) high that ranges 11–37 m (35–120 ft) wide. The Mystic River flows eastward from the Pyramid room for approximately 762 m (2500 ft) in passage that ranges 4–20 m (15–20 ft) high and 3–6 m (10–20 ft) wide, respectively, with several low air spaces before terminating the beneath breakdown.

The Crystal Cave represents a remnant of a lower portion of the Wonder Cave, which was also developed from the Monteagle Limestone. The Historic Section of the cave extends from the historic entrance (E1 white in Figure 1) for 229 m (750 ft) to the main stream passage. Steps were constructed at the entrance, and several areas in the historic section were excavated to provide access for visitors. Much of this section of the cave averages 1.5 m (5 ft) high and 3 m (10 ft) wide, respectively, with a small stream before lowering to a 0.6 m (2 ft) high crawl in water before connecting to the main stream passage. The main stream passage contains the same water that exits the Wonder Cave. From the junction of the historic section and main stream passage, the passage continues upstream following the Mystic River but gradually lowers in height, becoming a crawl in the stream and also over the breakdown toward the insurgence entrance (E4 white in Figure 1). Downstream of the junction, the main stream passage enlarges to 18 m (60 ft) wide and 2.4 m (8 ft) high, respectively, to a breakdown on the right leading to the quarry entrance (E3 white in Figure 1). The passage continues downstream past the quarry entrance to another junction, with the right passage continuing for ca. 91 m (300 ft) and the left continuing as a water-filled tube to the spring entrance (E2 white in Figure 1). The spring entrance is 8 m (26 ft) wide and 2.7 m (9 ft) high, respectively.

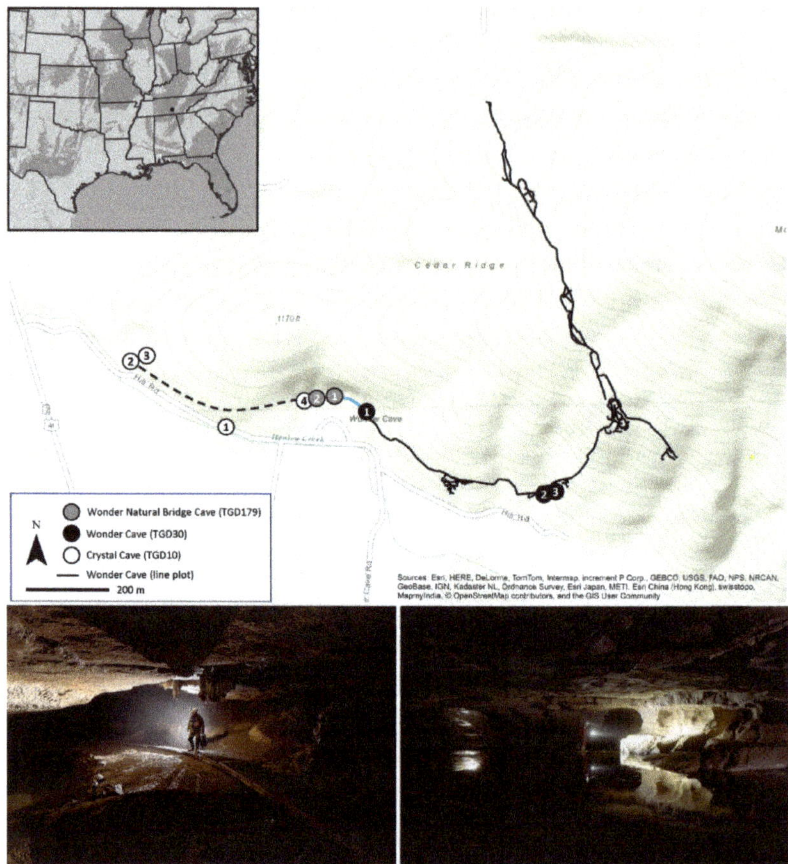

Figure 1. Line map and location of the Crystal-Wonder Cave System (top) in southwestern Grundy County, Tennessee, USA. Entrances to the three caves of the Crystal-Wonder Cave System are numbered and colored as follows: the Crystal Cave (TGD10) in white, the Wonder Cave (TGD30) in black, and the Wonder Natural Bridge Cave (TGD179) in gray, respectively. A map of the Crystal Cave is not available. The stream in the Wonder Cave (i.e., Mystic River) flows out of the Wonder Cave (E1 black) and on the surface briefly (blue line) before sinking into the insurgence entrance (E1 gray) of the Wonder Natural Bridge Cave, and then quickly emerging again (E2 gray) to briefly flow on the surface before sinking into the insurgence entrance (E4 white) of the Crystal Cave (TGD10). The stream flows through the Crystal Cave (general flow path shown as a dashed line) and finally resurges at the spring entrance (E2 white) of the Crystal Cave. The bottom photographs show the Mystic River along the former commercial cave tour in the Wonder Cave (left) and in the Crystal Cave near the connection of the Mystic River passage and Historical Section (right). Photographs by Amata Hinkle.

1.2. Discovery, Exploration, and History of the Crystal-Wonder Cave System

Descriptions of the Wonder Cave and details on the history of exploration, dye tracing, and commercial operations can be found in [11–15]. Native Americans likely camped at the spring based on the presence of artifacts nearby, while the Wonder Cave itself was named by the Vanderbilt University students Robert Nelson, Melville Anderson, and Will Fitzgerald when they "discovered" it in 1897 [11]. It was then quickly purchased by the local businessman R.M. Payne in 1898 as a potential attraction for his hotel—the Monteagle Hotel. Payne enlarged the entrance to allow for boat tours and constructed a new rock wall and walking trails. In addition, a large stream pump was installed to pump spring water from the cave to the hotel. The first commercial operation offered flat-bottom boat tours via the Mystic River which runs through the cave and lantern tours along the walking trails. After more than a decade of boat tours, Payne created a new tunnel entrance into the cave in 1914 before passing away in 1917. A significant moonshine alcohol operation was located a few hundred meters inside the cave during the Prohibition era [16]. In 1929, Jefferson Jones Raulston, Payne's grandson-in-law and granddaughter Mary, took over the operations of the Wonder Cave, constructing a stone entry façade leading into the cave, and a log house near the entrance that contained a gift shop and ticket office. This cave was Tennessee's most popular commercial cave at that time, being ideally located just 0.4 km (0.25 mi) off of U.S. Route 41. After the construction of Interstate 24 in the 1960s, visitation plummeted by nearly 90 percent, as it was an 11 km (7 mi) detour from the interstate. The cave remained in the family until 1987 when it was sold to the Born family, who operated it commercially for a short time during the summer months until 2000. The Wonder Cave was one of the best-known historical commercial caves in the United States, with over two million people visiting between 1897 and 2000 and attracting more than 40,000 visitors a year at its peak [17].

The first partial map of the Wonder Cave was published by the State Geological Survey of Tennessee in 1912. Thomas Barr published a complete map of the known passage of the Wonder Cave in his book titled the "Caves of Tennessee" [12] based on a trip with Bill Cuddington, Roy Davis, Frank Raulston, and others in December 1953. In 2014, Jason Hardy began a resurvey of the Wonder Cave, and his map was archived with the Tennessee Cave Survey in 2020. Despite its proximity to and hydrological connections with the Wonder Cave [15], a detailed map of the Crystal Cave remains to be published.

1.3. Biological Investigation of the Crystal-Wonder Cave System

The Crystal-Wonder Cave System has a long history of biological interest dating to the 1930s, when the first biological collections were conducted by J.M. Valentine, C. Mohr, K. Dearolf, and colleagues. Valentine collected the specimens of several troglobiotic species that would be described over the next three decades (e.g., [18–22]). Dearolf [23] summarized the observations and collections of several species from 75 caves visited in the United States, including records from both the Crystal and Wonder caves. Dearolf reported six cave-limited species and several additional records for non-troglobiotic fauna. Additional historical collections were made by T.C. Barr, L.R. Hubricht, L.P. Woods, J.G. Armstrong, and R.A. Brandon in the 1950s and 1960s that were included in later taxonomic studies (e.g., Woods and Inger [24], Malcolm and Chamberlin [25], Hubricht [26], Peck [27,28], and Shear [29]). Lewis [30] reported a list of cave-limited fauna from the Crystal Cave, which included 16 species. Finally, Niemiller and Zigler [2] identified the Crystal-Wonder Cave System as the most biodiverse cave system with respect to the cave-limited fauna in Tennessee. These authors reported 24 cave-limited species but did not include a faunal list for the cave system.

Herein we present an updated comprehensive list of the terrestrial and aquatic cave obligate fauna (i.e., troglobionts and stygobionts, respectively) for the Crystal-Wonder Cave System based on a comprehensive search of the scientific literature and museum records, and from recent biosurveys of the cave system conducted by the authors and colleagues between 2005 and 2022. In addition to the species list, we include a comprehensive bibliography on the cave obligate fauna of the Crystal-Wonder Cave System, discuss factors potentially driving its biodiversity, and comment on the conservation status of the exceptional biodiversity of this North American and global hotspot of subterranean biodiversity.

2. Materials and Methods

2.1. Ecological Classification of the Troglobionts and Stygobionts

We recognized troglobionts (i.e., troglobites; terrestrial species) and stygobionts (i.e., stygobites; aquatic species) as species that are permanent inhabitants of subterranean habitats [31–33], and that are unable to complete their life cycles outside of such habitats [34]. Troglobionts and stygobionts have source populations in subterranean habitats but may have sink populations in surface habitats from a metapopulation perspective [32]. While the use of morphology alone cannot definitively classify species ecologically [33], we used the presence of traits often observed in troglobiotic and stygobiotic fauna (i.e., troglomorphisms), such as reduced eyes and pigmentation, or hypertrophy of nonvisual sensory structures, but not found in surface-dwelling relatives, as evidence for isolation in subterranean habitats.

2.2. Cave Biosurveys

We conducted faunal bio-inventories of the Crystal-Wonder Cave System on six occasions since 2005 in association with other projects: the Crystal Cave on 12 May 2005, 21 November 2006, 21 June 2015, 22 April 2022, and 22 August 2022; and the Wonder Cave on 20 June 2015 and 22 April 2022, respectively. Bio-inventories consisted of time-constrained visual encounter surveys for cave life in terrestrial and aquatic habitats, including entrance areas and twilight zones, walls and ceilings, mud banks, rimstone pools, streams, and talus slopes. We searched underneath rocks and cover and within detritus and other organic debris, as well as searching through stream cobble. Each survey was conducted by two to seven researchers.

We identified common vertebrate and invertebrate species in the field. For many vertebrates, we field-identified taxa by direct observations without capture, or through taxonomically reliable indirect observations, such as the visual identification of mammal scat or footprints left in the mud. For many invertebrates, we collected specimens and identified them in the laboratory using the available taxonomic keys and literature. We outsourced identification to experts for taxa with which we had an insufficient taxonomic knowledge when possible. We took voucher photographs of the invertebrate and vertebrate taxa when possible.

2.3. Literature and Museum Searches

We conducted a search of the scientific literature to compile an updated list of troglobiont and stygobiont species for the Crystal-Wonder Cave System. Scientific literature sources included journal articles, book chapters, books, conference proceedings, theses and dissertations, and government reports. Searches of these scientific literature sources included keyword queries on the ISI Web of Science, Google Scholar, and zoological records. Keywords used in these searches included "Crystal Cave", "Wonder Cave", "Grundy County", "Tennessee", "Monteagle", "species", "troglobite", "stygobite", "troglobiont", "stygobiont", "troglobiotic", "stygobiotic", "groundwater", "subterranean", "salamander", "fish", "vertebrate", "snail", "mollusk", "insect", "fly", "beetle", "arthropod", "arachnid", "spider", "harvestman", "pseudoscorpion", "mite", "crustacean", "crayfish", "isopod", "amphipod", "copepod", "ostracod", and "flatworm". In addition, we also searched biodiversity databases, including the Global Biodiversity Information Facility (GBIF; https://gbif.org; accessed on 7 May 2023), VertNet (http://www.vertnet.org; accessed on 7 May 2023), Symbiota Collections of Arthropods Network (SCAN; https://scan-bugs.org/portal/; accessed on 7 May 2023, and InvertEBase (http://www.invertebase.org/portal/index.php; accessed on 7 May 2023). The list of cave obligate fauna includes the scientific name, authority, and conservation status of each species. Taxonomic nomenclature followed primarily the Integrated Taxonomic Information System (ITIS; http://itis.gov; accessed on 7 May 2023). For the conservation status, we included the International Union for Conservation of Nature (IUCN) red list of threatened species (http://www.iucnredlist.org; accessed on 8 May 2023) and NatureServe (http://www.natureserve.org); accessed on 8 May 2023) conservation statuses when they were available. The status of a species according to the United States list of threatened and endangered species under the U.S. Endangered Species Act was included (http://www.fws.gov/endangered; accessed on 8 May 2023), as well its conservation status in the state of Tennessee (Tennessee State Wildlife Action Plan; http://twraonline.org/2015swap.pdf; accessed on 8 May 2023).

3. Results

The list of cave-limited fauna documented within the Crystal-Wonder Cave System includes 31 species, with 23 troglobionts and eight stygobionts, respectively (Table 1; Figures 2 and 3). Twenty-eight species were known from the Crystal Cave, while eighteen species have been documented from the Wonder Cave, respectively. The Crystal Cave is the type locality for two cave-limited species (*Anillinus longiceps* and *Pseudanophthalmus humeralis*), while the Wonder Cave is the type locality for four species (Table 1): *Tolus appalachius*, *Chitrella archeri*, *Pseudanophthalmus intermedius*, and *Ptomaphagus hatchi*, respectively. *Anillinus longiceps* is known only from the Crystal-Wonder Cave System (Table 1). The cave-limited fauna represents five phyla, ten classes, nineteen orders, and twenty-six families.

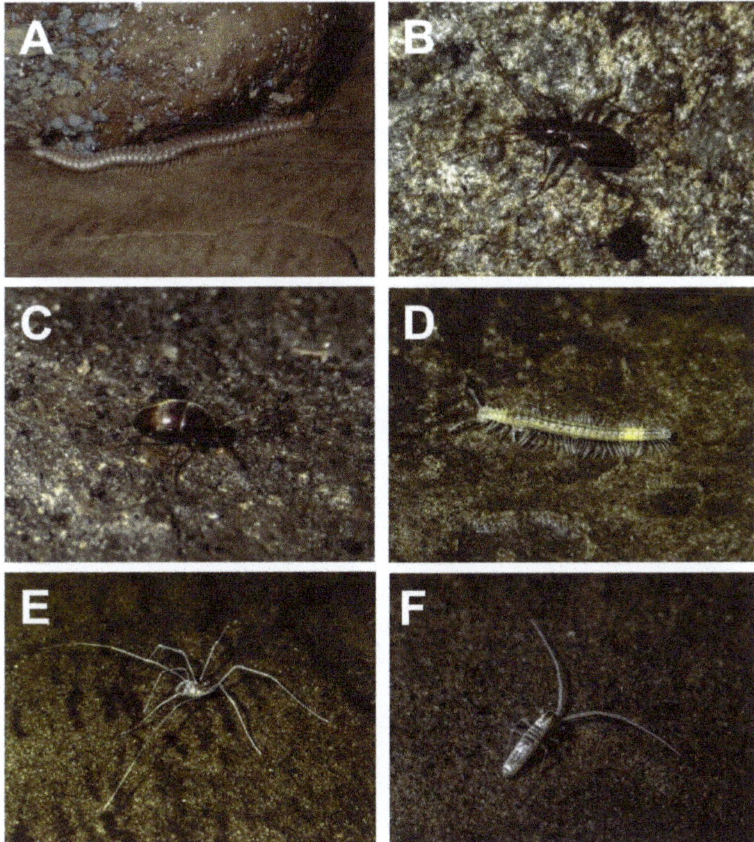

Figure 2. Representative terrestrial cave-limited fauna from the Crystal-Wonder Cave System, Alabama, USA: (**A**) *Tetracion tennesseensis*; (**B**) *Pseudanophthalmus intermedius*; (**C**) *Ptomaphagus hatchi*; (**D**) *Scoterpes ventus*; (**E**) *Tolus appalachius*; and (**F**) *Pseudosinella christianseni*. All photos were taken by Matthew L. Niemiller.

Table 1. Troglobionts and stygobionts of the Crystal-Wonder Cave System, Grundy County, Tennessee, USA. NatureServe conservation ranks include secure (G5), apparently secure (G4), vulnerable (G3), imperiled (G2), critically imperiled (G1), possibly extinct (GH), presumed extinct (GX), unranked (GNR), and unrankable (GU). T# is infraspecific taxon (i.e., subspecies) rank. A ? denotes an inexact numeric rank. State ranks for Tennessee are included in parentheses. IUCN red list categories include least concern (LE), near threatened (NT), vulnerable (VU), endangered (EN), critically endangered (CR), extinct in the wild (EW), and extinct (EX). Tennessee Wildlife Resources Agency state statuses include endangered (E), threatened (T), deemed in need of management (D), and special concern (S). Species of greatest conservation need in Tennessee are marked with an asterisk under State Status. No species are federally listed.

Taxon	Authority	NatureServe Status	IUCN Red List	State Status	Crystal Cave	Wonder Cave
TROGLOBIONTS						
Phylum Arthropoda						
Class Arachnida						
Order Araneae						
Family Linyphiidae						
Phanetta subterranea $	(Emerton, 1875)	G5 (S4)			X	X
Family Zoropsidae						
Liocranoides archeri $	Platnick, 1999	G2 (S2)			X	X
Order Opiliones						
Family Phalangodidae						
Tolus appalachius T,$	Goodnight and Goodnight, 1942	G3G4 (S3)		*	X	X
Order Pseudoscorpiones						
Family Chernetidae						
Hesperochernes mirabilis $	(Banks, 1895)	G5 (S3)			X	
Family Syarinidae						
Chitrella archeri T,$	Malcolm & Chamberlin, 1960	G1G2 (S1S2)		*		X
Order Acari						
Family Rhagidiidae						
Unidentified genus and species $					X	X
Class Collembola						
Order Entomobryomorpha						
Family Entomobryidae						
Pseudosinella christianseni $	Salmon, 1964	G5 (S2)		*	X	X
Pseudosinella spinosa $	(Delamare DeBoutteville, 1949)	G5 (S2)		*	X	X
Order Symphypleona						
Family Arrhopalitidae						
Pygmarrhopalites pavo $	(Christiansen and Bellinger, 1996)	G3? (S1S2)			X	
Class Diplura						
Order Rhabdura						
Family Campodeidae						
Litocampa cookei	(Packard, 1871)	G5 (S3)				X
Litocampa valentinei $	(Conde, 1949)	G5 (S2)		*	X	
Class Diplopoda						
Order Callipodida						
Family Abacionidae						
Tetracion tennesseensis $	Causey, 1959	G2G3 (S2S3)		*	X	X
Order Chordeumatida						
Family Cleidogonidae						
Pseudotremia barri $	Lewis, 2005	G2 (S2)			X	X
Family Trichopetalidae						
Scoterpes ventus $	Shear, 1972	G3 (S1)		*	X	X
Class Insecta						
Order Coleoptera						
Family Carabidae						

Table 1. *Cont.*

Taxon	Authority	NatureServe Status	IUCN Red List	State Status	Crystal Cave	Wonder Cave
TROGLOBIONTS						
Anillinus longiceps [T,E]	Jeannel, 1963				X	
Pseudanophthalmus humeralis [T,$]	Valentine, 1931	G2 (S2)		*	X	X
Pseudanophthalmus intermedius [T, $]	(Valentine, 1931)	G2 (S2)			X	
Family Leiodidae						
Ptomaphagus hatchi [T,$]	Jeannel, 1933	G3 (S3?)			X	X
Family Staphylinidae						
Subfamily Pselaphinae						
Batrisodes valentinei	Park, 1951	G2G4 (S1?)		*	X	
Order Diptera						
Family Sphaeroceridae						
Spelobia tenebrarum [$]	(Aldrich, 1897)	G5 (S4,S5)			X	X
Phylum Mollusca						
Class Gastropoda						
Order Ellobiida						
Family Ellobiidae						
Carychium stygium	Call, 1897	G3 (S2)		*	X	
Order Stylommatophora						
Family Helicodiscidae						
Helicodiscus notius specus	Hubricht, 1962	G5T2 (S1?)		*	X	
Family Zonitidae						
Glyphyalinia specus	Hubricht, 1965	G4 (S3)			X	
STYGOBIONTS						
Phylum Platyhelminthes						
Class Turbellaria						
Order Tricladida						
Family Kenkiidae						
Sphalloplana percoeca [$]	(Packard, 1879)	G5 (S3?)			X	
Phylum Arthropoda						
Class Malacostraca						
Order Amphipoda						
Family Crangonyctidae						
Stygobromus vitreus	Cope, 1872	G4 (S2)			X	
Stygobromus nov. sp. 1						X
Stygobromus nov. sp. 2					X	
Order Decapoda						
Family Cambaridae						
Orconectes australis [$]	(Rhoades, 1941)	G5 (S3)	LC		X	X
Order Isopoda						
Family Asellidae						
Caecidotea bicrenata [$]	(Steeves, 1963)	G5 (na)			X	X
Phylum Chordata						
Class Actinopterygii						
Order Percopsiformes						
Family Amblyopsidae						
Typhlichthys subterraneus [$]	Girard, 1859	G4 (S3)	NT	D *	X	X
Class Amphibia						
Order Caudata						
Family Plethodontidae						
Gyrinophilus palleucus [$]	McCrady, 1954	G2,G3 (S2)	VU B2ab(ii,v)	T *	X	

[T] Type locality in the Crystal-Wonder Cave System; [E] Endemic to the Crystal-Wonder Cave System; [$] Observed since 2015.

Figure 3. Representative aquatic cave-limited fauna from the Crystal-Wonder Cave System, Grundy County, Tennessee, USA: (**A**) *Gyrinophilus palleucus*; (**B**) *Orconectes australis*; (**C**) *Typhlichthys subterraneus*; and (**D**) *Caecidotea bicrenata*. All photos were taken by Matthew L. Niemiller.

3.1. Terrestrial Fauna

Troglobiotic spiders known from the Crystal-Wonder Cave System include linyphiids and zoropsids. The cave linyphiid *Phanetta subterranea* has one of the largest distributions of any troglobiont in North America [35,36]. Lewis [30] reported this species from the Crystal Cave, while we observed the spider at the Wonder Cave in 2022. *Liocranoides archeri* is known from several caves along the Western Escarpment of the Cumberland Plateau from southern Warren County, Tennessee southward into northeastern Alabama [30,37]. This species is pale in coloration but does not possess other obvious troglomorphic characters [37]; however, it has only been reported from caves [37] and has been treated as a troglobiont by other authors [2,3]. Platnick [37] reported this species from the Crystal Cave from a collection by Valentine and Beakley in 1935. We observed this spider in both the Crystal and Wonder caves in 2022.

The Wonder Cave is the type locality of the cave harvestman *Tolus appalachius* [19]. The holotype and paratypes have previously been collected from the Wonder Cave, with additional paratypes collected from the Crystal Cave by Valentine and Beakley in 1935. *Tolus appalachius* is a small, highly troglomorphic harvestman that occurs in several caves along the Western Escarpment of the Cumberland Plateau from Overton County, Tennessee southward into Jackson County, Alabama [3]. This species has also been reported from the Crystal-Wonder Cave System by Peck [38], Lewis [30], and Hedin and Thomas [39].

Two troglobiotic pseudoscorpions occur in the Crystal-Wonder Cave System. *Hesperochernes mirabilis* is a widely distributed chernetid species that is most abundant near entrances where it is associated with bat guano, rodent nests, and mammal scat [8,40,41]. We observed this species at the Crystal Cave in 2022 in the vicinity of raccoon scat near the main entrance. The Wonder Cave is the type locality for the syarinid *Chitrella archeri*. The male holotype was collected in 1938, while an allotype male and female paratype were collected in 1957 [25]. We collected a female from the top of the breakdown pile in the pyramid room of the Wonder Cave on 15 June 2015; this specimen represents the first reported occurrence of the species since its description in 1957. This species lacks eyes, has attenuated appendages and is only known from three caves: the Wonder Cave [25,30], the nearby Trussell Cave in Grundy County [30], and the Piper Cave on the Eastern Highland Rim in Smith County, Tennessee [25,30]. The record obtained from the Trussell Cave is based on a tentative identification [30] and may not represent *C. archeri*.

One unidentified troglobiotic rhagidiid mite is known from the Crystal-Wonder Cave System, where it was observed in both of these caves. This mite may be a species in the genus *Rhagidia* that has been reported from caves in northwestern Georgia [8,40].

Three troglobiotic millipedes have been documented in the Crystal-Wonder Cave System, including one callipodidan and two chordeumatids. *Tetracion tennesseensis* is a large cave millipede (up to 8 cm in length) known from several caves along the Western Escarpment of the Cumberland Plateau from southern White County southward into southwestern Grundy County and northeastern Franklin County [30,42]. This species has previously been reported from the Crystal Cave [30,42], and we observed *T. tennesseensis* in the Crystal and Wonder caves in 2022. *Pseudotremia barri* occurs in caves along the Western Escarpment of the Cumberland Plateau from southern Warren County into northeastern Franklin County [43]. Lewis [43] reported collecting specimens in stream detritus and pitfall traps in riparian mudbanks in the Crystal Cave. We observed *P. barri* on mudbanks along the main streams in both the Crystal and Wonder caves in 2022. We attributed an early report of *Pseudotremia* sp. by Dearolf [23] from the Wonder Cave to *P. barri*. Lewis [30] also reported this species (as *Pseudotremia* sp. nov. 7) from the Crystal Cave. *Scoterpes ventus* has a broad distribution throughout the Cumberland Plateau and Eastern Highland Rim of Tennessee from the Jackson and Overton Counties southward to the Franklin, Grundy, and Marion Counties [29,30,44]. This small, blind, and unpigmented trichopetalid troglobiont likely represents a species complex [29]. *Scoterpes ventus* has been previously reported from both the Crystal [29,30] and Wonder caves [23,29]. We observed this species in both caves in 2022, where it is most often found in moist habitats with organic matter comprising rotting wood, debris, and cricket frass (guano).

Three species of cave-limited collembolans (i.e., springtails) have been documented from the Crystal-Wonder Cave System. Both *Pseudosinella christianseni* and *P. spinosa* are broadly distributed across the Interior Low Plateau [45]. *Pseudosinella spinosa* is the largest of the troglobiotic *Pseudosinella* in North America, and lacks eyes and pigmentation [45]. *Pseudosinella christianseni* also lacks eyes and pigmentation and may be a species complex [45]. Both species have been previously collected from the Crystal Cave, while *P. spinosa* has been collected from the Wonder Cave (Christiansen Springtail Collection), respectively. We observed both species in the Crystal and Wonder caves in 2022. Dearolf [23] reported *Pseudosinella* sp. from the Wonder Cave, which may be either or both of *P. christianseni* and *P. spinosa*. Lewis [30] reported *Pygmarrhopalites pavo* from the Crystal Cave. This small globular springtail is known from caves observed in Virginia [46], as well as caves from the Grundy and Overton Counties, Tennessee [30].

Two troglobiotic diplurans have been reported from the Crystal-Wonder Cave System. Lewis [30] reported *Litocampa valentinei* from the Crystal Cave. This dipluran is known from several caves in northeastern Alabama and south-central Tennessee along the escarpments of the Cumberland Plateau [30,47], including several caves near the Crystal-Wonder Cave System [30]. Dearolf [23] reported *L. cookei* from the Wonder Cave. This species has the broadest distribution of any troglobiotic dipluran in the eastern United States, ranging from western Kentucky to southwestern Virginia, and southward into south-central Tennessee [47]. The occurrence of two *Litocampa* species is notable, as syntopy of *Litocampa* is rare [47]. *Litocama cookei* co-occurs with an undescribed species at the Goodmans Cave in Hancock County, Tennessee [47]. Ferguson [47] did not examine specimens from the Wonder Cave, but there is a nearby record at the Wet Cave in Franklin County [30]. We included both species in our list of troglobiotic taxa, but also noted that additional survey efforts and a comprehensive phylogenetic study are needed to ascertain the species limits in this complex genus.

The troglobiotic beetle fauna of the Crystal-Wonder Cave System is diverse, and includes three carabids, one leiodid, and one staphylinid. All three cave carabid species are blind and wingless. Two species of *Pseudanophthalmus* trechine cave beetles have been previously reported from both the Crystal and Wonder caves [18,30,48–54]. The Crystal Cave is the type locality of *P. humeralis* [18] of the *engelhardti* species group [54], while the Wonder Cave is the type locality of *P. intermedius* [18] of the *intermedius* species group [54]. Both species were first collected in 1931 from the Crystal-Wonder Cave System [18,48] and are known from caves along the Western Escarpment of the Cumberland Plateau from the Franklin and Grundy Counties [18,30,48,54]. *Pseudanophthalmus intermedius* is larger at 5–6 mm compared to 3.9–4.5 mm for *P. humeralis*, respectively [18,48]. Valentine [48,49] recognized two subspecies of *P. humeralis*—*P. h. humeralis* from the Crystal Cave and *P. h. brevis* from the Wonder Cave—but these subspecies were later synonymized by Jeannel [51]. Both species have been found primarily in association with rotting wood [48]. The Crystal Cave is the type locality for the bembidiine ground beetle *Anillinus longiceps* [22,55,56]. The holotype male was collected in 1931 by J.M. Valentine. Although many species in the genus are endogean, *A. longiceps* is considered as a troglobiont [56].

The Wonder Cave is the type locality of the round fungus beetle *Ptomaphagus hatchi*. This beetle has been reported previously from both the Crytal and Wonder caves [21,27,28,51,52,57–59]. This species is the most broadly distributed troglobiotic *Ptomaphagus* in the southern Cumberland Plateau [27,28]. The staphylinid cave ant beetle *Batrisodes valentinei* is known from the Crystal Cave [20,60,61]. The range of this troglobiont is primarily in northern Alabama, with the presumably isolated population from the Crystal Cave. However, Park [62] noted that this specimen from the Crystal Cave should be reexamined, as it might represent a new subspecies or species. We observed both species of *Pseudanophthalmus* and *Ptomaphagus hatchi* during biosurveys in 2022; in contrast, *A. longiceps* and *B. valentinei* have not been observed in the Crystal-Wonder Cave System since 1931.

The only other troglobiotic insect documented from the Crystal-Wonder Cave System is the cave dung fly *Spelobia tenebrarum*, which has been reported from many caves in the eastern United States [8,30,53,63,64], where it has been associated with scat. This species has reduced eyes and is the only known troglobiotic fly in the United States [63,64]. This species was reported from the Crystal Cave by Lewis [30], and we observed this fly in both caves in 2022.

Three troglobiotic snails occur in the Crystal-Wonder Cave System. *Carychium stygium* is a minute (<2 mm) terrestrial snail known from >75 caves throughout the Interior Low Plateau karst region of Kentucky and Tennessee [30,65,66], where it is often found in association with cricket guano [67]. Lewis [30] reported *C. stygium* from the Crystal Cave. Weigand et al. [68,69] suggested that *C. stygium* may be a morph of the troglophile *C. exile* based on mitochondrial COI sequence data. *Glyphyalinia specus* is a wide-ranging troglobiotic glyph known from twenty-seven caves in five states [66], including the Crystal Cave [26], in association with cricket guano [65]. *Helicodiscus notius specus* also has a broad distribution but is known from just four caves in Kentucky and Tennessee [66], including the Crystal Cave [30], where it is associated with cricket guano [65].

3.2. Aquatic Fauna

One cave flatworm—*Sphalloplana percoeca*—has been previously reported from the Crystal-Wonder Cave System [28]. We observed this flatworm on occasion in isolated drip pools. This species has been reported from 34 caves in the TAG region, including several nearby caves in Grundy County [2,30,70].

The cave-limited crustacean fauna includes one crayfish, one isopod, and three amphipods. The stygobiotic crayfish *Orconectes australis* is common in pools in the cave streams in the Crystal-Wonder Cave System. This cave crayfish also has been reported previously by Mohr [71] (as *Cambarus hamulatus*), Hobbs and Barr [72], Hobbs et al. [73], Buhay and Crandall [74], and Lewis [30]. *Orconectes australis* is the most wide-ranging and common stygobiotic crayfish in the southern Cumberland Plateau, occurring in >250 caves from Overton County in Tennessee southward to the Madison and Jackson Counties in Alabama [2,30,74]. We observed as many as 30 *O. australis* in the Crystal Cave in 2005 and 19 crayfish in the Wonder Cave in 2022, respectively.

The stygobiotic isopod *Caecidotea bicrenata* occurs throughout the Crystal-Wonder Cave System in several habitats, including stream riffles and pools, rimstone pools, and drip pools. *Caecidotea bicrenata* is widely distributed across a variety of subterranean habitats throughout the Interior Low Plateau [75] but may represent a cryptic species complex. This stygobiotic asellid was reported previously from the Crystal and Wonder caves by Dearolf [23] (as *C. nickajackensis*) and from the Crystal Cave by Lewis [30,75]. *Caecidotea bicrenata* was found to be common in both caves in 2022.

Dearolf [23] reported the cave amphipod *Stygobromus vitreus* from the Crystal Cave based on a collection in 1937. This species is known primarily from the Mammoth Cave region in central Kentucky with scattered occurrences in south-central Tennessee and northern Alabama [23,76]. Holsinger [76] noted that populations from Alabama morphologically differed slightly from populations in Kentucky, and potentially may represent a distinct species. An undescribed *Stygobromus* amphipod is known from the Wonder Cave [30]. This species apparently occurs in the southern Cumberland Plateau of south-central Tennessee and northern Alabama but extends southward to the north of Birmingham (Holsinger, pers. comm. in Lewis, [30]). Another undescribed *Stygobromus* amphipod is known from the Crystal Cave that is distinct from the taxon observed in the Wonder Cave. This species occurs in several caves along the Cumberland Plateau from the White County southwestward into the Franklin County (Holsinger, pers. comm. in Lewis [30]). Additional specimens have been collected from the Crystal Cave, which may be either of the two undescribed taxa, *S. vitreus*, or another species, such as *S. exilis*, which also has a broad distribution in the Interior Low Plateau karst region, including the southern Cumberland Plateau [30,76].

Two cave-limited vertebrates are known from the Crystal-Wonder Cave System. The southern cavefish *Typhlichthys subterraneus* has been observed in both the Crystal [24,30,77–80] and Wonder caves [30,77,78]. This cavefish is abundant in the main stream and tributaries in the Crystal Cave [78] and is considered as a top aquatic predator. *Typhlichthys subterraneus* is a cryptic species complex [79], with lineage B identified from the Upper Elk River watershed [79,80]. From our biosurveys of the Crystal Cave we observed six cavefish in 2005 and as many as 30 cavefish in 2006.

The Tennessee cave salamander *Gyrinophilus palleucus* was first observed in the Crystal Cave in 2006 [81,82]. This neotenic plethodontid salamander is considered as a top predator of cave streams and is known from several caves along the escarpments of the southern Cumberland Plateau in Tennessee, Alabama, and Georgia [81,83,84]; however, it has been presumed to be rare in the Crystal-Wonder Cave System. Only one individual has been observed on three occasions, most recently in 2022. *Gyrinophilus palleucus* has yet to be observed in the Wonder Cave.

4. Discussion

The cave-limited fauna of the Crystal-Wonder Cave System is remarkably diverse with 31 troglobionts and stygobionts, making it the most diverse cave system in the southern Cumberland Plateau region, and one of the most diverse cave systems in all of North America. With 31 cave-limited species, the Crystal-Wonder Cave System trails only the San Marcos artesian Well in central Texas (55 species; [85]) and the Mammoth Cave System in central Kentucky (49 species; [86]) and ranks ahead of the Fern Cave System in Alabama (27 species; [10]) and Sistema Huautla in Oaxaca, Mexico (27 species; [87]). In particular, the terrestrial fauna is exceptionally rich with 23 species, trailing only the Mammoth Cave System (32 troglobionts; [86]) and Sistema Huautla (27 species; [87]) in North America. The stygofauna of the Crystal-Wonder Cave System is also diverse (8 species), but not remarkable compared to other hotspot subterranean communities in North America, such as the San Marcos Artesian Well in Texas (55 species; [85]), Mammoth Cave System in Kentucky (17 species; [86]), and Shelta Cave in Alabama (12 species, [9,88]). The high number of troglobionts relative to the stygobionts in the Crystal-Wonder Cave System is not surprising given the cave's location in Tennessee, where troglobionts outnumber stygobionts by roughly four to one [2]. The Crystal-Wonder Cave System is the most speciose cave system in the southern Cumberland Plateau region ahead of the Fern Cave (27 species; [10]), the Shelta Cave (24 species; [9]), and the Big Mouth-Big Room System nearby in Grundy County, Tennessee (20 species; [2]).

The remarkable level of cave biodiversity observed in the Crystal-Wonder Cave System can largely be attributed by its location. The cave system sits at the junction of the Cumberland Plateau escarpment and the Eastern Highland Rim, the two ecoregions supporting the most cave biodiversity in Tennessee [2]. The cave's location in southern Tennessee also contributes, as cave biodiversity increases towards the southern interface of the Cumberland Plateau escarpment and the Eastern Highland Rim [2]. An additional factor is its history as a highly visited commercial cave, which attracted early cave biologists and increased the likelihood of the detection of exceptionally rare species (e.g., the single-cave endemic *Anillinus longiceps*), and explains how the cave came to be the type locality for six troglobionts.

The physical structure of the Crystal-Wonder Cave System also contributes to its remarkable troglobiont community. The cave system is relatively large (5.8 km), and hosts a variety of terrestrial and aquatic habitats, including a large cave stream that flows through much of the cave system. This stream, along with the multiple entrances, increases the opportunity for nutrients to enter and disperse throughout the cave system, thereby supporting a diverse troglobiont community. Although it has a long history as a commercial cave, the Wonder Cave has been closed to the public for more than two decades, reducing recent human impacts. Although the cave's watershed has not been mapped, the slopes of the adjacent Cumberland Plateau escarpment are forested, and the areas around the cave entrances are largely undeveloped, which benefits the cave community.

Of the thirty-one cave-limited species known from the Crystal-Wonder Cave System, eighteen species (sixteen troglobionts and two stygobionts) are of conservation concern (i.e., G1–G3 NatureServe conservation rank, state status). None of the species have federal status. Many of these species are at an increased risk of extinction due to their restricted distributions, or are known from few occurrences, such as *Anillinus longiceps*, which is known only from the Crystal-Wonder Cave System. The cave-limited fauna of the Crystal-Wonder Cave System are also facing potential threats related to changes in land use within the cave's watershed, particularly logging, home building, and a proposed sand quarry, as all of the land near the cave system is privately owned. Changes in the land use within the cave's watershed could increase sedimentation and otherwise negatively impact the water quality and quantity in the cave stream. In addition, the Crystal-Wonder Cave System is privately owned, and the various entrances to the cave system are owned by different individuals, thereby limiting the public's ability to monitor and manage the cave system with biodiversity in mind.

Despite its exceptional cave biodiversity, the list of cave-limited fauna that occur in the Crystal-Wonder Cave System is likely incomplete. Several sections, particularly within the Wonder Cave, remain to be comprehensively bio-inventoried, and several habitats, such as the epikarst and stream sediments, have not been adequately sampled and may harbor additional species. Several taxonomic groups are notably absent from the cave-limited fauna of the Crystal-Wonder Cave System, including stygobiotic copepods and troglobiotic woodlice, all of which may be discovered during future biosurveys. For example, two species of copepods—*Diacyclops yeatmani* and *D. indianensis*—are known from nearby the Big Mouth Cave [30] and may occur in the Crystal-Wonder Cave System. Other taxonomic groups have not been particularly well-studied in the Crystal-Wonder Cave System, including mites, spiders, pseudoscorpions, and springtails. More intensive biosurvey efforts on these groups may uncover additional taxa, as several species yet to be documented in the Crystal-Wonder Cave System have reported from other nearby caves in Grundy County [30]. Finally, comprehensive sampling within the Crystal-Wonder System coupled with genetic analyzes has the potential to uncover cryptic diversity, which is an increasingly common discovery of phylogenetic studies in cave-limited taxa [79,89]. Phylogeographic studies have incorporated specimens from the Crystal-Wonder Cave System, including the studies of *Orconectes australis* [74], *Gyrinophilus palleucus* [82], *Tetracion tennesseensis* [42], *Typhlichthys subterraneus* [79], and *Ptomaphagus hatchi* [59], but several taxonomic groups, such as stygobiotic isopods and amphipods, remain to be studied from a genetic perspective.

Author Contributions: Conceptualization: M.L.N.; methodology and analysis: M.L.N.; data acquisition: M.L.N., B.T.M., C.D.R.S., B.C., J.H. (Jared Higgs), A.H., N.M., K.D.K.N., K.S.Z., K.S. and J.H. (Jason Hardy); original draft preparation: M.L.N., C.D.R.S., A.H., K.S.Z. and K.S.; review and editing: M.L.N., B.T.M., C.D.R.S., B.C., A.H., N.M., K.D.K.N., K.S.Z., K.S. and J.H. (Jason Hardy). All authors have read and agreed to the published version of the manuscript.

Funding: Support for this project included grants from the National Science Foundation (award no. 2047939 to M.L.N.) and the Tennessee Wildlife Resources Agency (Tennessee Wildlife Resources Agency contract nos. ED-04-01467-00 and ED-06-02149-00 to B.T.M.; 328.01 to M.L.N.) as well as support from the Department of Biology at Middle Tennessee State University.

Institutional Review Board Statement: All research was conducted under approved Institutional Animal Care and Use Committee protocols at the University of Alabama in Huntsville (protocol no. 2017.R005) and Middle Tennessee State University (protocol no. 04-006).

Data Availability Statement: No new data were created or analyzed in this manuscript. Data sharing is not applicable for this manuscript.

Acknowledgments: We thank the Born family for allowing access to the cave system and supporting this study. We thank J. Todd and Pamela B. Hart for their assistance with the fieldwork. We thank Marshal Hedin, Marc Milne, and Karen Ober for identifying the specimens. Collection of specimens was authorized under the Tennessee Wildlife Resources Agency scientific permit no. 1385.

Conflicts of Interest: The authors declare no conflict of interest.

References

1. Culver, D.C.; Master, L.L.; Christman, M.C.; Hobbs, H.H. Obligate cave fauna of the 48 contiguous United States. *Conserv. Biol.* **2000**, *14*, 386–401. [CrossRef]
2. Niemiller, M.L.; Zigler, K.S. Patterns of cave biodiversity and endemism in the Appalachians and Interior Plateau of Tennessee, USA. *PLoS ONE* **2013**, *8*, e64177. [CrossRef] [PubMed]
3. Niemiller, M.L.; Zigler, K.S.; Fenolio, D.B. *Cave Life of TAG: A Guide to Commonly Encountered Species in Tennessee, Alabama and Georgia*; Biology Section of the National Speleological Society: Huntsville, AL, USA, 2013.
4. Niemiller, M.L.; Taylor, S.J.; Slay, M.E.; Hobbs, H.H., III. Biodiversity in the United States and Canada. In *Encyclopedia of Caves*, 3rd ed.; Culver, D.C., White, W.B., Pipan, T., Eds.; Elsevier: Amsterdam, The Netherlands, 2019; pp. 163–176.
5. Zigler, K.S.; Niemiller, M.L.; Fenolio, D.B. Cave Biodiversity of the Southern Cumberland Plateau. In *National Speleological Society Convention Guidebook*; National Speleological Society: Huntsville, TX, USA, 2014; pp. 159–163.
6. Christman, M.C.; Culver, D.C. The relationship between cave biodiversity and available habitat. *J. Biogeogr.* **2001**, *28*, 367–380. [CrossRef]
7. Culver, D.C.; Deharveng, L.; Bedos, A.; Lewis, J.J.; Madden, M.; Reddell, J.R.; Sket, B.; Trontelj, P.; White, D. The mid-latitude biodiversity ridge in terrestrial cave fauna. *Ecography* **2006**, *29*, 120–128. [CrossRef]
8. Zigler, K.S.; Niemiller, M.L.; Stephen, C.D.R.; Ayala, B.N.; Milne, M.A.; Gladstone, N.S.; Engel, A.S.; Jensen, J.B.; Camp, C.D.; Ozier, J.C.; et al. Biodiversity from caves and other subterranean habitats of Georgia, USA. *J. Cave Karst Stud.* **2020**, *82*, 125–167. [CrossRef]
9. Culver, D.C.; Sket, B. Hotspots of subterranean biodiversity in caves and wells. *J. Cave Karst Stud.* **2000**, *62*, 11–17.
10. Niemiller, M.L.; Slay, M.E.; Inebnit, T.; Miller, B.; Tobin, B.; Cramphorn, B.; Hinkle, A.; Jones, B.D.; Mann, N.; Niemiller, K.D.K.; et al. Fern Cave: A Hotspot of Subterranean Biodiversity in the Interior Low Plateau Karst Region of Alabama in the Southeastern United States. *Diversity* **2023**, *15*, 633. [CrossRef]
11. Nelson, W.A. The Monteagle Wonder Cave. *Resour. Tenn.* **1912**, *2*, 294–306.
12. Barr, T.C. Caves of Tennessee. Nashville: Tennessee Department of Environment and Conservation, Division of Geology. *Bulletin* **1961**, *64*, 567.
13. Matthews, L.E. Wonder Cave. In *Caves of Chattanooga*; Matthews, L.E., Ed.; National Speleological Society: Huntsville, AL, USA, 2007; pp. 161–178.
14. Smallwood, K.M. *Wonder Cave, Grundy County, Tennessee*; Spelean History Series No. 17; Speece Publications: Altoona, PA, USA, 2013.

15. Smallwood, K.M. The dye tracing of Hurricane Creek, Layne Cove and Smith Hollow Cove. *Tenn. Caver. in press.*
16. Douglas, J.C. Miners and moonshiners: Historic industrial uses of Tennessee caves. *Midcont. J. Archaeol.* **2001**, *26*, 251–267.
17. Fraser, M. Down in the Hole: Outlaw Country and Outlaw Culture. *South. Cult.* **2018**, *24*, 83–100. [CrossRef]
18. Valentine, J.M. New cavernicole Carabidae of the subfamily Trechinae Jeannel. *J. Elisha Mitchell Sci. Soc.* **1931**, *46*, 247–258.
19. Goodnight, C.J.; Goodnight, M.L. New Phalangodidae (Phalangida) from the United States. *Am. Mus. Novit.* **1942**, *1188*, 1–18.
20. Park, O. Cavernicolous pselaphid beetles of Alabama and Tennessee. *Geol. Surv. Alabama Mus. Pap.* **1951**, *31*, 1–107.
21. Jeannel, R. Trois *Adelops* nouveaux de l'Amerique du Nord. *Bull. Soc. Entomol. France* **1933**, *38*, 251–253. [CrossRef]
22. Jeannel, R. Supplément a la monographie des Anillini (1). Sur quelques espèces Nouvelles de l'Amérique du Nord. *Rev. Française D'entomologie* **1963**, *30*, 145–152.
23. Dearolf, K. The invertebrates of 75 caves in the United States. *Proc. Pa. Acad. Sci.* **1953**, *27*, 225–241.
24. Woods, L.P.; Inger, R.F. The cave, spring, and swamp fishes of the family Amblyopsidae of central and eastern United States. *Am. Midl. Nat.* **1957**, *58*, 232–256. [CrossRef]
25. Malcolm, D.R.; Chamberlin, J.C. The pseudoscorpion genus *Chitrella* (Chelonethida, Syarinidae). *Am. Mus. Novit.* **1960**, *1989*, 1–19.
26. Hubricht, L. Four new land snails from the southeastern United States. *Nautilus* **1965**, *79*, 4–7.
27. Peck, S.B. A systematic revision and evolutionary biology of the *Ptomaphagus adelops*. *Bull. Mus. Comp. Zool.* **1973**, *145*, 29–162.
28. Peck, S.B. The distribution and evolution of cavernicolous *Ptomaphagus* beetles in the southeastern United States (Coleoptera; Leiodidae; Cholevinae) with new species and records. *Canad. J. Zool.* **1984**, *62*, 730–740. [CrossRef]
29. Shear, W.A. The milliped family Trichopetalidae, Part 2: The genera *Trichopetalum*, *Zygonopus* and *Scoterpes* (Diplopoda: Chordeumatida, Cleidogonoidea). *Zootaxa* **2010**, *2385*, 1–62. [CrossRef]
30. Lewis, J.J. *Bioinventory of Caves of the Cumberland Escarpment Area of Tennessee. Final Report to Tennessee Wildlife Resources Agency & The Nature Conservancy of Tennessee*; Lewis & Associates: Borden, IN, USA, 2005.
31. Sket, B. Can we agree on an ecological classification of subterranean animals? *J. Nat. Hist.* **2008**, *42*, 1549–1563. [CrossRef]
32. Trajano, E. Ecological classification of subterranean organisms. In *Encyclopedia of Caves*, 2nd ed.; White, W.B., Culver, D.C., Eds.; Academic/Elsevier Press: Amsterdam, The Netherlands, 2012; pp. 275–277.
33. Trajano, E.; de Carvalho, M.R. Towards a biologically meaningful classification of subterranean organisms: A critical analysis of the Schiner-Racovitza system from a historical perspective, difficulties of its application and implications for conservation. *Subterr. Biol.* **2017**, *22*, 1–26. [CrossRef]
34. Culver, D.C.; Pipan, T. Ecological and evolutionary classifications of subterranean organisms. In *Encyclopedia of Caves*, 3rd ed.; Culver, D.C., White, W.B., Pipan, T., Eds.; Elsevier: Amsterdam, The Netherlands, 2019; pp. 376–379.
35. Barr, T.C., Jr.; Holsinger, J.R. Speciation in cave faunas. *Ann. Rev. Ecol. Syst.* **1985**, *16*, 313–337. [CrossRef]
36. Peck, S.B. A summary of diversity and distribution of the obligate cave-inhabiting faunas of the United States and Canada. *J. Caves Karst Stud.* **1998**, *60*, 18–26.
37. Platnick, N.I. A revision of the Appalachian spider genus *Liocranoides* (Araneae: Tengellidae). *Amer. Mus. Nov.* **1999**, *3285*, 1–13.
38. Peck, S.B. The cave fauna of Alabama: Part I. The terrestrial invertebrates (excluding insects). *Nat. Speleolog. Soc. Bull.* **1989**, *51*, 11–33.
39. Hedin, M.; Thomas, S.M. Molecular systematics of eastern North American Phalangodidae (Arachnida: Opiliones: Laniatores), demonstrating convergent morphological evolution in caves. *Mol. Phylogenet. Evol.* **2010**, *54*, 107–121. [CrossRef]
40. Holsinger, J.R.; Peck, S.B. The invertebrate cave fauna of Georgia. *Nat. Speleolog. Soc. Bull.* **1971**, *33*, 23–44.
41. Reeves, W.K.; Jensen, J.B.; Ozier, J.C. New faunal and fungal records from caves in Georgia, USA. *J. Cave Karst Stud.* **2000**, *62*, 169–179.
42. Loria, S.F.; Zigler, K.S.; Lewis, J.J. Molecular phylogeography of the troglobiotic millipede *Tetracion* Hoffman, 1956 (Diplopoda, Callipodida, Abacionidae). *Int. J. Myriap.* **2011**, *5*, 35–48.
43. Lewis, J.J. Six new species of *Pseudotremia* from caves of the Tennessee Cumberland Plateau (Diplopoda: Chordeumatida: Cleidogonidae). *Zootaxa* **2005**, *1080*, 17–31. [CrossRef]
44. Shear, W.A. Studies in the milliped order Chordeumida (Diplopoda): A revision of the family Cleidogonidae and a reclassification of the order Chordeumida in the New World. *Bull. Mus. Comparat. Zool.* **1972**, *144*, 151–352.
45. Christiansen, K.; Bellinger, P. *The Collembola of North America North of the Rio Grande: A Taxonomic Analysis*, 2nd ed.; Grinnell College: Grinnell, IA, USA, 1998.
46. Christiansen, K.; Bellinger, P. Cave *Arrhopalites*: New to science. *J. Cave Karst Stud.* **1996**, *58*, 168–180.
47. Ferguson, L.M. Systematics, Evolution, and Zoogeography of the Cavernicolous Campodeids of the Genus *Litocampa* (Diplura: Campodeidae) in the United States. Ph.D. Dissertation, Virginia Polytechnic Institute and State University, Blacksburg, VA, USA, 1981.
48. Valentine, J.M. A classification of the genus *Pseudanophthalmus* Jeannel (fam. Carabidae) with descriptions of new species and notes on distribution. *J. Elisha Mitchell Sci. Soc.* **1932**, *48*, 261–280.
49. Valentine, J.M. Anophthalmid beetles (fam. Carabidae) from Tennessee caves. *J. Elisha Mitchell Sci. Soc.* **1937**, *53*, 93–100.
50. Valentine, J.M. New anophthalmid beetles from the Appalachian region. *Geol. Surv. Alabama, Mus. Pap.* **1948**, *27*, 1–20.
51. Jeannel, R. Les coleopteres cavernicoles de la region des Appalaches. III. Etude systematique. *Notes Biospeol.* **1949**, *4*, 37–104.
52. Nicholas, B.G. Checklist of macroscopic troglobitic organisms of the United States. *Am. Midl. Nat.* **1960**, *64*, 123–160. [CrossRef]

53. Peck, S.B. The cave fauna of Alabama. Part II: The insects. *Nat. Speleolog. Soc. Bull.* **1995**, *57*, 1–19.
54. Barr, T.C., Jr. *A Classification and Checklist of the Genus Pseudanophthalmus Jeannel (Coleoptera: Carabidae: Trechinae)*; Special Publication 11; Virginia Museum of Natural History: Martinsville, VA, USA, 2004.
55. Erwin, T.L.; House, G.N. A catalogue of the primary types of Carabidae (incl. Cicindelinae) in the collections of the United States National Museum of Natural History (USNM) (Coleoptera). *Coleopt. Bull.* **1978**, *32*, 231–256.
56. Sokolov, I.M.; Carlton, C.; Cornell, J.F. Review of *Anillinus*, with descriptions of 17 new species and a key to soil and litter species (Coleoptera: Carabidae: Trechinae: Bembidiini). *Coleopt. Bull.* **2004**, *58*, 185–233. [CrossRef] [PubMed]
57. Jeannel, R. Monographie des Catopidae. *Mem. Mus. Natl. Hist. Natur.* **1936**, *1*, 1–433.
58. Barr, T.C., Jr. Studies on the cavernicole *Ptomaphagus* of the United States (Coleoptera; Catopidae). *Psyche* **1963**, *70*, 50–58. [CrossRef]
59. Leray, V.L.; Caravas, J.; Friedrich, M.; Zigler, K.S. Mitochondrial sequence data indicate "Vicariance by Erosion" as a mechanism of species diversification in North American *Ptomaphagus* (Coleoptera, Leiodidae, Cholevinae) cave beetles. *Subterr. Biol.* **2019**, *29*, 35–57. [CrossRef]
60. Park, O. New or little known species of pselaphid beetles from southeastern United States. *J. Tenn. Acad. Sci.* **1956**, *31*, 54–100.
61. Park, O. New or little-known species of pselaphid beetles, chiefly from southeastern United States. *J. Tenn. Acad. Sci.* **1958**, *33*, 39–74.
62. Park, O. Cavernicolous pselaphid beetles of the United States. *Amer. Midl. Nat.* **1960**, *64*, 66–104. [CrossRef]
63. Marshall, S.A.; Peck, S.B. Distribution of cave-dwelling Sphaeroceridae (Diptera) of eastern North America. *Proc. Entomol. Soc. Ontario* **1984**, *115*, 37–41.
64. Marshall, S.A.; Peck, S.B. The origin and relationships of *Spelobia tenebrarum* Aldrich, a troglobitic, eastern North American sphaerocerid fly. *Can. Entomol.* **1985**, *117*, 1013–1015. [CrossRef]
65. Hubricht, L. The distributions of the native land mollusks of the eastern United States. *Fieldiana* **1985**, *24*, 1–191.
66. Gladstone, N.S.; Carter, E.T.; McKinney, M.L.; Niemiller, M.L. Status and distribution of the cave-obligate land snails in the Appalachians and Interior Low Plateau of the eastern United States. *Am. Malacol. Bull.* **2018**, *36*, 62–78. [CrossRef]
67. Dourson, D.C. *Kentucky's Land Snails and Their Ecological Communities*; Goatslug Publications: Bakersville, NC, USA, 2010.
68. Weigand, A.M.; Jochum, A.; Pfenninger, M.; Steinke, D.; Klussmann-Kolb, A. A new approach to an old conundrum—DNA barcoding sheds new light on phenotypic plasticity and morphological stasis in microsnails (Gastropoda, Pulmonata, Carychiidae). *Mol. Ecol. Res.* **2011**, *11*, 255–265. [CrossRef]
69. Weigand, A.M.; Jochum, A.; Slapnik, R.; Schnitzler, J.; Zarza, E.; Klussmann-Kolb, A. Evolution of microgastropods (Ellobioidea, Carychiidae): Integrating taxonomic, phylogenetic and evolutionary hypotheses. *BMC Evol. Biol.* **2013**, *13*, 18. [CrossRef]
70. Kenk, R. Freshwater triclads (Turbellaria) of North America, IX. The genus *Sphalloplana*. *Smithson. Contrib. Zool.* **1977**, *246*, 1–38. [CrossRef]
71. Mohr, C.E. I Explore Cave. *Nat. Hist.* **1939**, *43*, 190–204.
72. Hobbs, H.H., Jr.; Barr, T.C., Jr. Origins and affinities of the troglobitic crayfishes of North America (Decapoda: Astacidae), II: Genus Orconectes. *Smithson. Contrib. Zool.* **1972**, *105*, 1–84. [CrossRef]
73. Hobbs, H.H., Jr.; Hobbs, H.H., III; Daniel, M.A. A review of the troglobitic decapod crustaceans of the Americas. *Smithson. Contrib. Zool.* **1977**, *244*, 1–83. [CrossRef]
74. Buhay, J.E.; Crandall, K.A. Subterranean phylogeography of freshwater crayfishes shows extensive gene flow and surprisingly large population sizes. *Mol. Ecol.* **2005**, *14*, 4259–4273. [CrossRef] [PubMed]
75. Lewis, J.J. The systematics, Zoogeography and Life History of the Troglobitic Isopods of the Interior Plateaus of the Eastern United States. Ph.D. Dissertation, University of Louisville, Louisville, KY, USA, 1988.
76. Holsinger, J.R. *Freshwater Amphipod Crustaceans (Gammaridae) of North America, Biota of Freshwater Ecosystems, Identification Manual No. 5*; U.S. Environmental Protection Agency: Washington, DC, USA, 1972.
77. Armstrong, J.G.; Williams, J.D. Cave and spring fishes of the southern bend of the Tennessee River. *J. Tenn. Acad. Sci.* **1971**, *46*, 107–115.
78. Niemiller, M.L.; Miller, B.T.; Fitzpatrick, B.M. *Status and Distribution of the Amblyopsid Fishes Forbesichthys agassizii and Typhlichthys subterraneus in Tennessee*; Tennessee Wildlife Resources Agency: Nashville, TN, USA, 2010.
79. Niemiller, M.L.; Near, T.J.; Fitzpatrick, B.M. Delimiting species using multilocus data: Diagnosing cryptic diversity in the southern cavefish, *Typhlichthys subterraneus* (Teleostei: Amblyopsidae). *Evolution* **2012**, *66*, 846–866. [CrossRef]
80. Niemiller, M.L.; Graening, G.O.; Fenolio, D.B.; Godwin, J.C.; Cooley, J.R.; Pearson, W.D.; Fitzpatrick, B.M.; Near, T.J. Doomed before they are described? The need for conservation assessments of cryptic species complexes using an amblyopsid cavefish (Amblyopsidae: *Typhlichthys*) as a case study. *Biodivers. Conserv.* **2013**, *22*, 1799–1820. [CrossRef]
81. Miller, B.T.; Niemiller, M.L. Distribution and relative abundance of Tennessee cave salamanders (*Gyrinophilus palleucus* and *G. gulolineatus*) with an emphasis on Tennessee populations. *Herpetol. Conserv. Biol.* **2008**, *3*, 1–20.
82. Niemiller, M.L.; Fitzpatrick, B.M.; Miller, B.T. Recent divergence with gene flow in Tennessee cave salamanders (Plethodontidae: *Gyrinophilus*) inferred from gene genealogies. *Mol. Ecol.* **2008**, *17*, 2258–2275. [CrossRef] [PubMed]
83. Miller, B.T.; Niemiller, M.L. Gyrinophilus palleucus. *Cat. Am. Amphib. Reptiles* **2012**, *884*, 1–7.
84. Niemiller, M.L.; Niemiller, K.D.K. *Species Status Assessment for the Tennessee Cave Salamander (Gyrinophilus palleucus) McCrady, 1954, Version 1.0.*; Tennessee Wildlife Resources Agency: Nashville, TN, USA, 2020; p. 59.

85. Hutchins, B.T.; Gibson, J.R.; Diaz, P.H.; Schwartz, B.F. Stygobiont diversity in the San Marcos Artesian Well and Edwards Aquifer groundwater ecosystem, Texas, USA. *Diversity* **2021**, *13*, 234. [CrossRef]
86. Niemiller, M.L.; Helf, K.; Toomey, R.S. Mammoth Cave: A hotspot of subterranean biodiversity in the United States. *Diversity* **2021**, *13*, 373. [CrossRef]
87. Francke, O.F.; Monjaraz-Ruedes, R.; Cruz-Lopez, J.A. Biodiversity of the Huautla cave system, Oaxaca, Mexico. *Diversity* **2021**, *13*, 429. [CrossRef]
88. Cooper, J.E. Ecological and Behavioral Studies in Shelta Cave, Alabama, with Emphasis on Decapod Crustaceans. Ph.D. Thesis, University of Kentucky, Lexington, KY, USA, 1975.
89. Ethridge, J.Z.; Gibson, J.R.; Nice, C.C. Cryptic diversity within and amongst spring-associated *Stygobromus* amphipods (Amphipoda: Crangonyctidae). *Zool. J. Linn. Soc.* **2013**, *167*, 227–242. [CrossRef]

diversity

Article

Fern Cave: A Hotspot of Subterranean Biodiversity in the Interior Low Plateau Karst Region of Alabama in the Southeastern United States

Matthew L. Niemiller [1,*], Michael E. Slay [2], Thomas Inebnit [3], Benjamin Miller [4], Benjamin Tobin [5], Brendan Cramphorn [1], Amata Hinkle [1], Bradley D. Jones [6], Nathaniel Mann [7], K. Denise Kendall Niemiller [1] and Steve Pitts [8]

[1] Department of Biological Sciences, The University of Alabama in Huntsville, 301 Sparkman Dr NW, Huntsville, AL 35899, USA; btc0011@uah.edu (B.C.); amatahinkle@gmail.com (A.H.); dk0047@uah.edu (K.D.K.N.)
[2] Ozark Highlands Office, The Nature Conservancy, Fayetteville, AR 72701, USA; mslay@tnc.org
[3] Arkansas Ecological Services Field Office, U.S. Fish & Wildlife Service, Conway, AR 72032, USA; thomas_inebnit@fws.gov
[4] Lower Mississippi-Gulf Water Science Center, U.S. Geological Survey, Nashville, TN 37211, USA; bvmiller@usgs.gov
[5] Kentucky Geological Survey, University of Kentucky, Lexington, KY 40506, USA; benjamin.tobin@uky.edu
[6] Alabama Cave Survey, Huntsville, AL 35810, USA; nss63915@gmail.com
[7] Spencer Mountain Grotto, Spencer, TN 38585, USA; thany@blomand.net
[8] Southeastern Cave Conservancy, Inc., Signal Mountain, TN 37377, USA; stevepitts597@gmail.com
* Correspondence: matthew.niemiller@uah.edu or cavemander17@gmail.com

Citation: Niemiller, M.L.; Slay, M.E.; Inebnit, T.; Miller, B.; Tobin, B.; Cramphorn, B.; Hinkle, A.; Jones, B.D.; Mann, N.; Niemiller, K.D.K.; et al. Fern Cave: A Hotspot of Subterranean Biodiversity in the Interior Low Plateau Karst Region of Alabama in the Southeastern United States. *Diversity* 2023, 15, 633. https://doi.org/10.3390/d15050633

Academic Editors: Tanja Pipan, David C. Culver and Louis Deharveng

Received: 31 March 2023
Revised: 27 April 2023
Accepted: 28 April 2023
Published: 6 May 2023

Abstract: The Fern Cave System, developed in the western escarpment of the Southern Cumberland Plateau of the Interior Low Plateau karst region in Northeastern Alabama, USA, is a global hotspot of cave-limited biodiversity as well as home to the largest winter hibernaculum for the federally endangered Gray Bat (*Myotis grisescens*). We combined the existing literature, museum accessions, and database occurrences with new observations from bioinventory efforts conducted in 2018–2022 to generate an updated list of troglobiotic and stygobiotic species for the Fern Cave System. Our list of cave-limited fauna totals twenty-seven species, including nineteen troglobionts and eight stygobionts. Two pseudoscorpions are endemic to the Fern Cave System: *Tyrannochthonius torodei* and *Alabamocreagris mortis*. The exceptional diversity at Fern Cave is likely associated with several factors, such as the high dispersal potential of cave fauna associated with expansive karst exposures along the Southern Cumberland Plateau, high surface productivity, organic input from a large bat colony, favorable climate throughout the Pleistocene, and location within a larger regional hotspot of subterranean biodiversity. Nine species are of conservation concern, including the recently discovered Alabama cave shrimp *Palaemonias alabamae*, because of their small range sizes, few occurrences, and several potential threats.

Keywords: checklist; karst; species richness; stygobiont; troglobiont

1. Introduction

The Fern Cave System in Jackson County, Northeastern Alabama, USA, is the most extensive cave system in the state of Alabama, with over 25 km (15.6 miles) of mapped passage [1], including 163 m (536 ft) of vertical extent and five entrances (Figure 1). The cave system is managed by the U.S. Fish and Wildlife Service (USFWS) and Southeastern Cave Conservancy, Inc. (SCCi). The largest colony of Gray Bats (*Myotis grisescens*) hibernates in sections of Fern Cave [2]. The 80.5 hectare Fern Cave National Wildlife Refuge was established in 1981 to protect this federally endangered species. Five entrances are known, with four located on Fern Cave National Wildlife Refuge and another entrance owned and managed by SCCi on the 32.4 hectare Kay Hill Deen Fern Cave Preserve. Fern Cave National Wildlife Refuge is managed as part of the Wheeler National Wildlife Refuge Complex.

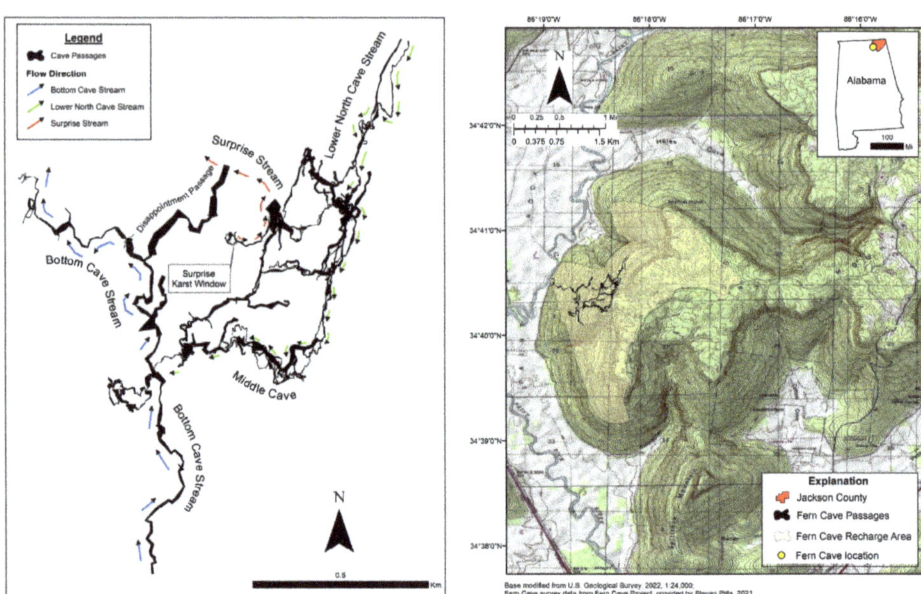

Figure 1. Line map of the Fern Cave System (**left**) and location of the cave system on Nat Mountain, Jackson County, Alabama, USA (**right**).

Fern Cave is developed along the western margin of Nat Mountain in the Paint Rock River Valley of Jackson County, Alabama. Nat Mountain is a highly dissected lobe of the Southern Cumberland Plateau that is bounded to the north and west by the Paint Rock River and to the south by Yellow Branch in Peter Cove [3]. The geology of Nat Mountain was mapped and described recently by Osbourne et al. [4] and is briefly described herein. The vertical relief of Nat Mountain is 305 m from 180 to 488 m in elevation. Nat Mountain is capped by the Lower Pennsylvanian-aged Pottsville Formation (61 m thick quartzose sandstone with occasional conglomeratic interbeds of shale). Beneath the Pottsville Formation is the Upper Mississippian-aged Pennington Formation (91 m thick), which consists of interbeds of sandstone, limestones, chert, dolomites, and shales. The presence of carbonate interbeds within the Pennington Formation creates karstified intervals, which result in springs discharging from the base of the formation at the contact with the underlying Upper Mississippian-aged Bangor Limestone (91 m thick), which contains interbeds of chert in the upper part and shale in the lower part. The Upper Mississippian-aged Hartselle Sandstone underlies the Bangor Limestone but is absent or so thin in the study area that it is typically unmapped. However, there does appear to be some hydrologic control from the Hartselle Sandstone at locations along the flanks of Nat Mountain where its presence and limited vertical permeability may create small springs that issue from the upper contact of the formation and sink underground a short distance from the point of issuance [3]. The Upper Mississippian-aged Monteagle Limestone (55–67 mm thick) underlies the Hartselle Sandstone and is extensively karstified throughout the study area. Below the Monteagle Limestone is the Middle Mississippian-aged Tuscumbia Limestone (6–12 m thick) that forms the base of Nat Mountain and the valley floor of Hales Cove.

The Fern Cave System is developed in four major limestone layers—Pennington Formation, Bangor Limestone, Monteagle Limestone, and Tuscumbia Limestone—and is known for two primary entrances: a 133 m (437 ft) deep, voluminous pit (Surprise Pit) popular with recreational cavers [1] and the largest known federally endangered Gray Bat (*Myotis grisescens*) hibernaculum in the world, with over 1.3 million bats [2]; [T. Inebnit, unpub. data]. The cave is characterized by a diversity of passage morphologies, providing a range of subterranean habitats. The Fern Cave System consists of a complex network of vadose and phreatic passages with more than 12 horizontal levels of passages representing distinct stages of development associated with past periods of water table stability and intervening periods of downcutting of the Paint Rock River through the resistant caprock into the soluble limestone layers below. The hydrology of the cave system was recently described by Miller and Tobin [3]. While much of the cave is dry, at least three distinct subterranean streams occur within the cave systems: Lower North Cave Stream, Surprise Stream, and the Bottom Cave Stream. These cave streams originate as surface streams at different locations atop Nat Mountain that sink into Fern Cave and ultimately issue into springs along the eastern bank of the Paint Rock River. Many of the upper-level passages in Fern Cave have floors with steep gradients; however, the lowest level—Bottom Cave—is characterized by a relatively flat floor with associated stream and flood debris indicative of back-flooding from the Paint Rock River. A recent dye tracing study by Miller and Tobin [3] delineated a recharge area of 6.7 km^2 (2.6 mi^2). The Fern Cave System is fed largely by recharge through a combination of surface-water runoff from atop Nat Mountain and discharge of groundwater in the overlying Pennington Formation sinking into the Monteagle Limestone. The recharge area for the Fern Cave System lies primarily along the western escarpment of Nat Mountain and drains to multiple springs along the Paint Rock River. The recharge area is bounded to the southeast by the Kennamer Cave System. There is evidence of some hydrological connection between the Fern and Kennamer cave systems. On the north side, the Fern Cave recharge area is bounded by recharge areas for Roadside Spring and Big Spring that discharge into Hales Cove. Current land use within the recharge area of the cave system is predominantly mixed deciduous forest with <1% shrub/scrub and pasture [3].

Unlike other hotspot caves in North America with a long history of biological studies (e.g., Mammoth Cave in Kentucky, reviewed in [5]; San Marcos artesian well in Texas, reviewed in [6]; and Shelta Cave in Alabama, ref. [7]), much of our knowledge of the biodiversity of the Fern Cave System is derived from a recent two year biological inventory presented herein. However, early knowledge of the Fern Cave fauna is based on visits and studies by biologists from the 1950s to the 1990s. Peck [8,9] summarized the terrestrial cave life of Alabama caves and reported on eight species from the Fern Cave System, including six troglobionts. Additional significant publications on the fauna of the Fern Cave System include Muchmore [10,11], Carpenter [12], Fleming [13], Holt [14], Peck [15,16], Hart and Hart [17], Hobbs et al. [18], Kenk [19], Hubbell and Norton [20], Ferguson [21], Gertsch [22], Lewis [23], McGregor et al. [24], Martin [2], Niemiller et al. [25], and Hedin and Milne [26].

Herein we present the first comprehensive list of terrestrial and aquatic cave obligate fauna (i.e., troglobionts and stygobionts, respectively) of the Fern Cave System based on the results of a recent two-year bioinventory of the cave system in 2018–2020 and a thorough search of the scientific literature and museum records. In addition to the species list, we include a comprehensive bibliography on the cave obligate fauna of Fern Cave, discuss factors associated with its exceptional biodiversity, and comment on the conservation status of the exceptional biodiversity of this North American and global hotspot of subterranean biodiversity.

2. Materials and Methods

2.1. Ecological Classification of Troglobionts and Stygobionts

We follow past authors in recognizing troglobionts (i.e., troglobites) as species that are permanent inhabitants of subterranean habitats [27–30] and cannot complete their life cycle outside of such habitats [30]. From a metapopulation perspective, troglobionts have source populations in subterranean habitats but may have sink populations in surface habitats [28].

While morphology alone cannot be used to definitively classify species ecologically [29], we used the presence of traits often observed in troglobiotic fauna, i.e., troglomorphisms such as reduced eyes, pigmentation, and hypertrophy of nonvisual sensory structures, but not found in presumed surface relatives, as evidence for isolation in subterranean habitats. We use the terms troglobiont and stygobiont in reference to species that occur in terrestrial and aquatic habitats, respectively.

2.2. Cave Biosurveys

We conducted faunal bioinventories in several areas throughout the Fern Cave System as well as three additional caves that are hydrologically connected to Fern Cave located on Fern Cave National Wildlife Refuge between June 2018 and December 2020. Bioinventories primarily consisted of time-constrained visual encounter surveys for cave life in terrestrial, riparian, and aquatic habitats, including entrance areas and the twilight zone starting at the drip line, walls and ceilings, ledges, mud banks, rimstone pools, streams, and talus slopes. The search effort included examining and overturning rocks, detritus, organic debris, and other cover, as well as searching through stream cobble. Surveys were conducted by two to seven researchers per cave visit. In the West Passage, we supplemented visual encounter surveys with baited traps in December 2020.

We field-identified common vertebrate and invertebrate species. In other cases, we collected invertebrate specimens and identified them in the laboratory using available taxonomic keys and the literature. We outsourced identification to experts for taxa for which we had insufficient taxonomic knowledge when possible. For many vertebrates, we field-identified taxa by direct observation without capture or through taxonomically reliable indirect observations, such as visual identification of mammal scat or footprints left in mud. Where possible, we took voucher photographs of invertebrate and vertebrate taxa. For some salamanders and decapods, we collected tissue samples and voucher specimens.

2.3. Literature and Museum Searches

We conducted a search of the scientific literature to compile an updated list of troglobiont and stygobiont species for the Fern Cave System. Scientific literature sources included journal articles, book chapters, books, conference proceedings, theses and dissertations, and government reports. Searches of literature sources included keyword queries on ISI Web of Science, Google Scholar, and Zoological Record. In addition, we also searched biodiversity databases, including the Global Biodiversity Information Facility (GBIF; https://gbif.org; accessed on 24 June 2022), VertNet (http://www.vertnet.org; accessed on 24 June 2022), Symbiota Collections of Arthropods Network (SCAN; https://scan-bugs.org/portal/; accessed on 24 June 2022), and InvertEBase (http://www.invertebase.org/portal/; accessed on 24 June 2022). The list of cave-obligate fauna includes the scientific name, authority, and conservation status of each species. Taxonomic nomenclature followed primarily the Integrated Taxonomic Information System (ITIS; http://itis.gov; accessed on 15 September 2022). For conservation status, we include the International Union for Conservation of Nature (IUCN) Red List of Threatened Species (http://www.iucnredlist.org; accessed 28 March 2023) and NatureServe (http://www.natureserve.org; accessed 28 March 2023) conservation statuses when available. The status of a species according to the United States list of threatened and endangered species under the U.S. Endangered Species Act is included (http://www.fws.gov/endangered; accessed on 28 March 2023), as is its conservation status in the state of Alabama [31].

3. Results

Our list of cave-limited fauna documented within the Fern Cave System includes twenty-seven species, including nineteen troglobionts and eight stygobionts (Table 1; Figures 2 and 3). Fern Cave is the type locality for two cave-limited species, and three species are endemic to the Fern Cave System (Table 1). The cave-limited fauna represents four phyla, ten classes, eighteen orders, and twenty-six families.

Table 1. Troglobionts and stygobionts of the Fern Cave System, Jackson County, Alabama, USA. NatureServe conservation ranks include Secure (G5), Apparently Secure (G4), Vulnerable (G3), Imperiled (G2), Critically Imperiled (G1), Possibly Extinct (GH), Presumed Extinct (GX), Unranked (GNR), and Unrankable (GU). IUCN Red List categories include Least Concern (LE), Near Threatened (NT), Vulnerable (VU), Endangered (EN), Critically Endangered (CR), Extinct in the Wild (EW), and Extinct (EX). Federal conservation status under the U.S. Endangered Species Act includes Listed Endangered (LE) and Listed Threatened (LT). Alabama Department of Conservation and Natural Resources statuses include Highest Conservation Concern (Priority 1), High Conservation Concern (Priority 2), Moderate Conservation Concern (Priority 3), Low Conservation Concern (Priority 4), and Lowest Conservation Concern (Priority 5). Abbreviations: na—conservation status is not available for the species; X—species has been reported historically or during biosurveys conducted during the current study.

Taxon	Authority	NatureServe Status	IUCN Red List	Federal Status	State Status	Historical	This Study
TROGLOBIONTS							
Phylum Arthropoda							
Class Arachnida							
Order Araneae							
Family Linyphiidae							
Phanetta subterranea	(Emerton, 1875)	G5	na				X
Family Nesticidae							
Nesticus barri	Gertsch, 1984	G3	na			X	X
Family Zoropsidae							
Liocranoides unicolor	Keyserling, 1881	GU	na				X
Order Pseudoscorpiones							
Family Chernetidae							
Hesperochernes mirabilis	(Banks, 1895)	G5	na				X
Family Chthoniidae							
Tyrannochthonius torodei [T,E]	Muchmore, 1996	G1	na			X	X
Family Neobisiidae							
Alabamocreagris mortis [T,E]	(Muchmore, 1969)	G1G2	na			X	X
Order Acari							
Family Rhagidiidae							
Unidentified genus and species		na	na				X
Class Collembola							
Order Entomobryomorpha							
Family Entomobryidae							
Pseudosinella hirsuta	(Delmare DeBoutteville, 1949)	G5	na			X	X
Pseudosinella spinosa	(Delmare DeBoutteville, 1949)	G5	na				X
Order Symphypleona							
Family Arrhopalitidae							
Pygmarrhopalites sp.		na	na				X
Class Diplura							
Order Rhabdura							
Family Campodeidae							
Litocampa valentinei	(Conde, 1949)	G3	na			X	X

Table 1. *Cont.*

Taxon	Authority	NatureServe Status	IUCN Red List	Federal Status	State Status	Historical	This Study
Class Diplopoda							
Order Callipodida							
Family Abacionidae							
Tetracion jonesi	Hoffman, 1956	G3G4	na				X
Order Chordeumatida							
Family Cleidogonidae							
Pseudotremia sp.		na	na				X
Family Trichopetalidae							
Scoterpes sp.		na	na				X
Order Polydesmida							
Family Xystomdesmidae							
Gyalostethus sp. nov. [E]		na	na				X
Class Insecta							
Order Coleoptera							
Family Carabidae							
Pseudanophthalmus sp.		na	na				X
Family Leiodidae							
Ptomaphagus hatchi	Jeannel, 1933	G3	na			X	X
Family Staphylinidae							
Subfamily Pselaphinae							
Unidentified genus and species		na	na				X
Order Diptera							
Family Sphaeroceridae							
Spelobia tenebrarum	(Aldrich, 1897)	G5	na				X
STYGOBIONTS							
Phylum Platyhelminthes							
Class Turbellaria							
Order Tricladida							
Family Kenkiidae							
Sphalloplana percoeca	(Packard, 1879)	G5	na			X	X
Phylum Annelida							
Class Oligochaeta							
Order Branchiobdellida							
Family Branchiobdellidae							
Cambarincola sheltensis	Holt, 1973	G1G2	na			X	

Table 1. *Cont.*

Taxon	Authority	NatureServe Status	IUCN Red List	Federal Status	State Status	Historical	This Study
Phylum Arthropoda							
Class Malacostraca							
Order Amphipoda							
Family Crangonyctidae							
Stygobromus sp.		na	na				X
Order Decapoda							
Family Atyidae							
Palaemonias alabamae	Smalley, 1961	G2G3	EN	LE	Priority 1		X
Family Cambaridae							
Orconectes australis	(Rhoades, 1941)	G5	LC		Priority 3	X	X
Order Isopoda							
Family Asellidae							
Caecidotea bicrenata	(Steeves, 1963)	G5	na			X	X
Class Ostracoda							
Order Podocopida							
Family Entocytheridae							
Sagittocythere barri	(Hart and Hobbs, 1961)	G5	na			X	
Phylum Chordata							
Class Actinopterygii							
Order Percopsiformes							
Family Amblyopsidae							
Typhlichthys subterraneus	Girard, 1859	G4	NT		Priority 3	X	X

[T] Type locality in the Fern Cave System; [E] Endemic to the Fern Cave System.

Figure 2. Representative terrestrial cave-limited fauna from the Fern Cave System, Alabama, USA: (**A**) *Pseudosinella hirsuta* (photo by Michael E. Slay); (**B**) *Liocranoides unicolor* (photo by Matthew L. Niemiller); (**C**) *Litocampa valentinei* (photo by Matthew L. Niemiller); (**D**) *Tyrannochthonius torodei* (photo by Michael E. Slay); (**E**) *Alabamocreagris mortis* (photo by Matthew L. Niemiller); (**F**) *Tetracion jonesi* (photo by Matthew L. Niemiller); (**G**) *Nesticus barri* (photo by Michael E. Slay); and (**H**) *Gyalostethus* sp. nov. (photo by Michael E. Slay).

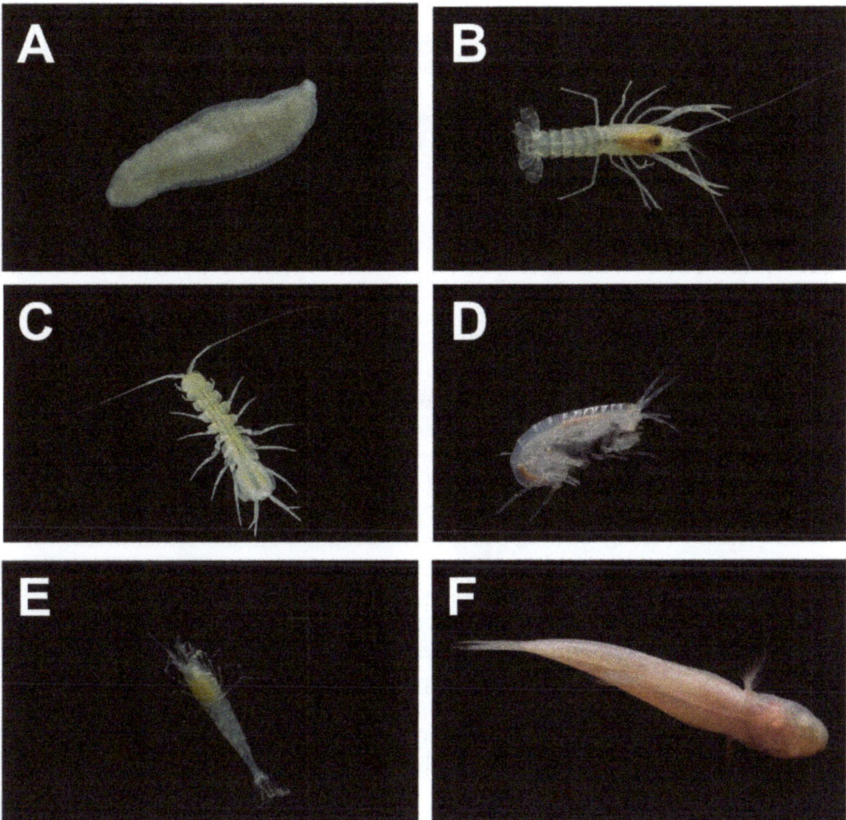

Figure 3. Representative aquatic cave-limited fauna from the Fern Cave System, Alabama, USA: (**A**) *Sphalloplana percoeca* (photo by Matthew L. Niemiller); (**B**) *Orconectes australis* (photo by Matthew L. Niemiller); (**C**) *Caecidotea bicrenata* (photo by Matthew L. Niemiller); (**D**) *Stygobromus* sp. (photo by Michael E. Slay); (**E**) *Palaemonias alabamae* (photo by Matthew L. Niemiller); and (**F**) *Typhlichthys subterraneus* (photo by Matthew L. Niemiller).

3.1. Terrestrial Fauna

Troglobiotic spiders documented in the Fern Cave System include one linyphiid, one nesticid, and one zoropsid. *Phanetta subterranea* has one of the largest distributions of any troglobiont in North America [32,33]. *Nesticus barri* is one of the most common troglobionts in dozens if not hundreds of caves along the escarpments of the Southern Cumberland Plateau in South-Central Tennessee and Northeastern Alabama, reaching its southwestern range limit in the Fern Cave area [22,26,34,35]. *Liocranoides unicolor* has a broad distribution throughout much of the Interior Low Plateau, including several caves in Northeastern Alabama [8]. This species is pale in coloration but does not possess other troglomorphic characters [36,37].

Three troglobiotic pseudoscorpions occur in the Fern Cave System. *Hesperochernes mirabilis* is a widely distributed species most abundant near entrances, where it is associated with bat guano, rodent nests, and other mammal scat [38–40]. The other two species are typically associated with deep cave habitats. *Tyrannochthonius torodei* was described from Fern Cave and named after William Torode, who first collected specimens in 1968 [11]. *Alabamocreagris mortis* is the largest of three troglobiotic pseudoscorpions in the Fern Cave System. It was described from specimens collected near the Morgue Entrance by Muchmore [10] and is the most widely distributed pseudoscorpion within the Fern Cave System.

At least one unidentified troglobiotic rhagidiid mite is known from the Fern Cave System, which was observed from several sections of the cave. This mite may be a species in the genus *Rhagidia*, which has been reported from caves in Northwestern Georgia [38,40].

Four troglobiotic millipedes have been documented in the Fern Cave System, including one callipodidan, two chordeumatids, and one polydesmid. *Tetracion jonesi* is the most widely distributed and frequently observed millipede in the Fern Cave System. This large cave-limited millipede has been reported from more than 85 caves along the Cumberland Plateau in Northeastern Alabama [8]. *Pseudotremia* is a diverse genus with several described troglobionts, including three from Alabama [41]. Troglobiotic individuals have been observed in multiple locations within the Fern Cave System and may represent one of the three described species in Alabama or an undescribed taxon. Millipedes of the genus *Scoterpes* have been observed throughout the Fern Cave System, often found in moist habitats with organic matter (rotting wood, debris, and cricket guano). The distributions of these two species—*S. syntheticus* and *S. stewartpecki*—overlap in the area, and two species are known to co-occur at nearby Crossings Cave [42]. Preliminary genetic analyses indicate that both forms likely occur in the Fern Cave System, but additional morphological analyses are required for confirmation. An undescribed troglobiotic species of *Gyalostethus* was discovered in cracks and under rocks on mud banks near the sump in the Bottom Cave section. This new species appears to be a relative of *Gyalostethus monitcolens*, the lone species currently described in this genus, which ranges from Southwestern Virginia to Northern Georgia and Northeastern Alabama, including Jackson and Morgan counties [43].

At least three species of cave-limited collembolans (i.e., springtails) have been documented from the Fern Cave System. Both *Pseudosinella hirsuta* and *P. spinosa* are broadly distributed in the Interior Low Plateau and were observed in several locations throughout the cave system. *Pseudosinella hirsuta* was reported previously from Fern Cave [9]. Troglomorphic individuals of the genus *Pygmarrhopalites* were collected from several locations in the Fern Cave System. These individuals have extremely reduced pigment and a troglomorphic foot complex. No troglobiotic *Pygmarrhopalites* have been reported from Alabama to date, but several species have been described from the Interior Low Plateau and Appalachians Karst regions in the Eastern United States [44–47]. The troglophile *Pygmarrhopalites pygmaeus*, a cosmopolitan species, was also observed and has been reported from several caves in Jackson, Madison, and Marshall counties [9].

A single troglobiotic dipluran occurs in the Fern Cave System. *Litocampa valentinei* is known from several caves in Northeastern Alabama and South-Central Tennessee along the escarpments of the Cumberland Plateau [21]. This cave dipluran has been reported previously from the Fern Cave System [9,21].

The troglobiotic beetle fauna of the Fern Cave System includes one carabid, one leiodid, and one staphylinid. Five carabids of the troglobiotic genus *Pseudanophthalmus* were collected from the Fern Cave System. These specimens may be *P. profundus*, which is known from nearby Crossings, Nat, and Paint Rock caves, and/or an undescribed species, which is known from nearby Pig Pen Cave [9]. The round fungus beetle *Ptomaphagus hatchi* was observed throughout the Fern Cave System, often in great abundance. This species was previously reported by Peck [9,15,16]. An unidentified cave ant beetle (subfamily Pselaphinae) was collected from the West Passage and likely belongs to either the genus *Batrisodes* or *Speleochus*. Several species of both genera are known from caves in Alabama [9,48–50].

The only other troglobiotic insect documented from the Fern Cave System is the dipteran *Spelobia tenebrarum*, which has been reported from many caves in the Eastern United States [9,40,51–53], where it is associated with scat. This species has reduced eyes and is the only known troglobiotic fly in the United States [51,52].

3.2. Aquatic Fauna

One cave flatworm—*Sphalloplana percoeca*—has been reported previously from the Fern Cave System [12,19]. *Sphalloplana percoeca* occurs primarily in epikarst-fed drip pools in upper-level passages, often in great numbers, but has also been observed in lower-level passages. A medium-sized branchiobdellid (2.5 mm), *Cambarincola sheltensis*, is an ectosymbiont of the stygobiotic crayfish *Orconectes australis*. This species has been confirmed from the type locality of Shelta Cave in Madison County, but identifications also include specimens taken from *O. australis* from Fern Cave collected by John E. and Martha R. Cooper [14].

The cave-limited crustacean fauna includes one shrimp, one crayfish, one isopod, one amphipod, and one ostracod. The federally endangered stygobiotic shrimp *Palaemonias alabamae* was discovered in August 2018 when four individuals were observed in an isolated pool near the Davison Entrance in the Bottom Cave section. Two cave shrimp were observed in July 2019: one in the same isolated pool and a second shrimp in a pool in the main stream. A single cave shrimp was again observed in the isolated pool in September 2020. Cave shrimp have yet to be observed upstream in Bottom Cave. Morphological and genetic analyses confirmed that this population is *P. alabamae* [25]. Preliminary environmental DNA analyses detected *P. alabamae* eDNA from water samples collected from Haley Spring Cave in addition to the water samples collected from pools near the Davison Entrance. This discovery represents the first new occurrence for this federally endangered species in 14 years and just the fifth population discovered to date, extending the geographic range into the Paint Rock River watershed [25]. The stygobiotic crayfish *Orconectes australis* is common in streams in both upper- and lower-level passages of the Fern Cave System. McGregor et al. [24] observed 34 individuals, including a female with eggs, in the Davison section of Bottom Cave in September 1993. This cave crayfish has also been reported from the Fern Cave System by Holt [14] and Hobbs et al. [18].

The isopod *Caecidotea bicrenata* was found throughout the Fern Cave System in several habitats, including stream riffles and pools, rimstone pools, and drip pools. *Caecidotea bicrenata* is widely distributed throughout the Interior Low Plateau [23] but may represent a cryptic species complex. This stygobiotic asellid was previously reported from the Fern Cave System by Fleming [13] and Lewis [23]. Cave amphipods of the genus *Stygobromus* were collected from isolated drip pools in the Morgue Pit area, West Passage, and Bottom Cave sections of the Fern Cave System. These specimens may be *S. vitreus* or *S. dicksoni*, which are known from Jackson and Madison counties [54], or an undescribed species. The ostracod *Sagittocythere barri* is an ectocommensal of the stygobiotic crayfish *Orconectes australis* and was reported from the Fern Cave System by Hart and Hart [17].

The only cave-limited vertebrate known from the Fern Cave System is the amblyopsid cavefish, *Typhlichthys subterraneus*. This cavefish was abundant in the stream in Bottom Cave and is considered a top aquatic predator. *Typhlichthys subterraneus* is a cryptic species complex [55] with two lineages contacting in Western Madison/Eastern Jackson counties. McGregor et al. [24] reported this cavefish previously.

4. Discussion

The Fern Cave obligate cave fauna is exceptionally rich with 27 troglobionts and stygobionts, making it one of the most diverse cave systems in North America. The terrestrial fauna is particularly diverse, with 19 species trailing only the Mammoth Cave System in North America (49 species overall, 32 troglobionts; ref. [5]). The stygofauna of the Fern Cave System is diverse (eight species) but not exceptional compared to other hotspot subterranean communities in North America, such as the San Marcos Artesian Well in Texas (55 species; ref. [6]), the Mammoth Cave System in Kentucky (17 species; ref. [5]), and Shelta Cave in Alabama (12 species; refs. [7,56]).

The exceptional cave-limited diversity within the Fern Cave System may be explained by several factors that operate in concert to influence patterns of subterranean biodiversity and endemism in the region. First, Fern Cave lies along the escarpments of the Southern Cumberland Plateau, which is highly dissected with numerous karst exposures and cave systems. The expansive karst and higher cave density in the region are expected to support greater species richness [57,58], but they may also offer greater dispersal opportunities [57]. Greater cave density may also provide increased opportunities for colonization of subterranean habitats [58]. The Fern Cave System lies within a hypothesized mid-latitude biodiversity ridge for terrestrial subterranean fauna, which is associated with long-term higher surface productivity and a favorable climate, particularly during the Pleistocene [58]. Within this region, cave systems likely have greater energy inputs from allochthonous sources to support more species, larger populations, and consequently lower extinct rates relative to other karst regions [57–59].

Of the 18 cave-limited species with a NatureServe conservation rank, half of the species (five troglobionts and four stygobionts) are of conservation concern (i.e., G1–G3 Nature-Serve conservation rank, federal or state status), highlighted by the federally endangered Alabama Cave Shrimp *Palaemonias alabamae* discovered in 2018 [25]. Most of these species are at an elevated risk of extinction due to their limited distributions and/or their few known occurrences. In particular, the cave pseudoscorpions *Tyrannochthonius torodei* and *Alabamocreagris mortis* are known only from the Fern Cave System. Cave-limited fauna face many threats, such as habitat loss and degradation, groundwater overexploitation and contamination, and climate change [60,61]. Although much of the Fern Cave System (and all five entrances) lies within the boundaries of the Fern Cave National Wildlife Refuge and the SCCi Kay Hill Deen Fern Cave Preserve, the cave system is not entirely immune to potential direct and indirect threats to its biodiversity. Nearly all (99%) of the 6.7 km^2 recharge area of the Fern Cave System is composed of mixed deciduous forest, which suggests a minimal risk of groundwater pollution [3]. However, future land use modifications, such as possible logging, could impact groundwater quality, although the risk is low at present. The Paint Rock River is known to backflood into the lowest level of the Fern Cave System [3]. Consequently, water quality in this section may be influenced by the water quality and flood stage of the Paint Rock River. Fortunately, the Paint Rock River has escaped most of the adverse anthropogenic impacts of other major tributaries of the Tennessee River; however, non-point source pollution of low to moderate magnitude primarily caused by nutrient enrichment has been identified within the Paint Rock watershed [62].

The list of cave-limited fauna within the Fern Cave System is likely incomplete, and there is great potential to discover new taxa. Much of the 25+ km of passage remains to be comprehensively bioinventoried, and some habitats, such as epikarst, are under-sampled and may harbor additional taxa. For example, an undescribed *Anillinus* beetle was collected from Magic City Cave on Fern Cave National Wildlife Refuge, which is likely hydrologically connected to the Fern Cave System. Most species in this genus are small (1–2 mm) litter or soil-dwelling inhabitants that are depigmented and lack eyes and wings, but several species are considered troglobionts, including some from Alabama [63]. In addition to the possibility of two *Scoterpes* millipedes co-occurring, it would be unsurprising if at least two species of *Pseudanophthalmus* cave beetles co-occur within the Fern Cave System, which is a common occurrence throughout much of the Interior Low Plateau and Appalachian Karst regions (e.g., refs. [64,65]). Several taxa are notably absent from the cave-limited fauna of Fern Cave, including terrestrial cavesnails, terrestrial woodlice, stygobiotic copepods, and stygobiotic salamanders, all of which may be discovered in the future. For example, three terrestrial cavesnails are known from Northern Alabama, including *Helicodiscus barri*, which has a broad distribution throughout the Interior Low Plateau karst region [66,67]. Twelve stygobiotic copepods occur in the Interior Low Plateau [68], but only one, *Diacyclops alabamensis*, has been reported from Alabama. *Gyrinophilus palleucus* is a top aquatic predator of many cave streams in Northern Alabama, including caves within the Paint Rock River watershed [69–71]. It is known to co-occur with *Orconectes australis*, *Typhlichthys subterraneus*, and *Palaemonias alabamae* [71], all of which occur within Fern Cave. Other taxonomic groups have not been particularly well studied in the Fern Cave System, including springtails and mites. More intensive biosurvey work on these groups may uncover additional taxa. Finally, few phylogenetic studies to date have incorporated specimens and samples from the Fern Cave System. Comprehensive sampling within the Fern Cave System has the potential to uncover cryptic diversity in some taxonomic groups, such as stygobiotic isopods and amphipods, which is an increasingly common discovery of phylogenetic studies in cave-limited taxa [55,72,73].

Author Contributions: Conceptualization: M.L.N.; methodology and analysis: M.L.N., M.E.S. and T.I.; data acquisition: M.L.N., M.E.S., T.I., B.M., B.T., B.C., A.H., B.D.J., N.M., K.D.K.N. and S.P.; original draft preparation: M.L.N., M.E.S. and T.I.; review and editing: M.L.N., M.E.S., T.I., B.M., B.T., B.C., A.H., B.D.J., N.M., K.D.K.N. and S.P. All authors have read and agreed to the published version of the manuscript.

Funding: The National Wildlife Refuge System (NWR) of the US Fish and Wildlife Service awarded funding for this project through the NWR's Inventory and Monitoring Program (Agreement no. F18AC00681). M.L. Niemiller was supported in part by the National Science Foundation (award no. 2047939), the Cave Conservancy Foundation, and the Alabama Department of Conservation and Natural Resources.

Institutional Review Board Statement: Not applicable.

Data Availability Statement: No new data were created or analyzed in this manuscript. Data sharing is not applicable for this manuscript.

Acknowledgments: We thank Pedro Ardapple, Reilly Blackwell, Forbes Boyle, Daniel Chapman, Kevin England, Mark Jones, Jacob Lieber, Hannah Lieffring, Pete Pattavina, Jennifer Pinkley, Dave Richardson, Mike Senn, Nick Sharp, Elliot Stahl, Matt Tomlinson, and Drew Westerman for assistance with fieldwork. We thank Marshal Hedin, Paul Marek, Marc Milne, Karen Ober, and Charles Stephen for identifying specimens. M.L. Niemiller and K.D.K. Niemiller thank E.R. Niemiller for his cooperation in utero during the study. The collection of specimens was authorized under ALDCNR scientific permit nos. 2018035450068680, 2018061776268680, 2018061777068680, 2019060225068680, 2019060224868680, 2020083527668680, and 2020083528068680. The views presented herein are those of the authors and do not necessarily represent those of the U.S. Fish and Wildlife Service or the U.S. Geological Survey.

Conflicts of Interest: Author S.P. was employed by the company Southeastern Cave Conservancy, Inc. The remaining authors declare that the research was conducted in the absence of any commercial or financial relationships that could be construed as a potential conflict of interest.

References

1. Pinkley, J.E. *Fern Cave: The Discovery, Exploration, and History of Alabama's Greatest Cave*; Blue Bat Books: Knoxville, TN, USA, 2014.
2. Martin, C.O. *Assessment of the Population Status of the Gray Bat (Myotis grisescens)*; Report ERDC/EL TR-07-22; U.S. Army Corps of Engineers: Vicksburg, MI, USA, 2007.
3. Miller, B.V.; Tobin, B. Mapping karst groundwater flow paths and delineating recharge areas for Fern Cave, Alabama through the use of dye tracing. U.S. Geological Survey Scientific Investigation Map; *in press*.
4. Osborne, W.E.; Ward, W.E., II; Irvin, G.D. Geologic Map and Cross Sections of the Paint Rock 7.5-Minute Quadrangle, Jackson and Madison Counties, Alabama. Geological Survey of Alabama Quadrangle Series Map QS59, Scale 1:24,000. 2013. Available online: https://www.gsa.state.al.us/img/Geological/Quads/QS59/QS59_Plate.pdf (accessed on 27 April 2023).
5. Niemiller, M.L.; Helf, K.; Toomey, R.S. Mammoth Cave: A hotspot of subterranean biodiversity in the United States. *Diversity* **2021**, *13*, 373. [CrossRef]
6. Hutchins, B.T.; Gibson, J.R.; Diaz, P.H.; Schwartz, B.F. Stygobiont diversity in the San Marcos Artesian Well and Edwards Aquifer groundwater ecosystem, Texas, USA. *Diversity* **2021**, *13*, 234. [CrossRef]
7. Culver, D.C.; Sket, B. Hotspots of subterranean biodiversity in caves and wells. *J. Cave Karst Stud.* **2000**, *62*, 11–17.
8. Peck, S.B. The cave fauna of Alabama: Part I. The terrestrial invertebrates (excluding insects). *Nat. Speleolog. Soc. Bull.* **1989**, *51*, 11–33.
9. Peck, S.B. The cave fauna of Alabama. Part II: The insects. *Nat. Speleolog. Soc. Bull.* **1995**, *40*, 39–63.
10. Muchmore, W.B. New species and records of cavernicolous pseudoscorpions of the genus *Microcreagris* (Arachnida, Chelonethida, Neobisiidae, Ideobisiinae). *Am. Mus. Novit.* **1969**, *2392*, 1–21.
11. Muchmore, W.B. The genus *Tyrannochthonius* in the eastern United States (Pseudoscorpionida: Chthoniidae). Part II. More recently described species. *Insecta Mundi* **1996**, *10*, 153–168.
12. Carpenter, J.H. Systematics and Ecology of Cave Planarians of the United States. Ph.D. Thesis, University of Kentucky, Lexington, KY, USA, 1970.

13. Fleming, L.E. The evolution of the eastern North American isopods of the genus *Asellus* (Crustacea: Asellidae). *Part I. Int. J. Speleol.* **1972**, *4*, 221–256. [CrossRef]
14. Holt, P.C. Branchiobdellids (Annelida: Clitellata) from some eastern North American caves, with descriptions of new species of the genus *Cambarincola*. *Int. J. Speleol.* **1973**, *5*, 219–256. [CrossRef]
15. Peck, S.B. A systematic revision and evolutionary biology of the *Ptomaphagus adelops*. *Bull. Mus. Comp. Zool.* **1973**, *145*, 29–162.
16. Peck, S.B. The distribution and evolution of cavernicolous *Ptomaphagus* beetles in the southeastern United States (Coleoptera; Leiodidae; Cholevinae) with new species and records. *Can. J. Zool.* **1984**, *62*, 730–740. [CrossRef]
17. Hart, D.G.; Hart, C.W., Jr. The ostracod family Entocytheridae. *Acad. Nat. Sci. Phila. Monogr.* **1974**, *18*, 1–239.
18. Hobbs, H.H., Jr.; Hobbs, H.H., III; Daniel, M.A. A review of the troglobitic decapod crustaceans of the Americas. *Smithson. Contrib. Zool.* **1977**, *244*, 1–83. [CrossRef]
19. Kenk, R. Freshwater triclads (Turbellaria) of North America, IX. The genus *Sphalloplana*. *Smithson. Contrib. Zool.* **1977**, *246*, 1–38. [CrossRef]
20. Hubbell, T.H.; Norton, R.M. The systematics and Biology of the cave crickets of the North American tribe Hadenoecini (Orthoptera), Saltatoria, Ensifera, Rhaphidophoridae, Dolichopodinae. *Misc. Publ. Mus. Zool. Univ. Mich.* **1978**, *156*, 1–124.
21. Ferguson, L.M. Systematics, Evolution, and Zoogeography of the Cavernicolous Campodeids of the Genus *Litocampa* (Diplura: Campodeidae) in the United States. Ph.D. Dissertation, Virginia Polytechnic Institute and State University, Blacksburg, VA, USA, 1981.
22. Gertsch, W.J. The spider family Nesticidae (Araneae) in North America, Central America, and the West Indies. *Tex. Mem. Mus. Bull.* **1984**, *31*, 1–91.
23. Lewis, J.J. The Systematics, Zoogeography and Life History of the Troglobitic Isopods of the Interior Plateaus of the Eastern United States. Ph.D. Thesis, University of Louisville, Louisville, KY, USA, 1988.
24. McGregor, S.W.; Rheams, K.F.; O'Neil, P.E.; Moser, P.H.; Blackwood, R. *Biological, Geological, and Hydrological Investigations in Bobcat, Matthews, and Shelta Caves and Other Selected Caves in North Alabama*; Open-File Report; Geological Survey of Alabama: Tuscaloosa, AL, USA, 1994.
25. Niemiller, M.L.; Inebnit, T.; Hinkle, A.; Jones, B.D.; Jones, M.; Lamb, J.; Mann, N.; Miller, B.; Pinkley, J.; Pitts, S.; et al. Discovery of a new population of the federally endangered Alabama Cave Shrimp, *Palaemonias alabamae* Smalley, 1961, in northern Alabama. *Subterr. Biol.* **2019**, *32*, 43–59. [CrossRef]
26. Hedin, M.; Milne, M.A. New species in old mountains: Integrative taxonomy reveals ten new species and extensive short-range endemism in *Nesticus* spiders (Araneae, Nesticidae) from the southern Appalachian Mountains. *ZooKeys* **2023**, *1145*, 1–30. [CrossRef]
27. Sket, B. Can we agree on an ecological classification of subterranean animals? *J. Nat. Hist.* **2008**, *42*, 1549–1563. [CrossRef]
28. Trajano, E. Ecological classification of subterranean organisms. In *Encyclopedia of Caves*, 2nd ed.; White, W.B., Culver, D.C., Eds.; Academic/Elsevier Press: Amsterdam, The Netherlands, 2012; pp. 275–277.
29. Trajano, E.; de Carvalho, M.R. Towards a biologically meaningful classification of subterranean organisms: A critical analysis of the Schiner-Racovitza system from a historical perspective, difficulties of its application and implications for conservation. *Subterr. Biol.* **2017**, *22*, 1–26. [CrossRef]
30. Culver, D.C.; Pipan, T. Ecological and evolutionary classifications of subterranean organisms. In *Encyclopedia of Caves*, 3rd ed.; Culver, D.C., White, W.B., Pipan, T., Eds.; Elsevier: Amsterdam, The Netherlands, 2019; pp. 376–379.
31. Alabama Department of Conservation and Natural Resources. *Alabama's Wildlife Action Plan 2015–2025*; Division of Wildlife and Freshwater Fisheries, Alabama Department of Conservation and Natural Resources: Montgomery, AL, USA, 2015.
32. Barr, T.C., Jr.; Holsinger, J.R. Speciation in cave faunas. *Ann. Rev. Ecol. Syst.* **1985**, *16*, 313–337. [CrossRef]
33. Peck, S.B. A summary of diversity and distribution of the obligate cave-inhabiting faunas of the United States and Canada. *J. Caves Karst Stud.* **1998**, *60*, 18–26.
34. Snowman, C.V.; Zigler, K.S.; Hedin, M. Caves as islands: Mitochondrial phylogeography of the cave-obligate spider species *Nesticus barri* (Araneae: Nesticidae). *J. Arachnol.* **2010**, *38*, 49–56. [CrossRef]
35. Niemiller, M.L.; Zigler, K.S.; Fenolio, D.B. *Cave Life of TAG: A Guide to Commonly Encountered Species in Tennessee, Alabama and Georgia*; Biology Section of the National Speleological Society: Huntsville, AL, USA, 2013.
36. Keyserling, E.G. Neue Spinnen aus Amerika, III. *Verh. Zool. Bot. Ges. Wien* **1881**, *31*, 269–314. [CrossRef]
37. Platnick, N.I. A revision of the Appalachian spider genus *Liocranoides* (Araneae: Tengellidae). *Am. Mus. Novit.* **1999**, *3285*, 1–13.
38. Holsinger, J.R.; Peck, S.B. The invertebrate cave fauna of Georgia. *Nat. Speleolog. Soc. Bull.* **1971**, *33*, 23–44.
39. Reeves, W.K.; Jensen, J.B.; Ozier, J.C. New faunal and fungal records from caves in Georgia, USA. *J. Cave Karst Stud.* **2000**, *62*, 169–179.
40. Zigler, K.S.; Niemiller, M.L.; Stephen, C.D.R.; Ayala, B.N.; Milne, M.A.; Gladstone, N.S.; Engel, A.S.; Jensen, J.B.; Camp, C.D.; Ozier, J.C.; et al. Biodiversity from caves and other subterranean habitats of Georgia, USA. *J. Cave Karst Stud.* **2020**, *82*, 125–167. [CrossRef]
41. Shear, W.A. Studies in the milliped order Chordeumida (Diplopoda): A revision of the family Cleidogonidae and a reclassification of the order Chordeumida in the New World. *Bull. Mus. Comparat. Zool.* **1972**, *144*, 151–352.

42. Shear, W.A. The milliped family Trichopetalidae, Part 2: The genera *Trichopetalum*, *Zygonopus* and *Scoterpes* (Diplopoda: Chordeumatida, Cleidogonoidea). *Zootaxa* **2010**, *2385*, 1–62. [CrossRef]

43. Hoffman, R.L. Revision of the milliped genera *Boraria* and *Gyalostethus* (Polydesmida: Xystodesmidae). *Proc. United States Natl. Mus.* **1965**, *117*, 305–348. [CrossRef]

44. Christiansen, K.; Bellinger, P. Cave *Arrhopalites*: New to science. *J. Cave Karst Stud.* **1996**, *58*, 168–180.

45. Christiansen, K.; Bellinger, P. *The Collembola of North America North of the Rio Grande. A Taxonomic Analysis*, 2nd ed.; Grinnell College: Grinnell, IA, USA, 1998.

46. Zeppelini, D.; Christiansen, K. *Arrhopalites* (Collembola: Arrhopalitidae) in U.S. caves with the description of seven new species. *J. Cave Karst Stud.* **2003**, *65*, 36–42.

47. Zeppelini, D.; Taylor, S.J.; Slay, M.E. Cave *Pygmarrhopalites* Vargovitsh, 2009 (Collembola, Symphypleona, Arrhopalitidae) in United States. *Zootaxa* **2009**, *2204*, 1–18. [CrossRef]

48. Park, O. Cavernicolous pselaphid beetles of Alabama and Tennessee. *Geol. Surv. Ala. Mus. Pap.* **1951**, *31*, 1–107.

49. Park, O. New or little-known species of pselaphid beetles, chiefly from southeastern United States. *J. Tenn. Acad. Sci.* **1958**, *33*, 39–74.

50. Park, O. Cavernicolous pselaphid beetles of the United States. *Am. Midl. Nat.* **1960**, *64*, 66–104. [CrossRef]

51. Marshall, S.A.; Peck, S.B. Distribution of cave-dwelling Sphaeroceridae (Diptera) of eastern North America. *Proc. Entomol. Soc. Ont.* **1984**, *115*, 37–41.

52. Marshall, S.A.; Peck, S.B. The origin and relationships of *Spelobia tenebrarum* Aldrich, a troglobitic, eastern North American sphaerocerid fly. *Can. Entomol.* **1985**, *117*, 1013–1015. [CrossRef]

53. Lewis, J.J. *Bioinventory of Caves of the Cumberland Escarpment Area of Tennessee*; Final Report to Tennessee Wildlife Resources Agency & The Nature Conservancy of Tennessee; Lewis & Associates, LLC: Clarksville, IN, USA, 2005.

54. Holsinger, J.R. Systematics of the subterranean amphipod genus *Stygobromus* (Crangonyctidae): Part II. Species of the eastern United States. *Smithson. Contrib. Zool.* **1978**, *266*, 1–144.

55. Niemiller, M.L.; Near, T.J.; Fitzpatrick, B.M. Delimiting species using multilocus data: Diagnosing cryptic diversity in the southern cavefish, *Typhlichthys subterraneus* (Teleostei: Amblyopsidae). *Evolution* **2012**, *66*, 846–866. [CrossRef]

56. Cooper, J.E. Ecological and Behavioral Studies in Shelta Cave, Alabama, with Emphasis on Decapod Crustaceans. Ph.D. Thesis, University of Kentucky, Lexington, KY, USA, 1975.

57. Niemiller, M.L.; Zigler, K.S. Patterns of cave biodiversity and endemism in the Appalachians and Interior Plateau of Tennessee, USA. *PLoS ONE* **2013**, *8*, e64177. [CrossRef] [PubMed]

58. Culver, D.C.; Deharveng, L.; Bedos, A.; Lewis, J.J.; Madden, M.; Reddell, J.R.; Sket, B.; Trontelj, P.; White, D. The mid-latitude biodiversity ridge in terrestrial cave fauna. *Ecography* **2006**, *29*, 120–128. [CrossRef]

59. Culver, D.C.; Master, L.L.; Christman, M.C.; Hobbs, H.H. Obligate cave fauna of the 48 contiguous United States. *Conserv. Biol.* **2000**, *14*, 386–401. [CrossRef]

60. Niemiller, M.L.; Bichuette, E.; Taylor, S.J. Conservation of cave fauna in Europe and the Americas. In *Ecological Studies: Cave Ecology*; Moldovan, O.T., Kovac, L., Halse, S., Eds.; Springer: Dordrecht, The Netherlands, 2018; pp. 451–478.

61. Mammola, S.; Cardoso, P.; Culver, D.C.; Deharveng, L.; Ferreira, R.L.; Fišer, C.; Galassi, D.M.P.; Griebler, C.; Halse, S.; Humphreys, W.F.; et al. Scientists' warning on the conservation of subterranean ecosystems. *BioScience* **2019**, *69*, 641–650. [CrossRef]

62. Barbour, M.S. *Paint Rock River Watershed Nonpoint Source Pollution*; Unpublished Report to Alabama Department of Environmental Management, Montgomery, Alabama; Alabama Natural Heritage Program: Montgomery, AL, USA, 2003; 184p.

63. Sokolov, I.M. Four new species of the genus *Anillinus* Casey (Coleoptera, Carabidae, Anillini) from Alabama, U.S.A., with a revised key to the Alabama species. *Zootaxa* **2020**, *4808*, 547–559. [CrossRef] [PubMed]

64. Barr, T.C., Jr. *A Classification and Checklist of the Genus Pseudanophthalmus Jeannel (Coleoptera: Carabidae: Trechinae)*; Special Publication 11; Virginia Museum of Natural History: Martinsville, VA, USA, 2004.

65. Ober, K.A.; Niemiller, M.L.; Philips, T.K. Cave trechine (Coleoptera: Carabidae) diversity and biogeography in North America. In *Cave Life—Drivers of Diversity and Diversification*; Wynne, J.J., Ed.; John Hopkins Press: Baltimore, MD, USA, 2022; pp. 192–223.

66. Gladstone, N.S.; Carter, E.T.; McKinney, M.L.; Niemiller, M.L. Status and distribution of the cave-obligate land snails in the Appalachians and Interior Low Plateau of the eastern United States. *Am. Malacol. Bull.* **2018**, *36*, 62–78. [CrossRef]

67. Gladstone, N.S.; Niemiller, M.L.; Pieper, E.B.; Dooley, K.E.; McKinney, M.L. Morphometrics and phylogeography of the cave-obligate land snail *Helicodiscus barri* (Gastropoda, Stylommatophora, Helicodiscidae). *Subterr. Biol.* **2019**, *30*, 1–32. [CrossRef]

68. Niemiller, M.L.; Taylor, S.J.; Slay, M.E.; Hobbs, H.H., III. Biodiversity in the United States and Canada. In *Encyclopedia of Caves*, 3rd ed.; Culver, D.C., White, W.B., Pipan, T., Eds.; Elsevier: Amsterdam, The Netherlands, 2019; pp. 163–176.

69. Miller, B.T.; Niemiller, M.L. Distribution and relative abundance of Tennessee cave salamanders (*Gyrinophilus palleucus* and *G. gulolineatus*) with an emphasis on Tennessee populations. *Herpetol. Conserv. Biol.* **2008**, *3*, 1–20.

70. Miller, B.T.; Niemiller, M.L. *Gyrinophilus palleucus*. *Cat. Am. Amphib. Reptiles* **2012**, *884*, 1–7.

71. Niemiller, M.L.; Niemiller, K.D.K. *Species Status Assessment for the Tennessee Cave Salamander (Gyrinophilus palleucus) McCrady, 1954. Version 1.0*; Tennessee Wildlife Resources Agency: Nashville, TN, USA, 2020; 59p.

72. Ethridge, J.Z.; Gibson, J.R.; Nice, C.C. Cryptic diversity within and amongst spring-associated *Stygobromus* amphipods (Amphipoda: Crangonyctidae). *Zool. J. Linn. Soc.* **2013**, *167*, 227–242. [CrossRef]
73. Devitt, T.J.; Wright, A.M.; Cannatella, D.C.; Hillis, D.M. Species delimitation in endangered groundwater salamanders: Implications for aquifer management and biodiversity conservation. *Proc. Natl. Acad. Sci. USA* **2019**, *116*, 2624–2633. [CrossRef] [PubMed]

Article

There and Back Again—The Igatu Hotspot Siliciclastic Caves: Expanding the Data for Subterranean Fauna in Brazil, Chapada Diamantina Region

Jonas Eduardo Gallão [1,2,*], Deyvison Bonfim Ribeiro [2], Jéssica Scaglione Gallo [1,2] and Maria Elina Bichuette [1,2]

1   Laboratório de Estudos Subterrâneos, Universidade Federal de São Carlos, Rodovia Washington Luís km 235, São Carlos 13565-905, São Paulo, Brazil; jessicasgallo@gmail.com (J.S.G.); lina.cave@gmail.com (M.E.B.)
2   Instituto Brasileiro de Estudos Subterrâneos, São Carlos 13565-545, São Paulo, Brazil; deyvisonribeiro@hotmail.com
*   Correspondence: jonasgallao@gmail.com

Abstract: The caves of Igatu, municipality of Andaraí, belonging to the region known as Chapada Diamantina represent a new hotspot of subterranean fauna. These caves are siliciclastic, which are sedimentary rocks where silica predominates, such as sandstones and (following metamorphism) quartzites, which makes them even more relevant from the point of view of subterranean diversity. For five caves, which we named Igatu Cave System (ICS), thirty-seven obligate cave species were found, of which thirty-five were troglobitic and two were stygobitic. The troglobitic taxa for ICS belong to three phyla, nine classes, 18 orders, and 32 families, representing a high phylogenetic diversity. Some taxa were, for the first time, reported as troglobitic in Brazil and even worldwide, such as Acari and scutigeromorphans (Chilopoda). We started the studies in 2009 and continue trough long-term monitoring projects. Some threats, severe in the past, such as "garimpo" (illegal small-scale artisanal mining) continue nowadays in an incipient way; however, the urban expansion due to the touristic appeal is also considered a threat. Our data ranked ICS as the Brazilian hotspot with the highest number of troglobitic/stygobitic species.

Keywords: subterranean biodiversity; conservation; Bahia state; Northeastern Brazil

Citation: Gallão, J.E.; Ribeiro, D.B.; Gallo, J.S.; Bichuette, M.E. There and Back Again—The Igatu Hotspot Siliciclastic Caves: Expanding the Data for Subterranean Fauna in Brazil, Chapada Diamantina Region. *Diversity* 2023, 15, 991. https://doi.org/10.3390/d15090991

Academic Editors: Tanja Pipan, David C. Culver and Louis Deharveng

Received: 1 July 2023
Revised: 28 August 2023
Accepted: 30 August 2023
Published: 4 September 2023

1. Introduction

The occurrence of karst areas in South America with high versus low troglobite diversity was predicted by Trajano [1,2], who considered paleoclimatic fluctuations during the Quaternary to explain this particular biodiversity, citing the Upper Ribeira region and the Campo Formoso region. Following this discussion, caves from the Upper Ribeira karst area, Southeastern Brazil, and the Campo Formoso region, Northeastern Brazil, were validated as hotspots [3,4]. However, other areas in Northeastern Brazil (Serra do Ramalho region, Chapada Diamantina region) were also considered hotspots and/or of high biodiversity for subterranean fauna [5,6].

Gallão and Bichuette [5] reported, for the first time, the high diversity of troglobites for caves in siliciclastic rocks of the Igatu region (Chapada Diamantina, Northeastern Brazil) and discussed why these caves could be considered remarkable, not only in terms of troglobite numbers but also in terms of phylogenetic diversity. To date, the subterranean fauna of the siliciclastic caves of Chapada Diamantina is remarkable as all, with the occurrence of the troglobitic scorpion *Troglorhopalurus translucidus* Lourenço, Baptista and Giupponi, 2004, from Gruta do Lapão Cave [7], the co-occurrence of troglobitic fishes, a rare event in siliciclastic caves: *Glaphyropoma spinosum* Bichuette, Pinna and Trajano 2008, and a new species of *Copionodon* [8,9].

In this work, we updated and reinforced the siliciclastic caves of Igatu, located in the Chapada Diamantina region, State of Bahia, Northeastern Brazil, as a troglobites/stygobites hotspot. In addition to the taxonomic richness, this region also contains indicators of phylogenetic diversity (presence of relict taxa), aspects that must be considered in the conceptualization of biodiversity hotspots [5,6]. Igatu is also a biogeographical region with a significant reservoir of biodiversity threatened by human activities [5,6]. We considered here five caves of Igatu, all connected by subterranean drainage, and forming a system: Gruna Rio de Pombos, Gruna Canal da Fumaça, Gruna Lava Pé, Gruna da Parede Vermelha, and Gruna Cantinho caves. We named this system as Igatu Cave System (ICS).

2. Material and Methods

2.1. Igatu Region and Their Caves

Igatu is located in the Chapada Diamantina National Park (CDNP) and is a district of the municipality of Andaraí, in the central part of the State of Bahia, Northeastern Brazil (Figures 1 and 2). It is part of the Serra do Sincorá and geologically belongs to the Tombador Formation [10]. The region has several streams (including subterranean drainages), tributaries of the Rio Coisa Boa and Rio Piabas rivers, part of the Upper Paraguaçu River basin, within the Northeast Atlantic Forest ecoregion, which presents high rates of endemism. The five caves considered in this work are crossed by the same subterranean drainage (tributary of the Rio Coisa Boa), and they present small galleries and low-ceiling conduits. The caves showed a small extent considering the passages, not surpassing 0.5 to 0.9 km each. In general, the conduits were formed by mechanical erosion caused by water allied to tectonism, with little evidence of chemical dissolution (Figures 2 and 3).

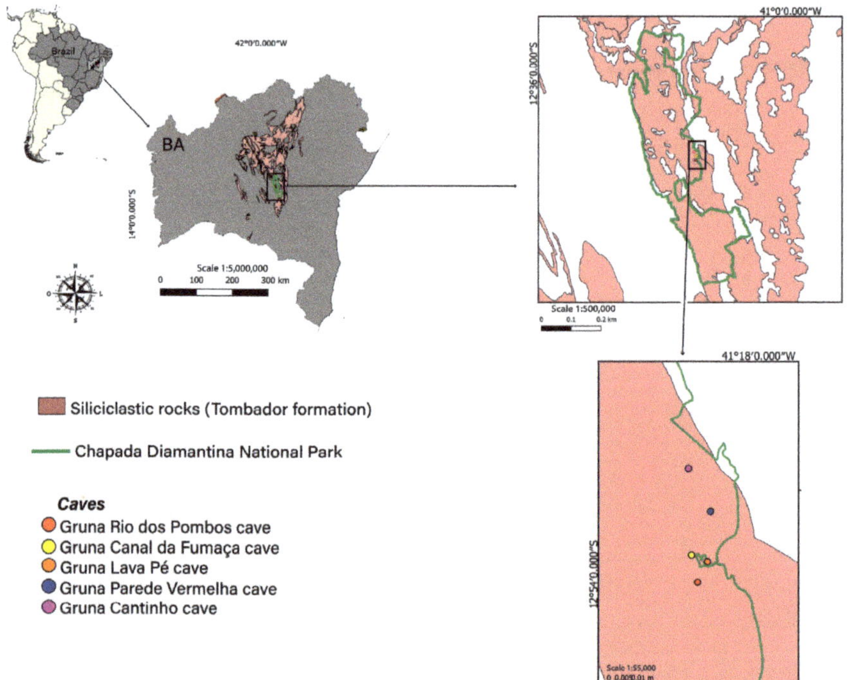

Figure 1. Map showing the region of Igatu Cave System (ICS), Chapada Diamantina region, Bahia state, Brazil. Developed in QGIS Development Team, QGIS Geographic Information System.

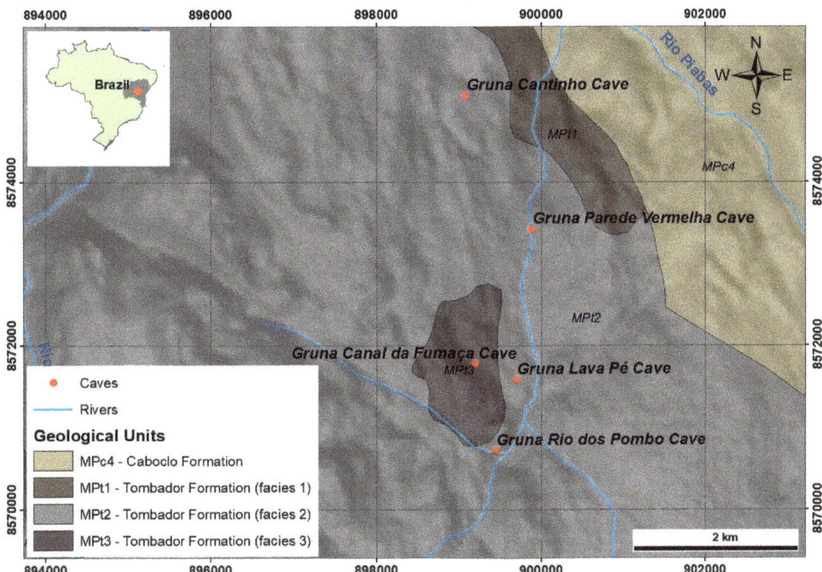

Figure 2. Geological map with details of Igatu Cave System (ICS) and surface drainages nearby. Blue lines: drainages. Developed in ArcGIS Desktop 10.6.1, version 10.6.1.9270; shapefiles for lithology: CPRM—Geological Map of Bahia, 1:1,000,000. 2003; shapefiles for drainages: ANA Metadata catalog (https://metadados.snirh.gov.br/geonetwork/srv/search?keyword=GEOFT_BHO_MASSA_DAGUA Accessed on 3 June 2023); shapefiles for relief: SRTM—Shuttle Radar Topography Mission.

The rocks exposed in Serra do Sincorá belong mainly to the Mesoproterozoic Tombador Formation [11] (Figure 2). In the Serra do Sincorá, the Tombador Formation is deposited on the Guiné Formation of the Paraguaçu Group. Its sandstones and conglomerates have the structure of a large anticlinorium with a wavy axis [11]. Severo Giudice [12] discussed that, geologically, the Chapada Diamantina is the product of a relief inversion, since it corresponds to the remnants of a sedimentary basin that settled over the São Francisco Craton about 1.8 billion years ago. The observed geological and geomorphologic elements of Igatu present themselves in different forms, such as mountains, tabular hills, waterfalls, caves (Figure 3), and rivers, and are responsible for a particular landscape, including its high number of caves (20+, ME Bichuette and JE Gallão, pers. obs.), with subterranean drainages and a rich fauna (Figure 3).

From 1846 to 1871, there was intensive diamond mining ("garimpo"—small-scale artisanal mining) in the region, and the waste from the old mines can still be seen along the Paraguaçu River and also inside the caves (Figure 3). After a golden age of about 25 years, diamond mining began to decline in 1871, and attempts were made to mechanize mining in the first half of the 20th century [12]. In the 1980s, mechanized mining was reintroduced in the Serra do Sincorá, installed in the riverbeds inside and outside the Chapada Diamantina National Park (CDNP). These "garimpos" were finally closed in March 1996. However, this activity continues today and is the main threat to the subterranean biodiversity of Igatu. Another threat in Igatu is the urban expansion, with many constructions over the outcrops (Figure 4).

2.2. Samplings, Determinations, Classification

We carried out inventories in several caves of the Igatu region between 2009 and 2016. These inventories were the first ones in Igatu siliciclastic caves. On those occasions, we discovered 11 caves with representative cave fauna, most of them with subterranean drainage. In this work, we considered five caves that represent the ICS (Gruna da Parede

Vermelha, Gruna Canal da Fumaça, Gruna Lava Pé, Gruna Cantinho, Gruna Rio dos Pombos), reaching ca. 5 km in a linear extension altogether (Figures 1 and 2).

Figure 3. The landscape of the Igatu region, and physical aspects of its caves: (**a**) view of the Igatu landscape with siliciclastic outcrops; (**b**) Rio Coisa Boa River, tributary of Upper Paraguaçu River basin; (**c**,**d**) Gruna da Parede Vermelha cave; (**e**) Gruna Cantinho cave; (**f**) Gruna Canal da Fumaça cave; (**g**) Gruna Rio dos Pombos cave; (**h**) Gruna Lava Pé cave.

Figure 4. Alterations and impacts observed in caves and landscape of Igatu: (**a**) dug walls in Gruna do Cantinho cave; (**b**) pebbles and gravels washed due "garimpo"; (**c**) general view of Rio Paraguaçu River with silting sand due to past activities of "garimpo" in Igatu (arrow), note the urbanization next to it. This activity was allowed until 1996 and is incipient nowadays.

We investigated several terrestrial and aquatic microhabitats by active search, without the installation of traps. The main observed substrates for the cave fauna were animal detritus (guano, etc.), vegetal debris, roots, rocky blocks, walls, and ceilings. The subterranean drainages consisted mainly of a soft bottom composed of sand and pebbles, in general, with lentic waters and few organic matter. Surveys were conducted by two to four researchers per cave, to avoid severe impacts by overcollecting. Specimens were identified in the laboratory using taxonomic keys, specific literature, and expert consultation/confirmation for some groups (Araneomorphae: A. Brescovit; Collembola: J. G. Palacios-Vargas and D. Zeppelini; Diplopoda: S. Golovatch; Chilopoda: A. Chagas-Jr.; Acari: M. Santos de Araújo;

Isopoda: I. S. Campos-Filho; Coleoptera: R. Bessi; Gastropoda: R. Salvador). Most of the taxa were confirmed as new and were also considered in the list.

For confirmation of troglobitic/stygobitic status, we also conducted several samplings in the epigean environment. We classified troglobites/stygobites as those species that did not occur in the epigean environment coupled with morphological clues (troglomorphisms). We used the presence of traits often observed in troglobitic fauna, such as reduced eyes, pigmentation, elongation of appendages, and hypertrophy of nonvisual sensory structures, but which are not found in presumed epigean relatives, as evidence for their long-term solation and evolution in subterranean habitats. To recognize these troglomorphisms, we performed comparisons with close epigean relatives, including those ones collected in the same region. We followed the classification proposed by Culver and Pipan [13] to classify troglobites: cave-obligate species that cannot complete their life cycle outside of subterranean habitats.

All material was deposited in scientific collections in Brazil, including Laboratório de Estudos Subterrâneos (LES), Museu de Zoologia da Universidade de São Paulo (MZUSP), Universidade Federal de Mato Grosso (UFMT), Universidade Estadual da Paraíba (UEPB), Museu Nacional do Rio de Janeiro (MNRJ), and Instituto Butantan (IB).

3. Results

The updated troglobitic/stygobitic species now counts with 37 troglobitic/stygobitic species in five caves (Table 1, Figures 5 and 6). Taxa are distributed in three phyla (Arthropoda, Mollusca, Chordata), nine classes, 18 orders, and 32 families, representing a high phylogenetic diversity. The five caves share part of the recorded species, with Gruna da Parede Vermelha being the richest one, with 19 troglobitic/stygobitic species.

Table 1. Troglobitic/stygobitic species recorded from ICS (Igatu Cave System), Brazil. Gen., genus; sp., species.

Taxonomic Group	Taxon	Cave
Diplopoda: Polydesmida: cf. Chelodesmidae	Gen. sp.	Gruna Cantinho
Diplopoda: Polydesmida: Oniscodesmidae	*Crypturodesmus* sp.	Gruna Cantinho, Gruna da Parede Vermelha
Chilopoda: Scutigeromorpha: Pselliodidae	*Sphendononema* sp.	Gruna da Parede Vermelha, Gruna Canal da Fumaça
Chilopoda: Scolopendromorpha: Scolopocryptopidae	*Scolopocryptops troglocaudatus* Chagas-Jr and Bichuette, 2015	Gruna Cantinho, Gruna Lava Pé
Chilopoda: Scolopendromorpha: Cryptopidae	*Cryptops* sp.	Gruna Lava Pé
Arachnida: Acari: Mesostigmata: Pachylaepidae	Gen. sp.	Gruna Cantinho
Arachnida: Acari: Mesostigmata: Dithinozerconidae	Gen. sp.	Gruna Rio dos Pombos
Arachnida: Acari: Sarcoptiformes: Oehserchestidae	Gen. sp.	Gruna da Parede Vermelha
Arachnida: Scorpiones: Buthidae	*Troglorhopalurus translucidus* Lourenço, Baptista and Giupponi, 2004	Gruna da Parede Vermelha, Gruna Canal da Fumaça, Gruna Cantinho, Gruna Lava Pé, Gruna Rio dos Pombos
Arachnida: Araneae: Theraphosidae	*Tmesiphantes hypogeus* Bertani, Bichuette and Pedroso, 2013	Gruna da Parede Vermelha
Arachnida: Araneae: Ctenidae	*Ctenus igatu* Polotow, Cizauskas and Brescovit, 2022	Gruna Canal da Fumaça

Table 1. *Cont.*

Taxonomic Group	Taxon	Cave
Arachnida: Araneae: Gnaphosidae: Prodidominae	Gen. sp.	Gruna Rio dos Pombos
Arachnida: Araneae: Ochyroceratidae	*Ochyrocera* sp.	Gruna Cantinho
Arachnida: Araneae: Pholcidae	*Metagonia* sp.	Gruna Rio dos Pombos, Gruna Cantinho
Arachnida: Araneae: Telemidae	Gen. sp.	Gruna da Parede Vermelha
Arachnida: Opiliones: Gonyleptidae	*Discocyrtus pedrosoi* Kury, 2008	Gruna da Parede Vermelha, Gruna Canal da Fumaça, Gruna Cantinho, Gruna Lava Pé, Gruna Rio dos Pombos
Arachnida: Opiliones: Tricommatidae	Gen. sp.	Gruna Cantinho
Arachnida: Pseudoscorpiones: Chernetidae	*Spelaeochernes* sp.	Gruna da Parede Vermelha
Arachnida: Pseudoscorpiones: Chthoniidae	*Pseudochthonius* sp.	Gruna da Parede Vermelha
Arachnida: Pseudoscorpiones: Syarinidae	Gen. sp.	Gruna da Parede Vermelha
Arachnida: Palpigradi: Eukoeneniidae	*Eukoenenia* sp.	Gruna Lava Pé, Gruna Cantinho
Malacostraca: Isopoda: Philosciidae	*Metaprosekia igatuensis* Campos-Filho, Fernandes and Bichuette, 2020	Gruna Rio dos Pombos
Malacostraca: Isopoda: Philosciidae	*Benthana xiquinhoi* Campo-Filho, Bichuette and Taiti, 2019	Gruna Lava Pé, Gruna da Parede Vermelha
Malacostraca: Isopoda: Philosciidae	Gen. sp.	Gruna da Parede Vermelha
Malacostraca: Isopoda: Plathyartridae	*Trichorhina* sp.	Gruna Rio dos Pombos, Gruna Lava Pé
Malacostraca: Isopoda: Platyarthridae	Gen. sp.	Gruna da Parede Vermelha, Gruna Rio dos Pombos
Collembola: Entomobryomorpha: Entomobryidae	*Verhoeffiella* sp.	Gruna da Parede Vermelha
Collembola: Entomobryomorpha: Entomobryidae: Heteromurinae: Heteromurini	Gen. sp.	Gruna Cantinho, Gruna Rio dos Pombos
Collembola: Entomobryomorpha: Paronellidae	*Troglopedetes* sp.	Gruna da Parede Vermelha, Gruna Cantinho
Diplura: Projapygidae	Gen. sp.	Gruna Rio dos Pombos
Insecta: Zygentoma: Nicoletiidae	Gen. sp.	Gruna Canal da Fumaça
Insecta: Blattaria: Blattellidae	Gen. sp.	Gruna da Parede Vermelha
Insecta: Coleoptera: Scydmaenidae	Gen. sp.	Gruna Cantinho
Insecta: Coleoptera: Staphylinidae: Pselaphinae	Gen. sp.	Gruna da Parede Vermelha
Gastropoda: Stylommatophora: Systrophiidae	*Happia* sp.	Gruna da Parede Vermelha, Gruna Canal da Fumaça, Gruna Lava Pé, Gruna Rio dos Pombos
Actinopterygii: Siluriformes: Trichomycteridae	*Copionodon* sp.	Gruna da Parede Vermelha, Gruna Canal da Fumaça, Gruna Cantinho, Gruna Lava Pé, Gruna Rio dos Pombos
Actinopterygii: Siluriformes: Trichomycteridae	*Glaphyropoma spinosum* Bichuette, de Pinna and Trajano, 2008	Gruna da Parede Vermelha, Gruna Canal da Fumaça, Gruna Cantinho, Gruna Lava Pé, Gruna Rio dos Pombos

Figure 5. Troglobitic/stygobitic fauna of ICS, Bahia State, Brazil: (**a**) *Scolopocryptops troglocaudatus* (Scolopendromorpha); (**b**) *Troglorhopalurus translucidus* (Scorpiones); (**c**) *Tmesiphantes hypogeus* (Mygalomorphae); (**d**) *Ctenus igatu* (Araneomorphae); (**e**) *Eukoenenia* sp. (Palpigradi); (**f**) *Happia* sp. (Stylommatophora); (**g**) *Copionodon* sp. (Siluriformes); (**h**) *Glaphyropoma spinosum* (Siluriformes).

Among the troglobites/stygobites recorded for ICS, eight were formally described: *Discocyrtus pedrosoi* Kury, 2008, *Glaphyropoma spinosum* Bichuette, de Pinna and Trajano 2008, *Troglorhopalurus translucidus* Lourenço, Baptista and Giupponi 2004, *Tmesiphantes hypogeus* Bertani, Bichuette and Pedroso 2013, *Metaprosekia igatuensis* Campos-Filho, Fernandes and Bichuette, 2020, *Benthana xiquinhoi* Campo-Filho, Bichuette and Taiti, 2019, *Ctenus igatu*

Polotow, Cizauskas and Brescovit, 2022, and *Scolopocryptops troglocaudatus* Chagas-Jr and Bichuette, 2015. ICS is the type-locality of seven species.

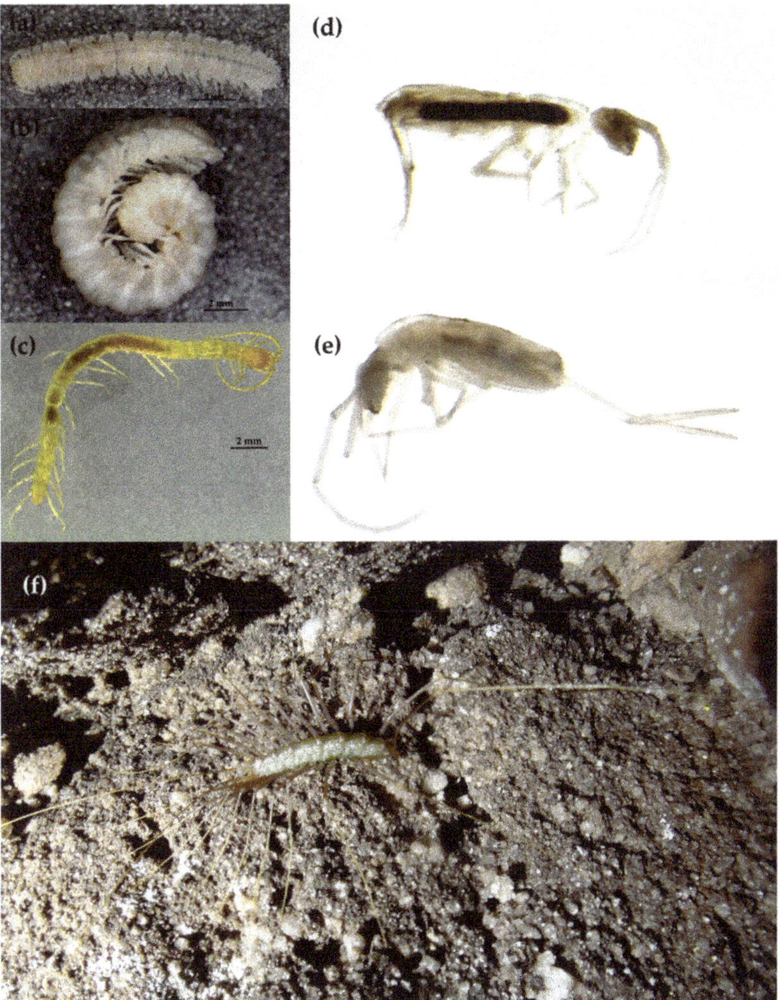

Figure 6. Troglobitic fauna of ICS, Bahia State, Brazil: (**a**) cf. *Chelodesmidae* sp. (Diplopoda: Polydesmida); (**b**) *Crypturodesmus* sp. (Diplopoda: Polydesmida); (**c**) *Cryptops* sp. (Chilopoda: Scolopendromorpha); (**d**) *Verhoeffiella* sp. (Collembola: Entomobryomorpha), 1.5 mm body size; (**e**) *Troglopedetes* sp. (Collembola: Entomobryomorpha), 1.1. mm body size; (**f**) *Sphendononema* sp. (Scutigeromorpha).

Considering the taxonomical records and some aspects of natural history, we can make some highlights.

For Myriapoda, most millipede species found in Brazilian subterranean habitats belong to the orders Polydesmida and Spirostreptida [14]. Polydesmida includes eight of 13 troglobitic species described for Brazil, all of which occur in limestone caves. For the Igatu region, two undescribed troglobitic Polydesmida are recorded: *Crypturodesmus* and one cf. Chelodesmidae. The genus *Crypturodesmus* (Oniscodesmidae) has been registered in Brazil and Mexico [15]. In the subterranean environment, the genus has been recorded in limestone caves in the states of Mato Grosso do Sul, São Paulo, and Paraná [14], and now for the ICS. In the family Chelodesmidae, five troglobitic species are known: two in

Brazilian limestone caves and three in Jamaica, Puerto Rico, and Spain, suggesting relict lineages [16]. This suggests that few, if any, radiations of chelodesmids have occurred within caves in the past [16].

Similarly, chilopods are representative of ICS; there are seven described troglobitic species for Brazil (two of the Order Geophilomorpha and five of the Order Scolopendromorpha). One of them occurs in the Igatu Cave System: *Scolopocryptops troglocaudatus*. Even more, a new species of the genus *Cryptops* (Scolopendromorpha) and a new species of the genus *Sphendononema* (Order Scutigeromorpha) also occur in ICS (A. Chagas-Jr., pers. comm.). *Cryptops* have a worldwide distribution, occurring in caves in Brazil, Europe, Australia, and Cuba [17]. The genus is common in Brazil, with three troglophilic and two troglobitic species described for limestone and iron ore caves [17]. Igatu caves have mainly exposed sandstone rock as a substrate, and their surroundings are mostly composed of outcrops; the discovery of a highly troglomorphic species of *Cryptops*, with appendages elongated in relation to the body, including antennae and anal legs (A. Chagas-Jr., pers. comm.), and the non-occurrence in the epigean environment reinforce its troglobitic status. *Scolopocryptops troglocaudatus* is the second troglobitic Scolopocryptopinae described and the first discovered in Brazil [18]. Additionally, this species is one of the most troglomorphic Scolopendromorpha known, with the anal leg reaching 2/3 of the body length [18]. Another relevant record for ICS caves is the new species of *Sphendononema* genus, representing the first troglobitic Scutigeromorpha worldwide; its legs, annal legs, and antennae are greatly elongated and the specimens showed low body sclerotization (comparatively with the widely distributed *S. guildingii*). These results corroborate the importance of ICS for Myriapoda taxonomic knowledge. In addition, these data reinforce the phylogenetic diversity of the ICS cave fauna.

There are about forty-three troglobitic species of Isopoda in Brazil and among them, two are known for Igatu caves, *Metaprosekia igatuensis* and *Benthana xiquinhoi*. In addition to these described species, three other new ones were also recorded (Table 1). The troglomorphisms were mainly a reduction in number of ocelli (or absence), body depigmentation, associated with low tolerance to dry conditions, also observed for other troglobitic isopods.

The fauna of Pseudoscorpiones are represented by 12 families and 22 genera in Brazilian caves [19], and the troglobitic fauna counts with 24 species, belonging to Chernetidae, Chthoniidae, Bochicidae, and Ideoroncidae families. Three undescribed species were recorded for ICS (Chernetidae, Chthoniidae, and Syarinidae families), the most modified (specialized) were Chthoniidae and Syarinidae, and the later one represents the first record for troglobitic species considering the family in Brazil (Table 1).

The scorpion *Troglorhopalurus translucidus* was discovered and described for Gruta do Lapão, in another region of Chapada Diamantina (municipality of Lençóis). This cave also belongs to the Espinhaço Supergroup, Tombador formation, however, at its northernmost point. Few specimens were recorded in the type-locality. On the contrary, in the Igatu caves, the abundance and distribution were greater, possibly representing the source population for the species. *Troglorhopalurus translucidus* is the most troglomorphic scorpion of the Buthidae family known and, together with *T. lacrau* (Lourenço and Pinto-da-Rocha, 1997), comprise the only two troglobitic scorpions known from Brazil. Some other subterranean scorpions in Brazilian caves are troglophiles, such as *Tityus blaseri* Mello-Leitão, 1931 and *T. spelaeus* Moreno-Gonzáles, Pinto-da-Rocha and Gallão, 2021, both species occur in caves and epigean habitats in the state of Goiás, *T. confluens* Borelli, 1899, in caves and epigean habitats in the states of Mato Grosso and Mato Grosso do Sul, *T. stigmurus* (Thorell, 1876), widely distributed in northeastern Brazil, with facultative cave populations in the state of Sergipe [20], and *T. obscurus* Gervais, 1843, with a well-established population in the caves of North Brazil (state of Pará) (J.E. Gallão, pers. obs.).

For Opiliones, the Brazilian subterranean fauna is remarkable with several representatives in trogloxenes and troglophiles species distributed in several families [21]. The updated number of described troglobitic opilionids counts with 14 species for Brazilian caves, most are Gonyleptidae. In addition to the described *Discocyrtus pedrosoi*, one undescribed troglomorphic Tricommatidae was recorded from Igatu caves.

To date, no troglobitic mite is known of from Brazil; however, several have been described as occurring in caves, as cave-dwellers. Three species recorded in the Igatu caves (Pachylaepidae, Dithinozerconidae, and Oehserchestidae families) presented troglomorphic characters when compared to the described species of these families (M. S. de Araújo, pers. comm.) such as elongated legs, as well as reduced sclerotization. In addition, surface collections did not reveal any mite species from these families, which justifies their troglobite status. Studies on the taxonomy of these taxa are urgently needed, which could corroborate the proposed category and also would provide important data on the biogeography of these families.

About the mygalomorphae spiders, *Tmesiphantes hypogeus* is the only known theraphosid troglobitic spider for Brazil. The species was discovered and described with females specimens only for Igatu caves. No male was found.

There are about 30 troglobitic Araneomorphae spiders for Brazil, with a dominance of Ochyroceratidae, Gnaphosidae, and Tetrablemmidae families, among others [22]. Igatu is remarkable due the occurrence of *Ctenus igatu*, a highly troglomorphic Ctenidae spider, in addition to three undescribed species of the families Ochyroceratidae, Pholcidae, and Gnaphosidae (Table 1).

There are at least 17 species of troglobitic Palpigradi for Brazil, all of which belong to the family Eukoeneniidae and most are from the genus *Eukoenenia*. In Igatu, there is one species of *Eukoenenia* that has not yet been described.

Brazil harbors 49 formally described troglobitic Collembola, none of which are from Igatu. The new records at ICS are of three undescribed species (*Verhoeffiella*, specimens of Heteromutini tribe, and *Troglopedetes* genus). It is worth noting that for these caves, we recorded the genus *Verhoeffiella*, which was previously recorded only in the Dinaric region of Europe. If confirmed by future detailed taxonomical studies, the presence of this genus would be a major discovery for Entomobryiidae biogeography. Even more, considering the records of *Troglopedetes* genus, this is the first record in South America of an European and Southern Asia genus, although there are many records of the related genus *Trogolaphysa* from the region. Like for *Verhoeffiella*, its discovery in Igatu raises an interesting and puzzling biogeographical problem.

In Brazilian caves, there are records of at least 24 troglobitic coleopterans, most of which are from the Carabidae family. None of the described species occur in the Igatu caves; however, there are two undescribed species belonging to the families Scydmaenidae and Staphylinidae, showing low abundance, and each one is restricted to only one cave.

For gastropods, there are currently 21 troglobitic species for Brazilian caves, but none are described for the caves of Igatu. In this region, there is only one troglobite, which remains undescribed, of the genus called *Happia* (Systrophiidae).

With regard to the stygobiotic fauna, we found that it was poor in Igatu. We recorded only two species, both of which were fishes. There are about 36 troglobitic fishes in Brazil [9], and two were found in Igatu: *Glaphyropoma spinosum* and an undescribed species of the genus *Copionodon*, both of which were widely distributed in Igatu caves. Both species belong to the subfamily Copionodontinae, endemic to the Chapada Diamantina region, and co-occur in the caves, which is a rare event in general. The wide distribution of these two Copionodontinae populations corroborates the connectivity of ICS caves.

The number of troglobites/stygobites for Igatu (37) does not include other relevant caves of the same geological supergroup (Espinhaço), such as Gruta do Lapão (municipality of Lençóis) and Gruta do Castelo (municipality of Mucugê). The total number of troglobitic species in the region rises up to 46 when these two caves are taken into account.

4. Discussion

Gallão and Bichuette [5] registered 162 cave-dwelling species in 11 caves from the Igatu region, with doubts about the possible connections between them. At that time, they considered 23 troglobitic species distributed in an area of 25 km^2. Now we reach 37 species for five caves, all connected by a subterranean drainage (part of the Rio Coisa river), covering a linear extension of 4.3 km. This extension is significantly smaller than that observed for other caves considered hotspots in Brazil and worldwide: the Areias Cave System in southeastern Brazil, formed by three connected caves, currently harbors more than 31 species and has about 8.5 km of mapped passages (ME Bichuette and JE Gallão, updated data); the Água Clara Cave System in northeastern Brazil, formed by four caves, harbors 31 species and has about 25.8 km of mapped passages [23]; the Huautla Cave System in Oaxaca, Mexico, harbors 27 species and has about 89 km of mapped passages [24]; the Fern Cave System in northeastern Alabama, USA, harbors 27 species and has over 25 km of mapped passages [25].

If we consider the troglobitic/stygobitic fauna of ICS in a phylogenetic context, we can realize the great diversity of these troglobites distributed in a variety of higher taxa. Currently, for ICS, nine classes, 18 orders, and 32 families are represented for 37 troglobitic/stygobitic species. Phylogenetic diversity also assists in choosing priorities for conservation. The extinction of species without close relatives is more damaging than extinction of species with close relatives [26,27], and so, the best conservation strategies are those that address the greatest possible phylogenetic diversity [27,28]. Although we did not perform any phylogenetic diversity test in this work, Gallão and Bichuette [5] performed tests for 11 caves of Igatu and in addition to calling attention to the troglobitic/stygobitic fauna, these authors demonstrated the relevance of one cave, the Gruna da Parede Vermelha, which, at the time, presented the greatest phylogenetic diversity considering subterranean fauna [5]. These comparisons emphasize the importance of Igatu with a greater potential for a higher number of troglobites/stygobites. In support of this idea, the Gruna da Parede Vermelha cave has about 0.7 km of mapped passages and harbors 18 troglobitic/stygobitic species at present.

We must also consider the lithology of the caves of Igatu (sandstone), which is generally neglected in inventories of subterranean fauna in general, not only cave-restricted species. In this sense, with the inclusion of troglophilic and trogloxene populations, we reached 184 species for the five caves considered here, which was clearly high for siliciclastic caves. When we compared with other studies conducted in siliciclastic areas in Brazil, we note how Igatu stood out in all relevant aspects considering biodiversity value, whether in number of subterranean species in general, of troglobitic/stygobitic species, and also phylogenetic diversity (Table 2).

Table 2. Comparison among siliciclastic regions with records of troglobitic/stygobitic fauna in Brazil. TR/STY: number of troglobitic/stygobitic species.

Region	Geomorphological Information	Number of Caves	TR/STY	Total of Species	References
Altamira and Medicilândia—North Brazil	Altamira—Itaituba	7	2	62	[29]
Altinópolis—Southeastern Brazil	Serra Geral, Botucatu Formation	9	0	83	[30]
Rurópolis—North Brazil	Altamira—Itaituba	1	0	16	[31]
Manoel Viana and São Pedro do Sul—South Brazil	Serra Geral, Botucatu Formation	3	0	30	[32]
Chapada Diamantina—Northeastern Brazil	Serra do Espinhaço, Tombador Formation	11	25	162	[5]
Altinópolis—Southeastern Brazil	Serra Geral, Botucatu Formation	8	0	131	[33]
Lima Duarte—Southeastern Brazil	Andrelândia geological group	20	6	469	[34]
Itirapina—Southeastern Brazil	Serra Geral, Botucatu Formation	1	3	67	E.L.B. Carvalho, undergraduated monograph (unpubl. data)
Altamira—North Brazil	Altamira—Itaituba	26	17	596	M.E. Bichuette, unpubl. data
Chapada Diamandina—Northeastern Brazil	Serra do Espinhaço, Tombador Formation	5	37	184	This study

There is currently a minimum threshold that counts only the number of troglobitic species to identify a cave or cave system as a hotspot. Culver and Sket [35] set this threshold at 20 species, and in a recent work, Culver et al. [36] increased this threshold to 25 species. However, Trajano et al. [6] discussed that caves and/or cave systems of Brazil can be considered as spots (or hotspots) not only based on the number of troglobitic/stygobitic species but also on phylogenetic diversity (such as the presence of relict taxa) as well as genetic diversity (such as the accumulation of autapomorphies). In this way, Trajano et al. [6] listed six sites: the Upper Ribeira karstic area in the state of São Paulo, the Serra da Bodoquena karst area in the state of Mato Grosso do Sul, the São Domingos karst area in the state of Goiás, in addition to the Chapada Diamantina karst area, the Serra do Ramalho karst area and the Chapada Diamantina siliciclastic area, the last three in the state of Bahia, and the last one considered in this work. It was noted that paleoclimatic fluctuations, in addition to geomorphological changes, have determined a high diversity of troglobites/stygobites in the state of Bahia as a whole [6], and the Igatu region clearly follows this pattern.

If we spread the number of siliciclastic caves of Chapada Diamantina from five to seven (five caves of Igatu region + Gruta do Lapão cave + Gruta do Castelo cave), we reach 46 species of troglobites/stygobites, some shared between them, representing an expressive subterranean biodiversity for a unique geological formation (Tombador Formation), with different facies, significantly increasing the relevance of the siliciclastic caves from Chapada Diamantina region.

The fauna of ICS is clearly remarkable, as previously stated by Gallão and Bichuette [5]. In contrast to the findings of Sousa Silva et al. [34], who considered the existence of caves with more than 30 troglobites/stygobites in Brazil, or even more, impossible, considering caves in siliciclastic rocks. This kind of affirmation, disregarding the existence of something not tested is clearly speculative and could threaten the decision on proposed areas particularly rich and unique in subterranean fauna, the case of Chapada Diamantina, which is one of the Brazilian regions with highest endemism rates [5,6,22], and a high rate of endemism to the subterranean fauna is expected too. Herein, we reinforced the hypothesis by Gallão and Bichuette [5], updating the number of troglobitic/stygobitic species to 37, in a small area covered by five sandstone caves, including beetles, centipedes, collembolans, acari, scorpions, spiders, gastropods, fish, and more. Most sandstone caves of Chapada Diamantina were heavily impacted by diamond mining ("garimpo") in the past, since 1846 and reaching until 1996. In recent years, the "garimpo" activity occurred in clandestine and residual ways. The five caves of Igatu Cave System were inserted in the Chapada Diamantina National Park (CDNP) and are legally protected. Even so, there are threats, such as the residual and clandestine "garimpo" in a small scale, and the urban expansion due the tourism in the region. Ecological long-term studies, allied to citizen science, are crucial to provide support in the effective protection of Igatu caves and its remarkable and particular fauna.

Author Contributions: Conceptualization, M.E.B. and J.E.G.; methodology, M.E.B. and J.E.G.; investigation, M.E.B., J.S.G., D.B.R. and J.E.G.; resources, M.E.B.; data curation, J.E.G. and J.S.G.; writing—original draft preparation, M.E.B., J.S.G., D.B.R. and J.E.G.; writing—review and editing, M.E.B., J.S.G. and J.E.G.; funding acquisition, M.E.B. All authors have read and agreed to the published version of the manuscript.

Funding: This research was funded by FAPESP (Fundação de Amparo à Pesquisa do Estado de São Paulo), grant number 2008/05678-7, and CNPq (Conselho Nacional de Desenvolvimento Científico e Tecnológico), grant number 303715/2011-1.

Institutional Review Board Statement: The study was conducted in accordance with the Environmental Brazilian Laws for fauna collection, permit number 20165-1 (SISBIO/ICMBio).

Data Availability Statement: Data sharing is not applicable for this manuscript.

Acknowledgments: We are very grateful to our guides in Igatu, Raimundo Cruz dos Santos ("Xiquinho de Igatu") and Rafael Pires de Souza (in memoriam); we also thank those who helped us in the several fieldwork trips to Igatu: Bianca Rantin, Luiza B. Simões, Tiago L. C. Scatolini, Camile S. Fernandes, Diego M. von Schimonsky, Tamires Zepon, Maria Rosendo, Ericson C. Igual; we also thank Leonardo de Assis for confection of the map in Figure 1, and Bruno Lenhare for confection of map in Figure 2; we thank Ericson C. Igual for the images of Figures 3c–e,h, 4a, 5a,f,g and 6f. J. Palacios-Vargas and D. Zeppelini revised the Collembola identifications and read the manuscript. The authors also thank the editors of this Special Issue of *Diversity*.

Conflicts of Interest: The authors declare no conflict of interest.

References

1. Trajano, E. Evolution of Tropical Troglobites: Applicability of the Model of Quaternary Climatic Fluctuations. *Mémoirs Biospéologie* **1995**, *22*, 203–209.
2. Trajano, E. Mapping Subterranean Biodiversity in Brazilian Karst Areas. In *Mapping Subterranean Biodiversity*; Culver, D.C., Deharveng, L., Gibert, J., Sasowsky, I., Eds.; Karst Waters Institute: Pennsylvania, PA, USA, 2001; pp. 67–70.
3. Deharveng, L. Diversity Patterns in the Tropics. In *Encyclopedia of Caves*; Culver, D.C., White, W.B., Eds.; Academic Elsevier Press: Amsterdan, The Netherlands, 2005; pp. 166–170.
4. Silva, M.S.; Ferreira, R.L. The First Two Hotspots of Subterranean Biodiversity in South America. *Subterr. Biol.* **2016**, *19*, 1–12. [CrossRef]
5. Gallão, J.E.; Bichuette, M.E. Taxonomic Distinctness and Conservation of a New High Biodiversity Subterranean Area in Brazil. *An. Acad. Bras. Cienc.* **2015**, *87*, 209–217. [CrossRef]
6. Trajano, E.; Gallão, J.E.; Bichuette, M.E. Spots of High Diversity of Troglobites in Brazil: The Challenge of Measuring Subterranean Diversity. *Biodivers. Conserv.* **2016**, *25*, 1805–1828. [CrossRef]
7. Lourenço, W.R.; Baptista, R.L.C.; de Leão Giupponi, A.P. Troglobitic Scorpions: A New Genus and Species from Brazil. *C. R. Biol.* **2004**, *327*, 1151–1156. [CrossRef] [PubMed]
8. Bichuette, M.E.; De Pinna, M.C.C.; Trajano, E. A New Species of Glaphyropoma: The First Subterranean Copionodontine Catfish and the First Occurrence of Opercular Odontodes in the Subfamily (Siluriformes: Trichomycteridae). *Neotrop. Ichthyol.* **2008**, *6*, 301–306. [CrossRef]
9. Bichuette, M.E.; Gallão, J.E. Under the Surface: What We Know about the Threats to Subterranean Fishes in Brazil. *Neotrop. Ichthyol.* **2021**, *19*, e210089. [CrossRef]
10. CPRM. *Projeto Chapada Diamantina: Parque Nacional da Chapada Diamantina—BA: Informações Básicas Para a Gestão Territorial: Diagnóstico do Meio Físico e da Vegetação*; CPRM: Brasília, Brazil, 1994.
11. Pedreira, A.J.; Serra Do Sincorá, B. *Sítios Geológicos e Paleontológicos do Brasil*; Schobbenhaus, C., Campos, D., Queiroz, E., Wing, M., Berbert-Born, M., Eds.; CPRM: Brasília, Brazil, 2001.
12. Severo Giudice, D.; Melo e Souza, R. Geologia E Geoturismo Na Chapada Diamantina. *Gestión Turística* **2010**, *2020*, 69–81. [CrossRef]
13. Culver, D.C.; Pipan, T. Ecological and Evolutionary Classifications of Subterranean Organisms. In *Encyclopedia of Caves*; White, W.B., Culver, D.C., Pipan, T., Eds.; Elsevier: Amsterdam, The Netherlands, 2019; pp. 376–379.
14. Trajano, E.; Golovatch, S.I.; Geoffroy, J.-J.; Pinto-da-Rocha, R.; Fontanetti, C.S. Synopsis of Brazilian Cave-Dwelling Millipedes (Diplopoda). *Pap. Avulsos de Zool. (São Paulo)* **2000**, *41*, 259–287.
15. Sierwald, P.; Spelda, J. MilliBase. Available online: https://www.millibase.org (accessed on 22 May 2023). [CrossRef]
16. Bouzan, R.S.; Means, J.C.; Ivanov, K.; Ferreira, R.L.; Brescovit, A.D.; Moretti Iniesta, L.F. Worldwide Distribution of Cave-Dwelling Chelodesmidae (Diplopoda, Polydesmida). *Int. J. Speleol.* **2022**, *51*, 235–248. [CrossRef]
17. Chagas, A.; Bichuette, M.E. A Synopsis of Centipedes in Brazilian Caves: Hidden Species Diversity That Needs Conservation (Myriapoda, Chilopoda). *Zookeys* **2018**, *2018*, 13–56. [CrossRef]
18. Chagas, A.; Bichuette, M.E. A New Species of Scolopocryptops Newport: A Troglobitic Scolopocryptopine Centipede from a Remarkable Siliciclastic Area of Eastern Brazil (Scolopendromorpha, Scolopocryptopidae, Scolopocryptopinae). *Zookeys* **2015**, *487*, 97–110. [CrossRef] [PubMed]
19. Von Schimonsky, D.M.; Bichuette, M.E. Distribution of Cave-Dwelling Pseudoscorpions (Arachnida) in Brazil. *J. Arachnol.* **2019**, *47*, 110–123. [CrossRef]
20. Moreno-González, J.A.; Pinto-da-Rocha, R.; Gallão, J.E. Bringing Order to a Complex System: Phenotypic and Genotypic Evidence Contribute to the Taxonomy of Tityus (Scorpiones, Buthidae) and Support the Description of a New Species. *Zookeys* **2021**, *1075*, 33–75. [CrossRef] [PubMed]
21. Pinto-da-Rocha, R. Sinopse Da Fauna Cavernícola Do Brasil (1907–1994). *Pap. Avulsos Zool.* **1994**, *39*, 61–173. [CrossRef]
22. Gallão, J.E.; Bichuette, M.E. Brazilian Obligatory Subterranean Fauna and Threats to the Hypogean Environment. *Zookeys* **2018**, *746*, 1–23. [CrossRef]
23. Ferreira, R.L.; Berbert-Born, M.; Souza-Silva, M. The Água Clara Cave System in Northeastern Brazil: The Richest Hotspot of Subterranean Biodiversity in South America. *Diversity* **2023**, *15*, 761. [CrossRef]

24. Francke, O.F.; Monjaraz-Ruedas, R.; Cruz-López, J.A. Biodiversity of the Huautla Cave System, Oaxaca, Mexico. *Diversity* **2021**, *13*, 429. [CrossRef]
25. Niemiller, M.L.; Slay, M.E.; Inebnit, T.; Miller, B.; Tobin, B.; Cramphorn, B.; Hinkle, A.; Jones, B.D.; Mann, N.; Niemiller, K.D.K.; et al. Fern Cave: A Hotspot of Subterranean Biodiversity in the Interior Low Plateau Karst Region of Alabama in the Southeastern United States. *Diversity* **2023**, *15*, 633. [CrossRef]
26. May, R.M. Taxonomy as Destiny. *Nature* **1990**, *347*, 129–130. [CrossRef]
27. Williams, P.H.; Humphries, C.J.; Vane-Wright, R.I. Measuring Biodiversity: Taxonomic Relatedness for Conservation Priorities. *Aust. Syst. Bot.* **1991**, *4*, 665–679. [CrossRef]
28. Vane-Wright, R.I.; Humphries, C.J.; Williams, P.H. What to Protect?—Systematics and the Agony of Choice. *Biol. Conserv.* **1991**, *55*, 235–254. [CrossRef]
29. Trajano, E.; Moreira, J.R.d.A. Estudo Da Fauna de Cavernas Da Província Espeleológica Arenítica Altamira—Itaituba, Pará. *Rev. Bras. Biol.* **1991**, *51*, 13–29.
30. Zeppelini Filho, D.; Ribeiro, A.C.; Ribeiro, G.C.; Fracasso, M.P.A.; Pavani, M.M.; Oliveira, O.M.P.; de Oliveira, S.A.; Marques, A.C. Faunistic Survey of Sandstone Caves from Altinópolis Region, São Paulo State, Brazil. *Pap. Avulsos Zool.* **2003**, *43*, 93–99. [CrossRef]
31. Reis, R.L.; Júnior, C.F.E.; Figueiredo, G.P.S.; Muriel-Cunha, J. Levantamento Preliminar da Biodiversidade da Caverna do Prudente, Província Espeleológica Arenitica Altamira-Itaituba, Rurópolis, Pará. In *ANAIS do 32º Congresso Brasileiro de Espeleologia*; Sociedade Brasileira de Espeleologia: Barreiras, BA, Brazil, 2013; pp. 115–119.
32. Sorbo, C.F.; Bichuette, M.E. Preliminary survey of invertebrates in three sandstone caves from rio do sul, Brazil. *Espeleo-Tema* **2013**, *24*, 41–47.
33. Gallo, J.S. Diversidade de Invertebrados Terrestres Em Cavernas Areníticas Do Estado de São Paulo, Com Ênfase Em Pseudonannolenidae (Diplopoda: Spirostreptida). Master's Dissertation, Universidade Federal de São Carlos, São Carlos, CA, USA, 2017.
34. Souza Silva, M.; Iniesta, L.F.M.; Lopes Ferreira, R. Invertebrates Diversity in Mountain Neotropical Quartzite Caves: Which Factors Can Influence the Composition, Richness, and Distribution of the Cave Communities? *Subterr. Biol.* **2020**, *33*, 23–43. [CrossRef]
35. Culver, D.C.; Sket, B. Hotspots of Subterranean Biodiversity in Caves and Wells. *J. Cave Karst Stud.* **2000**, *62*, 11–17.
36. Culver, D.C.; Deharveng, L.; Pipan, T.; Bedos, A. An Overview of Subterranean Biodiversity Hotspots. *Diversity* **2021**, *13*, 487. [CrossRef]

diversity

Article

The Água Clara Cave System in Northeastern Brazil: The Richest Hotspot of Subterranean Biodiversity in South America

Rodrigo Lopes Ferreira [1,2,*], Mylène Berbert-Born [3] and Marconi Souza-Silva [1,2]

1   Centro de Estudos em Biologia Subterrânea, Departamento de Ecologia e Conservação, Instituto de Ciências Naturais, Universidade Federal de Lavras, Campus Universitário, P.O. Box 3037, Lavras 37200-000, Minas Gerais, Brazil; marconisilva@ufla.br
2   Programa de Pós-Graduação em Ecologia Aplicada, Universidade Federal de Lavras, Lavras 37200-000, Minas Gerais, Brazil
3   Museu de Ciências da Terra—MCTer, Serviço Geológico do Brasil—SGB-CPRM, Av. Pasteur 404 Urca, Rio de Janeiro 22290-255, Rio de Janeiro, Brazil; mylene.berbertborn@cprm.gov.br
*   Correspondence: drops@ufla.br

Abstract: The Água Clara Cave System (ACCS) in Brazil is the richest hotspot of subterranean biodiversity in South America. In this study, we present an updated list of cave-restricted species in the ACCS and compare it with previously published hotspots in Brazil. Our list of cave-obligate fauna comprises 31 species, including 23 troglobionts and 8 stygobionts. The exceptional diversity of the ACCS can be attributed to factors related to the high dispersal potential of cave fauna within the system, high surface productivity, and the large size of the cave system size. Notably, we observed highly troglomorphic species in the ACCS, some of which are the most troglomorphic species in their respective groups in Brazil. The huge volume of galleries, high humidity, and trophic conditions prevailing in the ACCS may have played a role in shaping the strong troglomorphic traits observed in these species. However, all the obligate cave species in the ACCS require conservation attention and are at an elevated risk of extinction due to their limited ranges, few occurrences, and many potential threats. This study sheds light on the biodiversity and conservation status of cave-restricted fauna in the ACCS and highlights the importance of protecting these unique ecosystems.

Keywords: obligate cave fauna; conservation; species richness; stygobiont; troglobiont

Citation: Ferreira, R.L.; Berbert-Born, M.; Souza-Silva, M. The Água Clara Cave System in Northeastern Brazil: The Richest Hotspot of Subterranean Biodiversity in South America. *Diversity* **2023**, *15*, 761. https://doi.org/10.3390/d15060761

Academic Editors: Tanja Pipan, David C. Culver, Louis Deharveng and Michel Baguette

Received: 5 May 2023
Revised: 1 June 2023
Accepted: 7 June 2023
Published: 9 June 2023

1. Introduction

Subterranean environments are home to a distinct biodiversity that thrives under restricted food resources and stable environmental conditions, making them highly vulnerable to alterations in their pristine habitat characteristics [1]. However, our understanding of the diversity patterns of Neotropical cave fauna remains limited compared with other regions in the world [2], and the factors influencing their distribution are still largely unknown [3]. In South America, Brazil has the highest proportion of karst landscapes, and recent studies have identified many obligate cave species in these areas [4–6]. Karst landscapes in Brazil occur in different rock types, including granites, sandstone, iron ore, limestone, and dolomite, and are found in regions such as the Amazon Basin, the Atlantic Rain Forest, the Brazilian Savana, and the Caatinga under distinct climate conditions and geological ages [5–8].

Regrettably, industrial, economic, and human population growth in surrounding areas has had adverse effects on subterranean fauna and the karst landscape [1,9]. To safeguard these remarkable and delicate habitats and their fauna, some conservation strategies have been implemented globally to protect and preserve cave ecosystems [1]. The Red List of Threatened Species, established by the International Union for Conservation of Nature (IUCN), is one prominent initiative for indicating subterranean species that should be protected [2]. Furthermore, Culver and Sket [10] introduced the term "hotspots of subterranean biodiversity" to define subterranean habitats that have at least twenty or more obligate stygobitic and troglobitic species. However, they did not take into consideration the threats to biodiversity loss in those areas [4].

This approach of identifying hotspots based solely on species richness may not necessarily account for the ecological importance of these habitats or their vulnerability to anthropogenic impacts. Therefore, it is important to consider both the biodiversity value and the threats to the habitat in the process of identifying and prioritizing conservation areas [3,4]. Insufficient comprehension of subterranean systems has constrained the formulation of protective benchmarks, resulting in prioritization strategies that concentrate on identifying particular sites, such as caves, wells, or small aquifers, that display heightened levels of biodiversity. Nonetheless, these single-site methodologies are circumscribed by practical and financial limitations and neglect the interconnectivity among subterranean habitats as well as their mutual dependence with surface systems [11].

Despite the multifarious benefits that subterranean ecosystems and their biodiversity provide to humankind, they are infrequently incorporated into conservation plans on a large scale [12]. In order to safeguard cave biodiversity, it is imperative to not only preserve the intrinsic characteristics of subterranean habitats but also the pristine environmental conditions of the surrounding epigean environment [11]. Efforts focused on the development and implementation of conservation measures, as well as initiatives aimed at fostering collective action, are widely regarded as the most effective approach for identifying and safeguarding priority areas for conservation [13].

Despite persisting in a vision of conservation that remains constrained and fragmented in relation to subterranean habitats, identifying hotspots of subterranean biodiversity enables researchers and policymakers to optimize resource allocation and safeguard these exceptional and delicate ecosystems. Developing effective conservation strategies requires an understanding of the hazards that threaten these regions, including human development, pollution, and climate change. By investigating the taxonomic biodiversity, climate conditions, and organic resources present in these hotspots, researchers can improve their comprehension of the ecological mechanisms that govern these ecosystems and develop targeted conservation measures.

In Brazil, the speleological heritage is partially protected by a decree that requires that caves be classified according to their relevance degree prior to the installation of any activities that could potentially impact subterranean ecosystems. Currently, caves in Brazil are classified into four relevance categories (low, average, high, and maximum) through a multiparametric analysis. Only caves classified as having maximum relevance receive complete protection, while those in the remaining categories are susceptible to various degrees of impact, including complete cave suppression.

As part of the Diversity journal's Special Issue titled "Hotspots of Subterranean Biodiversity", we have compiled data on one of Brazil's most remarkable cave systems, utilizing both existing literature and original findings. This contribution presents the extensive biological diversity within the system and offers hypotheses on why such a rich cave-restricted community evolved in this particular cave system. Therefore, it aims to serve as a reference for future research on evolution and conservation, as well as to inspire cave exploration in the numerous unexplored cave systems in Brazil.

2. Regional Geology

About 2.5% of Brazil's territory is occupied by carbonate rocks, mostly in the central-eastern portion in an extensive strip of relatively continuous exposures between parallels 10° and 21° (Figure 1A). Some of the country's most significant karst areas are located in this wide region that is part of the São Francisco Craton (CSF). Eleven "speleological regions" are discriminated, taking into account the largest clusters of caves [14], one of which is the region of Serra do Ramalho karst (Figure 1B).

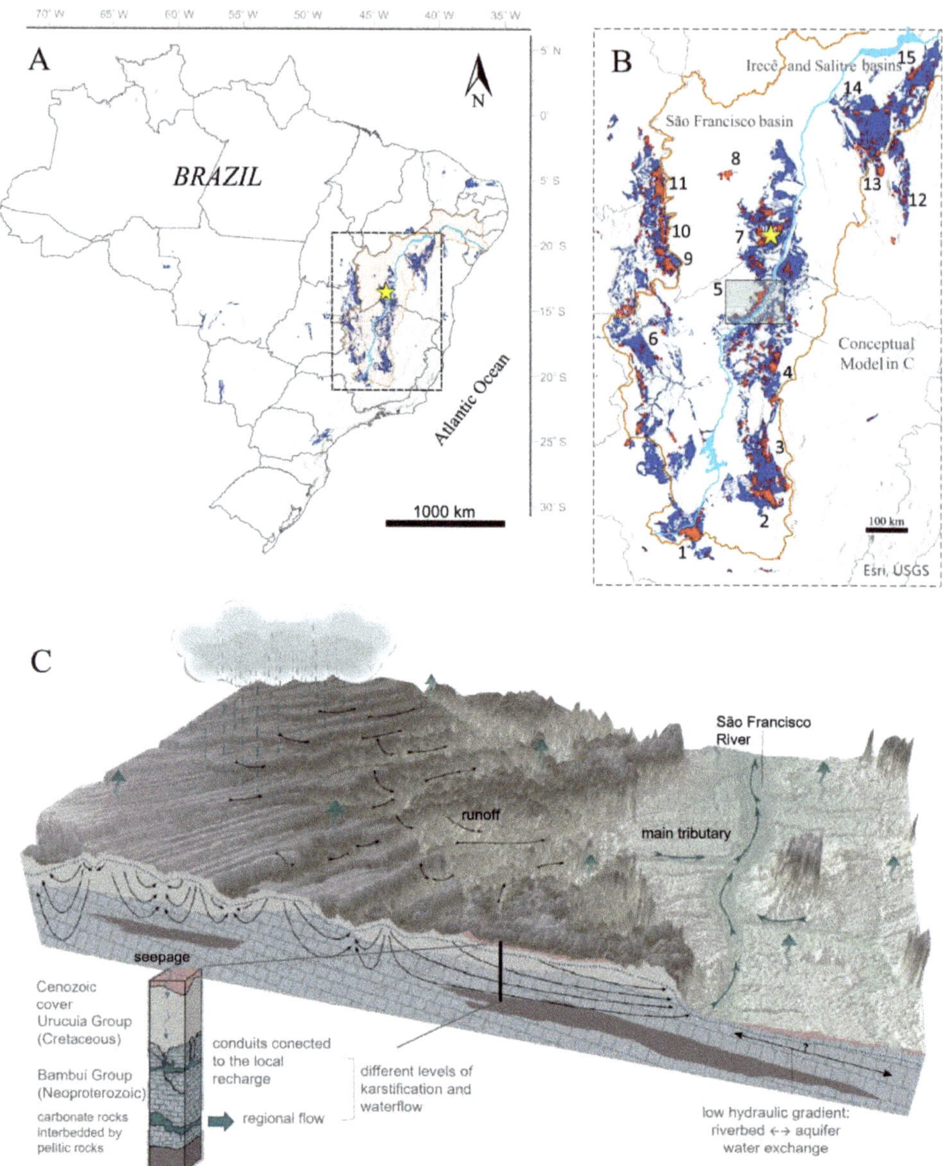

Figure 1. (**A**) Distribution of predominantly carbonate lithostratigraphic units in Brazil as dark blue areas; São Francisco River Basin (BHSF) as red shaded with São Francisco River in light blue; the Água Clara Cave System pointed out by the yellow star. (**B**) Neoproterozoic carbonate units that cover the São Francisco Craton: São Francisco Sedimentary Basin, Bambuí Group (central and western region); Irecê and Salitre Basins, Una Group (northeastern region). Numbers representing 11 Speleological Regions related to the carbonate rocks of the Bambuí Group plus 4 Regions involving the Una Group (sensu [14]), placed around the major concentrations of caves (red spots). "Serra do Ramalho and surroundings" is the region number 7. (**C**) Illustrative scheme of water dynamics in the São Francisco River basin in the northern portion of Minas Gerais/southwest Bahia states. The model can be considered representative of the hydrogeological Subdomain involving the Serra do Ramalho Karst (slightly modified from [15]).

Karstification in these regions occurs in limestones and dolomites of siliciclastic-carbonate sedimentary successions belonging to the Bambuí Group, more specifically involving the Sete Lagoas and Lagoa do Jacaré formations (basal and intermediate units, respectively). The Bambuí mega-sequence is also composed by the Serra de Santa Helena Formation, a thick interval of pelitic rocks between the two carbonate units, and the Serra da Saudade and Três Marias formations, which gather siltstones, mudstones, and arkose sandstones at the top of the classic stratigraphic column [16,17].

The entire package can reach 3000 thick [18], but given the complex interplay of lithology, depositional environment, and tectonic activity, the sedimentary successions of the Bambuí Group may exhibit different stratigraphic patterns in different locations along the São Francisco Sedimentary Basin [17,19–22]. Still under discussion, the age of the basin filling by Bambuí sediments comprises the end of the Neoproterozoic and the beginning of the Paleozoic (Ediacaran-Cambrian), between 635?–520 Ma [23–25].

The Phanerozoic sedimentary cover of the CSF is composed of Cretaceous sediments, including siltstones, volcaniclastic rocks, and sandstones from the Areado, Mata da Corda, and Urucuia groups, which partially overlay the rocks of the Bambuí Group in erosive unconformity [26,27]. These sediments form the continuous plateaus of the Serra Geral de Goiás, which span from the north of Minas Gerais to the south of Piauí [28]. Additionally, Cenozoic covers consist of ferruginous detritus-lateritic sediments and alluvial deposits.

3. Hydrogeology

The regional geology plays a crucial role in the development of karst systems, particularly with respect to the aquifer system and the hydraulic conditions related to the regional sedimentary and tectonic settings [29]. Some factors inherited from regional geological processes include: (i) relationships between lithologies with different permeabilities; (ii) reduced primary porosity vs. increased secondary porosity; (iii) the existence of large geological structures such as faults and lineaments; and (iv) potentially fluidized mineralizations. Together with climatological parameters, these aspects must be addressed from integrated perspectives at different scales.

The carbonate units of the Bambuí Group are widely distributed within the São Francisco River basin, where extensive exposures of its carbonate formations are found (Figure 1C). These units are karst aquifers or fissured-karst when interbedded with siliciclastic rocks (i.e., interfaces with pelitic rocks of the Serra de Santa Helena formation). They are of high environmental and socioeconomic importance within the geopolitical context of the hydrogeographic basin, as they play a significant role in preserving ecological streamflows and are the sole source for urban, industrial, and rural supplies in many areas [30–32]. Karst springs and resurgences are heavily exploited in diffuse rural settlements, often through rudimentary catchment systems providing water for small-scale agriculture, animals, and even direct human consumption.

The management of these aquifers is carried out based on differentiated hydrogeological, geomorphological, and climatic scenarios within the river basin (BHSF) as a whole. These "scenarios" refer to hydrogeological domains and subdomains that exhibit distinct hydrodynamic signatures. These include relationships between carbonate and non-carbonate units, the deformational conditions, the nature of non-karst coverings, and the karstification patterns in terms of the general arrangement, size, and types of morphogenetic structures found in both exokarst and endokarst environments. From this overview, fifteen hydrogeological units are recognized in the BHSF [33]. The Serra do Ramalho karst, where the ACCS is located, is part of the "Subdomain IIIe".

This subdomain features highly karstified areas in an exposure strip of undisturbed carbonate rocks located between the Urucuia plateaus and the middle reaches of the São Francisco River, spanning an elevation range of 450 to 800 m. The regional karstification in these areas is actively associated with the draining capabilities of the São Francisco River and large tributaries of its left bank, combined with high hydraulic gradient and permeability conditions, as illustrated in the general conceptual model applied to the

Itacarambi-Montalvânia region [15], north of the state of Minas Gerais, middle São Francisco (Figure 1C).

4. The Local Karst

The Serra do Ramalho karst system is delimited to the south by the Carinhanha River and to the north by the Corrente River, two large perennial tributaries that are very important for the São Francisco River flow [34]. These rivers are formed on the Urucuia plateaus (Cerrado biome [35]) and cross in a W–E direction over 100 km of dissected karst areas of the so-called "Sanfranciscana depression" (Figure 2A).

The interfluve of these rivers covers the broader region of Serra do Ramalho karst (approximately 12,000 km^2), including the areas and potential areas of more direct karst recharge closer to the edge of Serra do Urucuia, where the sandstone cover is less thick. The wider karst also includes the discharge areas at local base levels and the discharge areas at the regional base level represented by the São Francisco River. The municipalities of Cocos, Feira da Mata, Coribe, Carinhanha, Serra do Ramalho, and São Félix do Coribe are all located entirely within this region.

A significant portion of the "broad karst system" of Serra do Ramalho corresponds to a large elevated block configuring an already quite incised extension of the Urucuia plateau, where the main karst structures that characterize the local karst are found. The essentially horizontal lithostratigraphic units of the Bambuí Group that support this block [36,37] gain geomorphological expressions in the terraces and erosive steps towards the plateau boundaries, as well as in the locally developed "erosional windows" inside the plateau. The karst features are notably associated with the two carbonate units depicted by the Sete Lagoas and Lagoa do Jacaré formations (calcilutites, calcarenites, calcirudites, and dolomites), although other lithostratigraphic units exert some kind of influence on the active karst processes [38].

In the descriptive model for the local karst system [39], the highest levels of the block are at the interface with the sandstone plateaus and present less evolved dissolution features. In intermediate elevations, the karst relief has strongly undulating surfaces with many dolines and limestone towers in conformation or exhumation. There is progressive dissection towards the margins of the block, gradually inverting the dominance between negative and positive forms. The endokarst is characterized by several underground river systems that, observed on a macro scale, constitute multiple cave conduits organized in a predominantly dendritic and regionally centrifugal pattern, which can extend for a few tens of kilometers. In general, these systems are linked to the multiple autochthonous recharge points associated with polygonal depressions and karst windows located in the central portions of the high and middle plateaus. They present convergent flow towards punctual discharges in several locations along the entire perimeter of the large rocky block. Recharge within the block also occurs diffusely in non-karstic coverings (e.g., residual sandstones from Urucuia; interbedded fissured pelitic rocks; soils). The contour of the karst block is marked by deep incisions that form large canyons and steep valleys. Although they are large, most of these marginal valleys are currently associated with intermittent or ephemeral drainage. Marginal groundwater outlets are perennial springs (phreatic level) or resurgences of highly seasonal flow associated with conduit systems.

Based on these hydrogeological, geomorphological, and speleogenetic aspects, one may distinguish, from the center to the edges of the plateau, karst domains with their own morphological patterns and hydrological and sedimentary dynamics [39].

Among them, the eastern margin of the block is exceptionally prominent in the local geomorphology, rising abruptly from the low levels of the São Francisco plains (460–480 m). This boundary extends for over 70 km in a very peculiar, strongly indented path, with contours delimited by cliffs tens of meters high, under which several springs and resurgences are located. On this face are the main discharges of the local karst system, and it is in this section that the ACCS is housed. This extensive discharge zone is also marked by the increased expression of the epikarst and, consequently, the movement of clastic materials that cover the karstified rocks. Under these conditions, soil volumes can be more easily injected into and later removed from the dissolution networks by water activity. Cyclical phases of aggradation and erosion are expressed in the underground and interstitial environments, while karren fields are progressively formed on the surface of the massifs.

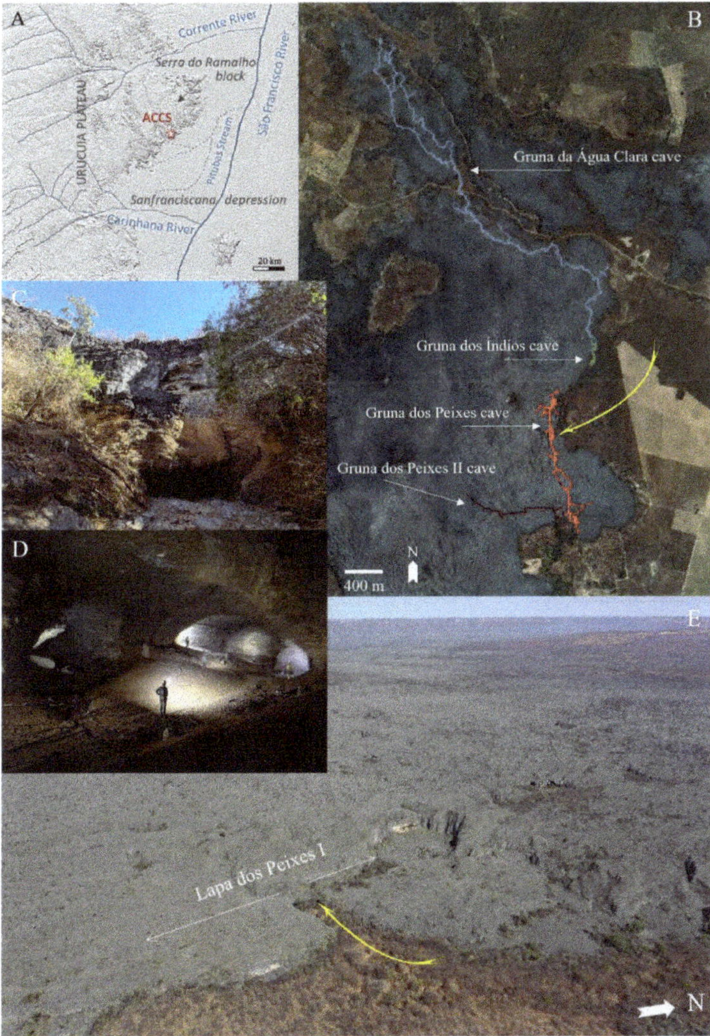

Figure 2. Geomorphological-hydrological context of the Serra do Ramalho Karst and the Agua Clara Cave System (ACCS), highlighting the overall interrelationship with the São Francisco River: (**A**) Detached karst block in the interfluve of the Carinhanha (south) and Corrente (north) rivers; the ACCS placed on the eastern edge of the karst block; (**B**) Spatial distribution of the caves Gruna da Água Clara, Gruna dos Índios, Lapa dos Peixes I and Lapa dos Peixes II; the yellow arrow indicates some entrances of the Lapa dos Peixes cave; (**C**) One of the entrances of the Água Clara cave, sink of the seasonalstream that drains sections of the ACCS; (**D**) Inner conduit of the Água Clara cave; (**E**) Aerial view of the exposed massif at the edge of the Serra do Ramalho block, where the ACCS is located; the yellow arrow indicates some entrances of the Lapa dos Peixes cave. Bambuí Speleological group provides the cave maps (https://bambuiespeleo.wordpress.com/, accessed on 4 May 2023).

All the outflows from this margin converge on the São Francisco plain, forming the Pitubas stream basin, a direct tributary of the São Francisco River that flows about 60 km along its alluvial plain. The tributaries in this "terminal basin of the karst system" have a strongly meandering pattern, with generations of wandering meanders characteristic of alluvial plains.

5. The Água Clara Cave System (ACCS)

The ACCS consists of a sequence of four large sets of underground conduits developed in the Sete Lagoas calcarenites. This "cave complex" is hydrologically connected by one of the multiple fluvial outlets located on the eastern periphery of the Serra do Ramalho karst block. From upstream to downstream, it involves the caves Gruna da Água Clara (13,880 m), Gruna dos Índios (510 m), Lapa dos Peixes I (9320 m), and Lapa dos Peixes II (2100 m). Linear development measurements and plan projections are indicated by Franco-Brazilian speleotographies carried out by Grupo Bambuí de Pesquisas Espeleológicas (GBPE) and Groupe Spéléo Bagnols Marcoule (GSBM) between 1998 and 2001 (Figure 2B) [14,40–43].

By direct observation of the relief and the surface drainage pattern, a maximum area of ca. 130 km^2 is inferred for the possible catchment basin corresponding to the ACCS outflow, a tributary of the Riacho das Pitubas (Figure 2A). This calculation does not consider possible aquifer interactions of greater extension that may be associated with the regional stratigraphic and geomorphological framework.

The conduit system (Figure 2D) extends along a linear axis of about 8 km and a different level of ca. 80 m from the innermost end in the massif ("distal", upstream) to the last resurgence ("proximal", downstream), where the massif contour projects towards the alluvial plain of the São Francisco River. The stream that drains the basin at this discharge point has an intermittent flow and runs only 3.5 km underground, along two sections interspersed with the riverbed that surrounds the massif externally (Figure 2E). Some entrances also receive a seasonal water contribution (Figure 2C).

Considering the position, geometry, morphology, and sedimentary content, some hydrologically inactive subterranean segments may possibly depict ancient tracings of the main course (i.e., Gruna dos Índios) or secondary flow pathways active only under overflow conditions. Other segments correspond to smaller underground tributary channels, or palaeochannels. Floodwater mazes are especially close to the sinks and next to the final outlet of the system.

Overall, the system comprises the following subcompartments: (i) a sinuous main conduit with intermittent flow, with permanently inactive sections (fossil compartments); local anastomoses possibly related to overflow regimes, with eventual diversion of the main course to new parallel flow paths; channels with a markedly elliptical cross-section resulting from pressure flow, locally evolved to the "keyhole" type and tending towards quadrangular and irregular (polygonal) in the vicinity of entrances (sinks and resurgences) and gallery intersections; (ii) secondary conduits of ephemeral tributaries, or perennial drainages of low flow associated with runoffs of diffusely stored water; they form sinuous to straight sections of smaller diameter with diverse morphology; (iii) sections with lateral "accessory galleries" in a reticulated pattern or isolated fissures, associated with gravitational infiltrations from fissures connected to the exposed surface of the rocky mass (epikarst with higher karstification degree); they typically develop or reach the highest levels of the system; (iv) maze sectors lateral to the main conduit located near entrances (sink and resurgences), with galleries of diversified morphology in a reticular or braided arrangement, frequency of collapsed blocks and sedimentary cones (talus) injected from the surface, partially rearranged by backflows from the main channel, phreatic oscillations, and/or supervening floods; (v) lower proto-levels of punctual runoff to which fine sediments previously deposited are being selectively relocated.

The morphological pattern of the ACCS may represent a general speleogenetic model valid for other flow systems in similar transects, encompassing the upper portions of autochthonous recharges (diffuse and concentrated), the front steps of erosive retreat of non-karst coverings, the frank exposure of the carbonate rocks, and the peripheral gradient of karstification, ending with the arrangement of the discharges in the São Francisco plain that is linked to the São Francisco river flow dynamics [39].

Observing the currently prevailing hydrological conditions, the ACCS fits into a mixed condition of convergent surface flow towards a short underground transit channel, seasonally enhanced by a regime of rapid rainfall infiltration through an intensely fissured rocky substrate. Such conditions favor seasonal overflows in the system as well as water table oscillations, mainly in the lowest terminations of the system.

Considering the diameter of the fossil conduit that constitutes the distal (upstream) portion of the Gruna da Água Clara cave, it is plausible that the system extends into the inner massif after the collapses and sedimentary blockages. Sedimentary and morphogenetic studies may contribute to elucidating past flow regimes and related sources.

6. Compiling the Data (Overview of Invertebrate Sampling)

The database of the obligate cave fauna was compiled through a combination of published literature [44–55] and visits to the ACCS, with 10 visits conducted in total. To determine potentially troglobitic or stygobitic species, we identified 'troglomorphisms' in unknown sampled specimens and consulted with specialist researchers (specialists are acknowledged further on) or referred to previously identified and described species from the literature. All collected organisms were preserved in 70% ethanol, identified to an accessible taxonomic level, and deposited in the Subterranean Invertebrate Collection (ISLA) of the Center for Studies on Subterranean Biology at the Federal University of Lavras (CEBS-UFLA).

The cartographic sources for Figure 1A,B were obtained through ArcGIS Pro 3.1 geoprocessing, specifically from the Geologic Maps of Brazil at a scale of 1:1,000,000 provided by the Serviço Geológico do Brasil (https://geoportal.cprm.gov.br/server/rest/services/geologia/litoestratigrafia_1000000/MapServer, accessed on 4 May 2023) and the Cadastro Nacional de Informações Espeleológicas (https://www.gov.br/icmbio/pt-br/assuntos/centros-de-pesquisa/cecav/cadastro-nacional-de-informacoes-espeleologicas/canie, 19 December 2022 updated data, accessed on 4 May 2023). The BHSF data in Figure 1C was sourced from open data provided by the Agência Nacional de Águas (ANA) (https://dadosabertos.ana.gov.br/, 29 March 2022 update, accessed on 4 May 2023). The shaded relief in Figure 2A was obtained from the Instituto de Pesquisas Espaciais—INPE (Brazilian Space Research Institute), specifically from the Topodata Project maps 14S45 and 13S45 (http://www.webmapit.com.br/inpe/topodata/, accessed on 4 May 2023). The cave maps in Figure 2B were provided by the Bambuí Speleological Group and can be accessed via their website at https://bambuiespeleo.wordpress.com/, accessed on 4 May 2023. The photographs of living specimens in Figure 3 were captured using a Canon EOS 80D digital camera (Canon, Tokyo, Japan). Lastly, the aerial photographs of the external landscape in Figure 2E were taken using a DJI Mavic 2 Pro drone (DJI, Shenzhen, China).

Figure 3. Some of the cave-restricted species found in the Água Clara cave system (ACCS), Brazil. (**A**) Ochyroceratidae sp.1; (**B**) *Charinus troglobius* Baptista & Giupponi 2002; (**C**) *Pseudochthonius koinopoliteia* Prado & Ferreira 2023; (**D**) *Giupponia chagasi* Perez & Kury 2002; (**E**) Eukoenenia sp.1; (**F**) *Pectenoniscus carinhanhensis* Cardoso, Bastos-Pereira, Souza & Ferreira 2020; (**G**) *Xangoniscus aganju* Campos-Filho, Araujo & Taiti 2014; (**H**) Styloniscidae sp.1; (**I**) Styloniscidae sp.2; (**J**) *Trichorhina* sp.; (**K**) *Troglobentosminthurus luridus* Souza, Medeiros & Bellini 2022; (**L**) Blattodea sp.1; (**M**) *Nylanderia* sp.1; (**N**) *Endecous infernalis* Carvalho, Junta, Castro-Souza & Ferreira 2023; (**O**) *Mesodiplatys falcifer* Kamimura 2018; (**P**) Oniscodesmidae sp.1; (**Q**) *Phaneromerium* sp.1; (**R**) Chelodesmidae sp.1; (**S**) Pyrgodesmidae sp.1; (**T**) Geophilomorpha sp.1; (**U**) *Girardia spelaea*; (**V**) *Spiripockia punctata*; (**W**) *Trichomycterus rubbioli*.

7. The Checklist of Cave-Restricted Taxa in ACCS

The ACCS contains a total of 31 species that are restricted to caves, with distribution across Hexapoda (9 species), Arachnida (7 species), Crustacea (6 species), Myriapoda (5 species), Gastropoda (2 species), Turbellaria (1 species), and Siluriformes (1 species) (Figure 3). This makes the ACCS the South American cave system with the highest number of cave-restricted species. Terrestrial species were the most predominant (22 species), followed by amphibious (5 species) and aquatic species (3 species).

The low richness of aquatic species in the ACCS may be attributed to the intermittent nature of the drainage that traverses the system. The three aquatic species comprise a fish (*Trichomycterus rubiolli* Bichuette & Rizzato 2012), a snail (*Spiripockia punctata* Simone 2012), and a flatworm (*Girardia spelaea* Hellmann & Leal-Zanchet 2020), all associated with permanent water bodies (locally restricted within the system) that receive water input from epikarstic epigenic diffuse recharge. Consequently, populations of these species are only observed in specific areas along the ACCS, such as travertine pools or permanent ponds located in topographically lowered regions (as in the Peixes and Peixes II caves). However, during periods of rainfall, there can be dispersion events of organisms between different areas of the system as the main drainage becomes active. As an illustrative example, on a particular occasion, a solitary fish specimen (*T. rubiolli*) was observed in a minuscule travertine pool situated in an elevated area. It is plausible that this specimen was transported to that location during a flood event. It is worth noting that specific sampling methods targeting minute invertebrates, such as microcrustaceans, have not been employed. Therefore, the possibility of encountering additional stygobitic species in the ACCS in the future cannot be ruled out.

Among the caves in the ACCS, the Gruna da Água Clara cave was found to have the highest richness of troglobitic species, with a total of 23 species, including four species that were exclusively observed in this cave (Symphypleona sp.2, Rhagidiidae sp.1, Caponiidae sp.1, *Trichorhina* sp.1). The Lapa dos Peixes I cave had 19 species, while the Lapa dos Peixes II cave had 17 species, and the Gruna dos Índios cave had only five species (see Table 1). The Gruna da Água Clara and Lapa dos Peixes I caves had the largest number of shared species, even though their nearest entrances were approximately 3 km apart. It is noteworthy that the Gruna dos Índios cave, located in the intermediate portion of the ACCS, has a relatively dry main conduit due to the airflow that comes from the entrances on both sides of its main conduit. As a result, it is likely that the cave-restricted species shared by the Gruna da Água Clara and Lapa dos Peixes I caves are migrating through mesocaverns that connect these macrocaves. Only two species (Chelodesmidae sp.1 and *Endecous infernalis*) were found in all caves in the system (see Table 1). According to Souza-Silva et al. [51], the estimated troglobitic species richness suggests that the sampling effort reached adequate levels of completeness (obtained by Jack-Knife estimators), as the observed richness corresponded to over 78% of the estimated richness [56].

Table 1. Variation in the number of troglobitic and stygobitic species across different taxa in the three South American hotspots of subterranean biodiversity.

Taxon	Areias Cave System	Toca do Gonçalo Cave	Água Clara Cave System
Crustacea	5	5	7
Hexapoda	7	7	9
Arachnida	6	3	6
Myriapoda	5	5	5
Mollusca	1	1	2
Siluriformes	1	1	1
Platyhelminthes	2	0	1
Nemertea	1	0	0
Total	**28**	**22**	**31**

It should be noted that, in addition to the troglobitic species, the ACCS also harbors 142 non-troglomorphic invertebrate species, distributed among Hexapoda (85 spp.), Arachnida (43 spp.), Myriapoda (5 spp.), Annelida (3 spp.), Mollusca (3 spp.), Turbellaria (2 spp.), and Nematoda (1 sp.). This makes the ACCS one of the most biologically diverse cave systems in South America, with at least 173 species. Importantly, most of these species were found in areas far from the cave entrances, indicating that future sampling efforts in the cave, especially in para-epigean communities, could further increase this number.

Notable Cave Species

Among the troglobitic species observed in the ACCS, some are highly troglomorphic, such as the springtail *Troglobentosminthurus luridus* (Figure 3K), the whip spider *Charinus troglobius* (Figure 3B), and the harvestmen *Giupponia chagasi* (Figure 3D), all of which represent the most troglomorphic species of their respective groups in Brazil. It is possible that the voluminous galleries associated with the oligotrophic conditions prevailing in the system have played a role in shaping the strong troglomorphic traits observed in these species. The elongated appendages found in these species confer advantages for both ambulation and detection of organic resources in such an ample subterranean environment. This hypothesis is in accordance with Trontelj et al. [57], who have discussed the significance of pore size on the evolution of cave organisms using amphipods from the genus *Niphargus* as models.

The high richness of isopods is also outstanding. From the six troglobitic species registered, five belong to the Styloniscidae family. It is noteworthy that the occurrence of phylogenetically related species within the same set of habitats is typically precluded by the competitive exclusion principle [58]. However, the coexistence of isopod species in the ACCS may be due to niche displacement, as they occupy distinct microhabitats within the system. Some species, such as *Xangoniscus aganju* (Figure 3G), are amphibious, whereas others, like *Pectenoniscus carinhanensis* (Figure 3F), are strictly terrestrial. Additionally, preferences for trophic resources can vary, with *Trichorhina* sp. (Figure 3J) typically found in plant organic debris (decaying trunks) and *P. carinhanensis* being more attracted to bat guano.

Among the 31 species that are restricted to caves within the ACCS, two species in particular warrant further discussion regarding their cave-restricted status: the earwig *Mesodiplatys falcifer* and the ant *Nylanderia* sp. The earwig *M. falcifer* has only been observed within the ACCS, despite extensive biological inventories conducted in numerous other caves within the Serra do Ramalho region. Although possessing relatively developed eyes, this species displays weak pigmentation (even in non-teneral adults) and elongated appendages compared with other species within its genus. In the original species description, Kamimura and Ferreira [49] suggested the possibility of it being troglobitic but did not definitively confirm this diagnosis, mainly because a single specimen was found. However, subsequent surveys within the ACCS revealed the presence of immature specimens consistently located in deep sections of the caves, with no observations of individuals in external habitats. Hence, we are herein considering this species as troglobitic. The ant species *Nylanderia* sp. exhibits all the typical troglomorphic traits. While all species in this genus are dark-pigmented and present well-developed eyes, the species from the ACCS displays weak pigmentation and considerably reduced eyes. Furthermore, an entire colony was observed rather than a single or a few specimens. Additionally, an expert (R. Feitosa, pers. com.) confirmed the troglomorphic traits and the undescribed status of this species.

Finally, it is important to mention that of the 31 cave-restricted species occurring in the ACCS, 22 are endemic to this cave system. The remaining eight species, including *C. troglobius*, *G. chagasi*, and *X. aganju*, which have relatively wide distributions across the Serra do Ramalho area, may comprise cryptic species complexes. Ongoing studies of *C. troglobius* and *X. aganju* have identified morphological and genetic differences between populations in distinct caves, suggesting that each species is actually a set of cryptic species, each endemic to a single cave or cave system. Therefore, it is likely that the number of endemic species in the ACCS will increase in the future, underscoring the system's importance as a unique biodiversity hotspot.

8. Discussion

8.1. Importance of Continuous Surveys and Updated Checklists

Sampling subterranean environments can be a daunting task, especially in the case of invertebrates. This challenge is largely due to the difficulty of accessing crucial microhabitats such as fissures, mesocaves, and interstitial voids [1,57,59]. Thus, to document subterranean biodiversity effectively, it is crucial to conduct multiple collections over time. Unfortunately, Brazil has a history of insufficient long-term studies in caves, mainly due to two factors. Firstly, the continuous discovery of new caves and karst areas presents researchers with tempting opportunities to identify new species and ecological patterns, drawing them away from performing long-term studies in a same cave or cave system. Secondly, the legal framework for protecting Brazilian caves only requires two samplings of a given cave for environmental licensing purposes. While the allure of new karst areas attracts researchers, the minimal legal requirement for sampling also hinders long-term studies.

Therefore, it is highly probable that many undiscovered subterranean hotspots of biodiversity exist in tropical regions, and that the currently known hotspots represent only a small portion of the total. In Brazil, all identified hotspots consist of caves or cave systems that have undergone successive sampling, indicating that the scarcity of long-term studies may be a major obstacle to the discovery of new hotspots in tropical areas. As such, countries with cave conservation policies, like Brazil, should consider recommending a greater number of sampling events in caves to reveal their true diversity. Given that numerous troglobitic species are rare, it is improbable that the entire range of species present in a cave will be detected through just one or a few sampling efforts.

8.2. Taxonomic Impediment of Megadiverse Tropical Areas and the Challenge of Determining Cave-Restricted Species

The conservation of subterranean environments is often hindered by the fact that many of the species that inhabit these ecosystems remain undescribed and are therefore often overlooked in conservation efforts [60,61]. In tropical and subtropical regions, where most cave-restricted species are yet to be formally described [4,61], caves are typically considered to be relatively species-poor [60]. Therefore, describing the troglobitic species found in a given area is a crucial step towards their conservation.

As an example, among the 31 cave-restricted species observed in the ACCS, only 11 (35.5%) have been formally described (Table 1): *Girardia spelaea* (Platyhelminthes: Dugesiidae) [52], *Spiripockia punctata* (Mollusca: Caenogastropoda) [47], *G. chagasi* (Opiliones, Gonyleptidae) [45], *C. troglobius* (Amblypygi: Charinidae) [44], *Pseudochthonius koinopoliteia* (Pseudoscorpiones: Chthoniidae) [55], *X. aganju* (Isopoda, Styloniscidae) [48], *P. carinhanhensis* (Isopoda, Styloniscidae) [50], *T. luridus* (Collembola: Sminthuridae) [51], *E. infernalis* (Orthoptera: Phalangopsidae) [54], *Mesodiplatys falcifer* (Dermaptera: Diplatyidae) [49], and *Trichomycterus rubbioli* (Siluriformes: Trichomycteridae) [46]. Moreover, of the 11 species, six were only recently discovered during ecological surveys of the caves comprising the ACCS [53]. Prior to 2016, only six of the 24 known cave-restricted species in the region were formally described [62]. However, recent ecological studies conducted in 26 caves in the Serra do Ramalho region indicate the existence of at least 70 additional cave-restricted species [Ferreira et al. unpublished data], indicating the vast potential for discovering new species in this area.

The slow pace of species description of cave-dwelling organisms in Brazil highlights the lack of taxonomists who specialize in taxa commonly found in caves, as well as the difficulties encountered by foreign taxonomists in studying this fauna due to legal constraints. Despite some financial support being allocated towards the description of subterranean taxa, the number of species described in Brazil in recent years remains relatively low compared with the vast number of newly discovered species each year. Therefore, it is imperative not only to encourage species description but also to train new taxonomists, particularly for taxonomic groups with limited or no specialists capable of identifying and describing this unique, endemic, and threatened fauna.

Finally, identifying species that are exclusively restricted to caves is a difficult task in tropical areas. While the concept of troglobitic species is widely accepted, it can be challenging to determine this status with certainty. The only surefire way of confirming whether a species is exclusively restricted to subterranean habitats is to demonstrate its absence in surface habitats, which can be difficult, if not impossible, in mega-diverse tropical regions. To address this challenge, alternative approaches, such as the use of troglomorphic traits, have been employed to define such species. However, it should be noted that although typical troglomorphic traits, such as lack of pigmentation, eye reduction, and appendage elongation, are easily recognizable, there are several specific traits, especially for groups naturally devoid of pigment and eyes (e.g., Palpigradi), which can make identification challenging. Additionally, the use of troglomorphic traits can often lead to misdiagnosis. In some cases, a troglobitic species may not be recognized if it presents weak troglomorphisms, even though it is already restricted to caves. Conversely, an epigean troglomorphic species found in caves may be erroneously considered a troglobite.

8.3. Is 25 Cave-Restricted Species a "Magic" Number?

In their original proposal, Culver and Sket [10] introduced the term "hotspots of subterranean biodiversity" (HSB) to refer to subterranean habitats containing a minimum of twenty cave-restricted species. Although somewhat subjective, this threshold was most likely established based on the researchers' extensive expertise in cave faunas globally and their specific interest in investigating the biogeographic patterns that underlie variations in the number of cave-restricted species across different areas. However, more recently, there has been a movement to raise this threshold to at least 25 cave-restricted species for a cave or cave system to qualify as a hotspot. This increase in the cutoff is still arbitrary, nonetheless. Consequently, an inevitable question arises: does this higher cutoff reflect the global scenario or is it primarily based on the already recognized regions that are known for their richness in troglobitic and stygobitic faunas?

Cave ecosystems in tropical regions are unlikely to exhibit the same ratio of troglobitic to non-troglobitic species as those found in temperate regions. This disparity arises due to the greater temperature fluctuations that occur during glacial maximums in higher latitude areas, which have profoundly affected the isolation and evolution of troglobitic species, leading to a high species richness in temperate caves. In contrast, tropical caves have not experienced such severe temperature changes over their geological history, although they are usually located within highly diverse external landscapes. As a result, while only a small proportion of epigean species may become isolated and evolve to become cave-restricted, this number can still be relatively high in tropical caves. Nonetheless, it is important to emphasize that tropical caves will never attain the same ratios of troglobitic species observed in temperate caves. This effect is evident when comparing the proportions of hotspots of subterranean biodiversity (HSB) between temperate and tropical regions. While 81.6% of these hotspots occur in temperate areas, only 18.4% are located in the tropics [1,4,63,64].

Therefore, it is crucial to consider various parameters while defining subterranean hotspots. As an initial step, it is crucial to consider different scales when identifying hotspots, distinguishing between regional and global levels. Furthermore, the presence of natural breakpoints in datasets can provide valuable insights and serve as indicators of potential hotspots. Another significant criterion to be taken into account is the latitudinal range where the cave is located. A cave located in a high-latitude region with few cave-restricted species may still be considered a hotspot, given the extreme external climate conditions that would typically preclude the existence of any species. In contrast, in tropical areas, where external conditions are less severe, a smaller number of cave-restricted species (compared with temperate regions) should be taken into account when defining a cave as a hotspot. Another important factor to consider when defining HSB is the lithology associated with the cave. It has been consistently observed that iron-ore caves in Brazil generally display a higher average richness of cave-restricted species compared with caves associated with other lithologies [65]. Conversely, granite caves tend to exhibit a lower abundance of troglobitic fauna. Consequently, a HSB located in granite caves may potentially have a lower number of cave-restricted species in comparison to HSBs in iron-ore caves. An interesting example illustrating this pattern is the Wynberg Cave System (WCS) in South Africa. This cave system is situated within quartzite rocks and harbors an impressive diversity of 19 cave-restricted species [66]. This substantial number of species within a lithology typically considered less favorable for supporting cave-restricted organisms further confirms the designation of WCS as a HSB.

Consequently, it is essential to rethink the HSB concept and propose adaptations, as many countries can use or consider it for public policies regarding cave conservation. Hence, there is a risk of overlooking significant caves or systems on these lists if we always consider a high number of cave-restricted species. Nonetheless, further studies are necessary to determine the appropriate width of the latitudinal range and the proportion of troglobitic richness relative to the average that should be considered when defining a hotspot.

It is worth noting that Culver and Sket [10] did not take into account the level of threat to which subterranean habitats are exposed, as proposed by the hotspot model of Myers et al. [67]. Well-preserved landscapes can quickly turn into pastures or be destroyed by mining activities, as has happened in many karst regions around the world in recent years [68,69]. Therefore, relying solely on the number of cave-restricted species might not provide an accurate indication of the "health" of a given subterranean system, as this depends on the type of impact it has experienced, especially in recent times [4]. The Água Clara Cave System (ACCS) is a prime example, as it faces unprecedented threats from external factors (see Section 8.7). Thus, it is crucial to incorporate the level of threat to which a cave is exposed in this concept, as proposed by Myers et al. [67], particularly given that conservation policies often prioritize investment in areas of high conservation value [4].

Finally, every HSB is undoubtedly important, not only because of the richness of cave-restricted species it presents but mainly due to the high degree of endemism displayed by a large proportion of its species. Thus, another attribute that should always be taken into account when assessing the significance of a hotspot is the number of cave-restricted species exclusively found within that cave (or system) in relation to the total number of troglobitic/stygobitic species it presents.

8.4. Why Is ACCS So Rich in Cave Restricted Species?

In temperate regions, cave-restricted species are significantly influenced by epigean primary productivity since the amount of organic resources in surface environments potentially affects the availability of resources in subterranean ecosystems [1]. However, in contrast to temperate regions, external primary productivity was not found to have a significant influence on the richness of troglobitic species in Brazil [70]. This is possibly due to the high productivity observed in tropical regions, where even areas with relatively low productivity can provide sufficient resources for cave-restricted species [70]. As a result, factors other than external productivity are likely to have shaped the ecology and evolution of these species in tropical regions.

The intermediate disturbance hypothesis (IDH) posits that local species diversity can be increased when ecological disturbances occur at intermediate frequencies, neither too rare nor too frequent. At low disturbance levels, competitive organisms outcompete less competitive species, leading to their extinction and the dominance of the ecosystem by the more competitive species [71]. Conversely, at high disturbance levels, all species are at risk of extinction. The IDH suggests that at intermediate levels of disturbance, species that thrive at both early and late successional stages can coexist, promoting diversity. This hypothesis is based on three premises: (i) ecological disturbances have significant impacts on species richness within the disturbance area; (ii) interspecific competition results in one species driving a competitor to extinction, thus becoming dominant in the ecosystem; and (iii) moderate ecological scale disturbances prevent interspecific competition [72–74]. Despite criticism of this hypothesis [75], many authors continue to rely on its theoretical and empirical foundations [76]. Although the IDH is mostly used to explain ecological scenarios, the role of disturbances in evolutionary processes is widely accepted [75]. Therefore, species diversity in disturbance-mediated coexistence could be enhanced by the presence of a disturbance regime that resembles historical processes since species generally adapt to the level of disturbance in their ecosystem during their evolution.

The ACCS presents an intermittent drainage that flows along the system during the rainy periods. The cave system receives a vast amount of water from the external micro basin, which is significant due to intense and localized rainfall events common in the region. The flashflood pulses associated with these events can transport external materials, including large tree trunks, which often become lodged within the caves (see Figure 2D). Moreover, these flood pulses seasonally alter many cave substrates, modifying the cave floor and affecting numerous microhabitats (as bat guano piles, that can be washed away). These disturbances can be classified as intermediate since they partially and temporarily modify the cave's microhabitats. Therefore, flashflood pulses have not only shaped the invertebrate community structure of the ACCS but also likely influenced species evolution. By periodically changing the cave substrates, flashflood pulses prevent the establishment of dominant species, leading to the exploitation of various niches within the cave. Souza-Silva et al. [53], who analyzed the niches of ten troglobitic species from the ACCS, demonstrated that the most widespread troglobitic species could utilize microhabitats with distinct characteristics, thus avoiding niche overlap and promoting coexistence (Figure 4).

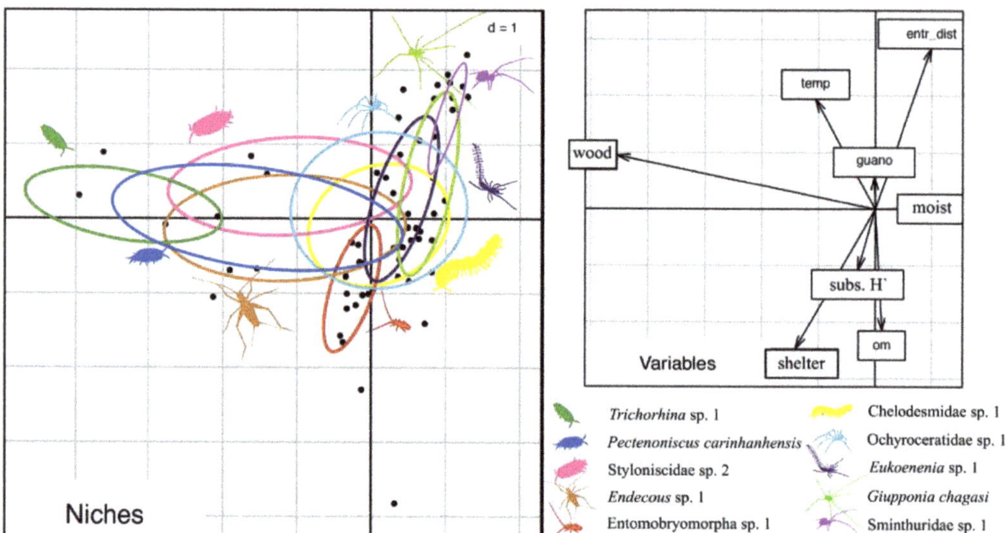

Figure 4. Niches of some cave-restricted species from the ACCS (modified from [53]). Results of the Outlying Mean Index (omi) analysis for the ten most widespread troglobitic species in ACCS that occupy the environmental niche according to the physical, trophic and microclimate characteristics of each transect.

Thus, despite cave-dwelling species often being classified as generalists, the presence of flashflood disturbances within the ACCS has likely resulted in reduced niche overlap. This reduction, in turn, has played a significant role in facilitating the coexistence of multiple troglobitic species within the system.

It is imperative to acknowledge that caves or cave systems harboring a high number of cave-restricted species typically correspond to oversized caves found in regions of high external productivity or with isolated water bodies from the surface [1,53]. The extent of a cave plays a significant role in fostering a greater diversity of cave-restricted species. Larger caves, by virtue of their size, offer a wider range of microhabitats, thereby providing the potential for accommodating a larger number of species [53,66]. The ACCS, situated in the Brazilian Tropical Dry Forest (specifically the Caatinga domain), is no exception. However, the remarkable richness observed in this system may be attributed not only to its extension, but also to both the IDH and the historical transformations undergone by the biome where the cave is located [53]. In particular, many caves in the Caatinga, especially those with perennial water sources, exhibit a remarkable diversity and endemism of cave-restricted fauna. It is noteworthy that the Caatinga biome had tropical rainforests spread over several areas during the Last Glacial Maximum (LGM) [77], which may have served as a refugia for the ancestors of current troglobitic species that were subsequently "trapped" inside caves following the retraction of the humid regions [78,79]. Supporting this hypothesis, three of the four known hotspots of subterranean biodiversity in Brazil, one of which remains undisclosed, are located in the Caatinga domain [4,53].

8.5. Comparison with Other Subterranean Hotspots in Brazil

South America is home to three recognized hotspots of subterranean biodiversity, all of which are located in Brazil. These include the Água Clara cave System and Toca do Gonçalo cave, located in the Bahia state (Northeastern Brazil), as well as the Areias cave systems situated in the São Paulo state (Southeastern Brazil). All these subterranean systems are associated with carbonate rocks. The Água Clara and Toca do Gonçalo caves are located in the semi-arid Caatinga Biome and harbor 31 and 22 species of obligate cave dwellers, respectively. The Areias Cave System, which is a unique Brazilian hotspot located within a conservation unit (Park Estadual Turístico do Alto Ribeira—PETAR), accommodates 28 species and is located in the Brazilian Atlantic Forest. These findings were reported in studies conducted by Souza-Silva and Ferreira [4] and Souza-Silva et al. [53].

The presence of either permanent or temporary water bodies provides a suitable habitat for troglobites and stygobites species in all three Brazilian hotspots of subterranean biodiversity. Additionally, the occurrence of siluriform fish species (*Trichomycterus rubbioli*, *Rhandiopsis* sp.n., *Pimelodella kronei* (Ribeiro, 1907)) is common in all three areas. These extensive subterranean systems (spanning over 500 m) are dependent on allochthonous donors for the input of organic resources into food chains, including percolation water, permanent runoff, streams, and bats. Among these hotspots, the Areias cave system is the only one that presents cave-restricted species from the epikarst zone compartments [4,53]. The occurrence of roots does not appear to be an essential resource for the maintenance of the fauna, as they are scarce in all three hotspots [4,53].

The Água Clara Cave System exhibits a predominance of terrestrial species (23 spp.), followed by amphibious species (5 spp.), and exclusively aquatic species (3 spp.). Similarly, in Toca do Gonçalo, terrestrial species are predominant (17 spp.), followed by species living in lentic aquatic habitats (5 spp.). The Areias Cave System also presents a predominance of terrestrial species (22 spp.), with the remaining species found in the aquatic lotic habitat (4 spp.) and epikarst zone (2 spp.) [4,53]. It is noteworthy that certain taxa such as Crustacea, Hexapoda, Arachnida, and Myriapoda consistently dominate as the richest taxa across all three hotspots, suggesting a pattern of successful colonization and establishment in subterranean habitats (Table 2).

Table 2. Taxonomic diversity and distribution of 31 obligate cave species within the Água Clara cave system, located in the western region of Bahia state, Brazil. The study includes observations from four specific caves within the cave system: Água Clara cave (AC), Índios cave (IN), Lapa dos Peixes I (LPI) and Lapa dos Peixes II (LPII). Additionally, the study distinguishes fauna between terrestrial (T) and aquatic (A) or both habitats.

Taxons	Taxon	Family	Species and Morphotypes	AC	IN	LP II	LP I	Habitat
Platyhelminthes	Tricladida	Dugesiidae	*Girardia spelaea*			+	+	T
Arachnida	Acari	Rhagidiidae	Rhagidiidae sp.1	+				T
	Amblypygi	Charinidae	*Charinus troglobius*	+			+	T
	Araneae	Caponiidae	Caponiidae sp.1	+				T
		Ochyroceratidade	Ochyroceratidae sp.1	+		+	+	T
	Opiliones	Gonyleptidae	*Giuponnia chagasi*	+		+		T
	Palpigradi	Eukoeneniidae	*Eukoenenia* sp.1	+		+	+	T
	Pseudoscorpiones	Ctoniidae	*Pseudochthonius koinopoliteia*	+			+	T
Collembola	Symphypleona	Sminthuridae	*Troglobentosminthurus luridus*	+			+	T
			Sminthuridae sp.1				+	T
		unidentified	Symphypleona sp.1	+				T
	Entomobryomorpha		Entomobryomorpha sp.1	+			+	T
			Entomobryomorpha sp.2	+			+	T
Hexapoda	Blattodea	unidentified	Blattodea sp.1	+		+	+	T
	Dermaptera	Diplatyidae	*Mesodiplatys falcifer*			+		T
	Ensifera	Phalangopsidae	*Endecous infernalis*	+	+	+	+	T
	Hymenoptera	Formicidae	*Nylanderia* sp.1	+		+	+	T
Crustacea	Isopoda	Styloniscidae	*Pectenoniscus carinhanhensis*	+	+			T
			Styloniscidae sp.2	+			+	T/A
			Styloniscidae sp.3	+		+	+	T/A
			Styloniscidae p.4			+		T/A
			Xangoniscus aganju	+		+	+	T/A
		Plathyarthridae	*Trichorhina* sp.1	+				T
Myriapoda	Geophilomorpha	unidentified	Geophilomorpha sp.1			+		T
	Polydesmida	Chelodesmidae	Chelodesmidae sp.1	+	+	+	+	T
		Trichopolidesmidae	*Phaneromerium* sp.1		+	+		T
		Pyrgodesmidae	Pyrgodesmidae sp.1	+		+	+	T
		Oniscodesmidae	Oniscodesmidae sp.1					T
Mollusca	Gastropoda	Pomatiopsidae	*Spiripockia punctata*			+	+	A
		unidentified	Eupulmonata sp.1	+			+	A
Osteichthyes	Siluriformes	Trichomycteridae	*Trichomycterus rubbioli*	+				A

It is crucial to emphasize the ecological significance of the three South American hotspots of subterranean biodiversity and to prioritize conservation efforts for their preservation. The presence of permanent or temporary water bodies seems to be a key factor in the occurrence of troglobitic and stygobitic species, particularly in those hotspots located in the semi-arid region of Brazil. Moreover, the dependence of these systems on allochthonous sources of organic resources underscores the importance of the surrounding landscapes in supporting subterranean life. The variation in species composition and diversity among the three Brazilian hotspots highlights their uniqueness and necessitates the development of tailored conservation strategies that account for the distinctive features of each ecosystem.

8.6. Global Relevance of the Brazilian Hotspots of Subterranean Biodiversity

Each year, more hotspots of subterranean biodiversity (HSB) are discovered, particularly in large cave systems, through long-term studies. However, tropical regions continue to be underrepresented in the discovery of new HSB. This lack of discovery in tropical areas is likely due to the relatively lower investment in research in these regions, particularly in Africa, the South and Central Americas, and Asia. Thus, the few known HSBs in tropical regions hold great contextual importance globally. It is crucial to note that although new HSB may be discovered in the future due to intensified research, it is improbable that they will be found as abundantly as those in some temperate regions. Therefore, existing HSB in tropical areas should be considered infrequent and protected, given their contextual rarity. Unfortunately, anthropogenic impacts in tropical regions have been rapidly increasing, particularly regarding deforestation in various biomes for agricultural expansion or logging purposes. This replacement of native vegetation with monocultures or pastures can severely disrupt the trophic webs of subterranean ecosystems that rely heavily on epigean organic sources. Therefore, the rarity of HSB in tropical regions, combined with the escalating human-induced impacts, raises concerns about the long-term continuity and viability of these ecosystems.

The HSB identified in Brazil carry a profound contextual significance, owing to the fact that three of the eight HSB that are currently known to occur in tropical regions are located in Brazil. Moreover, a recently discovered HSB in the country, although yet to be officially published (Ferreira et al. in prep.), further underscores the importance of the Brazilian Caatinga biome. The unearthing of this new HSB reveals the ongoing advancements in speleological research in Brazil, especially over the last two decades. It is thus anticipated that forthcoming research investments will certainly contribute to revealing additional HSB from other tropical regions in the future.

8.7. Challenges in Cave Conservation in Brazil

The conservation of caves in Brazil has been characterized by a turbulent history of policies that have oscillated between strengthening and weakening the protection of this invaluable natural heritage. This inconsistent and sometimes precarious state of protection has largely resulted from the ongoing conflicts between several productive sectors, notably the mining industry, and conservationists who are committed to preserving these unique ecosystems.

Prior to 1988, Brazilian caves received inadequate legal protection. However, since that year, they have been legally designated as "Assets of the Union" and have become increasingly recognized in conservation policies. In 1990, the enactment of Decree No. 99,556 granted complete legal protection to Brazilian caves, although, in practice, some caves have been subject to destruction even after the publication of this decree. In 2008, the publication of Decree No. 6640 (www.planalto.gov.br/ccivil_03/_Ato2007-2010/2008, accessed on 15 January 2010.) required the classification of Brazilian caves according to their degree of relevance for decision-making concerning the installation of commercial or industrial enterprises. Nonetheless, only caves deemed of utmost relevance received full legal protection, whereas others remained potentially vulnerable to destruction.

While Decree 6640 raised concerns by allowing the destruction of caves, it also spurred significant advancements in speleological research in Brazil. In comparison to the approximately 6000 registered caves prior to the decree's enactment, there are now more than 23,000. The number of described troglobitic species in the country has also risen from 75 in 2008 to 285 today. Despite these advancements, the permission to destroy caves in Brazil remains a subject of scrutiny, particularly with regards to the criteria used for determining their relevance. However, the situation has been significantly exacerbated by the January 2022 publication of Decree 10,935 (www.in.gov.br/en/web/dou/-/decreto-n-10.935-de-12 -de-janeiro-de-2022-373591582, accessed on 20 March 2023), which permits the destruction of even the most significant Brazilian caves [66]. Although the Brazilian Supreme Court has revoked parts of the decree, it is uncertain whether the country's most important caves will be protected going forward. Thus, it is imperative to use arguments beyond just those concerning HSB to advocate for the conservation of unique caves and those that could provide essential ecosystem services at various spatial scales.

Regarding the ACCS specifically, concerns are also severe, particularly in light of recent trends towards deforestation in the region where these caves are situated. This trend has led to the removal of significant portions of the original vegetation for the opening of arable lands (Figure 5C) and charcoal production (Figure 5D). Moreover, given the semi-arid nature of the region and the underground water sources accessible through the caves, pumps, frequently diesel-powered, are often employed to extract water for human or animal consumption as well as for irrigation purposes (Figure 5A,B). This practice, besides inducing a progressive reduction of the water table, often leads to contamination, especially when diesel pumps are utilized—the most common type in the area. Therefore, a highly recommended course of action is to establish a conservation unit with strict limitations in the region. Ideally, this unit should encompass the ACCS as well as all its catchment micro-basins and the corresponding area of influence.

Figure 5. Anthropic impacts in the caves and external surroundings of the ACCS: (**A**) Water capture in karstic resurgences, for diffuse rural supply; (**B**) Water capture in cave interior; (**C**) Deforestation near limestone outcrops; (**D**) Vegetal coal furnace.

8.8. ACCS Outreach and Public Awareness

The ACCS has recently received attention through various outreach and public awareness initiatives; however, these efforts remain in their early stages. The primary objective of these efforts is to increase public knowledge and appreciation of the unique subterranean biodiversity found in the cave system as well as promote the conservation of this valuable cave system.

Researchers, such as those from the Center of Studies on Subterranean Biology (CEBS/UFLA), extend an invitation to local community members to partake in most sampling activities and expeditions carried out within the ACCS, in addition to other caves located in the vicinity. The primary purpose of these opportunities is to elucidate the characteristics and significance of cave systems as well as engage local residents in the sampling activities. The overarching objective of this engagement is to mitigate any apprehension and misunderstandings concerning the cave ecosystem and its diverse fauna.

Educational lectures on the subject of cave systems were presented at schools within the region, and informative booklets detailing the topic of caves were freely distributed to students at the main school situated in the nearby small village of Agrovila 23. Additionally, informal conversations were held with local residents while conducting routine errands at markets, drugstores, and bakeries. The importance of engaging the local population and fostering their interest in preserving this unique natural heritage is recognized, as public support is pivotal for the conservation of cave fauna. Through technical visits, lectures, and informal talks, information regarding the subterranean environment can be subtly introduced, thereby expanding people's awareness of the subject matter. This approach helps to establish a strong link between knowledge production and dissemination while also enhancing the quality of work provided by local professionals [80].

Undoubtedly, these outreach and public awareness initiatives have a pivotal role in promoting the conservation of the ACCS and other subterranean habitats throughout Brazil. Through augmenting public knowledge and appreciation of these unique environments and engaging local communities in conservation efforts, significant strides can be taken towards ensuring that these invaluable resources are safeguarded for future generations.

Author Contributions: Conceptualization, R.L.F.; methodology and analysis, R.L.F.; data acquisition, R.L.F., M.B.-B. and M.S.-S.; original draft preparation, R.L.F.; review and editing, R.L.F., M.B.-B. and M.S.-S. All authors have read and agreed to the published version of the manuscript.

Funding: The authors would like to thank the Centro Nacional de Pesquisa e Conservação de Cavernas—CECAV and Instituto Brasileiro de Desenvolvimento e Sustentabilidade—IABS for the financial support (TCCE ICMBio/Vale 01/2018). We are also thankful to the CNPq (National Council for Scientific and Technological Development, grant n. 302925/2022-8) for the productivity scholarship provided to R.L.F., and to the team from the Center of Studies in Subterranean Biology (CEBS/UFLA) for the support in the field trips.

Institutional Review Board Statement: Not applicable.

Informed Consent Statement: Not applicable.

Data Availability Statement: No new data were created or analyzed in this manuscript. Data sharing is not applicable for this manuscript.

Acknowledgments: The authors would like to thank the Centro Nacional de Pesquisa e Conservação de Cavernas—CECAV and Instituto Brasileiro de Desenvolvimento e Sustentabilidade—IABS for the financial support (TCCE ICMBio/Vale 01/2018). We are also thankful to the CNPq (National Council for Scientific and Technological Development, grant n. 302925/2022-8) for the productivity scholarship provided to R.L.F., and to the team from the Center of Studies in Subterranean Biology (CEBS/UFLA) for the support in the field trips. M. Berbert-Born thanks to Serviço Geológico do Brasil-CPRM for the funding and logistics supported by the GEOKARST project—Geodiversity of the Sanfranciscana Depression. Rafael Costa da Silva, Rafael Henrique Grudka Barroso, Dandara Evangelista Ferreira Bustamante, Taís Novaes Santoro, Victor Scardua, Guilherme Neiva Rodrigues Oliveira, Livia Medeiros Cordeiro, Leda Zogbi, Allan Silas Calux and Nogueira for cooperation in field surveys.

Conflicts of Interest: The authors declare no conflict of interest.

References

1. Culver, D.C.; Pipan, T. *The Biology of Caves and Other Subterranean Habitats*; Oxford University Press: Oxford, UK, 2019; ISBN 978-0-19-882076-5.
2. Niemiller, M.L.; Taylor, S.J.; Bichuette, M.E. Conservation of cave fauna, with an emphasis on Europe and the Americas. In *Cave Ecology*; White, W.B., Culver, D.C., Eds.; Academic Press: Cambridge, MA, USA, 2018; pp. 451–478. ISBN 978-0-12-813036-2.
3. Mendes-Rabelo, L.; Souza-Silva, M.; Ferreira, R.L. Priority caves for biodiversity conservation in a key karst area of Brazil: Comparing the applicability of cave conservation indices. *Biodivers. Conserv.* **2018**, *27*, 2097–2129. [CrossRef]
4. Souza-Silva, M.; Ferreira, R.L. The first two hotspots of subterranean biodiversity in South America. *Subterr. Biol.* **2016**, *19*, 1–21. [CrossRef]
5. Ferreira, R.L.; de Oliveira, M.P.A.; Souza-Silva, M. Subterranean biodiversity in ferruginous landscapes. In *Cave Ecology*; White, W.B., Culver, D.C., Eds.; Academic Press: Cambridge, MA, USA, 2018; pp. 435–447. ISBN 978-0-12-813036-2.
6. Bento, M.D.; Souza-Silva, M.; Vasconcellos, A.C.; Bellini, B.C.; Prous, X.; Ferreira, R.L. Subterranean "oasis" in the Brazilian semiarid region: Neglected sources of biodiversity. *Biodivers. Conserv.* **2021**, *30*, 3837–3857. [CrossRef]
7. Auler, A.; Farrant, A.R. A brief introduction to karst and caves in Brazil. *Proc. Univ. Bristol Spelaeol. Soc.* **1996**, *20*, 187–200.
8. Souza-Silva, M.; Martins, R.P.; Ferreira, R.L. Cave conservation priority index to adopt a rapid protection strategy: A case study in Brazilian Atlantic rain forest. *Environ. Manag.* **2015**, *55*, 279–295. [CrossRef]
9. Lee, N.M.; Meisinger, D.B.; Aubrecht, R.; Kovacik, L.; Saiz-Jimenez, C.; Baskar, S.; Engel, A.S. Caves and karst environments. In *Life at Extremes: Environments, Organisms and Strategies for Survival*; CABI: Wallingford, UK, 2012; pp. 320–344.
10. Culver, D.C.; Sket, B. Hotspots of subterranean biodiversity in caves and wells. *J. Cave Karst Stud.* **2000**, *62*, 11–17.
11. Iannella, M.; Fiasca, B.; Di Lorenzo, T.; Di Cicco, M.; Biondi, M.; Mammola, S.; Galassi, D.M. Getting the 'most out of the hotspot' for practical conservation of groundwater biodiversity. *Glob. Ecol. Conserv.* **2021**, *31*, e01844. [CrossRef]
12. Canedoli, C.; Ficetola, G.F.; Corengia, D.; Tognini, P.; Ferrario, A.; Padoa-Schioppa, E. Integrating landscape ecology and the assessment of ecosystem services in the study of karst areas. *Landsc. Ecol.* **2022**, *37*, 347–365. [CrossRef]
13. Kukkala, A.S.; Moilanen, A. Core concepts of spatial prioritisation in systematic conservation planning. *Biol. Rev.* **2013**, *88*, 443–464. [CrossRef]
14. Rubbioli, E.; Auler, A.; Menin, D.; Brandi, R. Cavernas. In *Atlas do Brasil Subterrâneo*; ICMBio: Brasília, Brazil, 2019; 340p.
15. ANA—Agência Nacional de Águas. *Hidrogeologia dos Ambientes Cársticos da Bacia do Rio São Francisco Para a Gestão de Recursos Hídricos: Resumo Executivo*; ANA/TPF-Techne: Brasília, Brazil, 2018; p. 71.
16. Dardenne, M.A. Síntese sobre a estratigrafia do Grupo Bambuí no Brasil Central. In Proceedings of the Congresso Brasileiro de Geologia, Recife, Brazil, 30 November 1978; pp. 507–610.
17. Alkmim, F.F.; Martins-Neto, M.A. Proterozoic first-order sedimentary sequences of the São Francisco craton, eastern Brazil. *Mar. Pet. Geol.* **2012**, *33*, 127–139. [CrossRef]
18. Reis, H.L.S.; Suss, J.F. Mixed carbonate-siliciclastic sedimentation in forebulge grabens: An example from the Ediacaran Bambuí Group, São Francisco Basin, Brazil. *Sediment. Geol.* **2016**, *339*, 83–103. [CrossRef]
19. Brito Neves, B.B.; Campos, D.A.; Fuck, R.A.; Cordani, U.G. From Rodinia to Western Gondwana: An approach to the Brasiliano-Pan African Cycle and orogenic collage. *Epis. J. Int. Geosci.* **1999**, *22*, 155–166.
20. Reis, H.L.S.; Suss, J.F.; Fonseca, R.C.S.; Alkmim, F.F. Ediacaran forebulge grabens of the southern São Francisco basin, SE Brazil: Craton interior dynamics during West Gondwana assembly. *Precambrian Res.* **2017**, *302*, 150–170. [CrossRef]
21. Caetano-Filho, S.; Paula-Santos, G.M.; Guacaneme, C.; Babinski, M.; Bedoya-Rueda, C.; Peloso, M.; Amorim, K.; Afonso, J.; Kuchenbecker, M.; Reis, H.L.S.; et al. Sequence stratigraphy and chemostratigraphy of an Ediacaran-Cambrian foreland-related carbonate ramp (Bambuí Group, Brazil). *Precambrian Res.* **2019**, *331*, 105365. [CrossRef]
22. Uhlein, G.J.; Uhlein, A.; Pereira, E.; Caxito, F.A.; Okubo, J.; Warren, L.V.; Sial, A.N. Ediacaran paleoenvironmental changes recorded in the mixed carbonate-siliciclastic Bambuí Basin, Brazil. *Palaeogeogr. Palaeoclimatol. Palaeoecol.* **2019**, *517*, 39–51. [CrossRef]
23. Kuchenbecker, M.; Pedrosa-Soares, A.C.; Babinski, M.; Reis, H.L.S.; Atman, D.; Costa, R.D. Towards an integrated tectonic model for the interaction between the Bambuí basin and the adjoining orogenic belts: Evidences from the detrital zircon record of syn-orogenic units. *J. S. Am. Earth Sci.* **2020**, *104*, 102831. [CrossRef]
24. Da Silva, L.; Pufahl, P.; James, N.; Guimaraes, E.; Reis, C. Sequence stratigraphy and paleoenvironmental significance of the Neoproterozoic Bambui Group, Central Brazil. *Precambrian Res.* **2022**, *379*, 106710. [CrossRef]
25. Dantas, M.V.S.; Uhlein, A.; Uhlein, G.J.; Okubo, J.; Moura, S.A. Stratigraphy, isotope geochemistry, seismic stratigraphy and paleogeography of the Lagoa do Jacaré Formation, Bambuí Foreland Basin (Ediacaran-Cambrian), Southeast Brazil. *J. S. Am. Earth Sci.* **2023**, *121*, 104137. [CrossRef]
26. Campos, J.E.G.; Dardenne, M.A. Estratigrafia e Sedimentação da Bacia Sanfranciscana: Uma Revisão. *Rev. Bras. Geocienc.* **1997**, *27*, 269–282. [CrossRef]
27. Sgarbi, G.N.C.; Sgarbi, P.B.A.; Campos, J.E.G.; Dardenne, M.A.; Penha, U.C. Bacia Sanfranciscana: O registro Fanerozóico da Bacia do São Francisco. In *Bacia do São Francisco: Geologia e Recursos Naturais*; Pinto, C.P., Martins-Neto, M.A., Eds.; SBG: Belo Horizonte, Brazil, 2001; pp. 93–138.
28. Zalán, P.V.; Romeiro-Silva, P.C. Bacia do São Francisco. *Bol. Geociênc. Petrobrás.* **2007**, *15*, 561–571.
29. Ford, D.C.; Williams, P. *Karst Hydrogeology and Geomorphology*; John Wiley: Chichester, UK, 2007; p. 562.

30. ANA–Agência Nacional de Águas. Diagnóstico dos Meios Físico e Socioeconômico—Resumo Executivo. In *Hidrogeologia dos Ambientes Cársticos da Bacia do Rio São Francisco Para a Gestão de Recursos Hídricos: Relatório Final v.1*; ANA/TPF-Techne: Brasília, Brazil, 2018; p. 264.

31. Bhering, A.P.; Antunes, I.M.H.R.; Marques, A.G.; de Paula, R.S. Geological and hydrogeological review of a semi-arid region with conflicts to water availability (southeastern Brazil). *Environ. Res.* **2021**, *202*, 111756. [CrossRef] [PubMed]

32. Assunção, P.A.; Galvão, P.; Lucon, T.; Doi, B.; Fleming, P.M.; Marques, T.; Costa, F. Hydrodynamic and hydrodispersive behavior of a highly karstified neoproterozoic hydrosystem indicated by tracer tests and modeling approach. *J. Hydrol.* **2023**, *619*, 129300. [CrossRef]

33. ANA—Agência Nacional de Águas. Relatório técnico e temático dos Domínios e Subdomínios Hidrogeológicos. In *Hidrogeologia dos Ambientes Cársticos da Bacia do Rio São Francisco Para a Gestão de Recursos Hídricos*; ANA/TPF-Techne: Brasília, Brazil, 2018; p. 809.

34. ANA—Agência Nacional de Águas. *Hidrogeologia dos Ambientes Cársticos da Bacia do Rio São Francisco Para a Gestão de Recursos Hídricos: Relatório Final v.2*; Hidrogeologia; ANA/TPF-Techne: Brasília, Brazil, 2018; 217p.

35. IBGE—Instituto Brasileiro de Geografia e Estatística. Biomas e Sistema Costeiro—Marinho do Brasil—1:250,000. Available online: https://www.ibge.gov.br/geociencias/cartas-e-mapas/informacoes-ambientais/15842-biomas.html (accessed on 25 April 2023).

36. Bittencourt, A.L.V.; Rodet, J. Evolução morfológica do canion do Morro Furado no contexto dos calcários carstificados do Grupo Bambuí (Serra do Ramalho, Bahia, Brasil). *O Carste* **2002**, *14*, 224–237.

37. Santos, C.C.; Reis, C. *Geologia e Recursos Minerais da Folha Santa Maria da Vitória—SD.23-X-C-II: Escala 1:100,000, Estado da Bahia*; Nota Explicativa; Serviço Geológico do Brasil—CPRM: Salvador, Brazil, 2021; 75p.

38. Silva, R.C.; Berbert-Born, M.; Bustamante, D.E.F.; Santoro, T.N.; Sedor, F.; Avilla, L.S. Diversity and preservation of Pleistocene tetrapods from caves of southwestern Bahia, Brazil. *J. S. Am. Earth Sci.* **2019**, *90*, 233–254. [CrossRef]

39. Berbert-Born, M.; Silva, R.C. *Projeto Geokarst—Geodiversidade da Depressão Sanfranciscana: Relatório de Campo*; Serviço Geológico do Brasil—CPRM: Rio de Janeiro, Brazil, 2019; Unpublished.

40. Brandi, R. A descoberta da Gruta do Peixe. *O Carste* **2001**, *13*, 38–41.

41. Frágola, L. A Gruta dos Peixes II. *O Carste* **2001**, *13*, 56–62.

42. Frágola, L. O retorno à Gruta dos Peixes. *O Carste* **2001**, *13*, 66–67.

43. Jolivet, J. A Gruta de Água Clara—Ponto de encontro rue Mourfftard. *O Carste* **2001**, *13*, 28–31.

44. Baptista, R.L.C.; Giupponi, A.D.L. A new troglomorphic *Charinus* from Brazil (Arachnida: Amblypygi: Charinidae). *Rev. Ibér. Arac.* **2002**, *6*, 105–110.

45. Kury, A.B.; González, A.P. A new remarkable troglomorphic gonyleptid from Brazil (Arachnida, Opiliones, Laniatores). *Rev. Ibér. Arachnol.* **2002**, *246*, 43–50.

46. Rizzato, P.P.; Bichuette, M.E. A new species of cave catfish from Brazil, *Trichomycterus rubbioli* sp. n., from Serra do Ramalho karstic area, São Francisco River basin, Bahia State (Silurifomes: Trichomycteridae). *Zootaxa* **2012**, *3480*, 48–66.

47. Simone, L.R.L. A new genus and species of cavernicolous Pomatiopsidae (Mollusca, Caenogastropoda) in Bahia, Brazil. *Pap. Avulsos Zool.* **2012**, *52*, 515–524. [CrossRef]

48. Campos-Filho, I.S.; Araujo, P.B.; Bichuette, M.E.; Trajano, E.; Taiti, S. Terrestrial isopods (Crustacea: Isopoda: Oniscidea) from Brazilian caves. *Zool. J. Linn. Soc.* **2014**, *172*, 360–425. [CrossRef]

49. Kamimura, Y.; Ferreira, R.L. Description of a second South American species in the Malagasy earwig genus *Mesodiplatys* from a cave habitat, with notes on the definition of Haplodiplatyidae (Insecta, Dermaptera). *ZooKeys* **2018**, *790*, 87–100. [CrossRef]

50. Cardoso, G.M.; Bastos-Pereira, R.; Souza, L.A.; Ferreira, R.L. New cave species of *Pectenoniscus* Andersson, 1960 (Isopoda: Oniscidea: Styloniscidae) and an identification key for the genus. *Nauplius* **2020**, *28*, e2020039. [CrossRef]

51. Souza, P.G.C.; Medeiros, G.D.S.; Ferreira, R.L.; Souza-Silva, M.; Bellini, B.C. A highly troglomorphic new genus of Sminthuridae (Collembola, Symphypleona) from the Brazilian semiarid region. *Insects* **2022**, *13*, 650. [CrossRef]

52. Hellmann, L.; Ferreira, R.L.; Rabelo, L.; Leal-Zanchet, A.M. Enhancing the still scattered knowledge on the taxonomic diversity of freshwater triclads (Platyhelminthes: Dugesiidae) in caves from two Brazilian Biomes. *Stud. Neotrop. Fauna Envion.* **2022**, *57*, 148–163. [CrossRef]

53. Souza-Silva, M.; Cerqueira, R.F.V.; Pellegrini, T.G.; Ferreira, R.L. Habitat selection of cave-restricted fauna in a new hotspot of subterranean biodiversity in Neotropics. *Biodiv. Conserv.* **2021**, *30*, 4223–4250. [CrossRef]

54. Carvalho, P.H.M.; Junta, V.G.P.; Castro-Souza, R.A.; Ferreira, R.L. Three new cricket species and a new subgenus of *Endecous* Saussure, 1878 (Grylloidea: Phalangopsidae) from caves in northeastern Brazil. *Zootaxa* **2023**, *5263*, 001–039. [CrossRef] [PubMed]

55. Prado, G.C.; Ferreira, R.L. Three new troglobitic species of *Pseudochthonius* Balzan, 1892 (Pseudoscorpiones, Chthoniidae) from northeastern Brazil. *Zootaxa* **2023**, *5249*, 92–110. [CrossRef] [PubMed]

56. Ávila, A.C.; Pires, M.M.; Rodrigues, E.N.; Costi, J.A.; Stenert, C.; Maltchik, L. Drivers of the beta diversity of spider assemblages in southern Brazilian temporary wetlands. *Ecol. Entomol.* **2019**, *45*, 466–475. [CrossRef]

57. Trontelj, P.; Blejec, A.; Fišer, C. Ecomorphological convergence of cave communities. *Evolution* **2012**, *66*, 3852–3865. [CrossRef]

58. Den Boer, P.J. The present status of the competitive exclusion principle. *Trends Ecol. Evol.* **1986**, *1*, 25–28. [CrossRef] [PubMed]

59. Ortuño, V.M.; Gilgado, J.D.; Jiménez-Valverde, A.; Sendra, A.; Pérez-Suárez, G.; Herrero-Borgoñón, J.J. The "aluvial mesovoid shallow substratum", a new subterranean habitat. *PLoS ONE* **2013**, *8*, e76311. [CrossRef]

60. Cardoso, P.; Erwin, T.L.; Borges, P.A.; New, T.R. The seven impediments in invertebrate conservation and how to overcome them. *Biol. Conserv.* **2011**, *144*, 2647–2655. [CrossRef]

61. Pipan, T.; Deharveng, L.; Culver, D.C. Hotspots of subterranean biodiversity. *Diversity* **2020**, *12*, 209. [CrossRef]
62. Trajano, E.; Gallão, J.E.; Bichuette, M.E. Spots of high diversity of troglobites in Brazil: The challenge of measuring subterranean diversity. *Biodiv. Conserv.* **2016**, *25*, 1805–1828. [CrossRef]
63. Huang, S.; Wei, G.; Wang, H.; Liu, W.; Bedos, A.; Deharveng, L.; Tian, M. Ganxiao Dong: A hotspot of cave biodiversity in northern Guangxi, China. *Diversity* **2021**, *13*, 355. [CrossRef]
64. Iliffe, T.M.; Calderón-Gutiérrez, F. Bermuda's Walsingham Caves: A global hotspot for anchialine stygobionts. *Diversity* **2021**, *13*, 352. [CrossRef]
65. Souza-Silva, M.S.; Martins, R.P.; Ferreira, R.L. Cave lithology determining the structure of the invertebrate communities in the Brazilian Atlantic Rain Forest. *Biodivers. Conserv.* **2011**, *20*, 1713–1729. [CrossRef]
66. Ferreira, R.L.; Giribet, G.; Du Preez, G.; Ventouras, O.; Janion, C.; Souza-Silva, M. The Wynberg cave system, the most important site for cave fauna in South Africa at risk. *Subterr. Biol.* **2020**, *36*, 73–81. [CrossRef]
67. Myers, N.; Mittermeier, R.A.; Mittermeier, C.G.; Da Fonseca, G.A.; Kent, J. Biodiversity hotspots for conservation priorities. *Nature* **2020**, *403*, 853–858. [CrossRef] [PubMed]
68. Duan, Y.F.; Li, M.; Xu, K.W.; Zhang, L.; Zhang, L.B. Protect China's karst cave habitats. *Science* **2021**, *374*, 699. [CrossRef] [PubMed]
69. Ferreira, R.L.; Bernard, E.; da Cruz Júnior, F.W.; Piló, L.B.; Calux, A.; Souza-Silva, M.; Frick, W.F. Brazilian cave heritage under siege. *Science* **2020**, *375*, 1238–1239. [CrossRef]
70. Mendes-Rabelo, L.; Souza-Silva, M.; Ferreira, R.L. Epigean and hypogean drivers of Neotropical subterranean communities. *J. Biogeogr.* **2021**, *48*, 662–675. [CrossRef]
71. Dial, R.; Roughgarden, J. Theory of marine communities: The intermediate disturbance hypothesis. *Ecology* **1988**, *79*, 1412–1424. [CrossRef]
72. Wilkinson, D.M. The Disturbing History of Intermediate Disturbance. *Oikos* **1999**, *84*, 145–147. [CrossRef]
73. Kricher, J.C. *Tropical Ecology*; Princeton University Press: Princeton, NJ, USA, 2011.
74. Catford, J.A.; Daehler, C.C.; Murphy, H.T.; Sheppard, A.W.; Hardesty, B.D.; Westcott, D.A.; Rejmánek, M.; Bellingham, P.J. The intermediate disturbance hypothesis and plant invasions: Implications for species richness and management. Perspectives in Plant Ecology. *Evol. Syst.* **2012**, *14*, 231–241.
75. Fox, J.W. The intermediate disturbance hypothesis should be abandoned. *Trends Ecol. Evol.* **2013**, *28*, 86–92. [CrossRef]
76. Sheil, D.; Burslem, D.F.R.P. Defining and defending Connell's intermediate disturbance hypothesis: A response to Fox. *Trends Ecol. Evol.* **2013**, *28*, 571–572. [CrossRef]
77. Collevatti, R.G.; Terribile, L.C.; de Oliveira, G.; Lima-Ribeiro, M.S.; Nabout, J.C.; Rangel, T.F.; Diniz-Filho, J.A.F. Draw-backs to palaeodistribution modelling: The case of South American seasonally dry forests. *J. Biogeogr.* **2013**, *40*, 345–358. [CrossRef]
78. Wang, X.; Auler, A.S.; Edwards, L.; Cheng, H.; Cristalli, P.S.; Smart, P.L.; Richards, D.A.; Shen, C.C. Wet periods in northeastern Brazil over the past 210 kyr linked to distant climate anomalies. *Lett. Nat.* **2004**, *432*, 740–743. [CrossRef] [PubMed]
79. Polhemus, D.A.; Ferreira, R.L. Two unusual new genera of cavernicolous Hydrometridae (Insecta: Heteroptera) from Eastern Brazil. *Tijdschr. Voor Entomol.* **2018**, *161*, 25–38. [CrossRef]
80. Souza-Silva, M.S.; Ferreira, R.L.; Gonçalves, L.V. Public support as a key element for fauna conservation in Maquiné and Rei do Mato show caves. *Rev. Espeleo Tema* **2019**, *29*, 65–76.

 *diversity*

Interesting Images

Beyond Expectations: Recent Discovery of New Cave-Restricted Species Elevates the Água Clara Cave System to the Richest Hotspot of Subterranean Biodiversity in the Neotropics

Rodrigo Lopes Ferreira [1,2,*] and Marconi Souza-Silva [1,2]

[1] Centro de Estudos em Biologia Subterrânea, Departamento de Ecologia e Conservação, Instituto de Ciências Naturais, Universidade Federal de Lavras, Campus Universitário, P.O. Box 3037, Lavras CEP 37200-000, MG, Brazil; marconisilva@ufla.br

[2] Programa de Pós-Graduação em Ecologia Aplicada, Universidade Federal de Lavras, Lavras CEP 37200-000, MG, Brazil

* Correspondence: drops@ufla.br

Abstract: The Água Clara Cave System was previously recognized as a prominent hotspot of subterranean biodiversity in South America, harboring 31 cave-restricted species. However, a recent expedition conducted in September 2023, coinciding with an exceptionally dry period in the region, provided access to previously unexplored areas. Therefore, the objective of this research was to investigate the cave-restricted invertebrate species, extending the findings from a previous article on the Agua Clara Cave System published in June 2023, and emphasizing the significance of this system as one of the most crucial tropical biodiversity hotspots. This survey unveiled an additional 10 species, raising the count of cave-restricted species within the system to an impressive 41. This remarkable diversity not only solidifies the Água Clara Cave System's position as a paramount hotspot of subterranean biodiversity in the tropics but also serves as a stark warning about the imminent risks faced by these species. The escalating human-induced alterations in the region, notably deforestation, pose a significant risk to the survival of many of these unique and endemic species.

Keywords: obligate cave fauna; conservation; threats; species richness; stygobiont; troglobiont

Citation: Ferreira, R.L.; Souza-Silva, M. Beyond Expectations: Recent Discovery of New Cave-Restricted Species Elevates the Água Clara Cave System to the Richest Hotspot of Subterranean Biodiversity in the Neotropics. *Diversity* **2023**, *15*, 1215. https://doi.org/10.3390/d15121215

Academic Editors: David C. Culver, Louis Deharveng, Tanja Pipan and Ana Sofia Reboleira

Received: 21 November 2023
Revised: 4 December 2023
Accepted: 12 December 2023
Published: 14 December 2023

The Água Clara Cave System (ACCS), situated in northeastern Brazil, has gained recognition as a "Hotspot of Subterranean Biodiversity" (HSB) due to its remarkable richness of cave-restricted fauna [1,2]. This cave system comprises a network of extensive subterranean conduits interconnected through one of the numerous fluvial outlets found on the eastern periphery of the Serra do Ramalho karst area in the western Bahia state. The system encopasses four caves: Gruna da Água Clara (13,880 m), Gruna dos Índios (510 m), Lapa dos Peixes I (9320 m), and Lapa dos Peixes II (2100 m) [1]. It is noteworthy that the cumulative length of all caves within the Agua Clara Cave System (ACCS) stands at 25,810 m. However, when accounting for the inaccessible and unexplored areas existing between the caves, it is plausible that the system might extend to more than 27 km.

The concept of HSB was originally introduced by Culver and Sket [3] to designate subterranean environments that host a minimum of 20 or more troglobitic/stygobitic species. More recently, there has been a movement to raise this threshold to a minimum of 25 cave-restricted species. However, this change has faced some criticism due to concerns about the arbitrary nature of the cutoff and the necessity to consider various parameters when defining subterranean hotspots [1]. These parameters encompass the need to consider different scales when identifying hotspots, the cave's latitudinal location, the lithology associated with the cave, the level of threat to subterranean habitats, and the number of cave-restricted species exclusive to that cave (or system) relative to the total number of troglobitic/stygobitic species it supports [1].

Most of the HSBs are situated in temperate regions, with only a smaller portion located in sub-tropical or tropical areas [4–11]. Additionally, it is noteworthy that most HSBs within these sub-tropical and tropical regions have only recently come to the forefront, signifying an upsurge in cave fauna research in these regions. For instance, the discovery of the first hotspots of subterranean biodiversity (HSB) in South America occurred only in recent years, following extensive sampling in the Areias cave system located in the southern São Paulo state (southeastern Brazil) and the Toca do Gonçalo cave in the northern Bahia state (northeastern Brazil) [11]. Presently, four recognized HSBs exist in South America: the Toca do Gonçalo cave, housing 22 cave-restricted species [11]; the Areias Cave System, hosting 31 cave-restricted species [9,11]; the ACCS, with a previously documented 31 cave-restricted species [2,11]; and the Igatu caves, harboring 37 cave-restricted species. It is important to note that the Igatu caves do not constitute a conventional, functionally interconnected "system" in the traditional sense. Instead, they comprise a collection of caves situated in a relatively small geographical area. Nevertheless, the significant number of cave-restricted species inhabiting these caves underscores their importance and emphasizes the urgency of conservation efforts [9].

Prior investigations conducted at the ACCS have identified a total of 31 cave-restricted species, spanning multiple taxonomic groups, including Hexapoda (9 species), Arachnida (7 species), Crustacea (6 species), Myriapoda (5 species), Gastropoda (2 species), Turbellaria (1 species), and Siluriformes (1 species) [1]. These species predominantly occupy terrestrial (22 species) habitats, with a smaller representation of semiaquatic (5 species) and aquatic species (3 species) [1]. Notably, only a fraction of this assemblage (11 species) has been formally described [12–22]. A recent expedition conducted in September 2023 coincided with an exceptionally dry period in the region, allowing access to previously unexplored sections of the system. This survey unveiled an additional 10 new species, increasing the tally of cave-restricted species within the system to an impressive 41 (Table 1). This number makes the ACCS the cave system with the highest richness of cave-restricted species in the Neotropical region. These newly documented species were encountered in Lapa dos Peixes I cave, within a conduit featuring a permanent water body (Figure 1A). The extreme drought conditions likely prompted many previously undocumented species to relocate from their original microhabitats, congregating near this subterranean "oasis", where they became observable. It is worth highlighting that this small, moisture-rich area of the cave hosted 24 cave-restricted species, which were found in a single day, using direct intuitive research. This observation suggests that during periods of severe external drought, this cave section plays a pivotal role in safeguarding several of the cave-restricted species found in the system. Furthermore, within this limited space, a substantial root mat system is present, serving as a consistent organic resource for the cave invertebrates (Figure 1B).

Table 1. Taxonomic diversity and distribution of 41 obligate cave species within the Águas Claras cave system, located in the northeastern region of Bahia state, Brazil (the newly recorded taxa in this study are highlighted in bold). The study includes observations from four specific caves within the cave system: Águas Claras cave (AC), Índios cave (IN), and Lapa dos Peixes (LP). Additionally, the study distinguishes fauna between terrestrial (T) and aquatic (A) or both habitats.

Phylum/Classis	Order	Family	Species and Morphotypes	AC	IN	LP II	LP I	Habitat
Platyhelminthes	Tricladida	Dugesiidae	*Girardia spelaea*			+	+	A
Arachnida	Acari	Rhagidiidae	Rhagidiidae sp.1	+			+	T
	Amblypygi	Charinidae	*Charinus troglobius*	+			+	T
	Araneae	Caponiidae	Caponiidae sp.1	+			+	T
		Ochyroceratidade	Ochyroceratidae sp.1	+		+	+	T
		Ochyroceratidade	**Ochyroceratidae sp.2**				+	T
		Palpimanidae	**Palpimanidae sp.1**				+	T
		Tetrablemmidae	**Tetrablemmidae sp.1**				+	T
	Opiliones	Gonyleptidae	*Giuponnia chagasi*	+		+		T
	Palpigradi	Eukoeneniidae	*Eukoenenia* sp.1	+		+	+	T
	Pseudoscorpiones	Ctoniidae	*Pseudochthonius koinopoliteia*	+			+	T

Table 1. *Cont.*

Phylum/Classis	Order	Family	Species and Morphotypes	AC	IN	LP II	LP I	Habitat
Collembola	Symphypleona	Sminthuridae	*Troglobentosminthurus luridus*	+		+		T
		Sminthuridae	Sminthuridae sp.1				+	T
		unidentified	Symphypleona sp.1	+				T
	Entomobryomorpha		Entomobryomorpha sp.1	+			+	T
			Entomobryomorpha sp.2	+		+		T
Diplura		Projapygidae	**Projapygidae sp.1**				+	T
Insecta	Blattodea	unidentified	Blattodea sp.1	+		+	+	T
	Coleoptera	Carabidae	**Clivinina sp.1**			+	+	T
	Coleoptera	Carabidae	**Trechinae sp.1**				+	T
	Dermaptera	Diplatyidae	*Mesodiplatys falcifer*				+	T
	Ensifera	Phalangopsidae	*Endecous infernalis*	+	+	+	+	T
	Hemiptera	Delphacidae	**Delphacidae sp.1**				+	T
	Hemiptera	Hydrometridae	***Spelaeometra* sp.1**				+	T
	Hymenoptera	Formicidae	*Nylanderia* sp.1	+		+	+	T
Crustacea	Isopoda	Styloniscidae	*Pectenoniscus carinhanhensis*	+	+		+	T
		Styloniscidae	Styloniscidae sp.1	+			+	T/A
		Styloniscidae	Styloniscidae sp.2	+		+	+	T/A
		Styloniscidae	Styloniscidae sp.3			+	+	T/A
		Styloniscidae	*Xangoniscus aganju*	+		+	+	T/A
		Plathyarthridae	*Trichorhina* sp.1	+	+			T
Myriapoda	Geophilomorpha	unidentified	Geophilomorpha sp.1			+		T
	Polydesmida	Chelodesmidae	*Cayenniola* sp.1	+	+	+	+	T
		Trichopolidesmidae	*Phaneromerium* sp.1		+	+	+	T
		Trichopolidesmidae	**Trichopolidesmidae sp.1**				+	T
		Pyrgodesmidae	Pyrgodesmidae sp.1	+		+	+	T
		Oniscodesmidae	Oniscodesmidae sp.1				+	T
	Siphonophorida	Siphonophoridae	**Siphonophoridae sp.1**				+	T
Mollusca	Gastropoda	Pomatiopsidae	*Spiripockia punctata*			+	+	A
		unidentified	Eupulmonata sp.1	+			+	A
Osteichthyes	Siluriformes	Trichomycteridae	*Trichomycterus rubbioli*	+			+	A

The newly discovered species encompass a diverse array of taxa, all of which represent new taxa, with some already in the process of formal description. These include a palpimanid spider (Araneae: Palpimanidae—Figure 2A), an ochiroceratid spider (Araneae: Ochiroceratidae—Figure 2B), a tetrablemmid spider (Araneae: Tetrablemmidae: *Matta* sp.—Figure 2C), a siphonophorid millipede (Diplopoda: Siphonophorida: Siphonophoridae—Figure 2D), a trichopolydesmid millipede (Diplopoda: Polydesmida: Trichopolydesmidae), a delphacid planthopper (Hemiptera: Delphacidae—Figure 2E), two carabid beetles (Coleoptera: Carabidae: Clivinina—Figure 2F and Trechinae), a hydrometrid bug (Hemiptera: Hydrometridae: *Spelaeometra* sp.—Figure 2G), and a projapygid (Diplura: Projapygidae—Figure 2H).

All these species exhibited typical troglomorphic traits, including reduced or absent eyes and pigmentation. Additionally, specific troglomorphic characteristics were observed. The palpimanid spider displayed extremely reduced eyes and weak pigmentation, contrasting with the general morphology observed in the remaining species of this family. Similar traits were observed in the ochiroceratid and tetrablemmid spiders, which also exhibited an additional thickening of the cuticle and no eyes. Both the siphonophorid and the trichopolydesmid millipedes were completely unpigmented. They also showed an increase in sensory pits on the antennal segments and an unusually long tergal setae, considered troglomorphic traits in other millipede taxa [23–25]. The delphacid planthopper exhibited all the typical troglomorphic traits observed in other cave-restricted planthoppers, including the absence of eyes, pigment reduction, and wings reduction [26–30]. Both carabid beetles displayed typical troglomorphic traits, such as eye and pigment reduction, as well as wing reduction [31,32]. The hydrometrid bug (*Spelaeometra* sp.) exhibited all the troglomorphic traits observed in the other two known species of the genus. This included reduced eyes and pigmentation, along with elongated legs and antennae [33,34]. Finally, the projapygid (Diplura) exhibited the most traditional troglomorphic traits among diplurans, such as appendage elongation and an increase in sensory setae at the antennal segments [35]. Unfortunately, the specimen lost both cerci when discovered above a rock on the cave floor (Figure 2H), but one cercus was later recovered, indicating considerable elongation.

Figure 1. Água Clara cave system: (**A**) spatial distribution of the caves Gruna da Água Clara (1), Gruna dos Índios (2), Lapa dos Peixes I (3), and Lapa dos Peixes II (4); the yellow arrow indicates the region where the newly discovered cave-restricted species were found; (**B**) conduit where the newly discovered species were found (notice the water pond on the floor of the conduit); (**C**) root mats covering the cave floor; (**D**) a close-up view of a root mat with a troglobitic isopod (*Xangoniscus* sp.).

Figure 2. Newly discovered cave-restricted species from the ACCS: (**A**) Palpimanidae (Araneae); (**B**) Ochiroceratidae (Araneae); (**C**) Tetrablemmidae (Araneae); (**D**) Siphonophoridae (Diplopoda: Siphonophorida); (**E**) Delphacidae (Hemiptera); (**F**) Clivinina (Coleoptera: Carabidae); (**G**) *Spelaeometra* sp. (Hemiptera: Hydrometridae); (**H**) Projapygidae (Diplura).

While troglomorphisms can serve as valuable indicators of the potential status of these species, their analysis should always consider the contexts of the external ecosystems surrounding the caves. For instance, if a species, completely depigmented, blind, and with a reduced cuticle, is discovered in a cave within a humid forest (like the Amazon rainforest), these traits may not necessarily signify its restriction to that cave. This is because the surrounding forest provides numerous shaded and humid microhabitats (such as spaces under logs, leaf litter, etc.) that could easily accommodate individuals of this species. On the contrary, if a species with similar characteristics were found in a cave located in an arid or semi-arid region, these morphological traits would strongly indicate its restriction to subterranean habitats. This is because, in the surrounding epigean environments, such organisms would rarely encounter suitable microhabitats for their survival.

Thus, considering the highly xeric epigean environment surrounding the ACCS (Figure 1A), it is unlikely that these species (as well as the other 31 troglomorphic species previously registered in the system) can maintain viable populations on the surface. There-fore, not only were troglomorphic traits instrumental in identifying these species as troglobitic, but also the external surrounding conditions, which are highly restrictive, imposing physiological constraints and preventing the occurrence of these species in external habitats. Finally, it is noteworthy that some of the newly discovered species were examined by taxonomists who confirmed their status (as mentioned in the Acknowledgements section).

It is important to note that some of these newly discovered species hold particular significance, such as the palpimanid spider, which marks the first known troglobitic species within this family worldwide. Additionally, the presence of the delphacid planthopper in this region is noteworthy, as the three previously documented subterranean-restricted species from this family were exclusively recorded in New Caledonia [26,27].

Therefore, the Gruna da Água Clara cave, once recognized for harboring the highest troglobitic species richness among ACCS caves, with a total of 23 species, has now been surpassed by the Lapa dos Peixes I cave, which hosts 35 species. The Lapa dos Peixes II cave accommodates 17 species, whereas only 5 species have been found thus far in the Gruna dos Índios cave.

The recent discovery of ten additional species within the ACCS raises a red flag on two critical fronts. Firstly, it underscores the extraordinary diversity of this system, currently facing severe threats driven by an unprecedented increase in various anthropogenic impacts, particularly in recent decades [1]. Urgent actions, such as the establishment of a fully protected conservation unit, are imperative. Secondly, and perhaps more significantly, this discovery serves as a powerful reminder that limited samplings are insufficient when attempting to unveil the true extent of species richness within a cave or cave system. In Brazil, caves are increasingly at risk due to various industrial activities, such as mining and hydropower dam construction, among others. To assess which caves should be safeguarded, a mere two samplings are currently required to determine their relevance. Consequently, in Brazil, a cave may be deemed of low significance simply because of an inadequate sampling effort. Therefore, it is crucial to demand additional samplings during the environmental licensing processes to more accurately assess the relevance of a cave, especially given that Brazilian caves have never faced such high levels of threat as they do today [36].

Author Contributions: Conceptualization, R.L.F.; data acquisition, R.L.F. and M.S.-S.; original draft preparation, R.L.F.; review and editing, R.L.F. and M.S.-S. All authors have read and agreed to the published version of the manuscript.

Funding: The authors would like to thank the Centro Nacional de Pesquisa e Conservação de Cavernas (CECAV) and Instituto Brasileiro de Desenvolvimento e Sustentabilidade (IABS) for their financial support (TCCE ICMBio/Vale 01/2018). We are also thankful to the CNPq (National Council for Scientific and Technological Development, grant no. 302925/2022-8) for the productivity scholarship provided to R.L.F., and to the team from the Center of Studies in Subterranean Biology (CEBS/UFLA) for their support in the field trips.

Institutional Review Board Statement: Not applicable.

Informed Consent Statement: Not applicable.

Data Availability Statement: No new data were created or analyzed in this manuscript. Data sharing is not applicable for this manuscript.

Acknowledgments: The authors would like to thank the team from the Center of Studies in Subterranean Biology (CEBS/UFLA) for their support in the field trips. We also thank Gabriel Augusto Silva Vaz, Paulo César Reis Venâncio, and Priscila Emanuela Souza for their cooperation in field surveys. Finally, we would like to extend our sincere gratitude to the taxonomists who confirmed the cave-restricted status for some of the newly discovered species and who are currently describing some of them, including Leonardo Sousa Carvalho (Araneae), Leticia Vieira (Coleoptera), and Júlio César Do Carmo Vaz Santos (Auchenorrhynca).

Conflicts of Interest: The authors declare no conflict of interest.

References

1. Ferreira, R.L.; Berbert-Born, M.; Souza-Silva, M. The Água Clara Cave System in Northeastern Brazil: The Richest Hotspot of Subterranean Biodiversity in South America. *Diversity* **2023**, *15*, 761. [CrossRef]
2. Souza-Silva, M.; Cerqueira, R.F.V.; Pellegrini, T.G.; Ferreira, R.L. Habitat selection of cave-restricted fauna in a new hotspot of subterranean biodiversity in Neotropics. *Biodiv. Conserv.* **2021**, *30*, 4223–4250. [CrossRef]
3. Culver, D.C.; Sket, B. Hotspots of subterranean biodiversity in caves and wells. *J. Cave Karst Stud.* **2000**, *62*, 11–17.
4. Eberhard, S.M.; Howarth, F.G. Undara lava cave fauna in tropical Queensland with an annotated list of Australian subterranean biodiversity hotspots. *Diversity* **2021**, *13*, 326. [CrossRef]
5. Deharveng, L.; Le, C.K.; Bedos, A.; Judson, M.L.; Le, C.M.; Lukić, M.; Vermeulen, J. A hotspot of subterranean biodiversity on the brink: Mo So Cave and the Hon Chong Karst of Vietnam. *Diversity* **2023**, *15*, 1058. [CrossRef]
6. Francke, O.F.; Monjaraz-Ruedas, R.; Cruz-López, J.A. Biodiversity of the Huautla cave system, Oaxaca, Mexico. *Diversity* **2021**, *13*, 429. [CrossRef]
7. Deharveng, L.; Rahmadi, C.; Suhardjono, Y.R.; Bedos, A. The Towakkalak System, A hotspot of subterranean biodiversity in Sulawesi, Indonesia. *Diversity* **2021**, *13*, 392. [CrossRef]
8. Deharveng, L.; Ellis, M.; Bedos, A.; Jantarit, S. Tham Chiang Dao: A hotspot of subterranean biodiversity in Northern Thailand. *Diversity* **2023**, *15*, 1076. [CrossRef]
9. Gallão, J.E.; Ribeiro, D.B.; Gallo, J.S.; Bichuette, M.E. There and Back Again—The Igatu Hotspot Siliciclastic Caves: Expanding the data for subterranean fauna in Brazil, Chapada Diamantina Region. *Diversity* **2023**, *15*, 991. [CrossRef]
10. Clark, H.L.; Buzatto, B.A.; Halse, S.A. A hotspot of arid zone subterranean biodiversity: The Robe Valley in Western Australia. *Diversity* **2021**, *13*, 482. [CrossRef]
11. Silva, M.S.; Ferreira, R.L. The first two hotspots of subterranean biodiversity in South America. *Subterr. Biol.* **2016**, *19*, 1–21. [CrossRef]
12. Hellmann, L.; Ferreira, R.L.; Rabelo, L.; Leal-Zanchet, A.M. Enhancing the still scattered knowledge on the taxonomic diversity of freshwater triclads (Platyhelminthes: Dugesiidae) in caves from two Brazilian Biomes. *Stud. Neotrop. Fauna Envion.* **2022**, *57*, 148–163. [CrossRef]
13. Simone, L.R.L. A new genus and species of cavernicolous Pomatiopsidae (Mollusca, Caenogastropoda) in Bahia, Brazil. *Pap. Avulsos Zool.* **2012**, *52*, 515–524. [CrossRef]
14. Kury, A.B.; González, A.P. A new remarkable troglomorphic gonyleptid from Brazil (Arachnida, Opiliones, Laniatores). *Rev. Ibér. Arachnol.* **2002**, *246*, 43–50.
15. Baptista, R.L.C.; Giupponi, A.D.L. A new troglomorphic *Charinus* from Brazil (Arachnida: Amblypygi: Charinidae). *Rev. Ibér. Arac.* **2002**, *6*, 105–110.
16. Prado, G.C.; Ferreira, R.L. Three new troglobitic species of *Pseudochthonius* Balzan, 1892 (Pseudoscorpiones, Chthoniidae) from northeastern Brazil. *Zootaxa* **2023**, *5249*, 92–110. [CrossRef]
17. Campos-Filho, I.S.; Araujo, P.B.; Bichuette, M.E.; Trajano, E.; Taiti, S. Terrestrial isopods (Crustacea: Isopoda: Oniscidea) from Brazilian caves. *Zool. J. Linn. Soc.* **2014**, *172*, 360–425. [CrossRef]
18. Cardoso, G.M.; Bastos-Pereira, R.; Souza, L.A.; Ferreira, R.L. New cave species of *Pectenoniscus* Andersson, 1960 (Isopoda: Oniscidea: Styloniscidae) and an identification key for the genus. *Nauplius* **2020**, *28*, e2020039. [CrossRef]
19. Souza, P.G.C.; Medeiros, G.D.S.; Ferreira, R.L.; Souza-Silva, M.; Bellini, B.C. A highly troglomorphic new genus of Sminthuridae (Collembola, Symphypleona) from the Brazilian semiarid region. *Insects* **2022**, *13*, 650. [CrossRef]
20. Carvalho, P.H.M.; Junta, V.G.P.; Castro-Souza, R.A.; Ferreira, R.L. Three new cricket species and a new subgenus of *Endecous* Saussure, 1878 (Grylloidea: Phalangopsidae) from caves in northeastern Brazil. *Zootaxa* **2023**, *5263*, 001–039. [CrossRef]
21. Kamimura, Y.; Ferreira, R.L. Description of a second South American species in the Malagasy earwig genus *Mesodiplatys* from cave habitat, with notes on the definition of Haplodiplatyidae (Insecta, Dermaptera). *ZooKeys* **2018**, *790*, 87–100. [CrossRef] [PubMed]
22. Rizzato, P.P.; Bichuette, M.E. A new species of cave catfish from Brazil, *Trichomycterus rubbioli* sp. n., from Serra do Ramalho karstic area, São Francisco River basin, Bahia State (Silurifomes: Trichomycteridae). *Zootaxa* **2012**, *3480*, 48–66.
23. Iniesta, L.F.M.; Ferreira, R.L. The first troglobitic *Pseudonannolene* from Brazilian iron ore caves (Spirostreptida: Pseudonannolenidae). *Zootaxa* **2013**, *3669*, 85–95. [CrossRef] [PubMed]
24. Iniesta, L.F.M.; Ferreira, R.L. *Pseudonannolene lundi* n. sp., a new troglobitic millipede from a Brazilian limestone cave (Spirostreptida: Pseudonannolenidae). *Zootaxa* **2015**, *3949*, 123–128. [CrossRef] [PubMed]
25. Golovatch, S.I.; Wytwer, J. The South American millipede genus *Phaneromerium* Verhoeff, 1941, with the description of a new cavernicolous species from Brazil (Diplopoda: Polydesmida: Fuhrmannodesmidae). *Ann. Zool.* **2004**, *54*, 511–514.
26. Fennah, R.G. A cavernicolous new species of *Notuchus* from New Caledonia (Homoptera: Fulgoroidea: Delphacidae). *Rev. Suisse Zool.* **1980**, *87*, 757–759. [CrossRef]
27. Hoch, H.; Asche, M.; Burwell, C.; Monteith, G.M.; Wessel, A. Morphological alteration in response to endogeic habitat and ant association in two new planthopper species from New Caledonia (Hemiptera: Auchenorrhyncha: Fulgoromorpha: Delphacidae). *J. Nat. Hist.* **2006**, *40*, 1867–1886. [CrossRef]
28. Hoch, H.; Ferreira, R.L. *Ferricixius davidi* gen. n., sp. n.–the first cavernicolous planthopper from Brazil (Hemiptera, Fulgoromorpha, Cixiidae). *Dtsch. Entomol. Z.* **2012**, *59*, 201–206.

29. Hoch, H.; Ferreira, R.L. *Potiguara troglobia* gen. n., sp. n.–first record of a troglobitic Kinnaridae from Brazil (Hemiptera: Fulgoromorpha). *Dtsch. Entomol. Z.* **2013**, *60*, 33–40.
30. Hoch, H.; Ferreira, R.L. *Iuiuia caeca* gen. n., sp. n., a new troglobitic planthopper in the family Kinnaridae (Hemiptera, Fulgoromorpha) from Brazil. *Dtsch. Entomol. Z.* **2016**, *63*, 171–181. [CrossRef]
31. Balkenohl, M.; Pellegrini, T.G.; Zampaulo, R.D.A. A peculiar new beetle from Brazil associated with a cave habitat (Coleoptera, Carabidae, Clivinini). *Zootaxa* **2018**, *4497*, 398–410. [CrossRef] [PubMed]
32. Pellegrini, T.G.; Ferreira, R.L. Ultrastructural analysis and polymorphisms in *Coarazuphium caatinga* (Coleoptera: Carabidae, Zuphiini), a new Brazilian troglobitic beetle. *Zootaxa* **2014**, *3765*, 526–540. [CrossRef] [PubMed]
33. Polhemus, D.A.; Ferreira, R.L. Two unusual new genera of cavernicolous Hydrometridae (Insecta: Heteroptera) from eastern Brazil. *Tijdschr. Voor Entomol.* **2018**, *161*, 25–38. [CrossRef]
34. Cordeiro, I.R.; Bichuette, M.E.; Moreira, F.F. A New Species of *Spelaeometra* Polhemus & Ferreira, 2018 (Hemiptera: Heteroptera: Hydrometridae) from a Hotspot of Troglobites in Brazil, Serra do Ramalho Karst Area. *Animals* **2023**, *13*, 3199.
35. Sendra, A.; Yoshizawa, K.; Ferreira, R.L. New oversize troglobitic species of Campodeidae in Japan (Diplura). *Subterr. Biol.* **2018**, *27*, 53–73. [CrossRef]
36. Ferreira, R.L.; Bernard, E.; da Cruz Júnior, F.W.; Piló, L.B.; Calux, A.; Souza-Silva, M.; Barlow, J.; Pompeu, P.S.; Cardoso, P.; Mammola, S.; et al. Brazilian cave heritage under siege. *Science* **2022**, *375*, 1238–1239. [CrossRef]

 *diversity*

Article

The Subterranean Species of the Vjetrenica Cave System in Bosnia and Herzegovina

Teo Delić [1,*], Tanja Pipan [2], Roman Ozimec [3], David C. Culver [4] and Maja Zagmajster [1]

[1] SubBio Lab, Department of Biology, Biotechnical Faculty, University of Ljubljana, 1000 Ljubljana, Slovenia; maja.zagmajster@bf.uni-lj.si

[2] Karst Research Institute at Research Centre of the Slovenian Academy of Sciences and Arts, 6230 Postojna, Slovenia; tanja.pipan@zrc-sazu.si

[3] ADIPA: Society for Research & Conservation of Croatian Natural Diversity, 10000 Zagreb, Croatia; roman.ozimec@gmail.com

[4] Department of Environmental Science, American University, 4400 Massachusetts Ave. NW, Washington, DC 20016, USA; dculver@american.edu

* Correspondence: teo.delic@bf.uni-lj.si

Abstract: The Western Balkan's Vjetrenica Cave in southern Bosnia and Herzegovina is renowned for high richness of subterranean species. However, the data on its fauna have been published only in monographs printed in a small number of copies, making them hardly accessible to the wider scientific community. To overcome this issue, we compiled the data from published monographs with the data from our own recent field surveys. Further, as they are connected via water channels or small crevices in bedrock, we defined the Vjetrenica Cave System as a system comprising Vjetrenica and Bjelušica Caves and Lukavac Spring. Altogether, 93 troglobiotic, i.e., obligate subterranean aquatic (48) and terrestrial (45), taxa were reported for the system, verifying the Vjetrenica Cave System as the second richest locality in subterranean biodiversity in the world. The global uniqueness of the system is also reflected in the fact that as many as 40 troglobiotic species were described from the system. Finally, we reviewed the factors endangering this unique subterranean community and questioned whether it will withstand human-induced changes and pressures due to infrastructural development in southern Bosnia and Herzegovina.

Keywords: Dinaric Karst; Western Balkans; troglobiont; Vjetrenica; subterranean; hotspot; speleobiology

Citation: Delić, T.; Pipan, T.; Ozimec, R.; Culver, D.C.; Zagmajster, M. The Subterranean Species of the Vjetrenica Cave System in Bosnia and Herzegovina. *Diversity* **2023**, *15*, 912. https://doi.org/10.3390/d15080912

Academic Editors: Ana Sofia Reboleira and Michael Wink

Received: 5 July 2023
Revised: 31 July 2023
Accepted: 4 August 2023
Published: 6 August 2023

1. Introduction

The Western Balkan's Dinaric Karst is one of the global hotspots of subterranean biodiversity [1,2]. The long history of research in subterranean habitats [3] resulted in recognition of two geographically distant hotspots of species richness. The northwestern one, situated in southwestern Slovenia and northwestern Croatia, and the southeastern one, geographically settled at the territories bordering Bosnia and Herzegovina, Croatia and Montenegro [4–7]. Besides being exceptionally rich in subterranean taxa, each of the two bears its own unmistakable "crown gem" cave. While the updated subterranean species list of the northwestern gem, the Postojna-Planina Cave System (PPCS), was published fairly recently in the first special issue of *Subterranean Hotspots* [8], similar presentation of the southeastern gem, the Vjetrenica Cave, was already published 13 years ago [9]. The paper, however, did not include the species list. In addition to the paper, two extensive monographs have been published on Vjetrenica, both including data on its fauna but also paleontological and cultural heritage [10–12]. Yet, due to a limited number of copies and an outdated overview of fauna, there was a need to assemble and present the updated species list for the cave itself and the accompanying system.

In this contribution, we present the updated list of obligate subterranean taxa of the Vjetrenica Cave System, which includes not only Vjetrenica Cave but also Bjelušica Cave and Lukavac Spring; both are confirmed to be connected to Vjetrenica via water channels or crevices in bedrock. We mark the species that have been found in recent studies, and comment on dubious findings. We conclude with emphasizing the threats and conservation issues of the subterranean communities in southern Bosnia and Herzegovina.

History of Biological Studies of Vjetrenica Cave

The first document mentioning an undefined cave characterized by strong winds, similar to those occurring in Vjetrenica, was written 2000 years ago [13]. Pliny the Elder's (Plinius Senior) script mentions it in a way that it leaves us little doubt about the described cave being Vjetrenica. Archaeological artefacts demonstrate that the entrance parts of the cave were used by the poljes' settlers already in the Neolithic (7000–3000 BC) [12], while prehistoric animals, including leopards and hyenas, push its usage even further into past [14–16]. Up to 19th century, Vjetrenica was only occasionally mentioned in naturalists' manuscripts regarding Popovo Polje or the cave itself [13,17,18]. This largely changed with the annexation of modern Bosnia and Herzegovina territories, including Popovo Polje and Vjetrenica, by the Austro-Hungarian Empire (1878) [19–22]. Only a few decades before that, the first subterranean animal, *Leptodirus hochenwartii* Schmidt, 1832, from the Postojna Cave was described, and speleobiology—the biology of subterranean habitats—was born [23]. Southward extension of the empire suddenly enabled naturalists and admirers of subterranean caves to sample specialized fauna in Vjetrenica and other caves in Popovo Polje [24].

Thanks to its early recognition and proximity of the railway, Vjetrenica gained a lot of research interest in the early stages of speleobiology. By the end of the 19th and beginning of 20th century, some of the most eminent European scientists studied its fauna, transforming it into one of the most intensively sampled caves in the world [25–29]. Although preceded in sampling by K.W. Verhoeff [9], the earliest efforts to summarize its subterranean richness were made by Czech archeologist, geographer, paleontologist, and biologist Karel Absolon. Absolon [24] recognized Vjetrenica and the wider area of Popovo Polje as a hotspot of subterranean life and described some of Vjetrenica's outstanding life forms. The pace of discovery continued between the two World Wars [30–34], resulting in the cave's first species inventory by Wolf in 1937 [35]. As in other localities listed in his catalogues, Wolf did not pay attention to the "cave-adaptiveness" or ecology of animals occurring in Vjetrenica, fusing surface and subterranean taxa. Stanko Karaman [32,33,36,37] described a dozen specialized aquatic species from Vjetrenica and other caves in the vicinity, additionally emphasizing the uniqueness of area's aquatic fauna. Decades of sampling and numerous field excursions to Vjetrenica inspired Slovenian speleobiologist Boris Sket [11] to publish the first thorough overview, providing a special emphasis on troglobionts and stygobionts. In his comprehensive overview, he reported 40 stygobionts and 35 troglobionts, clearly placing Vjetrenica among the top ranked subterranean biodiversity hotspots [11,38]. Despite the exceptional results, the cave's inventory list did not stop at 75 species. Ozimec and Lučić [9] updated it and reported 101 troglobiotic species, however, without providing an actual list. The last in a series of inventories including specialists (stygobionts and troglobionts) and non-specialists (troglophiles and trogloxenes) was published by Ozimec and nearly 30 collaborators [12]. Their count comprised 41 troglobionts and 55 stygobionts for a total of 96 cave-dwelling species.

During a century and a half of systematic research in the area, Vjetrenica Cave received the majority of sampling efforts [9,11,12,39]. Herein, we chose a slightly different approach. In addition to carefully evaluating and updating Vjetrenica's subterranean fauna, we compiled an inventory list by combining it with the two nearby localities, the Lukavac Spring and the Bjelušica Cave (Figures 1 and 2). The main reason for their inclusion is their historical omission from similar inventories despite the fact they naturally contribute to the Vjetrenica Cave System [11]. In 2015, a simultaneous diving expedition into Lukavac and Donja Vjetrenica (lower part of the Vjetrenica Cave) resulted in confirmation of the connectedness of the two—divers from each side met under water (G. Balasz, personal communication). The other cave, Bjelušica, opens on a slope above the Vjetrenica Cave. It contains a small stream that disappears in the gravel floor. According to the spatial position of Bjelušica's main channel, and reappearance of the water flow in Vjetrenica's channel "Vilino gumno", we conclude that the two caves are connected (Figure 2).

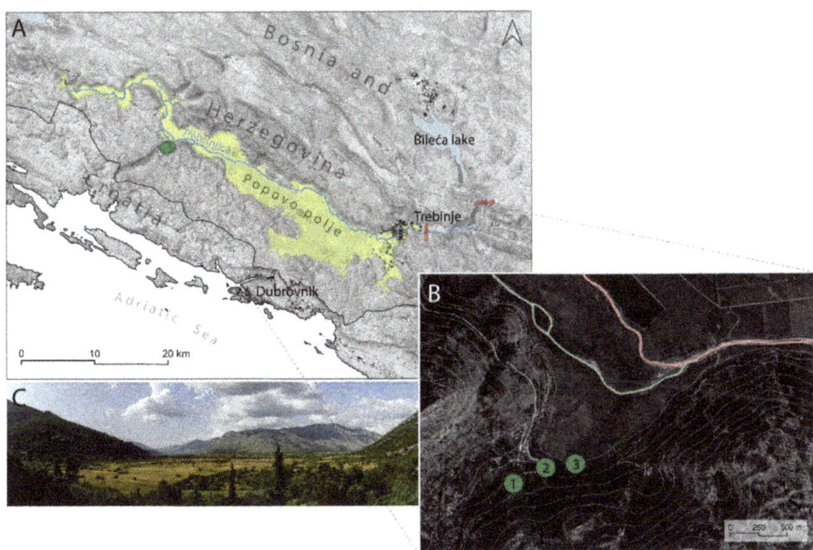

Figure 1. General position of the Vjetrenica Cave System in relation to the major landscape elements, defining the functioning of the Trebišnjica River and Popovo Polje (**A**). The hydropower plants Trebinje I and Trebinje II formed Bileća and Trebinje Lakes, respectively (dams marked by red lines). Downwards from the city of Trebinje, Trebišnjica is channelized on its way across Popovo Polje (presented in light green). Surface entrances to the Vjetrenica Cave System, marked in the satellite image (**B**), are situated in the northwestern part of the Popovo Polje; numbers refer to 1—Bjelušica Cave, 2—Vjetrenica Cave and 3—Lukavac Spring. The same image shows the natural (green) and artificial (red) course of the Trebišnjica River. View of the Popovo Polje from the Vjetrenica Cave's entrance (**C**) (Photo by T. Delić).

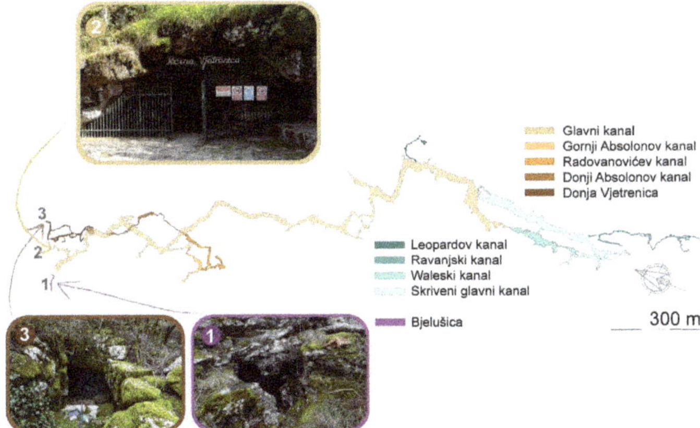

Figure 2. Three entrances to the Vjetrenica Cave System: 1—Bjelušica cave, 2—Vjetrenica cave, and 3—Lukavac spring (numbered as in Figure 1) and their relative positions on a simplified plan of the System (adapted from [12]). The main parts of the system are color coded on the right side. (Photo by E. Premate and T. Delić).

2. Geographical Setting and Description of the Vjetrenica Cave System

2.1. Geographic Setting

Due to its importance for the public recognition of the Dinaric Karst and the research fields of speleobiology, hydrology, and karstology [11,12,40–43], the Vjetrenica Cave System must be set into a wider context, which includes Popovo Polje and the sinking river feeding it, the Trebišnjica River. The Trebišnjica River drains from the boundary of the Black Sea and the Adriatic Sea drainages, first appearing at the surface below the ridge of Lebršnik (1985 m a.s.l.) and the area of Čemerno as the Mušnica River and its tributaries. Surface waters disappear in a series of ponors in the southwestern part of Gatačko Polje (930–950 m a.s.l.), reappear again in Cerničko (810 m a.s.l.) and Fatničko Polje (460–500 m a.s.l.), and finally occur as the Trebišnjica River in the resurgences beneath the town of Bileća. The two largest resurgences are Nikšička Vrela (325 m a.s.l.) and the now-submerged Dejanova Pećina (Dejan's Cave at 327 m a.s.l.) [44]. Before the alteration of its natural course, Trebišnjica flowed through Bilećko Polje and the city of Trebinje, across one of the largest karst poljes in the Dinaric Karst, Popovo Polje, on its way to the sinkhole Ponikva in Hutovo [45] (Figure 1). With 96.5 km of surface flow, Trebišnjica was the largest sinking river in Europe. In summers, it sank downstream from the city of Trebinje, making approximately 60 km of its flow seasonal [44]. Subterranean waters disappearing in Popovo Polje re-appear through resurgences in the Neretva River valley and a series of springs in the background of the city of Dubrovnik, the best known being the Ombla Spring (−15 m b.s.l.) [44,46].

The infrastructural works in the second half of the 20th century modified Trebišnjica's natural flow through several stages. The first stage included damming of Bilećko Polje, including the major springs of Trebišnjica, by changing it into a 20 km long artificial lake. Waters from the reservoir, which are accumulated behind a 120 m high dam, are used for the hydropower plant Trebinje I at Grančarevo (constructed between 1968–1975) [47,48]. With more than 1280 km^3 of water, Bilećko Jezero (Bileća Lake) is one of the largest lakes in the Dinaric Karst. Another dam, 35 m high and accumulating waters for the hydropower plant Trebinje II, was built in 1981. Along with it, a 60 km channel in the lower portions of the Popovo Polje was built to dispatch waters to the hydropower plant Čapljina. As a side effect, the channel prevented the polje's natural flooding and enabled its agricultural exploitation [44]. Consequently, interventions have had a large impact on the surface and subterranean watercourses in the area [44,49,50], decimating the local fauna and pushing some of the narrowly endemic species to the very edge of existence [51–53].

The largest portion of the Trebišnjica runs through the 65 km long Popovo Polje, one of the largest karstic fields in the Dinaric Karst. Due to its proximity to the Adriatic Sea (only 15 km airline distance), Popovo Polje is characterized by dry winters and mild, wet summers. The mean annual air temperature is around 11.4 °C, while the mean annual precipitation is approximately 1680 mm [12,54]. Both the polje and the major geomorphological elements, including locally more than 3 km thick Mesozoic limestones, are orientated NW–SE, in the so-called Dinaric direction. Based on its surface morphology, Popovo Polje is divided in two parts: the upper and lower Trebinjska Šuma and Popovo Polje, respectively. Trebinjska Šuma (šuma meaning woods) is a highly karstified area, dipping in the southeast–northwest (275–250 m a.s.l.) direction and extending from the city of Trebinje to Poljica [12,55]. Differently from the upper part of the polje, the lower part is covered in alluvial sediments, thickening towards its northwest end and reaching up to 25 m at the lowest part of the polje (220 m a.s.l.) [56]. Before the channelization, more than 500 ponors and estavelas were present in the polje, with the most impressive one being large caves (Plitica, Baba u Strujićima, Provalija, Doljašnica, Crnulja, Žira, and Ponikva) in the polje's lower part [44].

2.2. The Vjetrenica Cave System

For a cave whose entrance is characterized by winds reaching up to 89 km/h (Roman Ozimec, personal data), there is no wonder it bears a name meaning "a windy place" in local languages [9,10,12,23]. Similar to the polje, the Vjetrenica Cave System developed in the Mesozoic limestones, predominantly during the Cretaceous age, stretching in the NW–SE direction, and is situated in the outskirts of the Zavala Village in Popovo Polje, Bosnia and Herzegovina (42.8458, 17.9838) [55]. The whole system comprises three parts: the Vjetrenica and Bjelušica Caves and the Lukavac Spring (Figures 1 and 2).

Bjelušica (42.84538, 17.97794) is a rather simple, 80 m long cave linked to Vjetrenica by a water drip, reappearing in its "Vilino gumno" channel. Bjelušica opens on a slope westwards to Vjetrenica. Lukavac Spring (42.84646, 17.98456) lies northwards of Vjetrenica, 20 m lower than the cave's entrance, at the level of the polje (Figures 1 and 2). Although it has been long-hypothesized to be connected to Donja Vjetrenica, this was undoubtedly confirmed only recently by cave diving (G. Balasz, personal communication). Hydrologically, Lukavac Spring presents one of the outflows from the system [57].

The main part of the system, Vjetrenica Cave, is a relatively large and complex cave [12] (Figure 2), with the main channels reaching up to a couple of tens of meters in cross-section. The last topological surveys extended its length to 7324 m, with a vertical extent of 159 m [12]. Three quarters of Vjetrenica's length, the lowest point reaching 43 m in depth, are below the cave's entrance. Although the extant entrance is facing into Popovo Polje at the downstream end of the cave, 1500 m into the cave, there is a drainage divide [12], with the water past it presumably flowing towards the Adriatic Sea and the Neretva valley [44]. Due to the inclination of the layers and the overall topography of the cave, it comprises numerous syphons, occasional lakes, and streams of different sizes. Three levels can be recognized in the cave [12]. The most easily reachable and the most explored is the middle level, comprising predominantly horizontal passages—Glavni kanal, Radovanovićev kanal, Gornji Absolonov kanal, Leopardov kanal, Waleski kanal, Skriveni glavni kanal, and Ravanjski kanal (Figure 2). The lowest level of Vjetrenica consists of hydrologically active or even submerged passages, including Donja Vjetrenica, Donji Absolonov kanal, and Radovanovićev kanal. The potential third level, the uppermost, which is rich in domes and chimneys, extends along the whole cave and offers potential for future speleological surveys.

3. Compiling the List of Taxa

The herein presented list derives from the recently published monograph on Vjetrenica [12], additional records from the SubBio Lab (University of Ljubljana, Slovenia), and Jozef Grego. The existing list was critically evaluated, and species with dubious or not sufficiently known sampling origin were removed from the list. To provide support for the relevancy of the listed taxa, we supplemented the list with information on the year when the animal was last collected considering the period of the last 23 years. This information was retrieved from the database *SubBioDB*, which is managed by SubBio Lab (University of Ljubljana), as well as R. Ozimec's and J. Grego's field notes. In addition to the overview of the species, we provide the data on the species conservation statuses at national and international levels (Table 1).

Table 1. The list of troglobiotic taxa recorded in the Vjetrenica Cave System in Bosnia and Herzegovina. Ecological classification is marked with A—aquatic and T—terrestrial. Presence of the species in a specific part of the system, i.e., V—Vjetrenica Cave, B—Bjelušica Cave, and L—Lukavac Spring, is marked with "✓". The asterisk denotes if the cave/spring is a type locality for the respective species. The last two columns mark species conservation status: IUCN—categories according to IUCN Red List of Threatened Species; BIH—categories according to Red List of Threatened Species of Bosnia and Herzegovina. Categories refer to EN—endangered, VU—vulnerable, NT—near threatened, LC—least concern, and DD—data deficient. Data on species and the last collection year (labelled "Year" in the table) were retrieved from SubBio Lab's database, *SubBioDB* (2023), Ozimec et al. (2021), and J. Grego (Personal communication).

Higher Group	Family	Species	A/T	Cave/Spring			Year	Red List	
				V	B	L		IUCN	BIH
Rhabdocoella	Scutariellidae	*Scutariella stammeri* Matjašič, 1958	A	✓*			/		
		Stygodyticola hadzii Matjašič, 1958	A	✓*			/		
		Subtelsonia perianalis Matjašič, 1958	A	✓			/		
		Troglocaridicola spelaeocaridis Matjašič, 1958	A	✓			/		
		Troglocaridicola capreolaria herzegovinensis Matjašič, 1970	A	✓			/		
Tricladida	Geoplanidae	*Rhynchodemus* sp.	T		✓		2021		
Trematoda	Stenakridae	*Caudotestis protei* (Prudhoe, 1945) Yamaguti, 1958	A	✓*			/		
Nemertea	Prostomatidae	*Prostoma hercegovinense* Tarman, 1961	A	✓*			/		
Polychaeta	Serpulidae	*Marifugia cavatica* Absolon & Hrabe, 1930	A	✓		✓	2021		
Hirudinea	Erpobdellidae	*Dina absoloni* Johansson, 1913	A		✓*		2021		
Gastropoda	Cyclophoridae	*Pholeoteras euthrix* Sturany, 1904	T	✓	✓		2016	LC	
	Ellobiidae	*Zospeum troglobalcanicum* Absolon, 1916	T	✓			2016		
	Emmericiidae	*Emmericia ventricosa* Brusina, 1870	A			✓	/	VU	
	Hydrobiidae	*Kerkia briani* Rysiewska & Osikowski, 2020	A	✓		✓	2020		
		Narentiana vjetrenicae Radoman, 1973	A	✓*		✓*	/	EN	
		Pseudamnicola troglobia Bole, 1961	A	✓	✓	✓	/		
	Moitessieriidae	*Lanzaia vjetrenicae* Kuščer, 1933	A	✓*		✓	/	VU	
		Paladilhiopsis absoloni (Wagner, 1914)	A	✓		✓	/	LC	
		Iglicopsis butoti Falniowski & Hofman, 2021	A	✓			2010		
	Orientalinidae	*Radomaniola montana* (Radoman, 1973)	A	✓		✓	/		
	Pristilomatidae	*Vitrea spelaea* (Wagner, 1914)	T	✓			2016	EN	
		Gyralina candida (Wagner, 1909)	A	✓	✓		/		
	Spelaeoconchidae	*Spelaeoconcha paganettii polymorpha* Wagner, 1914	T	✓	✓*		2016	LC	
	Ferussaciidae	*Cecilioides spelaea* Wagner, 1914	T	✓					
	Agardhiellidae	*Agardhiella biarmata* (O.Boettger, 1880)	T	✓					
	Zonitidae	*Aegopis spelaeus* Wagner, 1914	T	✓	✓*		2016	NT	
Bivalvia	Dreissenidae	*Congeria kusceri* Bole, 1962	A	✓		✓	/	VU	
Diplopoda	Glomeridellidae	*Typhloglomeris coeca* Verhoeff, 1898	T	✓	✓	✓	2014		
	Julidae	*Typhloiulus edentulus* Attems, 1951	T	✓*			/		

Table 1. *Cont.*

Higher Group	Family	Species	A/T	Cave/Spring			Year	Red List	
				V	B	L		IUCN	BIH
	Polydesmidae	*Brachydesmus stygivagus* Verhoeff, 1899	T	✓	✓		2021		
	Trichopolydesmidae	*Verhoeffodesmus* sp.	T	✓			2021		
Chilopoda	Lithobiidae	*Lithobius matulici* Verhoeff, 1899	T	✓		✓	2021		
		Lithobius sketi Matic & Darabantu, 1968	T	✓*			2021		
		Eupolybothrus leostygis (Verhoeff, 1899)	T		✓		/		
Palpigradi	Eukoeneniidae	*Eukoenenia remyi* Conde, 1974	T	✓*			2014		
Acari	Labidostommatidae	*Labidostomma longipes* Willmann, 1940	T	✓	✓		2020		
Opiliones	Sironidae	*Cyphophthalmus* sp.	T	✓	✓		2021		
	Travuniidae	*Travunia vjetrenicae* Hadži, 1933	T	✓*			2021		
Pseudoscorpiones	Chtonidae	*Chthonius occultus* Beier, 1939	T		✓		2013		EN
		Neobisium vjetrenicae Hadži, 1932	T	✓*			/		EN
		Roncus anophthalmus (Ellingsen, 2013)	T	✓			/		
Araneae	Dysderidae	*Stalagtia hercegovinensis* (Nosek, 1905)	T	✓*	✓		2020		
		Stalitella noseki Absolon & Kratochvil, 1933	T	✓*			/		
	Linyphiidae	*Troglohyphantes salax* (Kulczynski, 1914)	T	✓	✓		/		
	Nesticidae	*Kryptonesticus fagei* (Kratochvil, 1933)	T		✓		/		
Copepoda	Cyclopidae	*Acanthocyclops troglophilus* (Kiefer, 1932)	A	✓*			/		
		Diacyclops charon (Kiefer, 1931)	A	✓			/		
		Diacyclops karamani (Kiefer, 1932)	A	✓*			/		
		Diacyclops tantalus (Kiefer, 1937)	A	✓*			/		
		Eucyclops inarmatus Kiefer, 1932	A	✓*	✓		/		
	Diaptomidae	*Troglodiaptomus sketi* Petkovski, 1978	A	✓			/		
Ostracoda	Cyprididae	*Pseudocypridopsis hartmanni* Petkovski et al., 2009	A	✓*			/		
		Pseudocypridopsis sywulai Petkovski et al., 2009	A	✓			/		
	Entocytheridae	*Sphaeromicola stammeri* Klie, 1930	A	✓			/		
Amphipoda	Hadziidae	*Hadzia fragilis* Karaman S., 1932	A	✓*			2021		DD
	Niphargidae	*Niphargus hercegovinensis* Karaman S., 1950	A	✓			2000		DD
		Niphargus kolombatovici Karaman S., 1950	A	✓			2003		DD
		Niphargus factor Karaman G. & Sket, 1990	A	✓*			2005		
		Niphargus boskovici Karaman S., 1952	A	✓*	✓		2021		DD
		Niphargus trullipes Sket, 1958	A	✓*			2021		DD
		Niphargus vjetrenicensis Karaman S., 1932	A	✓*		✓	2021		DD
		Niphargus balcanicus (Absolon, 1927)	A	✓*		✓	2021		DD

Table 1. *Cont.*

Higher Group	Family	Species	A/T	Cave/Spring			Year	Red List	
				V	B	L		IUCN	BIH
		Niphargus cvijici Karaman S., 1950	A	✓			/		DD
		Niphargus zavalanus Karaman S., 1950	A			✓*	/		DD
	Typhlogammaridae	*Typhlogammarus mrazeki* (Schäferna, 1907)	A	✓			2021		
Isopoda	Asellidae	*Proasellus hercegovinensis* (Karaman S., 1933)	A	✓*	✓	✓	2021		
		Proasellus anophtalmus (Karaman S., 1934)	A	✓	✓		/		
	Microparasellidae	*Microcharon* sp.	A	✓			/		
	Sphaeromatidae	*Monolistra hercegoviniensis* Absolon, 1916	A	✓*			2021		
	Trichoniscidae	*Alpioniscus heroldii* (Verhoeff, 1931)	T	✓		✓	2021		
		Cyphonethes herzegowinensis (Verhoeff, 1900)	T	✓	✓		2021		
Decapoda	Atyidae	*Spelaeocaris pretneri* Matjašič, 1956	A	✓			2003		
		Spelaeocaris hercegovinensis (Babić, 1922)	A	✓*		✓	2021		
		Troglocaris anophthalma periadriatica Jugovic et al., 2012	A	✓*		✓	2021		
Mysida	Mysidae	*Troglomysis vjetrenicensis* Stammer, 1933	A	✓*		✓	2000		
Collembola	Entomobryidae	*Verhoeffiella verdemontana* Lukić & Deharveng, 2018	T	✓		✓	2014		
		Verhoeffiella longicornis (Absolon, 1900)	T	✓			2021		
Diplura	Campodeidae	*Plusiocampa remyi* Condé, 1947	T	✓*			2021		
Thysanura	Nicoletiidae	*Coletinia* sp.	T	✓			/		
Coleoptera	Carabidae	*Neotrechus dalmatinus dalmatinus* (Miller L., 1861)	T	✓	✓	✓	2021		
		Scotoplanetes arenstorffianus Absolon, 1913	T	✓*			2021		EN
		Adriaphaenops pretneri Scheibel, 1935	T	✓*			/		EN
		Neotrechus suturalis otiosus (Obenberger, 1917)	T	✓	✓		/		
		Speluncarius anophthalmus (Reitter, 1886)	T		✓		/		
	Leiodidae	*Speonesiotes schweitzeri* Jeannel, 1941	T	✓	✓*		2014		
		Speonesiotes narentinus latitarsis (Apfelbeck, 1919)	T	✓	✓		/		
		Graciliella apfelbecki apfelbecki (Müller J., 1910)	T	✓*			2021		
		Hadesia vasiceki Müller J., 1911	T	✓*			2021		
		Nauticiella stygivaga Moravec & Mlejnek, 2002	T	✓*			2021		
		Anthroherpon primitivum (Absolon, 1913)	T	✓			/		
	Staphylinidae	*Troglamaurops ganglbaueri* (Winkler, 1925)	T	✓			2021		
		Nonveilleria sp.	T	✓			/		
Urodela	Proteidae	*Proteus anguinus* Laurenti, 1768	A	✓		✓	2021	VU	EN
				85	26	22			

4. The Overview of Troglobiotic Species in the Vjetrenica Cave System

4.1. General Overview

Altogether, 93 different subterranean species have been recorded and are considered as present in the Vjetrenica Cave System: 48 aquatic and 45 terrestrial (Table 1). Overall, 40 species have been scientifically described from the system: 35 from Vjetrenica, 4 from Bjelušica, and 2 from Lukavac (Table 1). Field surveys executed from the onset of the 21st century confirmed 50% of taxa previously reported from the system (Table 1).

Among a plethora of species, dozens of subterranean taxa inhabiting the system have been recognized as threatened. According to the IUCN Red List of Threatened Species (VIR), there are ten threatened taxa (Table 1): two endangered (EN), four vulnerable (VU), three of least concern (LC), and one near threatened (NT). According to the Red List of Bosnia and Herzegovina (VIR), there are fourteen threatened taxa: five endangered (EN) and nine data deficient (DD) (Table 1).

4.2. Comments to Selected Aquatic Taxonomic Groups

One of the most distinguishing characteristics of the Dinaric subterranean fauna is the presence of aquatic sessile and filtering species, deriving from marine or historically rich lacustrine fauna [58,59]. Three of these peculiar species were reported from the Vjetrenica Cave; the only subterranean tubeworm in the world, *Marifugia cavatica*; the only cave cnidarian, *Velkovrhia enigmatica*; and one of only a handful of subterranean clams, *Congeria kusceri* (Table 1; Figure 3C). *Marifugia cavatica* can be observed in the waters of the lower Vjetrenica's channels [60]. The presence of the other two species is highly questionable and needs additional confirmation. Even though *V. enigmatica* was reported from a cave in Croatia, it has been recently confirmed only in two caves from 500 km distant Ljubljanica River catchment in Slovenia [61]. Moreover, there are some indices that the data on *Velkovrhia* in Vjetrenica might be a result of an experimental error (Sket, personal communication). The second questionable species is *Congeria kusceri*, whose shell was presumably collected in an unknown part of Vjetrenica Cave [12]. Recent and intensive diving explorations in the lower parts of the cave did not result in finding live individuals (B. Jalžić and G. Balazs, personal communication). However, as it occurs in other caves in Popovo Polje, with the closest confirmed locality being the 1.7 km away Baba u Čvaljini Cave, its presence in the system cannot be completely ruled out.

The Vjetrenica Cave System harbors one of the most remarkable examples of single-genus diversity. There are as many as nine different species of the subterranean amphipod genus *Niphargus* [11,62] present in the system. To our knowledge, this exceptional richness is the highest number of subterranean congeners occurring in a single locality in the world, followed only by the community of six *Niphargus* species in the Postojna-Planina Cave System in Slovenia [8]. The co-occurring species largely differ in both general morphology and body size (ranging from the 3 mm large *N. factor* to the spiny and more than 30 mm long *N. balcanicus* (Figure 3A)). The species exploit a wide variety of habitats, including water drips, interstitial waters, and phreatic channels [62]. These characteristics have been related to the evolutionary effects of diminishing competition among closely related species [62,63]. In addition, more amphipod species were found in Vjetrenica, including *Hadzia fragilis* [32] and the largest and the bulkiest among all Dinaric amphipods, the monotypic *Typhlogammarus mrazeki* [64] (Figure 4C).

High species richness of "shrimp-like" crustaceans (Figure 3B) belonging to two different orders can be found within the system. Three species belong to the decapod family Atyiidae [65], which exhibited multiple transitions into the circum-Mediterranean subterranean habitats [66]. The fourth species is a monotypic Mysidae species found only in Vjetrenica's phreatic waters, *Troglomysis vjetrenicensis* [67].

The largest and the most outstanding animal of the subterranean habitats in the Dinaric Karst is the olm *Proteus anguinus* [68] (Figure 3D). Even though once commonly distributed in caves of Popovo Polje, it seems that its population largely disappeared from caves that had been cut off from the Trebišnjica River following to its channelization [51]. In

Vjetrenica Cave, the olm can be found in its lower parts in partly or completely submerged channels. Recently, olm populations from southern Dinarides, including those bound to the Trebišnjica River catchment, were recognized as a separate species-level lineage [69]. Considering the changes of the water regime in Popovo Polje, the olm's southern populations seem to be even more vulnerable than previously thought and highly threatened.

Figure 3. Diverse stygobionts reported from the Vjetrenica cave system: (**A**) the spiny amphipod *Niphargus balcanicus*, (**B**) cave shrimp *Spelaeocaris* sp., (**C**) the subterranean tubeworm *Marifugia cavatica*, along with the cave mussel *Congeria kusceri*, and (**D**) the olm *Proteus anguinus* (Photo: Teo Delić).

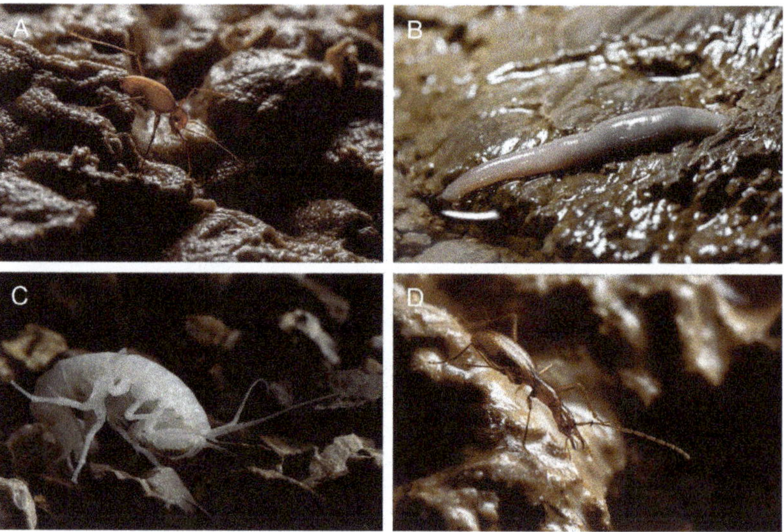

Figure 4. The cave hygropetric, specialized subterranean microhabitat was first described from Vjetrenica Cave [70]. Some of the specialized animals inhabiting it include (**A**) the semi-aquatic cave beetle *Hadesia vasiceki*; (**B**) the predatory cave leech *Dina absoloni*; (**C**) the bulkiest of all Dinaric subterranean amphipods, *Typhlogammarus mrazeki*; and (**D**) a highly troglomorphic and predatory Trechini beetle, *Scotoplanetes arenstorffianus* (Photo: Teo Delić).

4.3. Comments on Selected Terrestrial Taxonomic Groups

The most notable characteristics of Vjetrenica's terrestrial fauna is the existence of the species living in the special cave habitat, the so-called hygropetric [70]. The cave hygropetric is a specialized type of subterranean habitat, first recognized and described from Vjetrenica's depths. It refers to water flowing over the cave walls, forming a thin laminar flow or, sometimes, strong turbulent currents [70,71]. Organic matter dissolved in the water flowing down the vertical walls enables formation of microbial communities [72], which are scraped off the walls and used as nutrients by various groups of arthropods. Species or communities bound to this peculiar habitat are known only from the "hygropetricolous arc" spanning throughout the Dinaric Karst and Italian Prealps [71–74] and geographically distant Caucasus [75,76].

Probably the most known of all the hygropetricolous animals is the elusive beetle genus *Hadesia*, first to be recognized for its peculiar ecology and a semi-aquatic lifestyle [77,78]. Vjetrenica's *Hadesia vasiceki* (Figure 4A) bears some of the characteristics common to all terrestrial taxa inhabiting hygropetricolous habitats, including long claws, densely pubescent body, and mouthparts modified for scraping and grazing on organic matter [79]. The other hygropetricolous beetle in Vjetrenica, *Nauticiella stygivaga*, is rarely encountered. Following its description in 2002 [80] and despite many attempts, only two specimens were found in the cave's deeper sections in 2021 [81]. This semi-aquatic habitat is also exploited by the largest of Vjetrenica's amphipods, *Typhlogammarus mrazeki* [82] (Figure 4C), and the cave leech, *Dina absoloni* [83] (Figure 4B). Both species are known to climb the vertical walls and confront the hygropetric's waters in search of prey. In addition to the animals occurring in the water flow itself, a couple of them are known to occur at the edges of the hygropetric, presumably exploiting similar nutrient resources or preying on smaller invertebrates feeding on it. These include the millipede *Typhloiulus edentulus*, for which modified grazing mouthparts were also reported [84], and one of the most troglomorphic representatives of subterranean Trechini beetles in Europe, the predatory *Scotoplanetes arenstorffianus* [85,86] (Figure 4D).

Another remarkable characteristic of terrestrial fauna in the Vjetrenica Cave System is the high diversity of arachnids (Table 1), including mites (Acari), spiders (Araneae), harvestmen (Opiliones), palpigrades (Palpigradi), and pseudoscorpions (Pseudoscorpiones) [12]. The most recognizable among them are surely the large Dysderidae spiders, *Stalagtia hercegovinensis* (Figure 5A) and *Stalitella noseki*, which do not produce webs but freely walk and prey within the cave [87,88]. The predatory *Travunia vjetrenicae* (Figure 5B) is a member of a small opilionid family, Travuniidae, encompassing less than a dozen species worldwide. Despite its small size but due to its robust and spiny pedipalps, *Travunia* is considered a fierce predator of smaller invertebrates [89]. Some of the arachnids, including palpigrades, are rarely encountered due to its small size. Only a couple of millimeters long, *Eukoenenia remyi* [90] is, despite being a terrestrial animal, often found gliding on the calcite crusts on the surface of water pools (own observation).

Another species-rich group is the myriapods, including both diplopods and chilopods. Diplopods inhabit a wide variety of habitats, from the ones in transition to surface habitats to the already-mentioned cave hygropetric. Differences in their natural histories are well reflected onto morphologies, which range from the relatively short and round *Typhloglomeris coeca* (Figure 5C) to the elongated *Typhloiulus edentulus* [12,84]. The predatory chilopods are represented by three subterranean species, relatively small *Lithobius matulici*, and a large and highly troglomorphic *Lithobius sketi* [91,92] (Figure 5D).

High variability in size and ecology can also be noted in gastropods, whose representatives range in size from only two and a half millimeters to a centimeter and a half [12]. All of them except *Spelaeoconcha paganettii* (Figure 6A) are endemic to either Popovo Polje or the southeastern Dinaric Karst [12].

Figure 5. The examples of the striking terrestrial arthropod diversity in the Vjetrenica Cave System: (**A**) large and predatory spider *Stalagtia hercegovinensis*; (**B**) tiny opilionid *Travunia vjetrenicae*; (**C**) one of few subterranean representatives of Glomeridellidae family, *Typhloglomeris coeca*; and (**D**) the large and troglomorphic *Lithobius sketi*, named after late speleobiologist Boris Sket (1936–2023) (Photo: Teo Delić).

Figure 6. Additional diversity is brought into the system by (**A**) a rich molluscan community, including both aquatic and terrestrial species, such as *Speleaoconcha paganettii*, and (**B**) rich subterranean beetles fauna, including one of the largest leiodid beetles, *Graciliella apfelbecki*, (**C**) the tiny and elusive *Troglamaurops ganglbaueri*, and (**D**) the poorly studied collembolans, depicted by *Verhoeffiella longicornis* (Photo: Roman Ozimec and Teo Delić).

Besides hygropetric beetles, all three families with numerous subterranean representatives in the Balkans were recorded in Vjetrenica. The family Leiodidae is, besides *Hadesia* and *Nauticiella*, represented by *Graciliella apfelbecki* (Figure 6B), one of the largest (8 mm) and extremely troglomorphic leiodid species [93]. In addition to *Scotoplanetes*, the family Carabidae is represented by two congeners, *Neotrechus dalmatinus dalmatinus* and *Neotrechus suturalis otiosus* [12], and another presumably ecologically specialized species, *Adriaphaenops pretneri* [94]. Finally, the third family commonly distributed in the Balkan's subterranean habitats, Staphylinidae, is known by a yet undescribed species of *Nonveilleria* [12] and *Troglamaurops ganglbaueri* (SubBioDB) (Figure 6C).

5. Comments on Some of the Non-Troglobiotic Species Adding to the Conservation Importance of the Vjetrenica Cave System

Even though this paper is oriented towards presenting the list of obligate subterranean species, we need to bring forward some non-troglobiotic species that occur in the Vjetrenica Cave System. Three fish species found in subterranean waters of Vjetrenica Cave and are listed among endangered species. Two species, *Delminichthys ghetaldii* (Steindachner, 1882) and *Squalius svallize* Heckel & Kner, 1858, are declared as vulnerable under the IUCN criteria, while the third species, *Phoxinus lumaireul* (Schinz, 1840) is considered of least concern [95]. In addition, *Delminichthys ghetaldii* is considered endangered by the Red List of Bosnia and Herzegovina. Preceding the regulation of Trebišnjica (Figure 1), all three species were abundant in Popovo Polje. Moreover, the local inhabitants were exploiting them as a food source [13]. However, these customs gradually changed by the end of 1960s due to anthropogenic activities and the downfall of the limited habitats of fish species [96].

As for bats, five occasionally occurring species were recorded; all are listed as of least concern on the IUCN Red List. Additionally, are three Vespertilionidae species, namely *Myotis emarginatus* (E. Geoffroy Saint-Hilaire, 1806), *M. nattereri* (Kuhl, 1817), and *Plecotus* cf. *kolombatovici*, and two Rhinolophidae species, namely *Rhinolophus ferrumequinum* (Schreber, 1774) and *R. hipposideros* (Bechstein, 1800) [12,97,98]. In addition, three of these species have a higher threat status according to the Red List of Bosnia and Herzegovina; *M. emarginatus* and *R. ferrumequinum* are listed as vulnerable, while *R. hipposideros* is considered as endangered. Generally, the low number of bat species is presumably related to prevalent winds or limited size, which make Vjetrenica and Bjelušica, respectively, less suitable for hibernation or the establishment of nursery colonies.

6. Discussion

6.1. General Overview and Significance of the New Species List

Differently from most of the existing overviews of Vjetrenica's fauna, which focus only on the specialized fauna of the cave itself [9–13], we chose to broaden our scope by inclusion of the two nearby localities: Bjelušica Cave and Lukavac Spring. Their inclusion resulted in the listing of additional troglobiotic taxa and, finally, a higher number of troglobiotic species in the whole system than in the cave alone (Table 1). Due to morphological differences and the connectedness of the system's localities, not all of the listed species are found in all of them. Vjetrenica remains the richest locality with 85 troglobiotic species, followed by Bjelušica with 26 and, finally, Lukavac with 22 species.

Despite the long tradition of speleological surveying and high numbers of troglobiotic species, we are far from the final point of knowledge on both the Vjetrenica Cave System and its specialized fauna. Further increases in numbers of troglobiotic taxa may be expected by systematic sampling of overlooked microhabitats or taxa in addition to the usage of novel sampling and analytical techniques. Epikarst, which often includes its own specialized communities [99,100] and was never subjected to a thorough research in Vjetrenica, presents one of such habitats. Similarly, collembolans probably present the most illustrative example of overlooked taxa. Only two species of *Verhoeffiella* are listed for the whole system [12] (Figure 6D), although more species belonging to different genera and even families are known from it (Lukić M., personal communication). Finally, numbers might further increase by identification of morphologically cryptic species, which are repeatedly identified among specialized subterranean taxa, including Dinaric collembolans [101–103].

6.2. Monitoring of the Subterranean Communities

The proximity of the railway and infrastructural development along with the fascination about its size and accessibility changed Vjetrenica into a show cave more than half a century ago. E. Pretner questioned the rationality of this move already before its opening in 1960s. He proposed not to set the tourist needs ahead of the conservation of the cave and its peculiar fauna [39]. Tourism paved the way to educate visitors about the functioning and meaning of karst and karstic phenomena. At the same time, the arrangement of pathways and the growing number of visitors, as stressed already by Pretner, present a constant threat to fragile subterranean habitats [104]. Although relatively late, the monitoring scheme in Vjetrenica started in 2016, with an idea to detect changes in the physical and hydrological status of the cave and its microclimate, habitat conditions, the quantity of fauna, as well as modifications in its taxonomic composition. Both can serve as an alarm system for predicting potentially detrimental changes [105,106]. Along with the monitoring of fauna, special interest has been directed towards monitoring of the so-called lampenflora [107], the autotrophic communities developing near artificial light in caves. Algae, bryophytes, mosses, or plants otherwise absent from internal parts of the caves can alter the composition of subterranean communities by providing easily accessible nutrients to some of the species. Additional upgrades of monitoring practices will be assessed by the constant and long-term monitoring of physical parameters such as air and water temperature and the pH of the water and ground or air composition. Implementation of diverse and complementary monitoring practices is of crucial importance, as Vjetrenica and the whole area of Popovo Polje, due to its proximity to Dubrovnik (Figure 1), receive a growing number of tourists. In recent years, the number of tourists rose to more than 17,000 visitors in 2022, while the only exception was around 6000 visitors in 2020, which was heavily affected by the coronavirus pandemic [108]. Compared to the pre-Balkan war years, the number of visitors more than doubled after the cave's reopening. Such an increase presents additional pressure on subterranean ecosystems, calls for additional conservationist attention, and enhances the need for precise and thorough monitoring schemes.

6.3. Past, Present, and Future Threats and Conservation

Due to its geographical setting and connectedness to the Trebišnjica River, tourism does not present the largest issue for the Vjetrenica Cave System. This can be recognized in the progressing industrialization and engineering coupled with a growing need for agricultural land, which triggered construction of a series of dams over the course of the Trebišnjica, with its channelization and the transformation of the lower parts of the polje into agricultural land [44,109,110]. Before its damming and channelization, 155 sinkholes and estaveles existed in the polje [44]. Following the changes, Trebišnjica's hydrological networks, both surface and subterranean, were largely changed [43–47]. Excluding all of its natural meanders and overflowing areas caused the decimation of locally rich and endemic surface and subterranean fauna [51,52,69,111–113]. Despite its uniqueness on the world scale, destruction of Popovo Polje and Vjetrenica were only seldom documented by a couple series of publications, nature conservation actions, and scientific appeals for their conservation [43,46–51].

The effect of these anthropogenic alterations were never properly studied in Vjetrenica or in Popovo Polje. However, some hallmarks, like meters-thick layers of dry tubes of *M. cavatica* in Ponor Crnulja [114], testify to the irreversible changes. Iconic species such as *C. kusceri* and *M. cavatica* seem to have disappeared from some of most known localities in the Popovo Polje. Some papers report the catastrophic aftermath of these changes, resulting in extirpation of more than 99% of local populations [51]. In addition to changes of the water regimes, the land use also changed dramatically. Before channelization, the lower parts of the polje were flooded on average 240 (204–303) days per year [44].

Following channelization, approximately two-thirds of Popovo Polje was changed into agricultural land [109,110], coupling the changes in quantity of water with the potential changes in its quality. Although none of the available studies were executed on Vjetrenica's or Trebišnjica's subterranean fauna, increased concentrations of salt or nitrates were shown to have detrimental effects on subterranean communities [115–117].

As if not all of this was enough, the whole area of Popovo Polje suffered additional obstruction due to disintegration of Yugoslavia in the Balkan Wars during the 1990s [118,119]. The surroundings of the entrances to the Vjetrenica Cave System (Vjetrenica, Bjelušica, and Lukavac Spring) were literally changed into minefields. The wider area was demined in numerous actions following the war; still, some parts of the area may remain inapproachable—like the ridges above Vjetrenica's entrance.

Although the whole system, along with the Trebišnjica River, remains largely affected by the anthropogenic influence, not everything is being lost. Both Bosnia and Herzegovina and the Republic of Srpska proposed Vjetrenica as a future Natura 2000 site, and some of the species were listed on the IUCN's list of endangered and vulnerable species [95]. To further promote the uniqueness of the system and the accompanying Trebišnjica Basin, a Biospeleological Museum was founded in 2016 in close proximity to the system's entrances [120]. Finally, the attempts for conservation of these sites were crowned by an official application for the inclusion of Vjetrenica and the surrounding landscapes under the UNESCO's world heritage conservation scheme [121]. This might be a proper place to question how the possible inclusion of the Vjetrenica System onto UNESCO's list might help against the growing pressures, represented by the ambitious economic-developmental plans of Gornji Horizonti, which are already transforming landscapes in the Trebišnjica Basin. The Gornji Horizonti comprise an infrastructural plan for building additional series of dams and channels meant to feed a set of hydropower plants by draining waters from different, interconnected poljes or even drainages [44,45,122–125]. Despite the deluge of "green deals", "sustainability", and similar terms on the continental level [126] and the known effect of damming rivers on biodiversity [127,128], for now, it seems that nature and its conservation, along with the human wellbeing, are put aside.

6.4. Concluding Remarks

Only successful conservation attempts will enable further usage of Vjetrenica as a show cave, a touristic development of the area, and a scientific work, both in Vjetrenica and other parts of the system. For us, the scientific perspective is of a vital importance. Herein, we will list only two topics connected to evolutionary patterns and the mechanisms underlying them, wherein Vjetrenica's role cannot be overlooked. Vjetrenica's subterranean amphipod assemblage presents the richest subterranean amphipod community in the world. It comprises nine *Niphargus* congeners, largely differing in morphology and spatial use, and additional representatives of other amphipod families [11,62,63]. At least the *Niphargus* community was shown to originate through the mechanisms of adaptive radiation [129]. However, and despite the soundness of the topic, only the first steps towards understanding the mechanisms of community assembling have been made. In addition to this, Vjetrenica is renowned by its semi-aquatic hygropetricolous environment [70] and its peculiar inhabitants. The mechanisms and processes underlying the assembling of the hygropetricolous communities remain even less studied than those underlying the assembling of the niphargid communities.

Fieldworks expeditions after 2000 resulted in the repeated collection of approximately 50 percent of all the species recorded from the system. These are mostly larger taxa (Table 1), which are taxonomic groups that at least some of the authors, or their collaborators, are studying. Data on 50 percent of listed species, including hydrobiid snails or specialized trematodes, remain only literature-based and clearly demonstrate the lack of taxonomists in the scientific field. For most of the species, their presence in the system remains to be confirmed. Therefore, we found no better way to demonstrate how needy we are of both systematic sampling and thorough recording of the species occurring in the Vjetrenica Cave System.

The whole system, along with the Popovo Polje and the Trebišnjica River, present a unique combination of natural history and cultural heritage coupled with tourism and business opportunities. Long-term sustainability of the whole area is largely dependable on a wide variety of factors, including local inhabitants, scientists, farmers, decision makers and governmental agencies, employees in the tourism and energetic sectors, etc. With so many variable groups of interest, this is the right place to ask if we can cope with the burden and whether we, as a community, will be successful in attempts to preserve the second richest subterranean locality in the world?

Author Contributions: Conceptualization, D.C.C. and T.P.; original draft preparation, T.D.; review and editing, M.Z., R.O. and D.C.C.; visualization, T.D.; corresponding author T.D. All authors have read and agreed to the published version of the manuscript.

Funding: The study was partially funded by the Slovenian Research Agency (ARRS program P1-0184), the CEPF-funded project "SubBIOCODE—Developing new tools for rapid assessment of subterranean biodiversity in Bosnia and Herzegovina" to SubBio Lab and Biodiversa+, the European Biodiversity Partnership under the 2021–2022 BiodivProtect joint call for research proposals, co-funded by the European Commission (GA N°101052342), and with the funding organizations Ministry of Universities and Research (Italy), Agencia Estatal de Investigación—Fundación Biodiversidad (Spain), Fundo Regional para a Ciência e Tecnologia (Portugal), Suomen Akatemia—Ministry of the Environment (Finland), Belgian Science Policy Office (Belgium), Agence Nationale de la Recherche (France), Deutsche Forschungsgemeinschaft e.V.—BMBF-VDI/VDE INNOVATION + TECHNIK GMBH (Germany), Schweizerischer Nationalfonds zur Forderung der Wissenschaftlichen Forschung (Switzerland), Fonds zur Förderung der Wissenschaftlichen Forschung (Austria), and the Ministry of Higher Education, Science, and Innovation (Slovenia) and the Executive Agency for Higher Education, Research, Development, and Innovation Funding (Romania).

Data Availability Statement: No new data were created or analyzed in this study; all of the data is available in the study and the references cited.

Acknowledgments: The authors dedicate the paper to drivers of exploration in Vjetrenica: late professor Boris Sket (1936–2023), who's years of research in Vjetrenica and the wider area of Popovo Polje brought international recognition to it, and late Nikša Vuletić (1975–2022), director of the public enterprise Vjetrenica, who was a big supporter of research in Vjetrenica. In addition, the leading author would like to express gratitude to Ivo Lučić, for his generous help with the documentation on Popovo Polje and Vjetrenica, and my dear friends Martina Pavlek, Marko Lukić, Tvrtko Dražina, Branko Jalžić (all CBSS, Zagreb), Roman Lohaj, and Jozef Grego for their expertise in the selected taxonomic groups.

Conflicts of Interest: The authors declare no conflict of interest.

References

1. Culver, D.C.; Deharveng, L.; Bedos, A.; Lewis, J.J.; Madden, M.; Reddell, J.R.; Sket, B.; Trontelj, P.; White, D. The Mid-Latitude Biodiversity Ridge in Terrestrial Cave Fauna. *Ecography* **2006**, *29*, 120–128. [CrossRef]
2. Zagmajster, M.; Malard, F.; Eme, D.; Culver, D.C. Subterranean biodiversity patterns from global to regional scales. In *Cave Ecology*, 1st ed.; Moldovan, O.T., Kovač, L., Halse, S., Eds.; Ecological Studies; Springer International Publishing: Cham, Switzerland, 2018; pp. 195–227.
3. Schmidt, F. Beitrag zu Krain's Fauna. *Leptodirus Hochenwartii, n. g., n. sp. Illyrisches Blatt.* **1832**, *3*, 9–10.
4. Zagmajster, M.; Culver, D.C.; Sket, B. Species richness patterns of obligate subterranean beetles (Insecta: Coleoptera) in a global biodiversity hotspot–effect of scale and sampling intensity. *Divers. Distrib.* **2008**, *14*, 95–105. [CrossRef]
5. Sket, B. Diversity Patterns in Dinaric Karst. In *Encyclopedia of Caves*, 2nd ed.; White, W.B., Culver, D.C., Eds.; Elsevier: Amsterdam, The Netherlands, 2012; pp. 228–238.
6. Bregović, P.; Zagmajster, M. Understanding Hotspots within a Global Hotspot—Identifying the Drivers of Regional Species Richness Patterns in Terrestrial Subterranean Habitats. *Insect Conserv. Divers.* **2016**, *9*, 268–281. [CrossRef]
7. Borko, Š.; Altermatt, F.; Zagmajster, M.; Fišer, C. A Hotspot of Groundwater Amphipod Diversity on a Crossroad of Evolutionary Radiations. *Divers. Distrib.* **2022**, *28*, 2765–2777. [CrossRef]
8. Zagmajster, M.; Polak, S.; Fišer, C. Postojna-Planina Cave System in Slovenia, a Hotspot of Subterranean Biodiversity and a Cradle of Speleobiology. *Diversity* **2021**, *13*, 271. [CrossRef]

9.   Ozimec, R.; Lučić, I. The Vjetrenica cave (Bosnia & Herzegovina)–one of the world's most prominent biodiversity hotspots for cave-dwelling fauna. *Subterr. Biol.* **2009**, *7*, 17–24.
10.   Lučić, I. *Vjetrenica: Pogled u Dušu Zemelje*, 1st ed.; Lučić, I.: Zagreb, Croatia, 2003; p. 322.
11.   Sket, B. Životinjski svijet Vjetrenice. In *Vjetrenica: Pogled u Dušu Zemlje*, 1st ed.; Lučić, I., Ed.; Lučić, I.: Zagreb, Croatia, 2003; pp. 147–248.
12.   Ozimec, R.; Baković, N.; Bakšić, D.; Basara, D.; Bevanda, L.; Brajković, H.; Brancelj, A.; Erhard, C.; Gašić, Z.; Grego, J.; et al. *Vjetrenica—Cave Biodiversity Hotspot of the Dinarides*; Ozimec, R., Ed.; Public Enterprise Vjetrenica: Ravno, Bosnia and Herzegovina; ADIPA: Zagreb, Croatia, 2021; p. 356.
13.   Lučić, I. *Presvlačenje krša. Povijest Poznavanja Dinarskog Krša na Primjeru Popovog Polja*, 1st ed.; Synopsis: Zagreb, Croatia, 2018; p. 680.
14.   Malez, M.; Pepeonik, Z. Entdeckung des ganzen Skelettes eines fossilen Leoparden in der Vjetrenica—Höhle auf dem Popovo Polje (Herzegowina). *Bull. Sci. Que Sect. A* **1969**, *14*, 5–6.
15.   Miculinić, K. Fossil Remains of Leopard (*Panthera pardus*) from Vjetrenica Cave, Popovo Polje, BiH. Ph.D. Thesis, University of Zagreb, Zagreb, Croatia, 2012.
16.   Diedrich, C.G. Late Pleistocene leopards across Europe–northernmost European German population, highest elevated records in the Swiss Alps, complete skeletons in the Bosnia Herzegowina Dinarids and comparison to the Ice Age cave art. *Quat. Sci. Rev.* **2013**, *76*, 167–193. [CrossRef]
17.   Korać, J.V. *Trebinje. Istorijski Pregled. Knjiga I*; Zavičajni Muzej: Trebinje, Bosnia and Herzegovina, 1966; p. 240.
18.   Pamučina, J. Rizvanbegovića. In *Ljetopisi*; Čokorilo, P., Pamučina, J., Skenderova, S., Eds.; Veselin Masleša: Sarajevo, Bosnia and Herzegovina, 1976; pp. 83–88.
19.   Riedel, J. Eine Ventarole in der Herzegovina. *Mitteilungen Der Sect. Für Höhlenkunde* **1888**, *2*, 13–16.
20.   Cvijić, J. *Karst, Geografska Monografija*, 1st ed.; Državna štamparija: Beograd, Srbija, 1895; p. 176.
21.   Katzer, F. Geologischer Fuerer durch Bosnien und die Herzegovina. In *Herausgegeben Anlässlich des IX. Internationalen Geologencongresses von der Landesregierung in Sarajevo*, 1st ed.; Landesdruckerei: Sarajevo, Bosnia and Herzegovina, 1903; p. 280.
22.   Richter, E. Prilozi zemljopisu Bosne i Hercegovine. *Glas. Zemalj. Muzeja* **1905**, *7*, 257–414.
23.   Polak, S. Importance of discovery of the first cave beetle *Leptodirus hochenwartii* Schmidt, 1832. *Endins* **2005**, *28*, 71–80.
24.   Absolon, K. Výsledky Výskumných cest po Bálkaně. *Časopis Morav. Mus. Zemsk.* **1916**, *2*, 245–249.
25.   Nosek, A. Die Arachniden der herzegowinischen Höhlen. *Verh. Zool.-Bot. Ges. Wien.* **1905**, *55*, 212–221. [CrossRef]
26.   Schäferna, K. O novem slepem blesivci *Typhlogammarus* n. sbg. *Vestn. Kral. Ceske Spol. Nauk. Praha* **1906**, *26*, 1–25.
27.   Müller, J. Diagnosen neuer Höhlensilphiden. *Zool. Anz.* **1910**, *36*, 184–186.
28.   Müller, J. Zwei neue Höhlensilphiden aus den österreichischen Karstländern. *Wien. Entomol. Ztg.* **1911**, *30*, 175–176.
29.   Absolon, K. Über *Scotoplanetes arenstorffianus* nov. subg., nov. spec., eine neue Anophthalmentype (Coleoptera Carabidae) aus dem Ponor-Gebiete der Trebišnjica in Südosthercegovina. *Koleopterol. Rundsch.* **1913**, *2*, 93–100.
30.   Absolon, K. Les grandes amphipodes aveugles dans les grottes Balkaniques. *Compte Rendu de la 51e Session, Association Francaise Pour l'Avancement des Sciences* **1927**, *51*, 291–295.
31.   Hadži, J. Contribution to the knowledge of fauna of cave Vjetrenica. (Pseudoscorpionidea: *Neobisium* (*Blothus*) *vjetrenicae* sp. n., Opilionidea: *Travunia vjetrenicae* sp. n., *Nelima troglodytes* Roewer). *Glas. Srp. Kralj. Akad.* **1932**, *75*, 103–157.
32.   Karaman, S.L. Beitrag zur Kenntnis der Süsswasser-Amphipoden. *Prirodosl. Razpr.* **1932**, *2*, 179–232.
33.   Karaman, S. Über zwei neue Isopoden der Gruppe *Asellus* aus Jugoslavien. *Extr. Recl. Trav. Offer. A J. Georg.* **1933**, *1*, 103–113.
34.   Stammer, H.J. Einige seltene oder neue Höhlentiere. *Zool. Anz.* **1933**, *6*, 263–266.
35.   Wolf, B. *Animalium Cavernarum Catalogus. II.—Cavernarum Catalogus*; Junk Verl: Wien, Austria, 1937; p. 616.
36.   Karaman, S. Dve nove vrste podzemnih amfipoda Popova Polja u Hercegovini. O nekim amfipodima—Izopodima Balkana i. njihovoj sistematici [CLXIII]. In *Posebna Izdanja. Odjeljenje Prirodno-Matematickih Nauka*; Srpska Akademija Nauka: Belgrade, Yugoslavia, 1950; pp. 101–118.
37.   Karaman, S. Prilozi poznavanju nifarga Hercegovine i južne Dalmacije. *Prirodosl. Istraživanja* **1952**, *25*, 45–55.
38.   Culver, D.C.; Sket, B. Hotspots of subterranean biodiversity in caves and wells. *J. Cave Karst Stud.* **2000**, *62*, 11–17.
39.   Pretner, E. Kako zaštititi pećinsku faunu Vjetrenice kod Zavale? In Proceedings of the Treći Jugoslavenski Speleološki Kongres, Sarajevo, Jugoslavija, 21–27 June 1962; pp. 169–174.
40.   Cvijić, J. Das Karstphänomen. Versuch einer morphologischen Monographie. *Geogr. Abh.* **1898**, *5*, 217–329.
41.   Bonacci, O. *Karst Hydrology with Special References to the Dinaric Karst*, 1st ed.; Springer: Berlin, Germany, 1987; p. 194.
42.   Ford, D.; Williams, P. *Karst Hydrogeology and Geomorphology*; John Wiley & Sons Ltd.: West Sussex, UK, 2007; p. 562.
43.   Milanović, P. Karst of Eastern Herzegovina, the Dubrovnik Littoral and Western Montenegro. *Env. Earth Sci.* **2015**, *74*, 15–35. [CrossRef]
44.   Milanović, P. *Karst of East Herzegovina and Dubrovnik Littoral*, 1st ed.; Springer: Cham, Switzerland, 2023; p. 316.
45.   Bonacci, O. Hazards caused by natural and anthropogenic changes of catchment area in karst. *Nat. Hazards Earth Syst. Sci.* **2004**, *4*, 655–661. [CrossRef]
46.   Roje-Bonacci, T.; Bonacci, O. The possible negative consequences of underground dam and reservoir construction and operation in coastal karst areas: An example of the hydro-electric power plant (HEPP) Ombla near Dubrovnik (Croatia). *Nat. Hazards Earth Syst. Sci.* **2013**, *13*, 2041–2052. [CrossRef]

47. Milanović, P. *Hidrologija Karsta i Metode Istraživanja, HE 'Trebišnjica'*, 1st ed.; Institut za korišćenje i zaštitu voda na kršu: Trebinje, Jugoslavija, 1979; p. 302.
48. Milanović, P. Uticaj hidrosistema Trebišnjica na režim površinskih i podzemnih voda u Popovom polju. *Naš krš* **1983**, *IX/14-15*, 41–52.
49. Milanović, P. *Water Resources Engineering in Karst*, 1st ed.; CRC Press: Boca Raton, FL, USA, 2004; p. 328.
50. Stevanović, Z.; Eftimi, R. Karstic sources of water supply for large consumers in Southeastern Europe–sustainability, disputes and advantages. *Geol. Croat.* **2010**, *63*, 179–185. [CrossRef]
51. Čučković, S. The influence of the change in the water-course regime of the Trebišnjica water-system on the fauna of the karst underground regions. *Naš Krš* **1983**, *9*, 129–142.
52. Bilandžija, H.; Puljas, S.; Gerdol, M. Hidden from our sight, but not from our impact: The conservation issues of Cave Bivalves. *Preprints* **2021**, e2021050023. [CrossRef]
53. Reier, S.; Bogutskaya, N.; Palandačić, A. Comparative Phylogeography of *Phoxinus*, *Delminichthys*, *Phoxinellus* and *Telestes* in Dinaric Karst: Which factors have influenced their current distributions? *Diversity* **2022**, *14*, 526. [CrossRef]
54. Bakšić, D.; Paar, D.; Buzjak, N.; Lučić, I. Microclimatic monitoring in Vjetrenica Cave, B&H. In Proceedings of the International Scientific Symposium "Man and Karst", Bijakovići, Međugorje, Bosnia and Herzegovina, 13–16 October 2011; Lučić, I., Mulaomerović, J., Eds.; Centre for Karst and Speleology: Sarajevo, Bosnia and Herzegovina, 2011; pp. 13–14.
55. Čičić, S. Geološka građa terena šire okoline Popova polja i pećine Vjetrenica. *Naš krš* **2002**, *35*, 3–16.
56. Resulović, H.; Vlahinić, M. Characteristics of soil and specificities of drainage of karstic field Popovo Polje (Yugoslavia). *Poljopr. I Šumarstvo* **1983**, *29*, 65–79.
57. Spahić, M. Pećina Vjetrenica u Popovu polju–novo shvatanje speleogeneze. *Acta Geogr. Bosniae Et Herzegovinae* **2015**, *4*, 55–67.
58. Sket, B. The nature of biodiversity in hypogean waters and how it is endangered. *Biodivers. Conserv.* **1999**, *8*, 1319–1338. [CrossRef]
59. Bilandžija, H.; Morton, B.; Podnar, M.; Četković, H. Evolutionary history of relict *Congeria* (Bivalvia: Dreissenidae): Unearthing the subterranean biodiversity of the Dinaric Karst. *Front. Zool.* **2013**, *10*, 5. [CrossRef]
60. Kupriyanova, E.K.; Ten Tove, H.A.; Sket, B.; Zakšek, V.; Trontelj, P.; Rouse, G.W. Evolution of the unique freshwater cave dwelling tube worm *Marifugia cavatica* (Annelida: Serpulidae). *Syst. Biodivers.* **2009**, *7*, 389–401. [CrossRef]
61. Zagmajster, M.; Delić, T.; Prevorčnik, S.; Zakšek, V. New records and unusual morphology of the cave hydrozoan *Velkovrhia enigmatica* Matjašič & Sket, 1971 (Cnidaria: Hydrozoa: Bougainvilliidae). *Nat. Slov.* **2013**, *15*, 13–22.
62. Trontelj, P.; Blejec, A.; Fišer, C. Ecomorphological convergence of cave communities. *Evolution* **2012**, *66*, 3852–3865. [CrossRef] [PubMed]
63. Fišer, C.; Blejec, A.; Trontelj, P. Niche-based mechanisms operating within extreme habitats: A case study of subterranean amphipod communities. *Biol. Lett.* **2012**, *8*, 578–581. [CrossRef]
64. Karaman, G. XXXVIII. Contribution to the Knowledge of the Amphipoda. On the genus *Typhlogammarus* (Schäferna) (fam. Gammaridae) from Yugoslavia. *Fragm. Balc. Musei Maced. Sci. Nat.* **1972**, *9*, 21–34.
65. Zakšek, V.; Sket, B.; Trontelj, P. Phylogeny of the cave shrimp *Troglocaris*: Evidence of a young connection between Balkans and Caucasus. *Mol. Phylogenet. Evol.* **2007**, *42*, 223–235. [CrossRef]
66. Christodoulou, M.; Anastasiadou, C.; Jugovic, J.; Tzomos, T. Freshwater shrimps (Atyidae, Palaemonidae, Typhlocarididae) in the broader mediterranean region: Distribution, life strategies, threats, conservation challenges and taxonomic issues. In *A Global Overview of the Conservation of Freshwater Decapod Crustaceans*; Springer: Cham, Switzerland, 2016; pp. 199–236.
67. Wittmann, K.J.; Ariani, A.P.; Daneliya, M. The Mysidae (Crustacea: Peracarida: Mysida) in fresh and oligohaline waters of the Mediterranean. Taxonomy, biogeography, and bioinvasion. *Zootaxa* **2016**, *4142*, 1–70. [CrossRef]
68. Recknagel, H.; Trontelj, P. From cave dragons to genomics: Advancements in the study of subterranean tetrapods. *BioScience* **2022**, *72*, 54–266. [CrossRef] [PubMed]
69. Recknagel, H.; Zakšek, V.; Delić, T.; Gorički, Š.; Trontelj, P. Multiple transitions between realms shape relict lineages of *Proteus* cave salamanders. *Mol. Ecol.* **2023**, 1–16. [CrossRef]
70. Sket, B. The cave hygropetric-a little known habitat and its inhabitants. *Arch. Für Hydrobiol.* **2004**, *160*, 413–425. [CrossRef]
71. Giachino, P.M.; Vailati, D. *Kircheria beroni*, a new genus and new species of subterranean hygropetricolous Leptodirinae from Albania (Coleoptera, Cholevidae). *Subterr. Biol.* **2006**, *4*, 103–116.
72. Engel, A.S.; Paoletti, M.G.; Beggio, M.; Dorigo, L.; Pamio, A.; Gomiero, T.; Furlan, C.; Brilli, M.; Dreon, A.L.; Bertoni, R.; et al. Comparative microbial community composition from secondary carbonate (moonmilk) deposits: Implications for the *Cansiliella servadeii* cave hygropetric food web. *Int. J. Speleol.* **2013**, *42*, 181–192. [CrossRef]
73. Delić, T. First record of a specialized hygropetricolous cave beetle, genus *Croatodirus* (Coleoptera: Leiodidae), in Slovenia. *Nat. Slo.* **2017**, *19*, 55–61.
74. Lukić, M.; Fišer, C.; Delić, T.; Bilandžija, H.; Pavlek, M.; Komerički, A.; Dražina, T.; Jalžić, B.; Ozimec, R.; Slapnik, R.; et al. Subterranean Fauna of the Lukina Jama–Trojama Cave System in Croatia: The Deepest Cave in the Dinaric Karst. *Diversity* **2023**, *15*, 726. [CrossRef]
75. Vargovitsh, R.S. Cave Water Walker: An Extremely Troglomorphic *Troglaphorura gladiator* gen. et sp. nov. (Collembola, Onychiuridae) from Snezhnaya Cave in the Caucasus. *Zootaxa* **2019**, *4619*, 267–284. [CrossRef]
76. Vargovitsh, R.S. Deep troglomorphy: New arrhopalitidae (Collembola: Symphypleona) of different life forms from the Snezhnaya cave system in the Caucasus. *Diversity* **2022**, *14*, 678. [CrossRef]

77. Moldovan, O.T.; Jalžić, B.; Erichsen, E. Adaptation of the mouthparts in some subterranean Cholevinae (Coleoptera, Leiodidae). *Nat. Croat.* **2004**, *13*, 1–18.

78. Perreau, M.; Pavićević, D. The genus *Hadesia* Müller, 1911 and the phylogeny of Anthroherponina (Coleoptera, Leiodidae, Cholevinae, Leptodirini). In *Advances in the Studies of the Fauna of the Balkan Peninsula*, 1st ed.; Pavićević, D., Perreau, M., Eds.; Institute for Nature Conservation of Serbia: Beograd, Serbia, 2008; pp. 215–239.

79. Polak, S.; Delić, T.; Kostanjšek, R.; Trontelj, P. Molecular phylogeny of the cave beetle genus *Hadesia* (Coleoptera: Leiodidae: Cholevinae: Leptodirini), with a description of a new species from Montenegro. *Arthropod Syst. Phylogeny* **2016**, *74*, 241–254. [CrossRef]

80. Moravec, J.M.; Mlejnek, R.M. *Nauticiella stygivaga* gen. n. et sp. n., a new amphibiontic cavernicolous beetle from the Vjetrenica Cave, Herzegovina (Coleoptera: Leiodidae: Cholevinae: Leptodirini). *Acta Soc. Zool. Bohem.* **2002**, *66*, 293–302.

81. Delić, T.; Lohaj, R.; Brestovansky, J.; Čaha, D.; Jalžić, B. Questioning the monophyly of Anthroherponina (Coleoptera: Leiodidae: Cholevinae: Leptodirini) and description of three new, ecologically ultraspecialized subterranean species. *Zool. J. Linn. Soc.* **2023**, in press.

82. Karaman, G.S. On two new or interesting Amphipoda from Italy and Montenegro (contribution to the knowledge of the Amphipoda 290). *Ecol. Montenegrina* **2016**, *8*, 1–16. [CrossRef]

83. Sket, B.; Dovc, P.; Jalzic, B.; Kerovec, M.; Kucinic, M.; Trontelj, P. A cave leech (Hirudinea, Erpobdellidae) from Croatia with unique morphological features. *Zool. Scr.* **2001**, *30*, 223–229. [CrossRef]

84. Antić, D.; Dražina, T.; Rađa, T.; Lučić, L.; Makarov, S. Review of the genus *Typhloiulus* Latzel, 1884 in the Dinaric region, with a description of four new species and the first description of the male of *Typhloiulus insularis* Strasser, 1938 (Diplopoda: Julida: Julidae). *Zootaxa* **2018**, *4455*, 258–294. [CrossRef] [PubMed]

85. Lakota, J.; Lohaj, R.; Dunay, G. Taxonomical and ecological notes on the genus *Scotoplanetes* Absolon, with the description of a new species from Montenegro (Coleoptera: Carabidae: Trechini). *Nat. Cro.* **2010**, *19*, 99–110.

86. Pavićević, D.; Popović, M. A new species of *Scotoplanetes* Absolon, 1913 (Coleoptera; Carabidae; Trechini) from Durmitor Mt., Montenegro. *Biol. Serb.* **2023**, *45*, 1–5.

87. Absolon, K.; Kratochvíl, J. Zur Kenntnis der höhlenbewohnenden Araneae der illyrischen Karstgebiete. *Mitteilungen Über Höhlen Und Karstforschung* **1932**, *3*, 73–81.

88. Deeleman-Reinhold, C.L. The genus *Rhode* and the harpacteine genera *Stalagtia, Folkia, Minotauria* and *Kaemis* (Araneae, Dysteridae) of Yugoslavia and Crete, with remarks on the genus *Harpactea*. *Rev. Arachnol.* **1993**, *10*, 105–135.

89. Kury, A.B.; Mendes, A.C. Taxonomic status of the European genera of Travuniidae (Arachnida, Opiliones, Laniatores). *Mun. Ent. Zool.* **2007**, *2*, 1–14.

90. Condé, B. *Eukoenenia remyi* n. sp., palpigrade cavernicole d'Herzégovine. *Ann. Spéléol.* **1974**, *29*, 53–56.

91. Dányi, L.; Balázs, G.; Tuf, I.H. Taxonomic status and behavioural documentation of the troglobiont *Lithobius matulici* (Myriapoda, Chilopoda) from the Dinaric Alps: Are there semiaquatic centipedes in caves? *ZooKeys* **2019**, *848*, 1–20. [CrossRef]

92. Zapparoli, M. The present knowledge on the European fauna of Lithobiomorpha (Chilopoda). *Bull. Brit. Myr. Grp.* **2003**, *19*, 20–41.

93. Njunjić, I.; Perreau, M.; Hendriks, K.; Schilthuizen, M.; Deharveng, L. The cave beetle genus *Anthroherpon* is polyphyletic; molecular phylogenetics and description of *Graciliella* n. gen.(Leiodidae, Leptodirini). *Contrib. Zool.* **2016**, *85*, 337–359. [CrossRef]

94. Lohaj, R.; Lakota, J.; Quéinnec, E.; Pavićević, D.; Čeplík, D. Studies on *Adriaphaenops* Noesske with the description of five new species from the Dinarides (Coleoptera: Carabidae: Trechini). *Zootaxa* **2016**, *4205*, 501–531. [CrossRef]

95. IUCN. The IUCN Red List of Threatened Species. 2023. Available online: https://www.iucnredlist.org (accessed on 8 June 2023).

96. Palandačić, A.; Matschiner, M.; Zupančič, P.; Snoj, A. Fish migrate underground: The example of *Delminichthys adspersus* (Cyprinidae). *Mol. Ecol.* **2012**, *21*, 1658–1671. [CrossRef]

97. Mazija, M.; Rnjak, D. Rezultati istraživanja šišmiša u odabranim skloništima na dijeu Popovog polja u općini Ravno, Bosna i Hercegovina. *Hypsugo* **2016**, *1*, 20–29.

98. Zagmajster, M. Zanimivosti Vjetrenice na Popovem polju. *Glas podzemlja* **2006**, *1*, 66–67.

99. Pipan, T.; Culver, D.C. Epikarst communities: Biodiversity hotspots and potential water tracers. *Environ. Geol.* **2007**, *53*, 265–269. [CrossRef]

100. Culver, D.C.; Brancelj, A.; Pipan, T. Epikarst communities. In *Encyclopedia of Caves*, 3rd ed.; White, W.B., Culver, D.C., Pipan, T., Eds.; Academic Press: London, UK, 2019; pp. 399–406.

101. Trontelj, P.; Douady, C.J.; Fišer, C.; Gibert, J.; Gorički, Š.; Lefébure, T.; Zakšek, V. A molecular test for cryptic diversity in ground water: How large are the ranges of macro-stygobionts? *Freshw. Biol.* **2009**, *54*, 727–744. [CrossRef]

102. Lukić, M.; Delić, T.; Pavlek, M.; Deharveng, L.; Zagmajster, M. Distribution pattern and radiation of the European subterranean genus *Verhoeffiella* (Collembola, Entomobryidae). *Zool. Scr.* **2020**, *49*, 86–100. [CrossRef]

103. Hlebec, D.; Podnar, M.; Kučinić, M.; Harms, D. Molecular analyses of pseudoscorpions in a subterranean biodiversity hotspot reveal cryptic diversity and microendemism. *Sci. Rep.* **2023**, *13*, 430.

104. Sket, B. Prilog za zaštitu nekih speleobioloških značajnih objekata. *Naš Krš* **1983**, *9*, 123–127.

105. Piano, E.; Nicolosi, G.; Mammola, S.; Balestra, V.; Baroni, B.; Bellopede, R.; Cumino, E.; Muzzulini, N.; Piquet, A.; Isaia, M. A literature-based database of the natural heritage, the ecological status and tourism-related impacts in show caves worldwide. *Nat. Conserv.* **2020**, *50*, 159–174. [CrossRef]

106. Culver, D.C.; Sket, B. Biological monitoring in caves. *Acta Carsologica* **2002**, *31*, 55–64. [CrossRef]
107. Mulec, J. Lampenflora. In *Encyclopedia of Caves*, 3rd ed.; White, W.B., Culver, D.C., Pipan, T., Eds.; Academic Press: London, UK, 2019; pp. 635–641.
108. Davor Baković, V.D. Direktora Vjetrenice: Rekordna Prošla Godina, Optimistično Čekamo Novu Ljetnu Sezonu. Available online: https://visitbih.ba/davor-bakovic-v-d-direktora-vjetrenice-rekordna-prosla-godina-optimisticno-cekamo-novu-ljetnu-sezonu/ (accessed on 28 July 2023).
109. Paunović, A.S.; Bokić, Z.; Braeković, B. Investigations of the susceptibility of peach cvs, clingstone peaches and nectarines to powdery mildew (*Sphaerotheca pannosa* (Wall) Lev var *persicae*) in ecological conditions of Popovo Polje. *Jugosl. Vocar.* **1985**, *19*, 131–137.
110. Ćustovic, H.; Cero, M. Condition of Ameliorative Issues and Land Consolidation Measures in Area of Popovo Polje Area, Bosnia and Herzegovina; Unpublished Work. Available online: https://www.fao.org/fileadmin/user_upload/reu/europe/documents/LANDNET/2007/Bosnia.pdf. (accessed on 28 July 2023).
111. Darwall, W.; Carrizo, S.; Numa, C.; Barrios, V.; Freyhof, J.; Smith, K. *Freshwater Key Biodiversity Areas in the Mediterranean Basin Hotspot: Informing Species Conservation and Development Planning in Freshwater Ecosystems (Vol. 52)*, 1st ed.; IUCN: Gland, Switzerland, 2014; pp. 1–86.
112. Freyhof, J. Threatened Freshwater Fishes and Molluscs of the Balkan, Potential Impact of Hydropower Projects. ECA Watch Austria & EuroNatur; Unpublished Report. Available online: https://www.balkanrivers.net/uploads/legacy/Threatened-freshwater-fishes-and-molluscs_final.pdf (accessed on 23 June 2023).
113. Oikonomou, A.; Leprieur, F.; Leonardos, I.D. Biogeography of freshwater fishes of the Balkan Peninsula. *Hydrobiologia* **2014**, *738*, 205–220. [CrossRef]
114. Sket, B.; Paragamian, K.; Trontelj, P. A census of the obligate subterranean fauna of the Balkan Peninsula. In *Balkan Biodiversity*, 1st ed.; Griffiths, H.I., Kryštufek, B., Reed, J.M., Eds.; Springer: Amsterdam, The Netherlands, 2004; pp. 309–322.
115. Gerhardt, A. Sensitivity towards Nitrate: Comparison of groundwater versus surface water crustaceans. *J. Soil. Water. Sci.* **2020**, *4*, 112–121. [CrossRef]
116. Di Lorenzo, T.; Di Marzio, W.D.; Fiasca, B.; Galassi, D.M.P.; Korbel, K.; Iepure, S.; Pereira, J.L.; Reboleira, A.S.P.; Schmidt, S.I.; Hose, G.C. Recommendations for ecotoxicity testing with stygobiotic species in the framework of groundwater environmental risk assessment. *Sci. Total Environ.* **2019**, *681*, 292–304. [CrossRef]
117. Kokalj, A.J.; Fišer, Ž.; Dolar, A.; Novak, S.; Drobne, D.; Bračko, G.; Fišer, C. Screening of NaCl salinity sensitivity across eight species of subterranean amphipod genus *Niphargus*. *Ecotoxicol. Environ. Saf.* **2022**, *236*, 113456. [CrossRef] [PubMed]
118. Fagan, A.; Sircar, I. Environmental politics in the Western Balkans: River basin management and non-governmental organisation (NGO) activity in Herzegovina. *Env. Polit.* **2010**, *19*, 808–830. [CrossRef]
119. Naimark, N.M.; Case, H. *Yugoslavia and Its Historians: Understanding the Balkan Wars of the 1990s*, 1st ed.; Stanford University Press: Stanford, CA, USA, 2003; p. 296.
120. Ozimec, R. Otvoren prvi biospeleološki muzej u svijetu: Biospeleološki muzej Vjetrenica. *Subterranea Croat.* **2016**, *14*, 49–50.
121. World Heritage Convention Tentative List (Vjetrenica Cave). Available online: https://whc.unesco.org/en/tentativelists/1975/ (accessed on 15 June 2023).
122. Milanović, P.; Glišić, R.; Đorđević, B.; Dašić, T.; Sudar, N. Uticaj delimičnog prevođenja voda iz slivova Bune i Bregave u sliv Trebišnjice. *Vodoprivreda* **2012**, *44*, 255–257.
123. Mirković, U.B. Techno-economic analysis for the construction of the supply tunnel of hpp "Dabar" with the application of TBM technology. *Tehnika* **2019**, *74*, 359–366. [CrossRef]
124. Posljedice Megaprojekta "Gornji Horizonti": Ostaje li Hercegovina bez Vode i Kako Izbjeći Apokalipsu. Available online: https://www.klix.ba/vijesti/bih/posljedice-megaprojekta-gornji-horizonti-ostaje-li-hercegovina-bez-vode-i-kako-izbjeci-apokalipsu/200916105 (accessed on 28 July 2023).
125. Projekat Gornji Horizonti: Početak Kraja Neretve i Hercegovine Kakvu Poznajemo? Available online: https://abrasmedia.info/projekat-gornji-horizonti-pocetak-kraja-neretve-i-hercegovine-kakvu-poznajemo/ (accessed on 28 July 2023).
126. Fišer, C.; Borko, Š.; Delić, T.; Kos, A.; Premate, E.; Zagmajster, M.; Altermatt, F. The European green deal misses Europe's subterranean biodiversity hotspots. *Nat. Ecol. Evol.* **2022**, *6*, 1403–1404. [CrossRef] [PubMed]
127. Brunke, M. Colmation and depth filtration within streambeds: Retention of particles in hyporheic interstices. *Int. Rev. Hydrobiol.* **1999**, *84*, 99–117. [CrossRef]
128. Brenna, A.; Surian, N.; Mao, L. Alteration of gravel-bed river morphodynamics in response to multiple anthropogenic disturbances: Insights from the sediment-starved Parma River (northern Italy). *Geomorphology* **2021**, *389*, 107845. [CrossRef]
129. Borko, Š.; Trontelj, P.; Seehausen, O.; Moškrič, A.; Fišer, C. A subterranean adaptive radiation of amphipods in Europe. *Nat. Commun.* **2021**, *12*, 3688. [CrossRef]

Article

Subterranean Fauna of the Lukina Jama–Trojama Cave System in Croatia: The Deepest Cave in the Dinaric Karst

Marko Lukić [1,2], Cene Fišer [3], Teo Delić [3], Helena Bilandžija [1,2], Martina Pavlek [1,2], Ana Komerički [1], Tvrtko Dražina [1,4], Branko Jalžić [1], Roman Ozimec [2,5], Rajko Slapnik [1,6] and Jana Bedek [1,2,*]

1   Croatian Biospeleological Society, Rooseveltov trg 6, 10000 Zagreb, Croatia
2   Ruđer Bošković Institute, Bijenička cesta 54, 10000 Zagreb, Croatia
3   SubBio Lab, Department of Biology, Biotechnical Faculty, University of Ljubljana, Jamnikarjeva 101, 1000 Ljubljana, Slovenia
4   Faculty of Science, Department of Biology, Division of Zoology, University of Zagreb, Rooseveltov trg 6, 10000 Zagreb, Croatia
5   ADIPA: Society for Research & Conservation of Croatian Natural Diversity, Orehovečki ogranak 37, 10000 Zagreb, Croatia
6   ZOSPEUM, Molluscs, Cave & Karst Biological Consulting, Drnovškova pot 2, Mekinje, 1240 Kamnik, Slovenia
*   Correspondence: jana.bedek@hbsd.hr

Abstract: The Dinaric Karst is a global hotspot for subterranean diversity, with two distinct peaks of species richness in the northwest and southeast, and an area of a lower species richness in the central part. In this article, we present a species list and describe the ecological conditions of the Lukina jama–Trojama cave system, located in the central part of the Dinaric Karst. This cave system is the deepest and one of the most logistically challenging cave systems sampled so far in the Dinaric Karst. Repeated sampling resulted in a list of 45 species, including 25 troglobionts, 3 troglophiles, 16 stygobionts, and 1 stygophile. Most of the recorded species are endemic to the Velebit Mountain, while three species are endemic to the Lukina jama–Trojama cave system. Within the system, species richness peaks in the deepest third of the cave, most likely reflecting the harsh ecological conditions in the upper parts, including ice, cold winds, and occasional waterfalls. Milder and more stable deeper parts of the cave contain a rich subterranean species community, part of which is associated with two very distinct aquatic habitats, the cave hygropetric and the phreatic zone. The newly recognized hotspot of subterranean biodiversity in the central Dinaric Karst, which has emerged between the two known centers of biodiversity, further highlights the species richness in large cave systems, but also challenges the diversity patterns in the Dinaric Karst overall.

Keywords: Velebit Mt.; biospeleology; biodiversity; checklist; cave hygropetric; obligate cave species; troglobionts; stygobionts

Citation: Lukić, M.; Fišer, C.; Delić, T.; Bilandžija, H.; Pavlek, M.; Komerički, A.; Dražina, T.; Jalžić, B.; Ozimec, R.; Slapnik, R.; et al. Subterranean Fauna of the Lukina Jama–Trojama Cave System in Croatia: The Deepest Cave in the Dinaric Karst. *Diversity* **2023**, *15*, 726. https://doi.org/10.3390/d15060726

Academic Editor: Salvidio Sebastiano

Received: 30 March 2023
Revised: 16 May 2023
Accepted: 17 May 2023
Published: 31 May 2023

1. Introduction

Sampling subterranean fauna and describing the subterranean communities are challenging tasks. The majority of subterranean species live in habitats inaccessible to humans, such as permanently flooded zones or systems of narrow fissures, and voids in fractured rock [1]. Hence, species inventories remain incomplete even for the most well-known and best-explored cave systems. Restricted access to the subterranean environment, the so-called Racovitzan impediment [2], but see [3–5], remains the main obstacle in biospeleological research.

Caves are the easiest access points where humans can enter soluble rocky massifs and explore subterranean diversity. Compiling species lists demands coordinated actions of repeated collecting visits and taxonomic expertise. This is particularly challenging in large systems, especially those that traverse entire massifs and include deep vertical pits that require both technical expertise and psycho-physical preparedness. Large cave systems often contain high levels of habitat heterogeneity, which is an important predictor of species richness [6,7]. These systems can harbor a variety of terrestrial and aquatic habitats, including fissure systems, cave hygropetric or a permanently flooded phreatic zone. Moreover, deep cave systems experience strong depth-dependent environmental gradients in temperature, moisture, and food availability. Such gradients could, in theory, allow for species to spatially segregate at different altitudes [8–10]. Species inventories from such systems are particularly rare (see [8,11]); however, they remain the prime hotspot candidates, especially when located in a species-rich region.

The Dinaric Karst is a 650 km long limestone mountainous massif, that rises in the Western Balkans along the Eastern Adriatic coast. The entire region is recognized as one of the global hotspots of subterranean biodiversity [12–15]. Species richness along the Dinaric Karst is not evenly distributed and peaks in the northwest and southeast [4,5,16–18]. Four caves from Slovenia and Bosnia and Herzegovina are included in the list of subterranean biodiversity hotspots [19], while the area between the two centers, predominantly situated in Croatia, seems to be less species-rich [6,18,20].

Extensive sampling of Croatian caves over the last three decades, conducted mainly by researchers from the Croatian Biospeleological Society, has led to the recognition of several species-rich regions, e.g., the Ogulin–Plaški plateau dominated by subterranean aquatic species [18,21], and the Biokovo and Velebit Mountains, containing exceptionally rich terrestrial subterranean fauna [21]. However, the species inventories of these regions remain unpublished or are scattered across taxonomic papers, books, reports, or the so-called grey literature. For this reason, none of the caves in the central part of the Dinaric Karst have been listed as subterranean biodiversity hotspot, i.e., a cave inhabited by at least 25 aquatic and terrestrial obligate subterranean species [19].

Herein, we fill this gap by providing a species checklist for one of the deepest cave systems in the world, the Lukina jama–Trojama cave system (−1431 m), situated in the northern part of the Velebit Mt. In the years following its discovery, this cave system was the focus of both speleological and biospeleological research, resulting in numerous publications. However, while the speleological achievements were published in a comprehensive overview, e.g., [22], the biological data remained unpublished or scattered in narrowly focused scientific papers. In this paper, we compile both published and unpublished data, provide a comprehensive checklist of the fauna of the Lukina jama-Trojama cave system, and discuss its ecological and biogeographical significance.

2. Study Area

2.1. Northern Velebit

The Northern Velebit is the northernmost part of the 145 km long Velebit Mountain range, which stands out as the largest massif in the Dinaric Karst [23]. The uplifted karst plateau of the northern Velebit is mainly limestone and is composed of Jurassic carbonate rocks and massive calcareous breccias emerging from the Upper Paleogene to the Lower Neogene age [24]. The highest peaks of the plateau are 1600–1700 m a.s.l., while elevations of the karst poljes in the hinterland range from 400 to 500 m a.s.l. (Figure 1). The Lika and Gacka rivers flow across these karst poljes, and sink beneath the eastern Velebit foothills, resurging as coastal or even submarine springs along the Adriatic coast. Annual precipitation in the northern Velebit massif reaches up to 3000 mm [25], with a rapid water transfer through the limestone bedrock. The mean annual temperature on the plateau ranges from 3 to 8 °C, depending on geomorphological setting or elevation [25,26]. The distinctive surface relief with picturesque landscapes and the great diversity of karst features, including large collapsed dolines and extremely deep caves, together with the high biodiversity, were the reasons for the designation of the Northern Velebit National Park in 1999 [23].

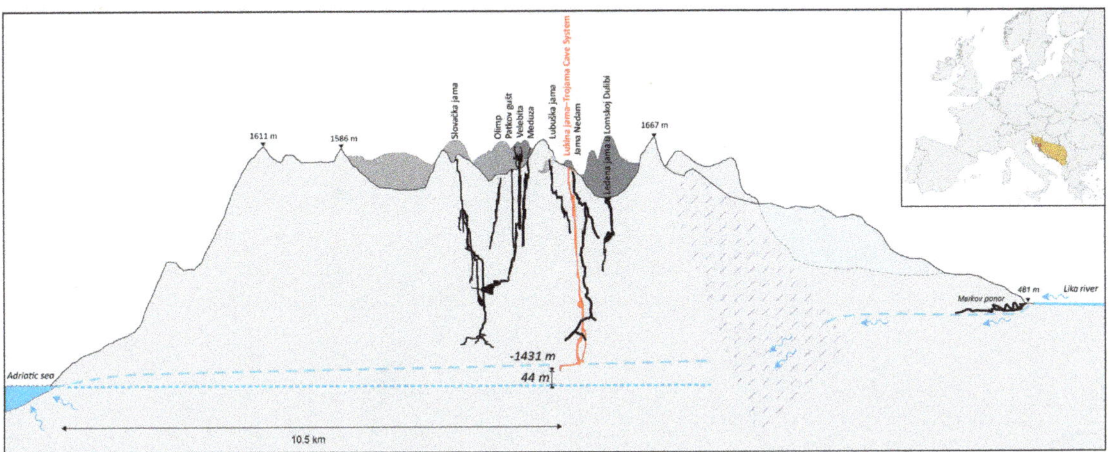

Figure 1. Cross-section of the northern Velebit, with sections of the deep caves and one of the sinkholes of the Lika river. The position of Lukina jama–Trojama cave system and its cross-section is emphasized in red; the area of the Dinaric Karst is marked in orange in the inset map. Modified from the original after Darko Bakšić. Used with permission.

More than 350 caves have been explored in the northern Velebit [27], but the caves for which the area is famous are its deep caves. Four caves exceed the depth of −1000 m and five others exceed the depth of −500 m (Figure 1). The vertical parts of the caves are predominantly of vadose origin, related to karstification processes of unbedded breccia [27], while only the deepest parts of the deepest caves might be of phreatic or epiphreatic origin. Large chambers (50–100 m in diameter) found in several caves at approximately 500 m a.s.l. [27] remain of uncertain origin and were, presumably, largely modified by collapse processes [26]. These chambers have a variety of terrestrial and aquatic habitats, and are important sampling sites for subterranean fauna. Among the deep caves of Velebit, the Lukina jama–Trojama cave system (−1431 m) stands out as the deepest cave in the whole Dinaric Karst.

2.2. The Lukina Jama, Trojama Cave System

The Lukina jama–Trojama cave system is located in the Hajdučki kukovi strict nature reserve in the Northern Velebit National Park. It was discovered in 1992 and explored in the following two years to a depth of −1392 m. At the time of its initial exploration, it was the 10th deepest cave in the world, and exploration of the sump at the bottom was one of the deepest dives in caves for a long time. It has been the subject of biospeleological, geological, hydrogeological, meteorological, and physical studies [22,26,28–33]. It is important from a historical point of view, as its discovery marked the initiation of intense speleological explorations of Velebit Mt., which resulted in the discovery of numerous impressive caves in the following decades.

The cave system is currently −1431 m deep and 3741 m long, and has two entrances (Figure 2): Trojama and Lukina jama, which open at 1475 and 1440 m a.s.l., respectively. The two cave channels join at about −550 m and form an extremely vertical cave morphology with a continuing sequence of shafts and very few chambers and ledges. The largest chamber, with a diameter of about 80 m, is located at −980 m (Figure 3). At the bottom, at a depth of about −1370 m, there is a small chamber with a lake and a sump (100 m a.s.l.). The phreatic channel was explored 120 m in length and a total depth of 60 m [34].

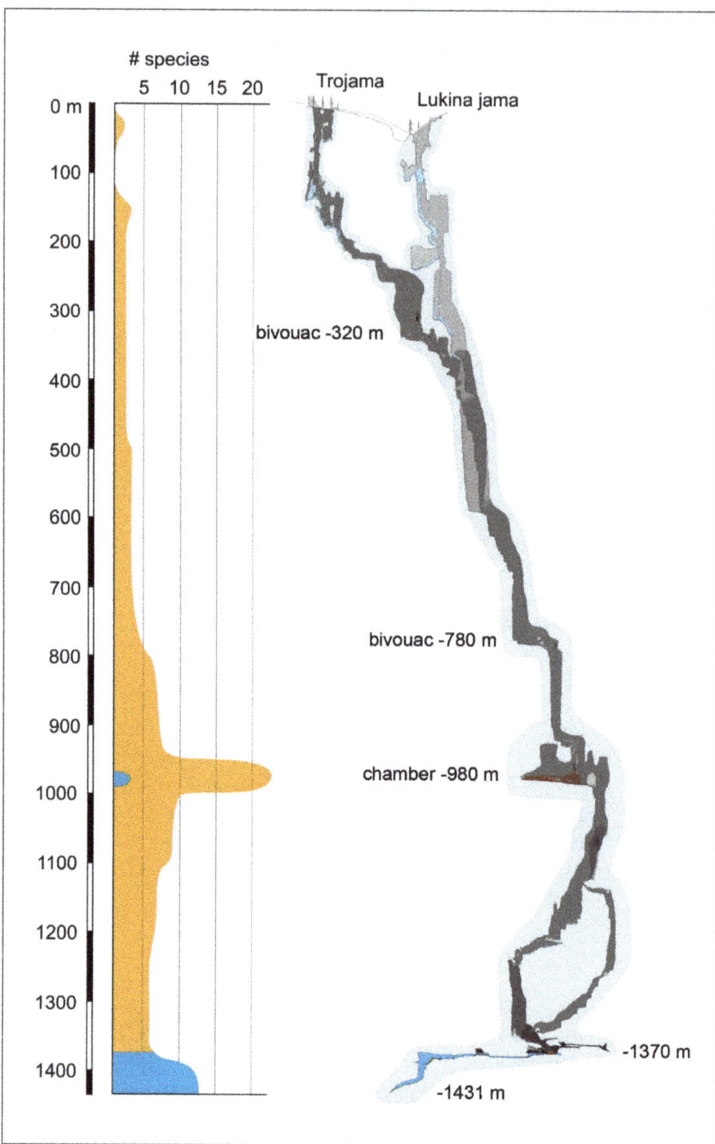

Figure 2. Cross-section of the Lukina jama–Trojama cave system with a histogram of vertical species richness distribution (orange, terrestrial species; blue, aquatic species). Cave map modified from the original after Darko Bakšić. Used with permission.

Studies of water dynamics and air circulation in the system suggest that they significantly alter the temperature profile of the system [26]. Two entrances and their morphology are open to a strong inflow of cold air in winter and a much weaker outflow in summer, resulting in the accumulation of ice in the upper parts of the cave. Intense air circulation ceases at a depth of 500 m. The snow, ice and ice crust accumulated in the passages reach a depth of 320 m in the Lukina jama branch, and 200 m in the Trojama branch (Figure 2). The cave system has a bimodal thermal gradient. From the top to a depth of 200 m, temperature gradually decreases from 4 °C to 0 °C in the summer, and from that point downward, rises again at the rate 0.39 °C/100 m, eventually reaching 5 °C at the bottom of the cave.

Figure 3. The large chamber at −980 m with bivouacs for researchers, the species-richest part of the system. Photo by Vedran Jalžić. Used with permission.

The upper parts of the system are fed mainly by percolating water, resulting in the first vadose streams just at −980 m and −1200 m. Therefore, the water regime of the system is mainly influenced by surface precipitation and snowmelt. During the rainstorms, some drip water areas turn into waterfalls, with a discharge of several tens to several hundreds of liters per second, especially in the lower parts of the cave [26]. The phreatic water in the siphon lake is under the influence of the vadose stream within the cave and broader watershed inflows [26]. Water tracing experiments in the northern Velebit revealed extremely complex groundwater drainage patterns [35]. The occasional inflow of groundwater from the broader area causes major floods in the bottom chamber and the channel above it. This was confirmed by year long water level monitoring, where the water rose rapidly by more than 100 m on one occasion, and by more than 20 m on another four occasions [26]. The water temperature of the vadose stream in the chamber at −980 m is constant and is around 3.1 °C, while in the siphon, it is between 4.5 and 7.5 °C [26]. Although drip water is present in many places throughout the cave, a permanent thin water film flowing over the walls, also known as the cave hygropetric [36], is present in the chamber at −980 m and near the syphone lake. Although these remain the only sites where it is reachable, it is possible that the cave hygropetric is also present in the upper parts, but remains inaccessible to researchers.

3. Materials and Methods

The biospeleological data presented in this paper have been accumulated over the course of three decades during caving expeditions in the Lukina Jama–Trojama cave system. In the 1990s, fauna was collected by a few cavers and biologists while they were exploring and mapping new channels. In more recent expeditions (2010, 2011, and 2013), dedicated teams of biospeleologists had each spent several days in the cave, collecting and surveying the subterranean biodiversity. Most of the data presented here were gathered during these expeditions.

Cave fauna was collected by hand using forceps in all the explored habitats. Baits were placed on the shaft walls near the ropes and in the horizontal parts of the channel. Baited pitfall traps were placed with an aqueous solution of sodium chloride as fixative. Adhesive tapes were used to collect the flying dipteran *Troglocladius hajdi*. Aquatic fauna was collected by hand using the plankton net, both in the vadose stream and phreatic water at the bottom. The fauna in the siphon lake was collected with traps for aquatic fauna and a Sket's bottle [37] during cave diving.

The collected material was identified to the lowest possible taxonomic level, and classified into the following categories: stygo-/troglobionts, stygo-/troglophiles, and stygo-/trogloxenes (Table 1). The categories were defined, following the definition of Sket [38], with a slight modification (we use the prefixes "stygo" instead of "aquatic troglo" and stygo-/troglophiles instead of eustygo-/eutroglophiles). Subtroglophiles and trogloxenes were scarce and not listed in the checklist.

Collection abbreviations: CBSS coll.—Croatian Biospeleological Society collection, Zagreb, Croatia; RO coll.—Roman Ozimec collection, Zagreb, Croatia; RS coll.—Rajko Slapnik collection, Slovenia; NHM coll.—Croatian Natural History Museum collection, Zagreb, Croatia; SubBio Lab coll.—SubBio Lab collection, Department for Biology, Biotechnical Faculty, University of Ljubljana, Slovenia.

4. Results

4.1. Checklist of the Lukina Jama–Trojama Cave System

The Lukina jama–Trojama cave system is both the best-studied and the most species-rich cave in the northern Velebit. The species list comprises 25 troglobionts, 16 stygobionts, 3 troglophiles, and 1 stygophile, listed in Table 1.

Table 1. Cave fauna of the Lukina jama–Trojama cave system. Ecology abbreviations: Sb–stygobiont; Sp–stygophile; Tb–troglobiont; Tp–troglophile. Species described from the Lukina jama–Trojama are marked with #. Undescribed species and genera are marked with *.

Taxonomic Group	Taxon	Ecology	Depth (m)	Distribution	Source
Porifera: Spongillidae	*Eunapius subterraneus* Sket and Velikonja, 1984 [39]	Sb	>1370	NW Dinarides	[32]
Gastropoda: Hydrobiidae	*Hauffenia* sp.	Sb	>1370	Velebit Mt.	[22]
Gastropoda: Hydrobiidae	*Lanzaia* sp.	Sb	>1370	Velebit Mt.	[22]
Gastropoda: Hydrobiidae	*Sadleriana* sp.	Sb	>1370	Velebit Mt.	RS coll.
Gastropoda: Acroloxidae	*Acroloxus* sp.	Sb?	>1370	NA	RS coll.
Gastropoda: Carychiidae	*Zospeum isselianum* Pollonera, 1887 [40]	Tb	800–1370	NW Dinarides	RS coll.
Gastropoda: Carychiidae	*Zospeum tholussum* Weigand, 2013 [31] #	Tb	980	northern Velebit	[31]
Gastropoda: Carychiidae	*Zospeum subobesum* Bole, 1974 [41]	Tb	800–1370	Dinarides	[22]
Gastropoda: Carychiidae	*Zospeum robustum* Inäbnit, Jochum & Neubert, 2019 [42]	Tb	800–1370	NW Dinarides	[42]
Bivalvia: Dreissenidae	*Congeria jalzici* Morton and Bilandžija, 2013 [43]	Sb	>1370	NW Dinarides	[43]
Polychaeta: Serpulidae	*Marifugia cavatica* Absolon and Hrabe, 1930 [44]	Sb	>1370	Dinarides	[22]
Clitellata: Erpobdellidae	*Croatobranchus mestrovi* Kerovec, Kučinić and Jalžić, 1999 [30] #	Sb	1370	northern Velebit	[30]
Clitellata: Erpobdellidae	*Dina.* sp. *	Sb	980	northern Velebit	SubBio Lab coll.
Clitellata: Haplotaxidae	*Haplotaxis* cf. *H. gordioides* (Hartmann, in Oken 1819) [45]	Sp	980	Holarctic	CBSS coll.
Palpigradi: Eukoeneniidae	*Eukoenenia* sp. *	Tb	980	northern Velebit	RO coll.
Acari: Rhagidiidae	*Rhagidia* sp. *	Tb	980	northern Velebit	[32]
Acari: Labidostommatidae	*Nicoletiella* sp. *	Tb	980	northern Velebit	[32]
Araneae: Dysderidae	cf. *Stalita* sp. *	Tb	980	Velebit Mt.	CBSS coll.

Table 1. *Cont.*

Taxonomic Group	Taxon	Ecology	Depth (m)	Distribution	Source
Pseudoscorpiones: Neobisiidae	*Neobisium* sp. *	Tb	980	northern Velebit	[32]
Opiliones: Nemastomatidae	*Hadzinia* sp. *	Tb	980	northern Velebit	RO coll.
Opiliones: Sironidae	*Cyphophthalmus* sp. *	Tb	980	northern Velebit	RO coll.
Isopoda: Trichoniscidae	*Androniscus* sp.	Tp	30	NA	CBSS coll.
Isopoda: Trichoniscidae	*Alpioniscus velebiticus* Bedek and Taiti, 2019 [46]	Tb	980–1370	Velebit	[46]
Isopoda: Trichoniscidae	Gen. * sp. *	Tb	800–980	Velebit	CBSS coll.
Isopoda: Asellidae	*Proasellus* cf. *P. slovenicus* (Sket, 1957) [47]	Sb	>1370	NW Dinarides	SubBio Lab coll.
Amphipoda: Niphargidae	*Niphargus arbiter* Karaman, 1984 [48]	Sb	>1370	Velebit Mt. and Lika	[32]
Amphipoda: Niphargidae	*Niphargus brevirostris* Sket, 1971 [49]	Sb	>1370	Velebit Mt. and Lika	[32]
Amphipoda: Niphargidae	*Niphargus croaticus* Jurinac, 1887 [50]	Sb	>1370	NW Dinarides	[22]
Amphipoda: Niphargidae	*Niphargus* sp. *	Sb	>1370	Lukina jama–Trojama	CBSS coll. & SubBio Lab coll.
Amphipoda: Niphargidae	*Chaetoniphargus lubuskensis* Karaman G.S. and Sket, 2019 [51]	Sb	980	northern Velebit	[51]
Decapoda: Atyidae	*Troglocaris* cf. *T. kapelana* Sket and Zakšek, 2009 [52]	Sb	>1370	Velebit Mt. and Lika	SubBio Lab coll.
Diplopoda: Polydesmidae	*Brachydesmus* sp.	Tp	30	NA	[32]
Diplopoda: Anthogonidae	*Haasia stenopodium* (Strasser, 1966) [53]	Tb	500–1370	NW Dinarides	[22]
Chilopoda: Geophilidae	*Geophilus hadesi* Stoev, Akkari, Komerički, Edgecombe and Bonato 2015 [32]	Tb	980–1100	Velebit Mt.	[32]
Collembola: Onychiuridae	Gen. sp. *	Tb	980–1370	northern Velebit	CBSS coll.
Collembola: Oncopoduridae	*Oncopodura* sp. *	Tb	980	Velebit Mt.	CBSS coll.
Collembola: Isotomidae	*Parisotoma* sp. *	Tb	980	northern Velebit	CBSS coll.
Collembola: Isotomidae	Gen. * sp. *	Tb	980	Lukina jama–Trojama	[32]
Collembola: Sminthuridae	*Disparrhopalites* sp. *	Tb	980	Velebit Mt.	[32]
Diplura: Campodeidae	*Plusiocampa* sp.	Tb	1370	northern Velebit	CBSS coll.
Coleoptera: Cholevidae	*Astagobius angustatus* (Schmidt, 1852) [54]	Tb	150–800	NW Dinarides	[22]
Coleoptera: Cholevidae	*Spelaeodromus pluto* (Reitter, 1881) [55]	Tb	150–800	Velebit Mt. and Lika	[32]
Coleoptera: Cholevidae	*Velebitodromus smidai* Casale, Giachino and Jalžić 2004 [56]	Tb	860–1200	northern Velebit	[56]
Diptera: Chironomidae	*Troglocladius hajdi* Andersen, Baranov and Hagenlund, 2016 [33] #	Tb	800–980	Lukina jama–Trojama	[33]
Diptera: Mycetophilidae	*Speolepta leptogaster* (Winnertz, 1863) [57]	Tp	980	Europe	CBSS coll.

4.2. The Fauna

The Lukina jama–Trojama is inhabited by a large number of endemic and evolutionary unique species (Figures 4–7), of which the ecological, phylogenetical and biogeographical significances are discussed below.

Scattered individuals of sponges were found in the middle part of the sump at the bottom of Lukina jama–Trojama. Using a molecular genetics approach, the sponges were identified as *Eunapius subterraneus* (Figure 6A). This stygobiotic species is distributed in the subterranean waters of the Dobra and Mrežnica rivers in the nearby Ogulin–Plaški valley and the adjacent regions [58]. The population in the underground parts of the Lika River basin, including Lukina jama–Trojama and Markov ponor, is the only one in the Adriatic Sea basin (all others are in the Black Sea basin). This species is known to occur in large aggregations and has a diverse habitus, ranging from small round to large flattened [58]. The sponges of Lukina jama–Trojama are up to 2 cm in diameter, spherical in shape, and have an irregular, wrinkled surface.

Figure 4. Troglobionts of the Lukina jama–Trojama cave system: (**A**) snail *Zospeum tholussum*; (**B**) Dysderidae spider; (**C**) harvestmen *Hadzinia* sp.; (**D**) mite *Rhagidia* sp.; (**E**) millipede *Haasia stenopodium*; and (**F**) centipede *Geophilus hadesi*. Photo by (**A,C,E,F**) Jana Bedek; (**B**) Branko Jalžić, and (**D**) Martina Pavlek.

Gastropods are the species-richest taxonomic group in this system. The empty shells of four stygobiotic species assigned to the genera *Hauffenia*, *Lanzaia*, *Sadleriana*, and *Acroloxus* were found in the sediment of the phreatic zone. In addition, four troglobiotic species of the family Carychiidae were recorded from −800 m to the bottom of the system: *Zospeum tholussum* (Figure 4A), which is also described in the system, *Z. isselianum*, *Z. subobesum*, and *Z. robustum*.

The cave-dwelling bivalve genus *Congeria* is a Miocene relict and endemic to the Dinaric Karst. It is one of the few cave-adapted bivalves found in the world [43,59]. Molecular phylogenetic analysis showed that the specimens found in the bottom sump belong to *C. jalzici* (Figure 6B), a species known from the sinkholes in the Lika River basin and a single submerged cave (spring) in Slovenia [43]. However, the morphological characteristics of the population from the Lukina jama–Trojama are surprising and unique. Their shell is thin and rounded in the ventral part, while all the other *Congeria* populations, regardless of species, have a much more robust and ventrally flattened shell. The particular morphology of the specimens is probably influenced by the specific environmental conditions in the sump. The absence of a turbulent water flow has possibly led to the development of a fragile, rounded shell.

Figure 5. Troglobionts of the Lukina jama–Trojama cave system: (**A**) springtail *Disparrhopalites* sp.; (**B**) springtail *Parisotoma* sp.; (**C**) beetle *Astagobius angustatus*; and (**D**) dipteran *Troglocladius hajdi*. Photo by (**A**) Marko Lukić; (**B,D**) Jana Bedek; and (**C**) Tin Rožman.

The cave tubeworm, *Marifuria cavatica*, is the only polychaete of the family Serpulidae adapted to freshwater caves (Figure 6C). *Marifugia* is a Pliocene relict and is widely distributed in the underground waters of the Dinaric Karst [60]; however, on the Velebit mountain, it is known only from two deep caves: Lukina jama–Trojama and Nedam. In certain areas, it forms massive colonies known as *Marifugia* deposits or Marifugia-tufa, which are classified as a special type of subterranean habitat, because it creates a substrate for many other species [61]. The population in Lukina jama–Trojama is not very large, but interesting because animals build their tubes perpendicular to the walls (Figure 6C). So, here they stick out into the water, unlike in most other places, where they are firmly attached to the walls with the entire length of the tube. This is another indication that the water flow in the sump is calm and steady.

Figure 6. Stygobionts of the Lukina jama–Trojama cave system: (**A**) sponge *Eunapius subterraneus*; (**B**) bivalve *Congeria jalzici*; (**C**) tubeworm *Marifugia cavatica*; (**D**) isopod *Proasellus* cf. *P. slovenicus*; (**E**) amphipod *Niphargus croaticus*; and (**F**) decapod *Troglocaris* cf. *T. kapelana*. Photo by (**A**) Vedran Jalžić; (**B**) Helena Bilandžija; (**C,F**) Jana Bedek; and (**D,E**) Tvrtko Dražina and Ana Komerički.

Two stygobiotic species of leeches belong to the family Erpobdellidae. *Croatobranchus mestrovi* was described from the Lukina jama–Trojama cave system (Figure 7D) [30] and is only known from four other deep caves of northern Velebit: Velebita cave system, Slovačka jama, Olimp, and Nedam. It is a peculiar species, easily recognizable by branchiae-like lateral processes, which made this leech a symbol of deep caves in northern Velebit. It has been found in the cave hygropetric or in drip pools with a water flow, with temperatures ranging from 4 to 6 °C [62]. They often move actively along the cave walls in a weak water flow, with their anterior part facing the water flow. The second species found in this system, based on preliminary data, is a new, yet undescribed species of the genus *Dina*.

Although spiders are a very species-rich group in the Dinaric caves, only one species is recorded for the system. The single specimen collected is actually a remnant of a dead spider, found in a chamber at −980 m. It belongs to the Dysderidae family, all of which are active hunters, meaning that they do not spin webs. This new, undescribed species is eyeless, and found in several other deep caves on Velebit Mt (Figure 4B).

Pseudoscorpions are highly diversified in the caves of Dinaric Karst; however, only one troglobiotic species of the genus *Neobisium* is recorded for the system. The species is also known from numerous caves on the high mountain plateau of the northern Velebit. It is probably a new, yet undescribed species morphologically similar to *N. svetovidi*. Recent molecular analyses from many caves of the Velebit Mt. revealed large intraspecific distances for this species [63].

Figure 7. Evolutionary and ecologically unique cave hygropetric specialists of the Lukina jama–Trojama cave system: (**A**) beetle *Velebitodromus smidai*; (**B**) amphipod *Chaetoniphargus lubuskensis*; (**C**) an undescribed isotomid springtail; (**D**) leech *Croatobranchus mestrovi*; (**E**) isopod *Alpioniscus velebiticus*; and (**F**) an undescribed trichoniscid isopod. Photo by (**A**) Branko Jalžić; (**B**) Ana Komerički and Tvrtko Dražina; and (**C–F**) Jana Bedek.

Harvestmen are represented by two new and undescribed species, attributed to the genera *Cyphophthalmus* and *Hadzinia* (Figure 4C). Both species are troglobiotic and troglomorphic. *Hadzinia* was also recorded in the caves on the northern slopes of the Velebit Mt.

Four isopod species from the families Trichoniscidae and Asellidae were recorded for the system. In the uppermost part of the cave, only females of the genus *Androniscus* were found, which probably belong to the species *A. roseus*, already recorded on the northern Velebit [64]. Two species endemic to the Velebit Mt., *Alpioniscus velebiticus* and a yet undescribed trichoniscid, were found in or near the cave hygropetric. *Alpioniscus velebiticus* (Figure 7E) is known across the Velebit Mt. mostly from the caves situated at higher altitudes [46]. The undescribed trichoniscid (Figure 7F) is known only from two caves in the northern Velebit, where it is associated to the cave hygropetric, and one cave in the southern Velebit, where it was recorded in the cave ponds. The preliminary identification of the aquatic *Proasellus*, suggests that it might belong to the species *P. slovenicus* (Figure 6D), which could extend the range of the species more than 100 km to the south.

Amphipods are represented by five species of the family Niphargidae. Two species, *Niphargus arbiter* and *N. croaticus* (Figure 6E), are large-bodied species, found in the lake at the bottom of the cave, and are distributed over a broader region [65–67]. Two other species, *N. brevirostris* and *Niphargus* sp., are smaller and more narrowly distributed. The former is a stouter species known from a broader region of Lika [49], while the latter is slender, presumably living in the flowing water, and may represent a new, yet undescribed species. The fifth species of the family has been described as *Chaetoniphargus lubuskensis* (Figure 7B) [51]; however, molecular phylogeny unambiguously places it within the genus *Niphargus* [68]. It was found in the cave hygropetric and in a small water pond in the chamber at −980 m. Its mouthparts imply it likely feeds on biofilm in percolating water or in the cave hygropetric, similarly to the genus *Niphargobates* [36].

Decapod crustaceans are represented by a single stygobiotic species, morphologically close to *Troglocaris kapelana*, found in the phreatic zone of the system (Figure 6F). This species has a wider distribution, including the areas of Velika Kapela Mt. [52] and the Lika region (unpublished data).

Millipedes are represented by two species. The most common is *Haasia stenopodium* (Figure 4E), found from −500 m to the bottom of the system. This genus, which encompasses only troglobiotic species, is endemic to NW Dinarides, with *H. stenopodium* having the largest range, from Nanos Mountain in western Slovenia, to the southeastern Velebit [69]. Juvenile specimens of the genus *Brachydesmus* were collected in the entrance part of Trojama, at a depth of −30 m.

The single centipede species recorded for the system is *Geophilus hadesi* (Figure 4F). This is only the second known troglobiotic species of the genus that is common in endogean habitats. It was collected in the chamber at −980 m, and also observed at −1100 m, but out of reach for collection. The species exhibits high troglomorphism and is known only from three deep caves on Velebit Mt. [32].

Five troglobiotic springtails, all new and yet undescribed species, are known from the system. Two species from the family Isotomidae are one of the few troglobiotic species of the family in the world. The first one is a new genus found in the cave hygropetric (Figure 7C). It has very thin and extremely elongated claws used for walking on the wet walls, a character shared with other hygropetricolous springtails, e.g., [70–73]. The second species belongs to the genus *Parisotoma* and is the first troglobiotic species of the genus that has a worldwide distribution (Figure 5B). All springtail species from the system exhibit a number of morphological traits typical for troglobiotic springtails (Figure 5A,B) [74]. Surprisingly, the species of the family Entomobryidae, otherwise species rich in the caves of the Dinaric Karst, were not found in the system.

The beetle fauna of the system, when set into a regional framework, is relatively poor. No subterranean representative of the tribe Trechini, which is otherwise very species-rich and -abundant in the Dinaric Karst, has ever been found in the system or any other cave in the vicinity. The other species-rich family, Leiodidae, is represented by three species. Two of them, *Astagobius angustatus* (Figure 5C) and *Spelaeodromus pluto*, are distributed in the northern part of the Dinaric Karst [75,76]. Both species are known exclusively from the caves located on high karstic plateaus, characterized by near-zero temperatures and often rich in ice formations [77]. The third species, the hygropetricolous *Velebitodromus smidai* (Figure 7A), is known only from a few caves in the vicinity of the system [56]. It was found in the deeper parts of the system, in places with a permanently existing cave hygropetric. Based on the present knowledge and the regional species pool [6,75], other troglobiotic species are expected within the system.

Two dipteran species were recorded at the depths between −800 and −980 m. The first is a chironomid *Troglocladius hajdi* (Figure 5D), endemic to the cave, and probably the only troglobiont in the world capable of flying [33]. Two specimens were collected by hand, while all other specimens were found trapped on the adhesive tapes placed on the walls of the −980 m chamber. Large and well-developed wings and halteres, as well as specimens that were found in the middle of the adhesive tape, suggest that the species is able to at least hover, if not actively fly. All specimens collected were females, indicating a possibility of parthenogenesis, a lifestyle not uncommon for chironomids living in extreme environments [78]. Little is known about its life cycle as no larvae were found, but shallow vadose streams with fine sediment in the chamber at −980 m that were not well sampled, are promising habitat for future sampling. The second dipteran species is a troglophile *Speolepta leptogaster*, which is widely distributed in Europe [79]. The obligate subterranean larvae spin silk nets on the cave walls, while short-lived adults are occasionally found in surface habitats [80]. Only larvae were collected in Lukina jama–Trojama, all within the chamber, at −980 m.

5. Threats and Conservation

The Lukina jama–Trojama cave system is located in a remote and difficult to access strict nature reserve Hajdučki and Rožanski kukovi, within the Northern Velebit National Park. Therefore, the entire surface area of the cave system and adjacent deep caves is well-protected from human impact. Thus, the habitats on the surface, below the surface and terrestrial deep cave habitats are in pristine condition. On the other hand, the phreatic zone, inhabited by a unique stygobiotic community, is threatened by the changes in the hydrological regime, caused by infrastructural and hydropower development. The most imminent threat is the construction of a dam for hydroaccumulation Kosinj, expected to be finished by 2028, which would build upon the existing Senj hydropower plant and the remaining hydropower potential of Lika river [81]. Velebit's underground aquifer is hydrologically connected to the Lika river, which sinks at its foothills. The hydrological regime of the Lika river has already been severely altered with construction of the first Kruščica reservoir for the same power plant. The extent of damage to the phreatic community of the Lukina jama–Trojama cave system is difficult to foresee, but it is likely that any further intervention in the hydrological regime of the Lika river could have lasting negative effects on this phreatic community, due to the change in the water regime, and lower nutrient input [82].

6. Discussion

The data presented in this paper reveal several peculiar features of the subterranean community of the Lukina jama–Trojama cave system: (i) a high proportion of endemic and obligate cave species; (ii) a distinct vertical distributional pattern and; and (iii) a high number of evolutionary and ecologically unique species associated with the cave hygropetric and the phreatic zone. Below, we discuss each of the three features and try to interpret them in light of the cave's morphology, ecology, and geographic position.

The fauna of the Lukina jama–Trojama cave system is characterized by both a high number of obligate cave species (41) and a high number of endemic species (Table 1). A relatively low share of troglophiles and trogloxenes can be partially attributed to sampling bias and harsh conditions in the upper parts of the cave (discussed below). Slightly less than half of the species are found within the broader region, namely 11 (27%) species in the broader Velebit–Lika region and 8 (20%) in the NW Dinaric Karst. Almost half of the recorded species (19 species, 46%) are endemic to the northern Velebit, and 3 of them have been reported so far only from this cave system, namely the amphipod *Niphargus* sp., dipteran *Troglocladius hajdi* [33], and an undescribed springtail genus. Nevertheless, we cannot rule out that these single-site endemics also live in other, insufficiently sampled caves of the northern Velebit. In accordance with the published data, stygobionts of the Lukina jama–Trojama cave system have larger ranges than troglobionts [12,16,43,58], largely corresponding with the hydrological connections of the deep groundwater aquifer beneath the Velebit Mt. and the Lika river [35].

The sampling revealed an easily discernible vertical distribution pattern of species richness. From the entrance to a depth of 800 m, we found a total of only five species (11%). The vast majority of the fauna (89%) was found in the deepest parts of the cave (Figure 2). As many as 26 species (58 %) were found in the zone between 800 and 1200 m, mostly in the large chamber at −980 m. Below −1200 m, 7 terrestrial and 14 aquatic species (47%) were found, the latter exclusively in the phreatic zone. Preliminary observations from nearby deep caves (Slovačka jama and Velebita cave system) indicate a similar depth-dependent increase in the species richness, although these caves harbor more species in the upper parts (unpublished data). These distributional patterns roughly resemble the vertical distribution of the species richness in deep caves in Slovenia, with more species found in the lower portions of the vertical caves [8]. This seems to be in contrast with the more than 2 km deep Caucasian Krubera Cave, where the highest number of species was found in the uppermost parts of the system (−70 m) [11], at least according to the currently available data for this large cave system.

The vertical distribution of the species richness in the Lukina jama–Trojama cave system could partially be attributed to the cave's morphology and its ecological characteristics. The upper parts are predominantly vertical, offering limited sampling possibilities. Considerable effort was made to search for fauna in these upper parts, especially in the areas in close vicinity of bivouacs, at the depths of 320 and 780 m, and by carefully examining the vertical walls during ascending. Despite these efforts, sampling success was negligible. It seems reasonable to assume that the low species richness reflects the ecological conditions in this part of the cave. The upper zone of the cave is ecologically harsh due to the confounding effects of the cold, wind, ice, and occasional waterfalls. The low mean annual temperature of the northern Velebit Mt. and high precipitation result in the accumulation of ice [83], the volume of which depends largely on the morphology of the cave entrances. The air circulation model proposed for the Lukina jama–Trojama cave system [26] explains how the temperature of the air and percolating water in the cave remains lower than the temperature of the karst massif itself. Up to a depth of approximately −200 m, the air temperature remains at about 0 °C, while a strong air circulation extends down to the depth of −500 m [28]. Moreover, heavy rainfall results in massive waterfalls with high discharge rates. The combination of low temperatures, ice accumulation and vertical water discharge makes the upper part of the system inhospitable for most of the subterranean species, and likely filters out the non-obligatory cave species. In contrast, environmental conditions stabilize in the large chamber at −980 m. Due to milder environmental conditions, but also a greater number of microhabitats comprising clay-like sediments, cave hygropetric, and vadose streams, the species richness increases in this zone. Finally, the chamber at −980 m has served as a main campsite, with several bivouacs for researchers, and was explored more thoroughly than other parts of the cave. In the deepest part of the cave, there is a strong species turnover from terrestrial to aquatic. The epiphreatic zone is mostly free of terrestrial species, possibly due to the devastating effects of periodic flooding, when the water level can rapidly rise over 100 m [26].

The third outstanding feature of the Lukina jama–Trojama cave system refers to a high number of evolutionary and ecologically unique taxa living in the two largely differing aquatic habitats, the cave hygropetric and the phreatic zone [84]. A combination of the geographic position and morphology of the cave is ideal for the development of a hygropetric habitat, characterized by rich microbial communities in the waters permanently flowing down the cave walls. The microbial communities are a sufficiently rich food source [85], and several species in this system have adapted to exploit them. Hygropetricolous species represent a case study of evolutionary convergence, as all of them share strongly modified mouthparts, used for the filtering or scraping of microbial communities, and claws [86,87], needed for standing and walking in a strong water current. The most remarkable cave hygropetric specialists in the Lukina jama–Trojama cave system are the leiodid beetle, *Velebitodromus smidai*, the amphipod, *Chaetoniphargus lubuskensis*, and the yet undescribed springtail genus. Additionally, several species are presumably bound to the cave hygropetric or habitats in its close proximity, e.g., the specialized cave leech, *Croatobranchus mestrovi*, the isopods, *Alpioniscus velebiticus*, and an undescribed trichoniscid, and several collembola species. In the northern Velebit, hygropetricolous species were usually found at depths greater than −500 m, indicating that this habitat develops only in the deep vadose caves, although this is not a general rule for hygropetricolous species [36]. In the deepest parts of the cave, permanently submerged channels of the phreatic zone are accessible only by cave diving. This habitat houses a unique subterranean community comprised of three filter-feeding species: (i) the cave clam, *Congeria jalzici*; (ii) the cave sponge, *Eunapius subterraneus*; and (iii) the cave serpulid, *Marifugia cavatica*. All three species represent the only (*M. cavatica*), or one of only a few subterranean lineages (*C. jalzici* and *E. subterraneus*) of the taxa otherwise rich in surface, marine, or freshwater representatives [40,88,89]. Moreover, filter-feeding animals are extremely rare among stygobionts, and Lukina jama–Trojama is one of only four caves in the world (unpublished data), all four located in the Dinaric Karst, where three filter-feeding cave-dwellers are known to coexist.

Despite the fact that the fauna of Lukina jama–Trojama is relatively well sampled compared to similar systems in the area, the list of species should not be considered complete, due to several reasons. The most obvious reason is the difficult access to the system, which allows for sampling only during logistically demanding speleological expeditions. Furthermore, the abundance of species in oligotrophic habitats is low [90] and the chances of finding a particular species are disproportionately low. An illustrative example is the results of the expedition in 2010. Although different taxa were found, all were collected in extremely low numbers, including the otherwise abundant Collembola. Five different Collembola species, represented by a total of only 29 specimens, were collected by hand in 32 working man-hours in the large chamber at −980 m, while not a single specimen was collected in baited traps. Some caveats in species inventory can also be attributed to taxonomy. Microcrustaceans such as copepods and ostracods were sampled, but not taxonomically studied. There are numerous undescribed (two new genera and 15 new species) or unidentified taxa awaiting further taxonomic evaluation, a process that may happen, if ever, with a substantial delay [91]. Nevertheless, additional species have been discovered for the system with each new expedition, and we can expect the species list to continue to grow. At the very least, there is a high probability that the species collected in neighboring deep caves, including a spider *Stalita pretneri* Deeleman-Reinhold, 1971 [92,93] a pseudoscorpion *Neobisium stygium* Beier, 1931 [63,94], and two hygropetricoulus species (a springtail *Tritomurus* sp. (unpublished data) and a beetle *Croatodirus casalei* Giachino and Jalžić, 2004 [56]), will eventually also be found in Lukina jama–Trojama.

This study did not confirm our hypothesis that species segregate at different altitudinal bands. Instead, most species were found in the deeper parts of the system. This might be attributed to the specific morphology of the cave and its ecological characteristics, and remains to be further tested in other deep cave systems. However, a great number of obligate cave species confirmed that the large cave systems, at least in the Dinaric Karst, indeed represent hotspot candidates. The results also reflect how advances in caving techniques and speleological research greatly increase our ability to sample much more complex and deeper subterranean habitats.

The comprehensive inventory of the species richness of the Lukina jama–Trojama cave system indicates that a seemingly lower species richness in the central part of the Dinaric Karst (see [6,16]) may be an artifact resulting from a limited access to caves in the mountainous area at higher altitudes and historical interest of biospeleologists the already-recognized cave biodiversity centers, such as the Vjetrenica cave [5] in the southeast and the Postojna–Planina cave system in the northwest [4]. The caves of central Dinaric mountain ridges, which might be as species-rich as the caves in the northwestern and southeastern parts of the Dinaric Karst, provide exciting possibilities and prospects for future discoveries.

Author Contributions: Conceptualization, M.L., J.B. and C.F.; original draft preparation, M.L. and J.B.; review and editing, M.L., J.B., T.D. (Teo Delić), H.B., M.P., A.K., T.D. (Tvrtko Dražina), B.J., R.O., R.S. and C.F.; visualization, M.L. and J.B.; taxonomic identification and interpretation, sponges, bivalves, polychaets H.B., gastropods R.S., leeches, milipedes T.D. (Tvrtko Dražina), centipedes A.K., spiders, dipterans M.P., pseudoscorpions, opilionids, mites R.O., isopods, decapods J.B., amphipods C.F. and T.D. (Teo Delić), springtails M.L., beetles T.D. (Teo Delić) and B.J.; corresponding author J.B. All authors have read and agreed to the published version of the manuscript.

Funding: M.L., J.B. and H.B. were supported by the Tenure Track Pilot Programme of the Croatian Science Foundation and the Ecole Polytechnique Fédérale de Lausanne and the Project TTP-2018-07-9675 EvoDark, with funds from the Croatian–Swiss Research Programme. C.F. and T.D. (Teo Delić) were supported by the Slovenian Research Agency through program P1-0184 and projects J1-2464 and J1-4391.

Institutional Review Board Statement: Fieldwork collecting was executed in accordance with the permits no. UP/I-612-07/10-33/750; 538-08-01-01/3-10-02 issued by the Ministry of culture on 24 June 2010; UP/I-612-07/11-33/0882; 532-08-01-01/1-11-02 issued by the Ministry of culture on 26 August 2011 and UP/I-612-07/12-48/30; 517-07-1-1-1-12-2 issued by the Ministry of Environment and Nature Protection on 18 September 2012 in accordance with the Croatian Nature Protection Act.

Data Availability Statement: All the data included in the analysis are available in the cited references, zoological collections of the Croatian Biospeleological Society and University of Ljubljana's SubBio Lab, and private collections of Roman Ozimec and Rajko Slapnik.

Acknowledgments: We are thankful to all cavers and speleological clubs that organized the expeditions to the Lukina jama–Trojama cave system; The Speleological Committee of the Croatian Mountaineering Association, Croatian Speleological Federation and the National Park Northern Velebit for the support of the expeditions; for the identification of fauna (Haplotaxidae, Mladen Kerovec; *Troglocaris*, Valerija Zakšek; *Proasellus*, Boris Sket; *Plusiocampa*, Kazimir Miculinić; *Speolepta*, Marija Ivković); Tin Rožman for image editing; Vedran Jalžić for providing the image of the chamber at -980 m; Darko Bakšić for providing the cross-section of northern Velebit and the cave map; anonymous reviewers for their comments that improved our manuscript.

Conflicts of Interest: The authors declare no conflict of interest.

References

1. Culver, D.C.; Pipan, T. *The Biology of Caves and Other Subterranean Habitats*, 2nd ed.; Oxford University Press: Oxford, UK, 2019; p. 301. ISBN 978-0-19-882076-5.
2. Ficetola, G.F.; Canedoli, C.; Stoch, F. The Racovitzan Impediment and the Hidden Biodiversity of Unexplored Environments. *Conserv. Biol.* **2019**, *33*, 214–216. [CrossRef] [PubMed]
3. Polak, S.; Pipan, T. The Subterranean Fauna of Križna jama, Slovenia. *Diversity* **2021**, *13*, 210. [CrossRef]
4. Zagmajster, M.; Polak, S.; Fišer, C. Postojna-Planina Cave System in Slovenia, a Hotspot of Subterranean Biodiversity and a Cradle of Speleobiology. *Diversity* **2021**, *13*, 271. [CrossRef]
5. Ozimec, R.; Baković, N.; Bakšić, D.; Basara, D.; Bevanda, L.; Brajković, H.; Brancelj, A.; Erhard, C.; Gašić, Z.; Grego, J.; et al. *Vjetrenica—Cave Biodiversity Hotspot of the Dinarides*; Ozimec, R., Ed.; Public Enterprise Vjetrenica: Ravno, Bosnia and Herzegovina; ADIPA: Zagreb, Croatia, 2021; p. 356.
6. Bregović, P.; Zagmajster, M. Understanding Hotspots within a Global Hotspot—Identifying the Drivers of Regional Species Richness Patterns in Terrestrial Subterranean Habitats. *Insect Conserv. Divers.* **2016**, *9*, 268–281. [CrossRef]
7. Reis-Venâncio, P.C.; Rabelo, L.M.; Pellegrini, T.G.; Ferreira, R.L. From Light to Darkness: The Duality of Influence of Habitat Heterogeneity on Neotropical Terrestrial Cave Invertebrate Communities. *Stud. Neotrop. Fauna Environ.* **2022**, 1–10. [CrossRef]
8. Trontelj, P.; Borko, Š.; Delić, T. Testing the Uniqueness of Deep Terrestrial Life. *Sci. Rep.* **2019**, *9*, 15188. [CrossRef]
9. Mammola, S. Finding Answers in the Dark: Caves as Models in Ecology Fifty Years after Poulson and White. *Ecography* **2019**, *42*, 1331–1351. [CrossRef]
10. Antić, D.Ž.; Reip, H.S. The Millipede Genus *Leucogeorgia* Verhoeff, 1930 in the Caucasus, with Descriptions of Eleven New Species, Erection of a New Monotypic Genus and Notes on the Tribe Leucogeorgiini (Diplopoda: Julida: Julidae). *EJT* **2020**, *713*, 1–106. [CrossRef]
11. Sendra, A.; Reboleira, A.S.P.S. The World's Deepest Subterranean Community—Krubera-Voronja Cave (Western Caucasus). *Int. J. Speleol.* **2012**, *41*, 221–230. [CrossRef]
12. Zagmajster, M.; Eme, D.; Fišer, C.; Galassi, D.; Marmonier, P.; Stoch, F.; Cornu, J.-F.; Malard, F. Geographic Variation in Range Size and Beta Diversity of Groundwater Crustaceans: Insights from Habitats with Low Thermal Seasonality: Range Size and Beta Diversity in Non-Seasonal Habitats. *Glob. Ecol. Biogeogr.* **2014**, *23*, 1135–1145. [CrossRef]
13. Culver, D.C.; Deharveng, L.; Bedos, A.; Lewis, J.J.; Madden, M.; Reddell, J.R.; Sket, B.; Trontelj, P.; White, D. The Mid-Latitude Biodiversity Ridge in Terrestrial Cave Fauna. *Ecography* **2006**, *29*, 120–128. [CrossRef]
14. Sket, B.; Paragamian, K.; Trontelj, P. A Census of the Obligate Subterranean Fauna of the Balkan Peninsula. In *Balkan Biodiversity—Pattern and Process in the European Hotspot*; Griffiths, H.I., Kryštufek, B., Reed, J.M., Eds.; Springer: Dordrecht, The Netherlands, 2004; pp. 309–322. ISBN 978-90-481-6732-6.
15. Sket, B. Diversity Patterns in Dinaric Karst. In *Encyclopedia of Caves*, 2nd ed.; White, W.B., Culver, D.C., Eds.; Elsevier: Amsterdam, The Netherlands, 2012; pp. 228–238.

16. Bregović, P.; Fišer, C.; Zagmajster, M. Contribution of Rare and Common Species to Subterranean Species Richness Patterns. *Ecol. Evol.* **2019**, *9*, 11606–11618. [CrossRef] [PubMed]
17. Zagmajster, M.; Culver, D.C.; Sket, B. Species Richness Patterns of Obligate Subterranean Beetles (Insecta: Coleoptera) in a Global Biodiversity Hotspot—Effect of Scale and Sampling Intensity. *Divers. Distrib.* **2008**, *14*, 95–105. [CrossRef]
18. Borko, Š.; Altermatt, F.; Zagmajster, M.; Fišer, C. A Hotspot of Groundwater Amphipod Diversity on a Crossroad of Evolutionary Radiations. *Divers. Distrib.* **2022**, *28*, 2765–2777. [CrossRef]
19. Culver, D.C.; Deharveng, L.; Pipan, T.; Bedos, A. An Overview of Subterranean Biodiversity Hotspots. *Diversity* **2021**, *13*, 487. [CrossRef]
20. Fišer, C.; Borko, Š.; Delić, T.; Kos, A.; Premate, E.; Zagmajster, M.; Zakšek, V.; Altermatt, F. The European Green Deal Misses Europe's Subterranean Biodiversity Hotspots. *Nat. Ecol. Evol.* **2022**, *6*, 1403–1404. [CrossRef]
21. Ozimec, R.; Bedek, J.; Gottstein, S.; Jalžić, B.; Slapnik, R.; Štamol, V.; Bilandžija, H.; Dražina, T.; Kletečki, E.; Komerički, A.; et al. *Crvena Knjiga Špiljske Faune Hrvatske. [Red Book of Croatian Cave Dwelling Fauna]*; Ozimec, R., Katušić, L., Eds.; Ministarstvo Culture: Zagreb, Croatia; Državni Zavod za Zaštitu Prirode: Zagreb, Croatia, 2009; p. 371. ISBN 978-953-7169-73-2.
22. Bakšić, D.; Barišić, T.; Božić, V.; Buzjak, S.; Čaplar, A.; Jalžić, B.; Lacković, D.; Stroj, A.; Šmida, B.; Vrbek, B.; et al. *Jamski Sustav Lukina Jama—Trojama*; Jalžić, B., Ed.; Hrvatski Planinarski Savez: Zagreb, Croatia, 2007; p. 143.
23. Glavaš, V.; Grlj, A.; Hočevar, G.; Novljan, Ž.; Stepišnik, U.; Stojilković, B.; Trenchovska, A.; Žebre, M. *Dinarski Kras*; Stepišnik, U., Ed.; Znanstvena Založba Filozofske Fakultete Univerze v Ljubljani: Ljubljana, Slovenia, 2018; p. 159. ISBN 978-961-06-0147-0.
24. Velić, I.; Velić, J. *Od Morskih Plićaka do Planine: Geološki Vodič Kroz Nacionalni Park Sjeverni Velebit (From Marine Shallows to the Mountain: Geological Guidebook through the Northern Velebit National Park)*; Kljajo, D., Ed.; Nacionalni Park Sjeverni Velebit: Krasno, Croatia, 2009; p. 143.
25. Perica, D.; Orešić, D. Klimatska Obilježja Velebita i Njihov Utjecaj na Oblikovanje Reljefa. *Senj. Zb.* **1999**, *26*, 1–50.
26. Stroj, A.; Paar, D. Water and Air Dynamics within a Deep Vadose Zone of a Karst Massif: Observations from the Lukina jama–Trojama Cave System (−1431 m) in Dinaric Karst (Croatia). *Hydrol. Process.* **2019**, *33*, 551–561. [CrossRef]
27. Bakšić, D.; Paar, D.; Stroj, A.; Lacković, D. Northern Velebit Deep Caves. In Proceedings of the 16th International Congress of Speleology, Brno, Czech Republic, 21–28 July 2013; Filippi, M., Bosak, P., Eds.; Czech Speleological Society and UIS: Brno, Czech Republic, 2013; Volume 2, pp. 24–29.
28. Paar, D.; Buzjak, N.; Bakšić, D.; Radolić, V. Physical Research in Croatia's Deepest Cave System: Lukina jama-Trojama, Mt. Velebit. In Proceedings of the 16th International Congress of Speleology, Brno, Czech Republic, 21–28 July 2013; Filippi, M., Bosak, P., Eds.; Czech Speleological Society and UIS: Brno, Czech Republic, 2013; Volume 2, pp. 442–446.
29. Paar, D.; Mance, D.; Stroj, A.; Pavić, M. Northern Velebit (Croatia) Karst Hydrological System: Results of a Preliminary 2H and 18O Stable Isotope Study. *Geol. Croat.* **2019**, *72*, 205–213. [CrossRef]
30. Kerovec, M.; Kučinić, M.; Jalžić, B. *Croatobranchus mestrovi* sp. n.—Predstavnik Nove Endemske Podzemne Vrste Pijavica (Hirudinea, Erpobdellidae). *Speleolog* **1999**, *44*, 35–36.
31. Weigand, A. New *Zospeum* Species (Gastropoda, Ellobioidea, Carychiidae) from 980 m Depth in the Lukina jama–Trojama Cave System (Velebit Mts., Croatia). *Subterr. Biol.* **2013**, *11*, 45–53. [CrossRef]
32. Stoev, P.; Akkari, N.; Komerički, A.; Edgecombe, G.; Bonato, L. At the End of the Rope: *Geophilus hadesi* sp. n.—The World's Deepest Cave-Dwelling Centipede (Chilopoda, Geophilomorpha, Geophilidae). *ZooKeys* **2015**, *510*, 95–114. [CrossRef] [PubMed]
33. Andersen, T.; Baranov, V.; Hagenlund, L.K.; Ivković, M.; Kvifte, G.M.; Pavlek, M. Blind Flight? A New Troglobiotic Orthoclad (Diptera, Chironomidae) from the Lukina jama—Trojama Cave in Croatia. *PLoS ONE* **2016**, *11*, e0152884. [CrossRef] [PubMed]
34. Mihoci, T.; Kovač Konrad, P. Ekspedicija Lukina jama—Sifon 2013. *Subterranea Croat.* **2014**, *12*, 3–12.
35. Stroj, A. Hidrogeološka Istraživanja Krških Vodonosnih Sustava. Primjer Istraživanja Podzemnih Tokova u Zaleđu Priobalnih Izvora Velebitskog Kanala. *Vijesti Hrvat. Geol. Druš.* **2011**, *48*, 20–34.
36. Sket, B. The Cave Hygropetric—A Little Known Habitat and Its Inhabitants. *Arch. Hydrobiol.* **2004**, *160*, 413–425. [CrossRef]
37. Calado, R.; Marschal, C.; Lejeusne, C.; Sket, B.; Chevaldonné, P. Improvements to the "Sket Bottle": A Simple Manual Device for Sampling Small Crustaceans from Marine Caves and Other Cryptic Habitats. *J. Crustac. Biol.* **2008**, *28*, 185–188. [CrossRef]
38. Sket, B. Can We Agree on an Ecological Classification of Subterranean Animals? *J. Nat. Hist.* **2008**, *42*, 1549–1563. [CrossRef]
39. Sket, B.; Velikonja, M. Troglobitic Freshwater Sponges (Porifera, Spongillidae) Found in Yugoslavia. *Stygologia* **1986**, *2*, 254–266.
40. Pollonera, C. Note Malacologiche. *Bull. Della Soc. Malacol. Ital.* **1887**, *12*, 204–223.
41. Bole, J. Rod *Zospeum* Bourguignat 1856 (Gastropoda, Ellobiidae) v Jugoslaviji. *Slov. Akad. Znan. Umet.* **1974**, *17*, 249–291.
42. Inäbnit, T.; Jochum, A.; Kampschulte, M.; Martels, G.; Ruthensteiner, B.; Slapnik, R.; Nesselhauf, C.; Neubert, E. An Integrative Taxonomic Study Reveals Carychiid Microsnails of the Troglobitic Genus *Zospeum* in the Eastern and Dinaric Alps (Gastropoda, Ellobioidea, Carychiinae). *Org. Divers. Evol.* **2019**, *19*, 135–177. [CrossRef]
43. Bilandžija, H.; Morton, B.; Podnar, M.; Ćetković, H. Evolutionary History of Relict *Congeria* (Bivalvia: Dreissenidae): Unearthing the Subterranean Biodiversity of the Dinaric Karst. *Front. Zool.* **2013**, *10*, 5. [CrossRef] [PubMed]
44. Absolon, K.; Hrabe, S. Über Einen Neuen Süsswasser-Polychaeten Aus Den Höhlengewässern Der Herzegowina. *Zool. Anz.* **1930**, *88*, 249–264.
45. Oken, L. *Isis, Oder Encyclopädische Zeitung von Oken*; Expedition der Isis: Wetter, Germany, 1819; Volume 4.
46. Bedek, J.; Taiti, S.; Bilandžija, H.; Ristori, E.; Baratti, M. Molecular and Taxonomic Analyses in Troglobiotic *Alpioniscus* (*Illyrionethes*) Species from the Dinaric Karst (Isopoda: Trichoniscidae). *Zool. J. Linn. Soc.* **2019**, *187*, 539–584. [CrossRef]

47. Sket, B. Einige Neue Formen Der Malacostraca (Crust.) Aus Jugoslawien. *Bull.Sci. Beogr.* **1957**, *3*, 70–71.
48. Karaman, G.S. Contribution to the Knowledge of the Amphipoda. Revision of the Niphargus Orcinus-Group, Part I. (Fam. Niphargidae). *Montenegrin Acad. Sci. Arts Glas. Sect. Nat. Sci.* **1984**, *4*, 7–79.
49. Sket, B. Vier Neue Aberrante *Niphargus*—Arten (Amphipoda, Gammaridae) und Einige Bemerkungen zur Taxonomie der *Niphargus*—Ähnlichen Gruppen. *Razpr. SAZU Diss. Acad. Sci. Artium Slov. Cl IV Hist. Nat. Med.* **1971**, *14*, 1–25.
50. Jurinac, A.E. Prilog Hrvatskoj Fauni Ogulinsko-Slunjske Okolice i Pećina. *Rad Jugosl. Akad. Znan. Umjet.* **1887**, *83*, 86–128.
51. Karaman, G.S.; Sket, B. New Genus and Species of the Family Niphargidae (Crustacea: Amphipoda: Senticaudata), *Chaetoniphargus lubuskensis* gen. nov., sp. nov. from Croatia. *Zootaxa* **2019**, *4545*, 249. [CrossRef]
52. Sket, B.; Zakšek, V. European Cave Shrimp Species (Decapoda: Caridea: Atyidae), Redefined after a Phylogenetic Study; Redefinition of Some Taxa, a New Genus and Four New *Troglocaris* Species. *Zool. J. Linn. Soc.* **2009**, *155*, 786–818. [CrossRef]
53. Strasser, K. Neue Diplopoden Aus Höhlen Jugoslawiens. *Senckenberg. Biol.* **1966**, *47*, 379–398.
54. Schmidt, F. Zwei Neue Arten von *Leptoderus*. *Stettin. Entomol. Ztg.* **1852**, *13*, 381–382.
55. Reitter, E. Neue Und Seltene Coleopteren, Im Jahre 1880 in Süddalmatien Und Montenegro Gesammelt Und Beschrieben. *Dtsch. Entomol. Z.* **1881**, *25*, 177–230. [CrossRef]
56. Casale, A.; Giachino, P.M.; Jalžić, B. Three New Species and One New Genus of Ultraspecialized Cave Dwelling Leptoderinae from Croatia (Coleoptera, Cholevidae). *Nat. Croat.* **2004**, *13*, 301–317.
57. Winnertz, J. Beitrag Zu Einer Monografie Der Pilzmücken (Mycetophilidae). *Verh. Der Zool. -Bot. Ges. Wien* **1863**, *13*, 637–964.
58. Bilandžija, H.; Bedek, J.; Jalžić, B.; Gottstein, S. The Morphological Variability, Distribution Patterns and Endangerment in the Ogulin Cave Sponge *Eunapius subterraneus* Sket & Velikonja, 1984 (Demospongiae, Spongillidae). *Nat. Croat.* **2007**, *16*, 1–17.
59. Simone, L.R.L.; Ferreira, R.L. *Eupera troglobia* sp. nov.: The First Troglobitic Bivalve from the Americas (Mollusca, Bivalvia, Sphaeriidae). *Subterr. Biol.* **2022**, *42*, 165–184. [CrossRef]
60. Jalžić, B.; Cukrov, M.; Bilandžija, H. Rasprostranjenost Dinarskog Špiljskog Cjevaša *Marifugia cavatica* Absolon i Hrabe, 1930. *Subterranea Croat.* **2021**, *19*, 38–46.
61. Matjašič, J. Marifugijska Favnula (Faune Marifugiale). In Proceedings of the Treći Jugoslavenski Speleološki Kongres, Sarajevo, Yugoslavia, 21–27 June 1962; Kanaet, T., Ed.; Speleološki Savez Jugoslavije: Sarajevo, Yugoslavia, 1962; pp. 155–156.
62. Sket, B.; Dovč, P.; Jalžić, B.; Kerovec, M.; Kučinić, M.; Trontelj, P. A Cave Leech (Hirudinea, Erpobdellidae) from Croatia with Unique Morphological Features. *Zool. Scr.* **2001**, *30*, 223–229. [CrossRef]
63. Hlebec, D.; Podnar, M.; Kučinić, M.; Harms, D. Molecular Analyses of Pseudoscorpions in a Subterranean Biodiversity Hotspot Reveal Cryptic Diversity and Microendemism. *Sci. Rep.* **2023**, *13*, 430. [CrossRef]
64. Bregović, P.; Rožman, T.; Bedek, J.; Dražina, T.; Jalžić, B.; Komerički, A.; Lukić, M.; Pavlek, M. Biospeleološka Istraživanja Špilja i Jama Nacionalnog Parka Sjeverni Velebit u 2018. Godini. *Senj. Zb.* **2019**, *46*, 37–60. [CrossRef]
65. Delić, T.; Trontelj, P.; Zakšek, V.; Fišer, C. Biotic and Abiotic Determinants of Appendage Length Evolution in a Cave Amphipod. *J. Zool.* **2016**, *299*, 42–50. [CrossRef]
66. Delić, T.; Švara, V.; Coleman Charles, O.; Trontelj, P.; Fišer, C. The Giant Cryptic Amphipod Species of the Subterranean Genus *Niphargus* (Crustacea, Amphipoda). *Zool. Scr.* **2017**, *46*, 740–752. [CrossRef]
67. Zakšek, V.; Delić, T.; Fišer, C.; Jalžić, B.; Trontelj, P. Emergence of Sympatry in a Radiation of Subterranean Amphipods. *J. Biogeogr.* **2019**, *46*, 657–669. [CrossRef]
68. Borko, Š.; Trontelj, P.; Seehausen, O.; Moškrič, A.; Fišer, C. A Subterranean Adaptive Radiation of Amphipods in Europe. *Nat. Commun.* **2021**, *12*, 3688. [CrossRef] [PubMed]
69. Antić, D.; Dražina, T.; Rađa, T.; Tomić, V.T.; Makarov, S.E. Review of the Family Anthogonidae (Diplopoda, Chordeumatida), with Descriptions of Three New Species from the Balkan Peninsula. *Zootaxa* **2015**, *3948*, 151. [CrossRef] [PubMed]
70. Thibaud, J.; Massoud, Z. Un Nouveau Genre d'Insectes Collemboles Onychiuridae Cavernicoles Des Picos de Europa (Espagne). *Bull. Mus. Hist. Nat. Paris* **1986**, *8*, 327–331. [CrossRef]
71. Deharveng, L.; Thibaud, J. *Bessoniella procera* n. g., n. sp., Nouvel Orchesellidae Cavernicole Relictues des Pyrenees. *Bull. Mus. Natl. Hist. Nat. Sect. A* **1989**, *11*, 397–405.
72. Lukić, M.; Houssin, C.; Deharveng, L. A New Relictual and Highly Troglomorphic Species of Tomoceridae (Collembola) from a Deep Croatian Cave. *ZooKeys* **2010**, *69*, 1–16. [CrossRef]
73. Vargovitsh, R.S. Cave Water Walker: An Extremely Troglomorphic *Troglaphorura gladiator* gen. et sp. nov. (Collembola, Onychiuridae) from Snezhnaya Cave in the Caucasus. *Zootaxa* **2019**, *4619*, 267–284. [CrossRef] [PubMed]
74. Lukić, M. Collembola. In *Encyclopedia of Caves*, 3rd ed.; White, W.B., Culver, D.C., Pipan, T., Eds.; Elsevier: Oxford, UK, 2019; pp. 308–319.
75. Pretner, E. Koleopterološka Fauna Pećina i Jama Hrvatske s Historijskim Pregledom Istraživanja (Fauna Coleopterologica Subterranea Croatiae mit einer Geschichtlichen Übersicht der Forschungen). *Krš. Jugosl. JAZU* **1973**, *8*, 101–239.
76. Jalžić, B. Über die Verbreitung der Hohlen-Gattung *Astagobius* Reitter (Col., Catopidae) in Velebit Gebirge (Kroatien, Jugoslawien) mit der Beschreibung von *A. angustatus vukusici* ssp. nov. *Acta Entomol. Jugosl.* **1982**, *18*, 15–20.
77. Polak, S. *Spelaeodromus sneznikensis* sp. nov. from Slovenia (Coleoptera: Cholevidae: Leptodirinae). *Acta Entomol. Slov.* **2002**, *10*, 5–12.
78. Armitage, P.D. Behaviour and Ecology of Adults. In *The Chironomidae: The Biology and Ecology of Non-Biting Midges*; Armitage, P.D., Cranston, P.S., Pinder, L.C.V., Eds.; Chapman & Hall: London, UK, 1995; pp. 194–224.

79. Ševčík, J.; Kjærandsen, J.; Marshall, S.A. Revision of *Speolepta* (Diptera: Mycetophilidae), with Descriptions of New Nearctic and Oriental Species. *Can. Entomol.* **2012**, *144*, 93–107. [CrossRef]
80. Dörge, D.; Zaenker, S.; Klussmann-Kolb, A.; Weigand, A. Traversing Worlds—Dispersal Potential and Ecological Classification of *Speolepta leptogaster* (Winnertz, 1863) (Diptera, Mycetophilidae). *Subterr. Biol.* **2014**, *13*, 1–16. [CrossRef]
81. HEP Hidroenergetski Sustav Senj 2. Available online: https://www.hep.hr/projekti/hidroenergetski-sustav-senj-2/247 (accessed on 29 January 2023).
82. Bilandžija, H.; Puljas, S.; Čuković, T. *Nacionalni Programi za Praćenje Stanja Očuvanosti Vrsta i Staništa u Hrvatskoj*; Congeria kusceri Bole, 1962 i *Congeria jalzici* Morton & Bilandžija, 2013; Državni Zavod za Zaštitu Prirode: Zagreb, Croatia, 2014; p. 31.
83. Buzjak, N.; Bočić, N.; Paar, D.; Bakšić, D.; Dubovečak, V. Ice Caves in Croatia. In *Ice Caves*; Elsevier: Oxford, UK, 2018; pp. 335–369. ISBN 978-0-12-811739-2.
84. Ford, D.; Williams, P. *Karst Hydrogeology and Geomorphology*; John Wiley & Sons Ltd.: West Sussex, UK, 2007; p. 562. ISBN 978-1-118-68498-6.
85. Paoletti, M.G.; Mazzon, L.; Martinez-Sañudo, I.; Simonato, M.; Beggio, M.; Dreon, A.L.; Pamio, A.; Brilli, M.; Dorigo, L.; Summers Engel, A.; et al. A Unique Midgut-Associated Bacterial Community Hosted by the Cave Beetle *Cansiliella servadeii* (Coleoptera: Leptodirini) Reveals Parallel Phylogenetic Divergences from Universal Gut-Specific Ancestors. *BMC Microbiol.* **2013**, *13*, 129. [CrossRef]
86. Giachino, P.M.; Vailati, D. Kircheria Beroni, a New Genus and New Species of Subterranean Hygropetricolous Leptodirinae from Albania (Coleoptera, Cholevidae). *Subterr. Biol.* **2006**, *4*, 103–116.
87. Polak, S.; Delić, T.; Kostanjšek, R.; Trontelj, P. Molecular Phylogeny of the Cave Beetle Genus *Hadesia* (Coleoptera: Leiodidae: Cholevinae: Leptodirini), with a Description of a New Species from Montenegro. *Arthropod Syst. Phylogeny* **2016**, *74*, 241–254.
88. Kupriyanova, E.K.; ten Hove, H.A.; Sket, B.; Zakšek, V.; Trontelj, P.; Rouse, G.W. Evolution of the Unique Freshwater Cave-dwelling Tube Worm *Marifugia cavatica* (Annelida: Serpulidae). *Syst. Biodivers.* **2009**, *7*, 389–401. [CrossRef]
89. Harcet, M.; Bilandžija, H.; Bruvo-Mađarić, B.; Ćetković, H. Taxonomic Position of *Eunapius subterraneus* (Porifera, Spongillidae) Inferred from Molecular Data—A Revised Classification Needed? *Mol. Phylogenet. Evol.* **2010**, *54*, 1021–1027. [CrossRef] [PubMed]
90. Kováč, L. Caves as Oligotrophic Ecosystems. In *Cave Ecology*; Moldovan, O.T., Kováč, Ľ., Halse, S., Eds.; Ecological Studies; Springer International Publishing: Cham, Switzerland, 2018; Volume 235, pp. 297–307. ISBN 978-3-319-98850-4.
91. Culver, D.C.; Trontelj, P.; Zagmajster, M.; Pipan, T. Paving the Way for Standardized and Comparable Subterranean Biodiversity Studies. *Subterr. Biol.* **2013**, *10*, 43–50. [CrossRef]
92. Deeleman-Reinhold, C.L. Beitrag Zur Kenntnis Höhlenbewohnender Dysderidae (Araneida) Aus Jugoslawien. *Slov. Akad. Znan.Umet. Razred Za Prirodosl. Vede Cl. IV Hist. Nat.* **1971**, *14*, 95–120.
93. Pavlek, M.; Gauthier, J.; Tonzo, V.; Bilat, J.; Arnedo, M.A.; Alvarez, N. Life-History Traits Drive Spatial Genetic Structuring in Dinaric Cave Spiders. *Front. Ecol. Evol.* **2022**, *10*, 910084. [CrossRef]
94. Beier, M. Zur Kenntnis Der Troglobionten Neobisien (Pseudoscorp.). *Eos Rev. Esp. Entomol.* **1931**, *7*, 9–23.

diversity

Article

The Baget Karstic System and the Interstitial Environment of Lachein, a Hotspot of Subterranean Biodiversity in the Pyrenees (France)

Franck Bréhier [1,*], Danielle Defaye [2], Arnaud Faille [3] and Anne Bedos [4]

[1] Independent Researcher, 9 Carrè Raou–Alas, 09800 Balaguères, France
[2] Independent Researcher, 27 rue Gaston Charlet, 87000 Limoges, France; defaye.danielle@orange.fr
[3] Department of Entomology, State Museum of Natural History, 70191 Stuttgart, Germany; arnaud.faille@smns-bw.de
[4] Institut de Systématique, Évolution, Biodiversité (ISYEB), UMR 7205 CNRS, Muséum National d'Histoire Naturelle, Sorbonne Université, 45 rue Buffon, 75005 Paris, France; bedosanne@yahoo.fr
* Correspondence: brehier-franck@orange.fr

Abstract: Located in Lestelas-Balaguères massif, central northern Pyrenees, France, the Baget catchment covers 13.25 km^2 and is highly karstified: so far, more than 80 caves have been recorded. The main outlet of the system, the exsurgence de Las Hountas, has an average flow of 550 L/s. Downstream, it is connected with the hyporheic of the Lachein stream. The Baget system, formed by both the karstic system and the hyporheic, has been intensively investigated by cave biologists and is known to be a hotspot for subterranean biodiversity. The synthesis provided here lists no less than 17 troglobionts and 40 stygobionts, with 3 single site endemics, making the Baget system the richest subterranean hotspot in the Pyrenees. This is notably due to the diversity of subterranean habitats and to the comprehensive knowledge of the stygofauna, likely unmatched at the European scale. Considering the significant speleological findings of the last 15 years that have not been yet biologically investigated, we can expect new discoveries, especially for the troglofauna.

Keywords: cave fauna; hyporheic; troglobiont; stygobiont

Citation: Bréhier, F.; Defaye, D.; Faille, A.; Bedos, A. The Baget Karstic System and the Interstitial Environment of Lachein, a Hotspot of Subterranean Biodiversity in the Pyrenees (France). *Diversity* 2024, 16, 62. https://doi.org/10.3390/d16010062

Academic Editor: Michel Baguette

Received: 19 December 2023
Revised: 12 January 2024
Accepted: 13 January 2024
Published: 18 January 2024

1. Introduction

The Pyrenean range, especially on the central and western area of its northern slope, is known to be one of the world hotspots for subterranean fauna [1]. Within this area, three sites stand out for their high diversity: the hyporheic of the Nert, the Coume Ouarnède system [2] and the subterranean Baget system, including both karstic and hyporheic habitats (Figure 1). The latter, which is still being explored by cavers, has been extensively studied and sampled over more than 50 years by biologists, benefiting from the close proximity of the CNRS laboratory in Moulis, and has become a global reference for the subterranean aquatic fauna of karst and hyporheic environments.

In this study, based on bibliographic data and unpublished observations, we gathered the most comprehensive list of troglobionts and stygobionts for the Baget system, considering both the karst environment and the hyporheic environment. The aim of this article is to present the general features of this system, incorporating some unpublished speleological and hydrological data, and to provide an overview of its fauna, put in its ecological and biogeographical contexts.

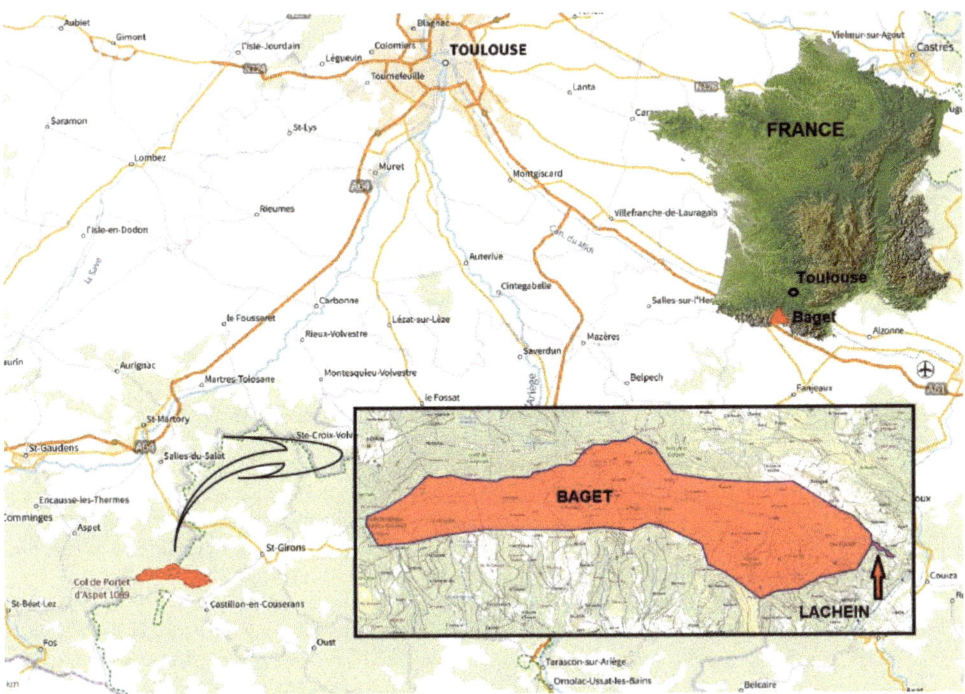

Figure 1. Location of the Baget system, central Pyrenees, South France.

2. Study Site

2.1. The Lestelas-Balaguères Massif

The Lestelas-Balaguères massif straddles the departments of Ariège (13 municipalities) and Haute-Garonne (4 municipalities). It covers an area of around 88 km^2. It is heavily karstified and attracted the interest of cavers very early on. The inventories, initiated by Georges Jauzion, taken over by Daniel Quettier and amended by the caving committees of Ariège and Haute-Garonne with the support of local caving groups, list more than 500 caves [3]; D. Quettier, pers com.

In this area, the Mesozoic series that covers the crystalline massif is represented by terrain ranging from the Triassic to the Turonian. The Jurassic and Lower Cretaceous are the two predominant carbonate formations, and karstification is most marked in the Jurassic-Neocomian and Upper Aptian with Urgonian facies [4].

The Lestelas-Balaguères massif is drained by several karstic systems. The most important are the Peillot system; the Bethmale/Hount Heredo system; the Pas du Loup/Teillèdes or Francazal system; the Belle/Cassagnous or underground Hider system; the Coume Ferrat/Aliou or Paloumé system; and the Papillon/Las Hountas or underground Baget system.

The Peillot system, located on the edge of the massif near the village of Cazavet, is apart as it is developed in sandstone marl from the Albian. The three subterranean systems of Hider, Paloumé and Baget share many morphological similarities. Their upper entrances are at similar altitudes (Belle, 1162 m; Bagagès, 1172 m; Papillon, 1150 m), with their respective exsurgences at 505 m (Cassagnous), 441 m (Aliou) and 493 m (Las Hountas). The upper entrances provide access to a series of shafts leading to an active stream with a low gradient. The gradient increases significantly in all three cases shortly before reaching the flooded zone, which is of wide extension.

For the Hider subterranean system, the junction between the upper entrance and the resurgence has been completed (exploration of Gouffre Belle by the Spéléo-Club de l'Epia and of the flooded zone by cave divers G. Tixier and F. Bréhier), [5,6]. For Paloumé and

Baget systems, a large gap remains. In all these systems, exploration is still in progress by local caving groups [3,7–9].

2.2. The Subterranean Baget System

The Baget catchment is located in the southern part of the Lestelas-Balaguères massif in Ariège. It adjoins the Arbas massif to the West at the level of the ridges passing through the Pic de Cornudère and the Tuc de Pissarelle. According to the limits proposed by A. Mangin [4,10–12], it extends from West to East over ten kilometers long with a fairly irregular width varying from 1 to 2 km. It is drained by the exsurgence of Las Hountas, located at its eastern end. The catchment covers 13.25 km^2, of which 4 km^2—i.e., around 30%—are of impermeable non-calcareous terrain.

It is made up of a homogenous structure of crystalline Mesozoic limestone [13]. The Alas fault forms its northern boundary. To the South, the limestone disappears beneath the clay-gravel formations of the Albo-Cenomanian. The western limit, in continuity with the Arbas massif, is more difficult to define, with metamorphosed limestones interspersed with dolomites. It is supposed to extend to the Col de la Croix de Guéret. Its surface area is mostly covered by forests and pastures. There is very little anthropic activity. Its average altitude is 955 m, ranging from 493 m (altitude of the Las Hountas resurgence) to 1417 m (altitude of the Tuc de Graué). Annual precipitation is around 1700 mm, the climate is oceanic with high water level ranging from November to April. The average temperature is 12.3 °C [10–12].

Not under threat and not under formal protection measures, the biological richness of the massif is nevertheless remarkable, and has been labeled as *Zone Naturelle d'Intérêt Écologique, Faunistique et Floristique* (ZNIEFF) Soulane de Balaguères au Char de Liqué (national ID: 730012100) and is included in a Natura 2000 site (Chars de Moulis et de Liqué, grotte d'Aubert, Soulane de Balaguères et de Sainte-Catherine, granges des vallées de Sour et d'Astien, national ID FR7300836).

The main outlet of the system is the perennial exsurgence de Las Hountas, made up of around ten impenetrable griffons, at an altitude of 493 m. Its average flow is 550 L/s, with a minimum of 50 L/s and floods approaching 11,000 L/s [10–12]. It gives rise to the perennial Lachein stream, which flows after 1.5 km into the Lez River at the village of Alas. At Saint-Girons, the Lez meets the Salat River, a tributary of the Garonne. Collections and studies of the hyporheic environment were carried out approximately halfway between the exsurgence of las Hountas and the confluence with the Lez, on a shallow slope.

A second exsurgence is located in the village of Alas, at an altitude of 483 m. It might be fed partly by sinkholes from the Lachein, and partly by seepage into the metamorphosed limestone on the right bank downstream of Las Hountas. The relationships between these two supposed exsurgences and systems are not clearly defined. Upstream of the exsurgence de Las Hountas, the stream only flows during episodes of rain. The caves Moulo del Jaur, Puits de la Hillère (Hillère sinkhole), Perte de la Hille (Hille sinkhole) and Perte de la Peyrère (Peyrère sinkhole) then function in turn as exsurgences, depending on the intensity of the flows.

2.3. Speleological Data

The Baget catchment area is largely karstified, and around 80 caves have been recorded by cavers [3], D. Quettier, pers com.

At present, four of these caves provide access to the system's perennial drain. These are from downstream to upstream Puits de la Hillère, Perte de la Hille, Gouffre de la Peyrère and Gouffre du Papillon.

Puits de la Hillère, Perte de la Hille and the gouffre de la Peyrère are located in the downstream part of the system, approximately 1 km from the outlet. Recent explorations by cave diving have allowed connecting these three caves together [6]. They form a network of passages more than 1500 m long, including 800 m of flooded conduits (Figure 2).

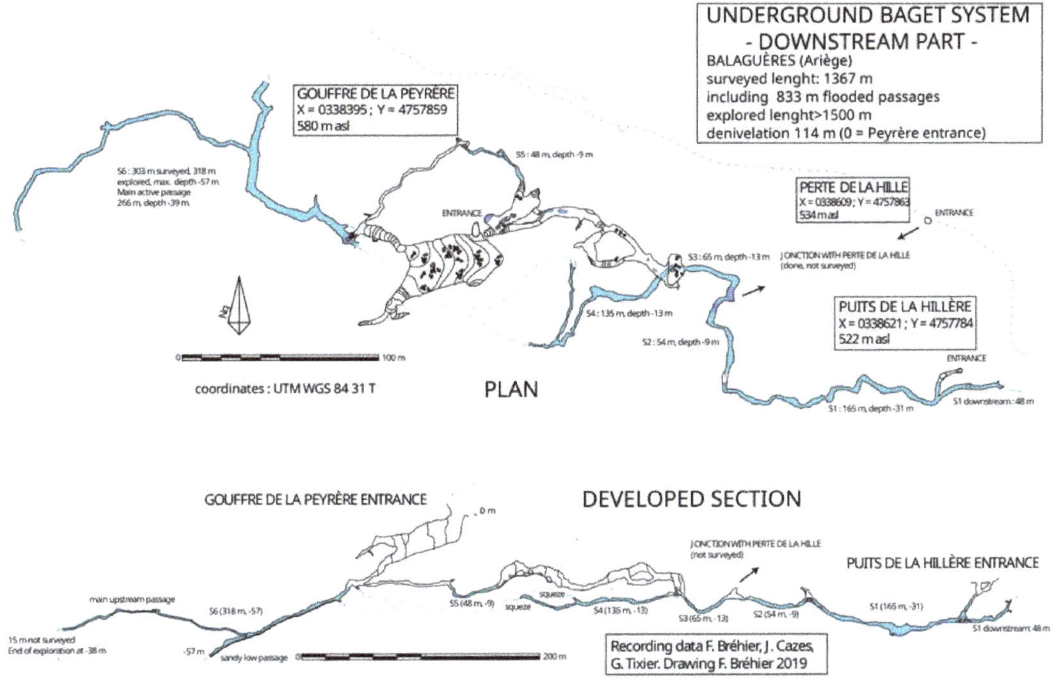

Figure 2. Map of the lower part of Baget karstic system, including Gouffre de la Peyrère, Perte de la Hille and Puits de la Hillère, connected by cave diving.

The Gouffre du Papillon, discovered in 1964 by Émile Bugat, is located more than 6 km from the outlet, making it the most upstream known section of the system. Recent speleological explorations by the Groupe Spéléologique du Couserans in 2007 permitted to extend its length to more than 6 km and its depth to more than 600 m (-604 m), and to reach the main drain. Upstream of the main drain, explorations have been temporarily stopped in sump 2, at a depth of -25 m, on an unstable sand slope. Downstream, at -563 m, a sump was dived to a depth of -43 m by Franck Bréhier. The passage continues with the same large dimensions and further explorations are needed. Between the far downstream end of the Gouffre du Papillon and the far upstream end of the Gouffre de la Peyrère, a large void 4.6 km long in a straight line remains (Figure 3). The vast majority of the network has yet to be discovered, but at present no other caves are known in the supposed route of the collector that could potentially join it. In addition, the small difference in level between the sumps in the Papillon and Peyrère suggests that a large part of the active system flows through a flooded zone, which greatly complicates exploration.

2.4. Hydrological Features

According to the model proposed by A. Mangin [10–12], the Baget system consists of a highly transmissive but low-capacity drain running beneath the Lachein valley talweg. Gouffre de la Peyrère, Puits de la Hillère and Grotte de Saint-Catherine, which are caves giving access to the flooded karst, would be ancillary systems to the drain, with low transmissivity but high capacitance. In this way, water reserves would be built up in large reservoirs that are not directly interconnected. This is the model that has been adopted to date [14]. In fact, recent underground diving explorations have shown that Puits de la Hillère and Gouffre de la Peyrère are directly linked and that the drain passes through both caves (Figure 2). This means that the drain is not located directly beneath the talweg, but rather, at least in the downstream part of the system, runs along its right bank. Perte de la Hille, supposed to be directly on the drain, is offset from it, and is connected to it by a narrow gallery inactive at low water. We can assume that Grotte de Sainte-Catherine, located on the left bank and further downstream, at the level of Las Hountas outlet, is disconnected from the drain, although it is not yet possible to confirm this.

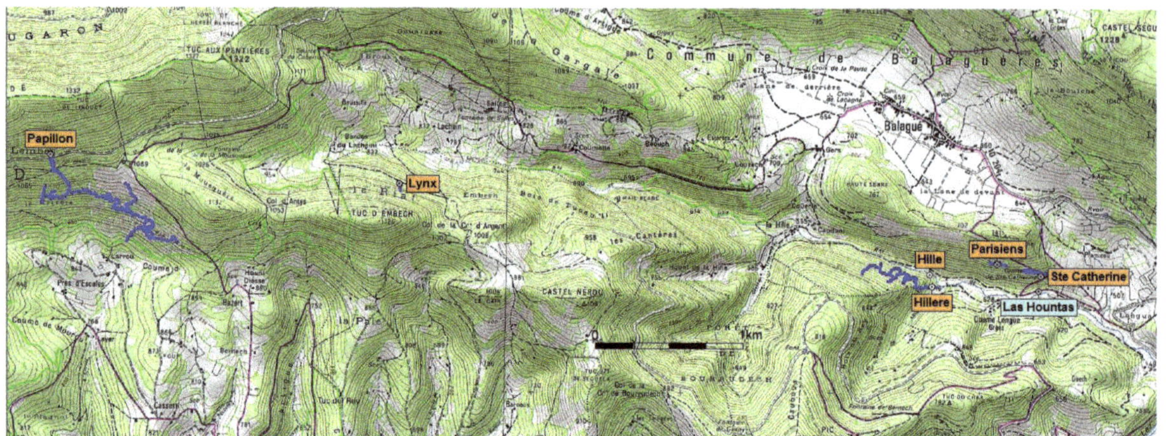

Figure 3. Relative positions of the main caves of the Baget system (see Figure 1 for location), from Gouffre du Papillon, upstream, to the exsurgence de Las Hountas. From CartoExplorer 3.

2.5. The Baget System in Terms of Subterranean Habitats

Despite its relatively small size, the subterranean Baget system presents a great diversity of habitats. For terrestrial fauna, the vast vertical entrances of Gouffre de la Peyrère and Grotte de Sainte-Catherine offer a scree area rich in organic matter, favorable to endogean species (Figure 4, top). The deposits of bat guano in Grotte de Sainte-Catherine favors the presence of guanophilic fauna. The variety of underground landscapes encountered through the numerous caves of the system and the kilometers of passages of Gouffre du Papillon, still untouched by any biospeleological surveys (fossil galleries, large wells, calcite flows, clay deposits, etc.) are all potential habitats.

For aquatic fauna, Grotte de Sainte-Catherine offers access to the vertical flows of the epikarst and to the fissural environment which was instrumented by the Moulis Laboratory. Gours, basins, and streams in the vadose zone are found in large numbers at Grotte de Sainte-Catherine, at Gouffre de la Peyrère and especially at Gouffre du Papillon. Flooded karst has been widely studied by filtration of the various perennial or temporary outlets of the system [15,16] or by pumping at Gouffre de la Peyrère [17]. If recent speleological discoveries have shown that the compartments of Exsurgence de Las Hountas, Puits de la Hillère and Gouffre de la Peyrère are not disconnected from each other, the fact remains that they shelter distinct biocenoses and that the flooded karst should not be considered as a single habitat at a system scale. Finally, the underflow of the Lachein downstream of Exsurgence de Las Hountas also offers a great diversity of habitats for hyporheic fauna [18].

2.6. History of Biological Research on the Subterranean Baget System

In 1948, R. Jeannel of the Museum National d'Histoire Naturelle (Paris) convinced the CNRS (National Center for Scientific Research) to set up an underground laboratory in the commune of Moulis, a few kilometers from Le Baget [19,20]. The Moulis site was chosen, among other criteria, for the rich underground fauna of the area, already well documented by Jeannel [21]. The aim was to carry out breeding of cave animals in a natural environment as well as to develop multidisciplinary studies of the subterranean environment including geology, hydrogeology and biology.

The Baget catchment was chosen as an experimental site for several reasons: its immediate proximity to the Moulis Laboratory and ease of access, its relatively small size and the richness of its fauna. It has been the subject of hydrological, climatic and hydrochemical monitoring for almost 50 years. Numerous interdisciplinary studies including geomorphology [22], population monitoring [23], climatology/population relationships [24,25], and hydrogeology [26], were carried out on this site and made it a national reference.

Figure 4. Different aspects of Baget karstic system. (**Top**): the main chamber of Gouffre de la Peyrère. (**Bottom left**): the main sump of Gouffre de la Peyrère. (**Bottom right**): caving exploration in Gouffre du Papillon. *Photos* F. Bréhier.

The site is particularly suited to an experimental ecological approach of the structures of communities of underground species. From 1967, R. Rouch carried out an in-depth study and sampled extensively the aquatic fauna of the Baget catchment. He began to sample the hyporheic fauna of Lachein in 1967 [27] and, from 1988, he led a comprehensive study of ground water communities linked to a karst aquifer [28–36]. His work has become a World reference.

Concerning terrestrial fauna, most of the published investigation were done at Grotte de Sainte-Catherine and concerned Coleoptera (see Section 4.1.7).

3. Material and Methods

The data presented in this article come from primary literature compilation as well as personal data.

Almost all of the investigations and publications on aquatic fauna are due to R. Rouch. Collections in the interstitial environment of the Lachein underflow were made by Bou-Rouch pumping, filtered through net with 100 μm mesh [18]. For the karstic stygobiotic fauna, the samples came from filtering the main outlet of Las Hountas; the temporary outlets of Moulo de Jaur, Puits de la Hillère and Perte de la Hille, during episodes of flooding. Data from the gouffre de la Peyrère come mainly from the pumping operation led by A. Mangin in 1991 [17]. Crustaceans were studied by Rouch, and identified by different specialists. Data on Oligochaeta, due to Route et al. [37], come from Rouch's sampling [36,37].

While aquatic sampling was made following strict protocols in the long term, terrestrial sampling was rather opportunistic. Data on terrestrial fauna are more scarce and compiled from various publications. They concern Gouffre de Tussau, Gouffre de la Peyrère, Grotte de Sainte-Catherine and Gouffre du Lynx. For our own investigations, we sampled by sight, by baiting and using pitfalls.

The names and validity of the species were checked using the taxonomic referential TaxRef of INPN (MNHN & OFB [Ed]. 2003–2023) [38]. Species ecological status were inferred from the taxonomic literature, particularly from R. Rouch publications. It covers all described species on our list. Only obligate subterranean species (stygobionts and troglobionts), and facultative subterranean species (stygophiles and eutroglophiles) have been considered here. The latter are often numerically dominant in subterranean communities and have been included here for this reason.

4. Results

We will separately consider aquatic and terrestrial subterranean fauna. For each, we will list in four tables (Tables 1–4, respectively) both the obligate and facultative subterranean species. Most relevant stygobionts and troglobionts will be commented on individually.

4.1. Aquatic Fauna

4.1.1. Oligochaeta

Delaya leruthi (Hrabe, 1958) is present in Grotte de Sainte-Catherine. This large species is known from central Pyrenees and especially common in caves of l'Estelas-Balaguères massif [39].

Cookidrilus ruffoi Giani, Martinez-Ansemill & Sambugar, 2004 is an endemic species of Lachein, whereas *Cookidrilus speluncaeus* Rodriguez & Giani, 1987 is known as well from grotte de Labouiche, some 50 km from Lachein [37].

Spiridion phreaticola (Juget, 1987) is a stygobiont with a larger distribution. It has been mentioned in the interstitial waters of the Rhône, the Neste d'Aure and the Dordogne, as well as in the underground river of Grotte de Labouiche [37].

Table 1. Stygobiotic species of the Baget system. K: Karst; HR: Hyporheic; **Species in bold:** single site endemics.

Oligochaeta	
Haplotaxidae	
Delaya leruthi (Hrabe, 1958)	K
Lumbriculidae	
***Cookidrilus ruffoi* Giani, Martinez-Ansemil, & Sambugar, 2004**	**HR**
Cookidrilus speluncaeus Rodriguez & Giani, 1987	HR
Tubificidae	
Spiridion phreaticola (Juget, 1987)	HR
Gastropoda	
Hydrobiidae	
Islamia moquiniana (Dupuy, 1851)	HR
Moitessieria simoniana (Saint Simon, 1848)	HR
Copepoda: Cyclopoida	
Cyclopidae	
Acanthocyclops sensitivus (Graeter & Chappuis, 1914)	HR
Diacyclops belgicus Kiefer, 1936	HR
Diacyclops clandestinus (Kiefer, 1926)	HR
Diacyclops languidoides ssp. (Lilljeborg, 1901)	HR, K
Graeteriella (Graeteriella) rouchi Lescher-Moutoué, 1968	HR
Graeteriella sp.	HR
Speocyclops anomalus Chappuis & Kiefer, 1952	K
Speocyclops kieferi Lescher-Moutoué, 1968	HR, K
Speocyclops racovitzai cf. *boscensis* Kiefer, 1954	K
Speocyclops racovitzai liquensis Chappuis & Kiefer, 1952	HR, K
Copepoda: Harpacticoida	
Ameiridae	
Nitocrella delayi Rouch, 1970	K
Nitocrella gracilis Chappuis, 1955	HR, K
Parapseudoleptomesochra subterranea subterranea (Chappuis, 1928)	HR, K
Canthocamptidae	
***Antrocamptus catherinae* Chappuis & Rouch, 1960**	**K**
Antrocamptus chappuisi Rouch, 1970	HR
Ceuthonectes gallicus Chappuis, 1928	HR, K
Elaphoidella bouilloni Rouch, 1965	HR, K
Elaphoidella coiffaiti Chappuis & Kiefer, 1952	HR, K
Moraria (Moraria) catalana Chappuis & Kiefer, 1952	HR, K
Parastenocarididae	
Parastenocaris dianae Chappuis, 1955	HR
Parastenocaris vandeli Rouch, 1988	HR
Proserpinicaris mangini (Rouch, 1992)	HR, K
Ostracoda	
Candonidae	
Pseudocandona rouchi Danielopol, 1973	HR
Pseudocandona sp.1	K
Pseudocandona sp.2	K
Pseudocandona sp.3	HR
Bathynellacea	
Bathynellidae	
Vandelibathynella vandeli (Delamare & Chappuis, 1954)	HR, K
Amphipoda	
Niphargidae	
Niphargus kochianus (Bate, 1859)	HR, K
Salentinellidae	
Parasalentinella rouchi Bou, 1971	HR, K
Salentinella petiti Coineau, 1963	HR, K

Table 1. *Cont.*

Ingolfiellida		
Ingolfiellidae		
	Ingolfiella thibaudi Coineau, 1968	HR, K
Isopoda		
Microparasellidae		
	Microcharon ariegensis (Coineau, 1968)	HR, K
Stenasellidae		
	Stenasellus virei boui Magniez, 1968	HR
	Stenasellus virei hussoni Magniez, 1968	K

Table 2. Stygophilic species of the Baget system.

Annelida: Polychaeta	
Aelosomatidae	
	Rheomorpha neizvestnovae (Lastochkin, 1935)
Copepoda: Cyclopoida	
Cyclopidae	
	Acanthocyclops venustus venustus (Norman & Scott, 1906)
	Eucyclops serrulatus (Fischer, 1851)
	Megacyclops viridis (Jurine, 1820)
	Paracyclops fimbriatus (Fischer, 1853)
Copepoda: Harpacticoida	
Canthocamptidae	
	Attheyella (Attheyella) crassa (Sars, 1863)
	Bryocamptus (Echinocamptus) echinatus (Mrazek, 1893)
	Bryocamptus (Rheocamptus) typhlops (Mrazek, 1893)
	Bryocamptus (Rheocamptus) zschokkei ssp. (Schmeil, 1893)
	Maraenobiotus vejdovskyi Mrazek, 1893
	Moraria (Moraria) varica (Graeter, 1911)
	Pesceus schmeili (Mrazek, 1893)
Ostracoda	
Candonidae	
	Cryptocandona vavrai Kaufmann, 1900
	Cypria sp.
	Fabaeformiscandona breuili (Paris, 1920)
	Pseudocandona marchica (Hartwig, 1899)
	Pseudocandona rostrata (Brady & Norman, 1899)
Cyprididae	
	Eucypris pigra (Fischer, 1851)
	Ilyocypris sp. Baltanas, Danielopol, Roca & Marmonier, 1993
	Psychrodromus betharrami
Cypridopsidae	
	Potamocypris zschokkei (Kaufmann, 1900)

4.1.2. Gastropoda

Bertrand (unpublished inf., 2002) mentions Moitessieria simoniana (Saint-Simon, 1848) and Neohoratia globulina Paladilhe, 1866), now considered as a synonym of Islamia moquiniana (Dupuy, 1851). *Moitessieria simoniana* has a distribution covering the Pyrenees and the South of the Massif Central. It is common in Ariège. *Islamia moquiniana*, also common in Ariège, has a wide distribution in the southern half of France. These species are present both in the hyporheic environment and in the karst [39].

4.1.3. Copepoda

Crustacea are considered as the most diversified stygobiotic taxon in Europe, and represent 70% of the groundwater species richness [40].

Table 3. Troglobiotic species of the Baget system. **Species in bold:** single site endemism.

Palpigradida Eukoeneniidae	
	***Eukoenenia pyrenaella* Condé, 1990**
Araneae Leptonetidae	
	Leptoneta convexa Simon, 1873
Chilopoda Lithobiidae	
	Lithobius cavernicola Fanzago, 1877
Julida Blaniulidae	
	Blaniulus lorifer consoranensis (Brölemann, 1921)
Isopoda Oniscida Trichoniscidae	
	Scotoniscus macromelos macromelos Racovitza, 1908
Collembola Onychiuridae	
	Micronychiurus n. sp.
Entomobryidae	
	Pseudosinella theodoridesi Gisin & Gama, 1969
Oncopoduridae	
	Oncopodura n. sp.
Tomoceridae	
	Tomocerus problematicus Cassagnau, 1964
Coleoptera Carabidae	
	Aphaenops (Hydraphaenops) sinester Coiffait, 1959
	Aphaenops (Hydraphaenops) cerberus bruneti Jeannel, 1926
	Aphaenops (Hydraphaenops) ehlersi ehlersi (Abeille de Perrin, 1872)
	Aphaenops (Hydraphaenops) tiresias (Piochard de La Brûlerie, 1872)
	Aphaenops (Hydraphaenops) bucephalus (Dieck, 1869)
	Aphaenops (Argonotrechus) orpheus consorranus (Dieck, 1870)
Leiodidae	
	Speonomus (Machaeroscelis) infernus infernus (Dieck, 1869)
Ascomycota Laboulbeniaceae	
	Rhachomyces aphaenopsis Thaxter, 1905

Table 4. Troglophilic species of the Baget system.

Araneae Leptonetidae	
	Leptoneta infuscata Simon, 1873
Diplura Campodeidae	
	Litocampa vandeli (Condé, 1947)
Collembola Onychiuridae	
	Onychiuroides pseudogranulosus Gisin, 1951
Arrhopalitidae	
	Pygmarrhopalites pygmaeus (Wankel, 1860)
Neelidae	
	Megalothorax cf incertus Börner, 1903
Coleoptera Carabidae	
	Geotrechus (Geotrechus) orpheus consorranus (Dieck, 1870)
	Laemostenus oblongus oblongus Dejean, 1828

Copepods constitute the dominant contingent of the subterranean aquatic fauna of the Baget. Taking into account both the hyporheic environment and the flooded karst, a total of 33 species have been documented, including 14 species of Cyclopoida Cyclopidae and 19 species of Harpacticoida distributed into Ameiridae (3), Canthocamptidae (13) and Parastenocarididae (3). This rich community includes many stygobionts: 10 cyclopid and 12 harpacticoid species, to which are added a number of stygophilous or stygoxenous forms that also occupy subterranean habitats to a greater or lesser extent.

For the hyporheic environment of the Lachein alone, over an area of 75 m^2, Rouch counted 22 species of harpacticoids including 10 stygobionts and 14 species of cyclopids including 7 stygobionts [41,42]. Although the spatial and physico-chemical heterogeneity of this environment leads to a heterogeneous distribution of the diversity, it remains stable over time and in its structure.

Studies of flooded karst by the filtration of outlets over long periods show a high degree of homogeneity of the populations in time and space, with each outlet having a taxonomic composition that is constant but original and different from the other outfalls [43–48]. Filtration of the exsurgence de Las Hountas yielded 11 species of cyclopids, including 4 stygobionts, and 21 species of harpacticids, including 8 stygobionts [15].

In addition to the large number of stygobionts, the copepods of the Baget system are characterized by a great number of taxa of higher order and a great number of endemic species.

Among the Cyclopidae, the majority are stygobionts. Several species of the genera *Acanthocyclops* and *Diacyclops*, common in underground environments, are found such as *Acanthocyclops sensitivus* (Graeter & Chappuis, 1914), *Diacyclops belgicus* Kiefer, 1936, *Diacyclops clandestinus* (Kiefer, 1926). Although we would rather consider *Diacyclops languidoides* (Lilljeborg, 1901) as a stygophilic species, we have listed it as a stygobiotic as it is often so-cited [2,40]. It is above all the genera *Graeteriella* and *Speocyclops*, typical of these environments and all endemic either to the Pyrenees or to underground habitats in France, that are the best represented, with *Graeteriella (Grateriella) rouchi* Lescher-Moutoué, 1968, *Speocyclops. anomalus* Chappuis & Kiefer, 1952, *S. kieferi* Lescher-Moutoué, 1968, *S. racovitzai boscensis* Kiefer 1954 and *S. racovitzai liquensis* Chappuis & Kiefer, 1952. Among the Harpacticoida, three families are represented, well known from underground environments: Ameiridae, Canthocamptidae and Parastenocarididae. The family Ameiridae is represented by three stygobionts, including two endemic ones: *Nitocrella delayi* Rouch, 1970 (microendemic in Ariège) and *N. gracilis* Chappuis, 1955, (endemic in Ariège and Haute-Garonne; the third one, *Parapseudoleptomesochra subterranea subterranea* (Chappuis, 1928) is known from many caves, hyporheic habitats and springs. The family Canthocamptidae, generally very common in such habitats, comprises six stygobitic representatives belonging to four genera: *Antrocamptus* with two species endemic in Ariège, *A. catherinae* Chappuis & Rouch, 1960 (cave) and *A. chappuisi* Rouch, 1970, the latter collected through filtering, *Ceuthonectes* with *C. gallicus* Chappuis, 1928 considered as the most abundant species in subterranean waters of Pyrenees, *Elaphoidella* with the microendemic *E. coiffaiti* Chappuis & Kiefer, 1952 preferring muddy bottoms and *E. bouilloni* Rouch, 1965, which prefers sites with fine sand; and *Moraria* with the endemic *M. (M.) catalana* Chappuis & Kiefer, 1952. Among the stygophile canthocamptid, *Bryocamptus* is the most diverse with three species: *B. (Echinocamptus) echinatus* (Mrazek, 1893), a cold-water stenotherm species with a wide palearctic distribution, *B. (Rheocamptus) typhlops* (Mrazek, 1893), spread across Europe, inhabiting fountains, springs and subterranean waters and the common *B. (R.) zschokkei* (Schmeil, 1893) living also in cold waters of mosses, springs of central Europe, Eurasia and North America.

Finally the large family Parastenocarididae, composed of more than 300 species highly specialized for life in groundwater habitats thanks to their physical ability to move in small spaces with their cylindrical and slender body and their small size (less than a millimeter), is represented in the Baget system by three species, all endemic: *Parastenocaris dianae*

Chappuis, 1955, *Parastenocaris vandeli* Rouch, 1988 and *Proserpinicaris mangini* (Rouch, 1992). It is likely that other representatives of this family will be collected in future investigations.

Among Copepoda, more than 40% of the species reported to date are endemics, illustrating the biological richness of the Baget system.

4.1.4. Ostracoda

The class Ostracoda occurs in almost all aquatic and even some terrestrial habitats. Several lineages have subterranean representatives or are exclusively living in groundwaters. The subterranean ostracod fauna is, however, still poorly known.

One of the most diverse family in freshwaters is the large family Candonidae (29% of all species in non-marine habitats [49]). In the Baget system, this family is represented by four species of the genus *Pseudocandona*, three of which undescribed [18]. Two undescribed species are found in the karstic system, whereas *Pseudocandona rouchi* Danielopol, 1973 and the third undescribed species occur in the hyporheic environment of Lachein.

Another species, *Fabaeformiscandona breuili* (Paris, 1920), previously often designated as *Candona hertzogi* Klie, 1937 was collected in the hyporheic of the Lachein. Considered as a stygobiont by Rouch, it is actually a styglophile.

4.1.5. Bathynellacea (Syncarida)

The genus *Vandelibathynella* is monospecific, represented by *Vandelibathynella vandeli* [50]. When Delamare and Chappuis described this species from Grotte de Font-Sainte, in Ariege, in 1954 under the genus *Bathynella*, they noted its special position within the group. It was later reported as being in the Baget karstic system [51] and from the hyporheic of Lachein [18]. Since then, it has only been found in two other localities, both in Ariège: Grotte de Passarolles [52], and in the hyporheic environment of the Nert stream [53]. Within the hyporheic environment of Lachein, it is found exclusively in areas of low permeability and low concentration in oxygen, which are also the least populated areas.

4.1.6. Amphipoda

Three species of amphipods inhabit the underground waters of the Baget system. One is a species of the genus *Niphargus*. It has been attributed to *Niphargus kochianus* Bate, 1859, but is actually a species of the *kochianus* group [54]. Its taxonomic status remains to be clarified. This species group is widely distributed throughout western Europe. Small for a *Niphargus* (5 to 6 mm), it prefers interstitial environments.

The other amphipods, which are also typical of interstitial environments, have a more restricted distribution.

Salentinella petiti Coineau, 1968 is found in North-West Spain and South-West France (Dordogne, Tarn, Tarn et Garonne, Lot, Pyrénées Atlantiques, Hautes-Pyrénées, Haute-Garonne and Ariège). Like the bathynellacea *Vandelibathynella vandeli*, it is typical of the least drained and the most oxygen-poor areas of the hyporheic environment.

Parasalentinella rouchi Bou,1971 is a small amphipod, less than 2 mm long with volvation abilities. Its distribution area is limited to the interstitial environments in the Pyrenees of Ariège and Haute-Garonne [55] (Figure 5).

4.1.7. Ingolfiellida

This peracarid order is represented in Baget by *Ingolfiella thibaudi* Coineau, 1968, a small anophtalmous Ingolfiellida, about 2 mm long and very elongated. It has a regional distribution in southern France covering Ariège, Hautes-Pyrénées, Tarn, Ardèche, Gard and Bouches-du-Rhône.

4.1.8. Isopoda

Two species have been recorded from the Baget: *Microcharon ariegensis* (Coineau, Boutin & Artheau, 2013) and *Stenasellus virei* Dollfus, 1897. *Microcharon* is a microparasellid isopod genus found in interstitial groundwater, river underflows and marine interstitial.

The species has a worldwide although highly fragmented distribution. It is well represented throughout the Mediterranean basin, with more than 70 known species and widespread in the Southern ground waters of France [56].

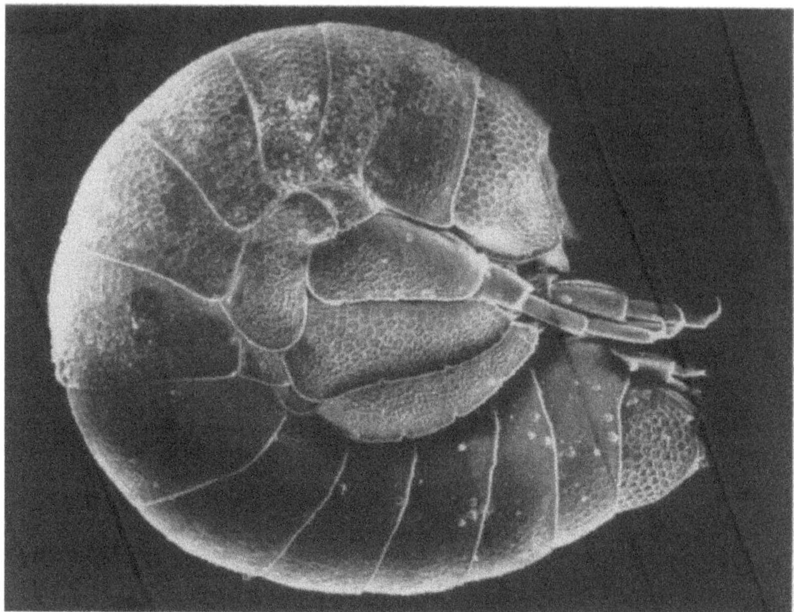

Figure 5. *Parasalentinella rouchi*. Photo C. Bou. From Danielopol, Rouch & Bou, 1999.

The species present in the Baget system was first described as *M. rouchi* (Coineau 1968), originally distributed in the subterranean waters of the Nive, Laran and Kakoueta rivers in the north of the Pays Basque, as well as in the subterranean waters of the Ariège river, its tributary the Hers and the Lachein. In 2013, the specimens of the Ariège, the Hers and the Lachein were shown to differ from the first description of *M. rouchi* and were assigned to a new species *M. ariegensis* (Coineau, Boutin & Artheau, 2013). In the Baget system, it was first collected at the Lachein hyporheic [57], and then found at Exsurgence de Las Hountas, Moulo de Jaur, Puits de la Hillère and Grotte de Sainte-Catherine [51].

Stenasellus virei Dollfus, 1897 is represented in the Baget system by two subspecies: *S. virei hussoni* Magniez, 1968 and *S. virei boui* Magniez, 1968 [58]. The specimens of the two subspecies are elongated, about 6 to 10 mm long, pinkish in color. *S.virei hussoni* is found in Exsurgence de Las Hountas, Moulo de Jaur, Puits de la Hillère, Gouffre de la Peyrère and Grotte de Saint-Catherine. The subspecies is widespread. *S. virei boui* is an endemic subspecies known only from a few localities in Ariège. Slightly thinner and whiter than *S. virei hussoni*, it is typical of interstitial environments. In the Baget system, it is found in the hyporheic of Lachein [59].

4.2. Terrestrial Fauna

Except for the Coleoptera, which have undergone more intense sampling (Table 5), all the specimens are cited from Grotte de Sainte-Catherine.

4.2.1. Palpigradi

All the palpigrades live either in the soil or in caves and are both anophtalmous and apigmented. Unlike tropical areas where endogenous species prevail, most of the European species live in caves. The very rare *Eukoenenia pyrenaella* Condé, 1990 has only been collected once and by a single specimen from the Sainte-Catherine cave [60].

4.2.2. Araneae

Among the many spiders found in the caves of the Baget system—seven common troglophilous species from the Sainte-Catherine cave (Déjean—com. pers.)—only one species, *Leptoneta convexa* Simon, 1873, can be considered troglobiotic [61]. Collected by Coiffait in 1950 in the Aven de Sainte-Catherine, it has never been recorded since [62].

4.2.3. Chilopoda

The only troglobiont known from the system, *Lithobius cavernicola* Fanzago, 1877 is endemic to the eastern Pyrenees [63].

4.2.4. Diplopoda

Blaniulus lorifer consoranensis (Brölemann, 1921) is a troglobitic species very common in the caves of Couserans (Ariège) and Comminges (Haute-Garonne).

4.2.5. Isopoda

Only one troglobitic Oniscoidae, *Scotoniscus macromelos macromelos* Racovitza, 1908 is known from the Baget karstic system. *Scotoniscus macromelos* is endemic of central Pyrenees.

4.2.6. Collembola

So far, four troglobiotic and three troglophilic species have been collected in the caves of the Baget system, mainly in Grotte de Sainte-Catherine. Among the troglobiotic species, a new species not yet described belongs to the genus *Micronychiurus*. Its troglobiotic status requires confirmation. The genus *Micronychiurus* is widespread along the Pyrenean range and highly diversified in caves and deep soils. Another new species belongs to *Oncopodura* of the *crassicornis* group and seems to be close to *O. pelissiei* from the karst of Quercy (North of Toulouse) [64]. *Pseudosinella theodoridesi* Gisin & Gama, 1969 and *Tomocerus problematicus* Cassagnau, 1964 are two endemic species, frequent in a few caves of the Couserans (western Ariège) and Comminges (Haute-Garonne) regions, where they often co-occur. Despite their status as strict troglobionts, they have retained reduced pigmentation and eyes.

4.2.7. Coleoptera (Table 5)

Only a few caves of the system were sampled for beetles, but the number of species is high: six species of the genus *Aphaenops*, genus endemic and emblematic of the Pyrenees, and one species of Leiodidae Leptodirini, *Speonomus infernus infernus* Coiffait, 1959, a species widespread in Ariège and Haute-Garonne.

The species composition is largely overlapping that of the Coume Ouarnède system, 5 of the 6 species beeing co-shared by the two systems. Only the species *A. sinester* Coiffait, 1959 is lacking from the Coume Ouarnède system.

Table 5. Coleoptera of the Baget system, with their location and bibliographical references.

Aphaenops (H) cerberus bruneti	Grotte de Sainte-Catherine [24,65–75] Gouffre du Papillon [67,68] Gouffre de la Peyrère [67] Gouffre du Lynx (new record)
Aphaenops (H) ehlersi ehlersi	Grotte de Sainte-Catherine [65,66,76–78]
Aphaenops (H) sinester	Grotte de Sainte-Catherine [75,77–80]
Aphaenops (H) tiresias tiresias	Gouffre de la Peyrère [67,81]
Aphaenops (H) bucephalus	Gouffre de la Peyrère [67,75,81]
Aphaenops (A) orpheus consorranus	Grotte de Sainte-Catherine [66,67,76,82,83]
Speonomus (M) infernus infernus	Gouffre de Tussau [84] Gouffre du Papillon (new record) Gouffre de la Peyrère [66] Grotte de Sainte-Catherine [23,66,85,86]

Carabidae

Aphaenops (Hydraphaenops) cerberus bruneti is a species widespread and common in the caves of the area (Figure 6). The population of the Baget system is not genetically distinct from the specimens from the rest of the massif of Lestelas-Balaguères and Arbas massifs [68].

Figure 6. *Aphaenops* (*Hydraphaenops*) *cerberus bruneti* from Grotte de l'Espugue. (Lestelas-Balaguères massif). Photo S. Huang.

- Aphaenops (Hydraphaenops) sinester

This enigmatic species, described from the Aven de Sainte-Catherine, known by very few specimens from this cave and the nearby Grotte de Liqué, is morphologically very close to *Aphaenops pluto* (Dieck, 1869), a species common on the other side of the Lez valley (Sourroque massif). *Aphaenops sinester* was described as a subspecies of *A. pluto* (Coiffait, 1959), but recently regarded as a distinct species [87]. More material would be required to confirm its validity.

Aphaenops (Hydraphaenops) tiresias tiresias and *A. bucephalus*, both uncommon, are nevertheless known from several caves of the Lestelas-Balaguères and Arbas massifs (Figure 7).

Surprisingly, none of those two species were found in Grotte de Sainte-Catherine. They are until today only known from Gouffre de la Peyrère. This disparity between Gouffre de la Peyrère and Grotte de Sainte-Catherine, although very close, roughly matches that observed for aquatic fauna, and raises questions about potential barriers to dispersion or ecological heterogeneity within the Baget system.

Aphaenops (Argonotrechus) orpheus ssp. *consorranus* is rather an endogean subspecies, not rare in caves or under stones in forest on the Lestelas-Balaguères and Arbas massifs. Being often mentioned as a troglobitic species (e.g., [81]), we listed it here as such.

Figure 7. *Aphaenops (Hydraphaenops) tiresias* feeding on a Diptera (Limoniidae). Grotte du Goueil-di-Her, Coume Ouarnède system. *Photo* A. Faille.

Leiodidae

The only species of cave Leiodidae, *Speonomus (Machaeroscelis) infernus*, is common and occurs in several caves of Ariège and Haute-Garonne (Figure 8).

Representatives of the endogean genus *Bathysciola* are also present, with many narrow endemics (e.g., *B. arcuatipes* Jeannel, 1924, *B. lapidicola* Saulcy, 1872, *B. liqueana* Fresneda, Bourdeau & Faille, 2010) in the area [88]. Their presence in the studied area remains to be confirmed.

Figure 8. *Speonomus (Machaeroscelis) infernus* from Grotte de l'Espugue, Lestelas-Balaguères massif. Photo C. Vanderbergh.

4.2.8. Laboulbeniales

Laboulbeniales are highly specialized parasitic fungi that are found exclusively on insects, arachnids and myriapods. The genus *Rhachomyces* parasitizes many species of different subgenera of *Aphaenops*. Among the five species paraziting *Aphaenops*, *Rhachomyces aphaenopsis* Thaxter, 1905 is the most common [75]. It has been found in Aven de Ste Catherine, on A. cerberus bruneti [65,75,89], on A. sinester [75] and on A. ehlersi [65,75]. In the Gouffre de la Peyrère, it was recorded on A. bucephalus [75]. Even if not specifically reported for the area, all the Trechini species (*A. bucephalus, A. cerberus, A. ehlersi, A. orpheus, A. sinester, A. tiresias*) present in the Baget system are known to host the species.

5. Discussion

In view of the taxa listed above, the Baget system deserves to be qualified as a hotspot for underground biodiversity. With 40 stygobiotic and 17 troglobiotic species, it is even to date the richest in the Pyrenees, ahead of Coume Ouarnède (21 stygobiotic and 17 troglobitic species) [2]. In addition to the number of species, its richness is also measured by the large number of higher-rank taxa and the presence of narrow endemic species (*Cookidrilus ruffoi* Giani, Martinez-Ansemil & Sambugar, 2004 for the hyporheic environment, *Antrocamptus catherinae* Chappuis & Rouch, 1960 for the karst groundwater, and *Eukoenenia pyrenaella* Condé, 1990 for the terrestrial environment).

Although the size of the catchment is relatively small (13.5 km^2), the system offers a large number of subterranean habitats, which can partly explain this richness. Above all, this strong biodiversity can be linked to its presence in a particularly rich sector, the Central Pyrenees. It may be interesting to compare the Baget biocenoses of the two neighboring Pyrenean hotspots: the Nert for the hyporheic fauna and the Coume Ouarnède system [2,90,91] (Figure 9 and Table 6).

For the terrestrial fauna, the Baget and the Coume Ouarnède are two contiguous systems, but with a faunistic assemblage very distinct from each other. Of 26 species present in total, only 8 are present on both sites. Richness is similar on both systems, with 17 troglobionts. In Coume Ouarnède system, it includes two remarkable relict species, the opilion *Arbasus caecus* (Simon, 1911) [92] and the springtail *Tritomurus falcifer* Cassagnau, 1958, which are absent in the Baget system. On the other hand, we find in Baget a strictly endemic species, *Eukoenenia pyrenaella* Condé, 1990.

For aquatic fauna, Coume Ouarnède system hosts fewer troglobionts (22) than Baget system (40). Two hypotheses may explain this. First, a less extended hyporheic habitat in Coume Ouarnède and second, a much more in-depth study of the aquatic fauna of the Baget catchment (Table 7). Rouch's work, conducted over a period of more than 30 years, and concerning both karst and hyporheic habitats, provides us with a detailed and complete knowledge of the stygofauna, probably unmatched at the European scale. Of a total of 49 to 50 species (uncertainty due to the unidentified *Salentinella* sp. in Coume Ouarnède), only 10 to 12 are present on both sites. If we compare the data of Baget with those of Nert [91], the number of species is again greater in Baget. Nevertheless, we obtain similar results if we consider the hyporheic environment only with 29 species in Nert and 31 in Baget. It can be noted that although the surface studied in the hyporheic of Nert is larger, the sampling was less intense. Of a total of 49 species on these two sites, 18 are present in both. Baget, Nert and Coume Ouarnède altogether host 56 stygobionts, among which only 10 are common to the three sites.

All these results lead us to believe that regarding aquatic fauna, the data collected in the Baget catchment reflect the reality of the population. No region of the Pyrenees has been the site of such an intense collection effort, and, given the fragmentary data available [20], it is very likely that more in-depth studies on other aquatic habitats, including Nert and Coume Ouarnède, would lead to a significant increase in the underground biodiversity of the central Pyrenees.

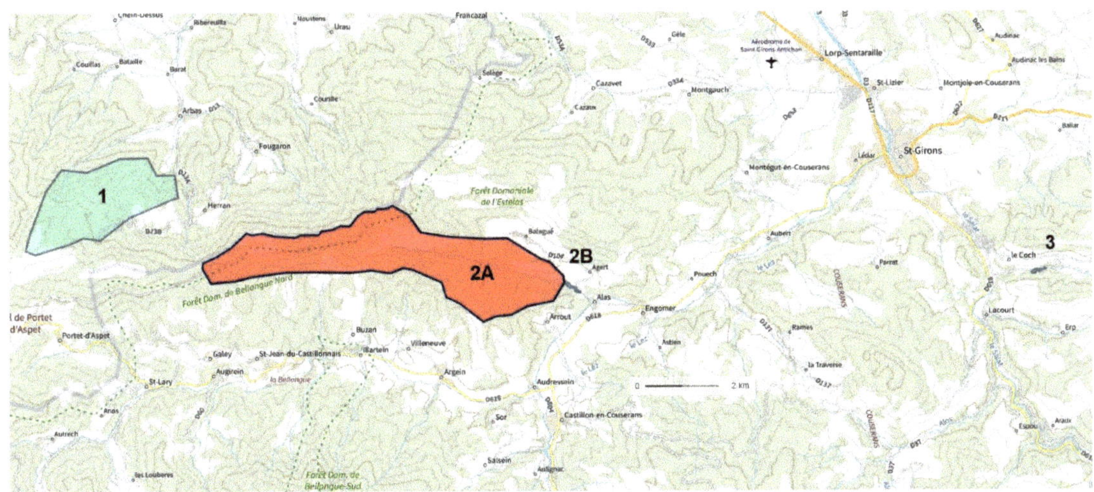

Figure 9. Location of the 3 Pyrenean subterranean hotspots. 1: Coume Ouarnède System; 2A: Baget System; 2B: Hyporheic of Lachein stream; 3: Hyporheic of Nert stream.

Table 6. Comparative biocenosis of Baget system, Coume Ouarnède system and the hyporheic of Nert stream. CO = Coume Ouarnède.

Stygobionts	BAGET	CO	NERT
Atrioplanaria delamarei			X
Plagnolia vandeli		X	X
Delaya leruthi	X	X	
Cookidrilus ruffoi	X		
Cookidrilus speluncaeus	X		
Spiridion phreaticola	X		
Islamia moquiniana	X	X	
Moitessieria simoniana	X	X	X
Acanthocyclops sensitivus	X		X
Diacyclops belgicus	X		
Diacyclops clandestinus	X		
Diacyclops languidoides ssp.	X	X	
Graeteriella (Graeteriella) rouchi	X		X
Graeteriella sp.	X		
Grateriella (Paragrateriella) sp.		X	
Speocyclops anomalus	X	X	
Speocyclops kieferi	X		X
Speocyclops racovitzai boscensis cf.	X		
Speocyclops racovitzai liquensis	X		
Speocyclops racovitzai ssp.		X	X
Nitocrella delayi	X		
Nitocrella gracilis	X	X	X
Parapseudoleptomesochra subterranea subterranea	X	X	X
Antrocamptus catherinae	X		
Antrocamptus chappuisi	X		X
Ceuthonectes gallicus	X	X	X

Troglobionts	BAGET	CO
Eukoenenia pyrenaella	X	
Troglocheles vandeli		X
Leptoneta convexa	X	
Leptoneta microphtalma		X
Arbasus caecus		X
Lithobius cavernicola	X	
Blaniulus lorifer consoranensis	X	X
Blaniulus troglobius gibbicollis		X
Spelaeoglomeris jeanneli		X
Scotoniscus macromelos macromelos	X	X
Micronychiurus n. sp.	X	
Pseudosinella theodoridesi	X	X
Oncopodura n. sp.	X	
Oncopodura tricuspidata		X
Tomocerus problematicus	X	
Tritomurus falcifer		X
Aphaenops (Hydraphaenops) sinester	X	
Aphaenops (Hydraphaenops) bucephalus	X	X
Aphaenops (Hydraphaenops) cerberus bruneti	X	X
Aphaenops (Hydraphaenops) crypticola		X
Aphaenops (Hydraphaenops) ehlersi	X	X
Aphaenops (Hydraphaenops) tiresias tiresias	X	X
Aphaenops (Argonotrechus) orpheus consorranus	X	
Speonomus (Machaeroscelis) infernus infernus	X	
Speonomus (Machaeroscelis) infernus arbasanus		X
Rhachomyces aphaenopsis	X	X

Table 6. *Cont.*

Stygobionts	BAGET	CO	NERT	Troglobionts		
					BAGET	C O
Elaphoidella bouilloni	X		X			
Elaphoidella coiffaiti	X					
Elaphoidella infernalis		X				
Moraria (Moraria) catalana	X					
Parastenocaris dianae	X	X	X			
Parastenocaris nertensis			X			
Parastenocaris vandeli	X		X			
Proserpinicaris mangini	X					
Proserpinicaris meridionalis			X			
Pseudocandona rouchi	X		X			
Pseudocandona sp.1	X					
Pseudocandona sp.2	X					
Pseudocandona sp.3	X					
Vandelibathynella vandeli	X		X			
Bathynella sp.		X				
Niphargus gineti			X			
Niphargus foreli		X				
Niphargus kochianus	X		X			
Niphargus pachypus		X	X			
Niphargus robustus		X	X			
Parasalentinella rouchi	X	X	X			
Salentinella petiti	X		X			
Salentinella sp.		X	X			
Coxosalentinella gineti			X			
Ingolfiella thibaudi	X		X			
Microcharon ariegensis	X		X			
Proasellus racovitzai		X				
Stenasellus virei boui	X		X			
Stenasellus virei hussoni	X	X				

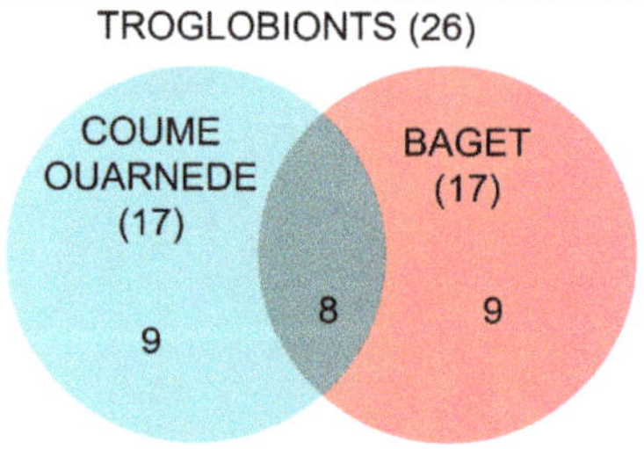

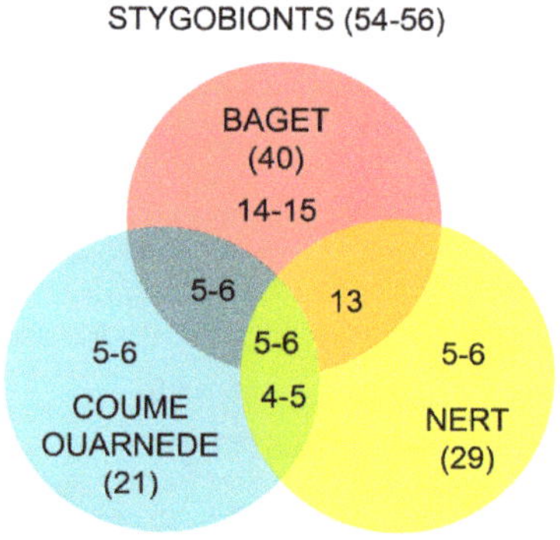

Table 7. Distribution of stygobionts between hyporheic and karstic habitats for the Baget system, the Coume Ouarnède system, and the hyporheic of the Nert.

Stygobionts	Karstic Habitat K	Hyporheic Habitat HR	K + HR
Baget system	23	31	40
Coume Ouarnède system	18	11	22
Hyporheic of Nert stream	0	29	29

Concerning the troglofauna of the system, the data come only from a few caves, mostly Grotte de Sainte-Catherine. Furthermore, many taxonomic groups such as Acari, Pseudoscorpiones, Oniscida, Diplura are poorly or not represented, probably due to insufficient sampling. The significant speleological developments of the last 15 years have opened up a vast field of study, which could lead, if biospeleologists get down to it, to new discoveries.

Author Contributions: All the authors participated in all phases of the realization and writing of the paper. All authors have read and agreed to the published version of the manuscript.

Funding: This research received no external funding.

Institutional Review Board Statement: Not applicable.

Data Availability Statement: Data are contained within the article.

Acknowledgments: We especially thank Louis Deharveng for his support all along the realization of the paper and providing access to his database. We thank Vincent Prié for his help on Gastropoda, Christian Vanderbergh and Sunbin Huang for providing pictures of Coleoptera. We are grateful to Daniel Quettier (Société Méridionale de Spéléologie et de Préhistoire) for his important inventory work of the Pyrenean caves and for giving us access to all his data. In general, we thank all the cavers (among them, Groupe Spéléologique du Couserans, Spéléo Club de l'Epia) with whom we continue to explore and study the caves of the Estelas-Balaguères massif.

Conflicts of Interest: The authors declare no conflicts of interest.

References

1. Culver, D.C.; Deharveng, L.; Pipan, T.; Bedos, A. An overview of subterranean biodiversity hotspots. *Diversity* **2021**, *13*, 487. [CrossRef]
2. Faille, A.; Deharveng, L. The Coume Ouarnède System, a Hotspot of Subterranean Biodiversity in Pyrenees (France). *Diversity* **2021**, *13*, 419. [CrossRef]
3. Société Méridionale de Spéléologie et de Préhistoire. Bulletins de la Section Spéléologie. Available online: http://smsp-speleo.blogspot.com/p/bibliotheque-pdf.html (accessed on 16 January 2024).
4. Mangin, A. Le système karstique du Baget (Ariège) (note préliminaire). *Ann. Spéléo.* **1970**, *25*, 561–580.
5. Weber, B.; Malard, A. Système souterrain de l'Hider. *Spéléo Mag.* **2019**, *108*, 20–25.
6. Bréhier, F.; Tixier, G. Reprise des explorations sur le réseau du Pas du Loup—Francazal (Haute-Garonne). *Le Fil* **2013**, *25*, 26–32.
7. Spéléo-Club EPIA. Le spéléo-club EPIA: Un club de la balle atomique! *Spelunca* **2014**, *133*, 41–44.
8. Cabanel, P. Aliou 2019. *Le Fil* **2020**, *30*, 43–55.
9. Alcamo, S.; Marguet, T.; Burle, C.; Gould, V. Le gouffre de la Lune—Haute-Garonne: Récit d'une exploration estélassienne. *Spéléo Mag.* **2022**, *118*, 6–13.
10. Mangin, A. Contribution à l'étude hydrodynamique des aquifères karstiques: Première partie Généralités sur le karst et les lois d'écoulement utilisées. *Ann. Spéléo.* **1974**, *29*, 283–332.
11. Mangin, A. Contribution à l'étude hydrodynamique des aquifères karstiques: Deuxième partie Concepts méthodologiques adoptés. Systèmes karstiques étudiés. *Ann. Spéléo.* **1974**, *29*, 495–601.
12. Mangin, A. Contribution à l'étude hydrodynamique des aquifères karstiques: Troisième partie Constitution et fonctionnement des aquifères karstiques. *Ann. Spéléo.* **1975**, *30*, 21–124.
13. Debroas, E.J. Géologie du bassin versant du Baget (zone nord-pyrénéenne, Ariège, France): Nouvelles observations et conséquences. *Strata* **2009**, *46*, 93.
14. Marsaud, B. *Structure et Fonctionnement de la Zone Noyée des Karst à Partir des Résultats Expérimentaux*; Editions BRGM: Orléans, France, 1997; 306p.
15. Rouch, R. Recherches sur les eaux souterraines. 17. Le système karstique du Baget. II. Etude des Harpacticides rejetés au niveau de Las Hountas au cours de plusieurs crues du cycle hydrologique 1970–1971. *Ann. Spéléo.* **1972**, *27*, 139–176.
16. Rouch, R. Recherches sur les eaux souterraines. 24. Le système karstique du Baget. III. Etude des sorties d'Harpacticides au niveau de Las Hountas lors de plusieurs crues des cycles hydrologiques 1971–1972 et 1972–1973. *Ann. Spéléo.* **1974**, *29*, 351–372.
17. Rouch, R.; Pitzalis, A.; Descouens, A. Effets de pompage à gros débit sur le peuplement des Crustacés d'un aquifère karstique. *Ann. Limn.* **1993**, *29*, 15–29. [CrossRef]
18. Rouch, R. Sur la répartition spatiale des Crustacés dans le sous-écoulement d'un ruisseau des Pyrénées. *Ann. Limn.* **1988**, *24*, 213–234. [CrossRef]
19. Jeannel, R. Quarante années d'explorations souterraines. *Notes Biospéologiques* **1950**, *6*, 1–93.
20. Juberthie, C.; Ginet, R. France. In *Encyclopaedia Biospeologica*; Tome 1; Juberthie, C., Decu, V., Eds.; Société de Biospéologie: Moulis, France, 1994; pp. 665–692.
21. Jeannel, R. *Faune Cavernicole de la France*; Lechevalier: Paris, France, 1926; 334p.
22. Renauld, P. Etude géomorphologique des grottes de Sainte-Catherine (Ariège). *Ann. Spéléo.* **1969**, *24*, 5–18.
23. Juberthie, C. Etude écologique des larves de *Speonomus infernus* dans la grotte de Sainte-Catherine. *Ann. Spéléo.* **1969**, *24*, 563–577.
24. Juberthie, C. Relation entre le climat, le microclimat et les *Aphaenops cerberus* de la grotte de Sainte-Catherine (Ariège). *Ann. Spéléo.* **1969**, *24*, 75–104.
25. Andrieux, C. Etude du climat de la grotte de Sainte-Catherine en Ariège selon le cycle de 1967. *Ann. Spéléo.* **1969**, *24*, 19–74.
26. Ulloa-Cedamanos, F.; Probst, J.L.; Binet, S.; Camboulive, T.; Payre-Suc, V.; Pautot, C.; Bakalowicz, M.; Beranger, S.; Probst, A. A forty-year karstic critical zone survey (Baget Catchment, Pyrenees-France): Lithologic and hydroclimatic controls on seasonal and inter-annual variations of stream water chemical composition, pCO_2, and carbonate equilibrium. *Water* **2020**, *12*, 1227. [CrossRef]
27. Bou, C.; Rouch, R. Un nouveau champ de recherches sur la faune aquatique souterraine. *Compte Rendu L'académie Sci. Paris* **1967**, *265*, 369–370.
28. Rouch, R. Contribution à la connaissance des Harpacticides hypogés (Crustacés Copépodes). *Ann. Spéléo.* **1968**, *23*, 5–167.
29. Rouch, R.; Bonnet, L. Le système karstique du Baget. IV. Premières données sur la structure et l'organisation de la communauté des Harpacticides. *Ann. Spéléo.* **1976**, *31*, 27–41.

30. Rouch, R.; Bonnet, L. Le système karstique du Baget. V. La communauté des Harpacticides. Sur la constance du peuplement. *Ann. Spéléo.* **1976**, *31*, 43–53.

31. Rouch, R. Le système karstique du Baget. XII. La communauté des Harpacticides. Sur l'interdépendance des nomocénoses épigée et hypogée. *Ann. Spéléo.* **1982**, *18*, 41–54. [CrossRef]

32. Rouch, R. Le système karstique du Baget. XIII. Comparaison de la dérive des Harpacticides à l'entrée et à la sortie de l'aquifère. *Ann. Limn.* **1982**, *18*, 133–150. [CrossRef]

33. Rouch, R.; Carlier, A. Le système karstique du Baget. XIV. La communauté des Harpacticides épigés. Evolution et comparaison des structures du peuplement épigé à l'entrée et à la sortie de l'aquifère. *Stygologia* **1985**, *1*, 71–91.

34. Rouch, R.; Carlier, A. Le système karstique du Baget. XV. Le peuplement des Harpacticides épigés. Analyse des prélèvements synchrones à l'entrée et à la sortie de l'aquifère. *Stygologia* **1985**, *1*, 224–238.

35. Rouch, R.; Bakalowicz, M.; Mangin, A.; D'hulst, D. Sur les caractéristiques chimiques du sous-écoulement d'un ruisseau des Pyrénées. *Ann. Limn.* **1989**, *25*, 3–16. [CrossRef]

36. Rouch, R.; Lescher-Moutoué, F. Structure du peuplement des Cyclopides (Cructacea: Copepoda) dans le milieu hyporhéique d'un ruisseau des Pyrénées. *Stygologia* **1992**, *7*, 197–211.

37. Route, N.; Martinez-Ansenil, E.; Sambugar, B.; Giani, N. On some interesting freshwater Annelida, mainly Oligochaeta, of the underground waters of southwestern France with the description of a new species. *Subteranea Biol.* **2004**, *2*, 1–5.

38. MNHN & OFB (Ed.). Inventaire National du Patrimoine Naturel (INPN). 2003–2023. Available online: https://inpn.mnhn.fr (accessed on 16 January 2024).

39. Bertrand, A.; Bréhier, F.; Dumas, P.; Juberthie, C.; Mangin, A.; Cabrol, P.; Grassaud, M. *Projet de Réserve Naturelle Souterraine de l'Ariège. Sites, Paysages, Nature*; Préfecture de l'Ariège & DREAL Midi-Pyrénées: Toulouse, France, 2002.

40. Zagmajster, M.; Malard, F.; Eme, D.; Culver, D.C. Subterranean Biodiversity Patterns from Global to Regional Scales. In *Cave Ecology*; Ecological Studies (Analysis and Synthesis); Moldovan, O., Kováč, L'., Halse, S., Eds.; Springer: Cham, Switzerland, 2018; Volume 235, pp. 195–227.

41. Rouch, R. Structure du peuplement des Harpacticides dans le milieu hyporhéique d'un ruisseau des Pyrénées. *Ann. Limn.* **1991**, *27*, 227–241. [CrossRef]

42. Rouch, R. Caractéristiques et conditions hydrodynamiques des écoulements dans les sédiments d'un ruisseau des Pyrénées. Implications écologiques. *Stygologia* **1992**, *7*, 13–25.

43. Rouch, R.; Bonnet, L. Le système karstique du Baget. VI. La communauté des Harpacticides. Signification des échantillons récoltés lors des crues au niveau de deux exutoires du système. *Ann. Limn.* **1977**, *13*, 227–249. [CrossRef]

44. Rouch, R.; Bonnet, L. Le système karstique du Baget. VII. La communauté des Harpacticides. Structure du peuplement hypogée pendant les crues. *Ann. Limn.* **1977**, *13*, 251–259. [CrossRef]

45. Rouch, R. Le systèmes karstique du Baget. VIII. La communauté des Harpacticides. Les apports au sein du système par dérive catastrophique. *Ann. Limn.* **1977**, *15*, 243–274. [CrossRef]

46. Rouch, R. Le systèmes karstique du Baget. X. La communauté des Harpacticides. Richesse spécifique, diversité et structures d'abondances de la nomocénose hypogée. *Ann. Limn.* **1980**, *16*, 1–20. [CrossRef]

47. Rouch, R. Le système karstique du Baget. XI. La communauté des Harpacticides. Sur l'évolution de la nomocénose épigée au sein de l'aquifère. *Ann. Limn.* **1980**, *16*, 299–314. [CrossRef]

48. Rouch, R. Le systèmes karstique du Baget. IX. La communauté des Harpacticides. Richesse spécifique, diversité et structure d'abondances des échantillons de dérive au niveau du ruissellement de surface. *Vie Milieu* **1980**, *30*, 229–241.

49. Meisch, C.; Smith, R.J.; Martens, K. A subjective global checklist of the extant non-marine Ostracoda (Crustacea). *Eur. J. Taxon.* **2019**, *492*, 1–135. [CrossRef]

50. Delamare Deboutteville, C.; Chappuis, P.A. Les Bathynelles de France et d'Espagne avec diagnoses d'espèces et de formes nouvelles. *Vie Milieu* **1954**, *4*, 114–115.

51. Rouch, R. Recherches sur les eaux souterraines. 12. Le système karstique du Baget. I. Le phénomène d'hémorragie au niveau de l'exutoire principal. *Ann. Spéléo.* **1970**, *25*, 667–709.

52. Serban, E.; Coineau, N.; Delamarre-Deboutteville, C. Recherches sur les Crustacés souterrains et mésopsamiques. I—Les Bathynellacés Malacostraca des régions méridionales de l'Europe occidentale. La sous-famille des Gallobathynellinae. *Mémoires Du Muséum Natl. D'histoire Nat. Série A Zool.* **1972**, *75*, 7–107.

53. Rouch, R. Peuplement des Crustacés dans la zone hyporhéique d'un ruisseau des Pyrénées. *Ann. Limn.* **1995**, *31*, 9–28. [CrossRef]

54. Ginet, R. Bilan systématique du genre *Niphargus* en France. *Soc. Linn. Lyon* **1996**, 1–237.

55. Bou, C. Recherches sur les eaux souterraines XVI. *Parasalentinella rouchi* n.gen.,n.sp. des eaux souterraines des Pyrénées francaises (Amphipoda, Gammaridea). *Ann. Spéléo.* **1971**, *26*, 481–494.

56. Coineau, N.; Boutin, C.; Artheau, M. Origin of the interstitial isopod Microcharon (Crustacea, Microparasellidae) from the western Languedoc and the northern Pyrenees (France) with the description of two new species. *Subterr. Biol.* **2013**, *10*, 1–16. [CrossRef]

57. Coineau, N. Contribution à l'étude de la faune interstitielle. Isopodes et Amphipodes. Mémoires du Muséum national d'Histoire naturelle Paris, N.S. *Ann. Zool.* **1968**, *55*, 147–216.

58. Magniez, G. L'espèce polytypique *Stenasellus virei* Dollfus, 1897 (Crustacé Isopode Hypogé). *Ann. Spéléo* **1968**, *23*, 363–407.

59. Magniez, G. Observations sur *Stenasellus virei* (Crustacea, Isopoda, Asellota) dans ses biotopes naturels des eaux souterraines. *Int. J. Speleol.* **1974**, *6*, 115–171. [CrossRef]

60. Condé, B. Palpigrades (Arachnida) de grottes d'Europe. *Rev. Suisse Zool.* **1989**, *96*, 823–840.
61. Le Peru, B. The spiders of Europe, a synthesis of data: Volume 1. Atypidae to Theridiidae. *Mém. Soc. Linn. Lyon* **2011**, *2*, 1–522.
62. Dresco, E. Etude des Leptoneta. Leptoneta convexa Sim. et microphtalma Sim. (Aranaeae Leptonetidae). *Bull. Soc. Hist. Nat. Toulouse* **1980**, *116*, 146–149.
63. Iorio, E. Catalogue biogéographique et taxonomique des Chilopodes (Chilopoda) de France métropolitaine. *Mém. Soc. Linn. Bordeaux* **2014**, *15*, 1–372.
64. Deharveng, L. Collemboles cavernicoles VIII. Contribution à l'étude des Oncopoduridae. *Bull. Soc. Ent. Fr.* **1988**, *92*, 133–147. [CrossRef]
65. Boyer-Lefèvre, N.H. Recherches sur les Laboulbéniales des Trechinae cavernicoles pyrénéens. *Ann. Spéléo.* **1965**, *20*, 117–131.
66. Coiffait, H. Biospeologica LXXVII. Enumération des grottes visitées (Neuvième série). *Arch. Zool. Exp. Gén.* **1959**, *97*, 209–465.
67. Faille, A. Endémisme et Adaptation à la Vie Cavernicole Chez les Trechinae Pyrénéens (Coleoptera: Carabidae). Approches Moléculaire et Morphométrique. Ph.D. Dissertation, Muséum National d'Histoire Naturelle, Paris, France, 2006.
68. Faille, A.; Tänzler, R.; Toussaint, E.F.A. On the way to speciation: Shedding light on the karstic phylogeography of the micro-endemic cave beetle *Aphaenops cerberus* in the Pyrenees. *J. Hered.* **2015**, *106*, 692–699.
69. Jauzion, G. Activités de la section spéléologie—1990–1991. *Bull. Soc. Mérid. Spéléo. Préhist.* **1991**, *31*, 101–107.
70. Jeannel, R. Les Trechinae de France (Deuxième Partie). *Ann. Soc. Entom. France* **1921**, *90*, 295–344. [CrossRef]
71. Jeannel, R. Monographie des Trechinae (Troisième livraison). Les Trechini cavernicoles. *L'Abeille* **1928**, *35*, 1–808.
72. Jeannel, R. *Faune de France. Coléoptères Carabiques I, Vol. 39*; P. Lechevalier: Paris, France, 1941; 571p.
73. Jeannel, R. *Faune de France. Vol. 51: Coléoptères Carabiques (Supplément)*; P. Lechevalier: Paris, France, 1949; 51p.
74. Juberthie, C.; Massoud, Z.; Piquemal, F. L'équipement sensoriel des Trechinae souterrains. I.—Les organes sensoriels de l'élytre. *Ann. Spéléo.* **1975**, *30*, 483–494.
75. Santamaria, S.; Faille, A. *Rhachomyces* (Ascomycota, Laboulbeniales) parasites on cave-inhabiting Carabidae beetles from Pyrenees. *Nova Hedwig.* **2007**, *85*, 159–186. [CrossRef]
76. Coiffait, H. Note sur les Trechidae cavernicoles et endogés de la région Pyrénéenne. *Notes Biospéologiques* **1958**, *XIII*, 7–23.
77. Coiffait, H. Localités nouvelles ou peu connues de Coléoptères cavernicoles (Pyrénées centrales). *Bull. Soc. Mérid. Spéléo. Préhist.* **1962**, *9*, 70–77.
78. Coiffait, H. Monographie des Trechinae cavernicoles des Pyrénées. *Ann. Spéléo.* **1962**, *17*, 119–170.
79. Coiffait, H. Nouveaux Trechini cavernicoles des Pyrénées françaises. *Ann. Spéléo.* **1959**, *XIV*, 343–350.
80. Déliot, P. Coléoptères Trechinae et Bathysciinae peuplant le domaine souterrain du département de l'Ariège. Liste des espèces, localisations et observations. *Ariège Nat.* **1995**, *5*, 14–31.
81. Faille, A.; Ribera, I.; Deharveng, L.; Bourdeau, C.; Garnery, L.; Queinnéc, E.; Deuve, T. A molecular phylogeny shows the single origin of the Pyrenean subterranean Trechini ground beetles (Coleoptera: Carabidae). *Mol. Phylogenetics Evol.* **2010**, *54*, 97–105. [CrossRef]
82. Coiffait, H. Note sur les premiers états de *Geotrechus orpheus* ssp. *consorranus* et sur la biologie larvaire de ce coléoptère. *Vie Milieu* **1951**, *II*, 461–469.
83. Fourès, H. Localités nouvelles ou peu connues de coléoptères cavernicoles des Pyrénées françaises. *Bull. Soc. Hist. Nat. Toulouse* **1947**, *82*, 252–256.
84. Mouriès, M. Fiche cavité. In *Fichier Spéléologique Du Comité Spéléologique Régional de Midi-Pyrénées*; 1978.
85. Jeannel, R. Révision des Bathysciinae (Coléoptères, Silphides). Morphologie, distribution géographique, Systématique. *Arch. Zool. Exp. Gén.* **1911**, *47*, 1–641.
86. Jeannel, R.; Racovitza, E. Enumération des grottes visitées, 1908–1909 (Troisième série). Biospeologica XVI. *Arch. Zool. Exp. Gén.* **1910**, *5*, 67–185.
87. Quéinnec, E.; Ollivier, E. Tribu Trechini. In *Faune de France 94. Coléoptères Carabiques, Compléments et Mise à Jour*; Coulon, J., Pupier, R., Quéinnec, E., Ollivier, E., Richoux, P., Eds.; Fédération Française des Sociétés de Sciences Naturelles: Paris, France, 2011; Volume 1, pp. 19–254.
88. Tronquet, M. *(Ed). Catalogue des Coléoptères de France*; Association Roussillonnaise d'Entomologie: Perpignan, France, 2014; 1052p.
89. Santamaria, S. Els *Rhachomyces* (Laboulbeniales, Ascomycotina) paràsits dels caràbids troglobionts als Pirineus. *Orsis* **1989**, *4*, 3–10.
90. Danielopol, D.L.; Rouch, R.; Bou, C. High Amphipoda Species Richness in the Nert Groundwater System (Southern France). *Crustaceana* **1999**, *72*, 863–882. [CrossRef]
91. Danielopol, D.L.; Rouch, R.; Baltanas, A. Taxonomic diversity of groundwater Harpacticoida (Copepoda, Crustacea) in Southern France. A contribution to characterise hot-spot diversity sites. *Vie Milieu* **2002**, *52*, 1–15.
92. Karaman, I. *Arbasus caecus* (Simon, 1911): A new member of the ancient family Buemarinoidae (Opiliones, Laniatores) and its relation to the known species. *Biol. Serbica* **2023**, *45*.

diversity

Article

The Cent Fonts Aquifer: An Overlooked Subterranean Biodiversity Hotspot in a Stygobiont-Rich Region

Vincent Prié [1,2,*], Cédric Alonso [3], Claude Bou [4], Diana Maria Paola Galassi [5], Pierre Marmonier [6] and Marie-José Dole-Olivier [7]

1   BIOPOLIS Program in Genomics, Biodiversity and Land Planning, CIBIO, Campus de Vairão, 4485-661 Vairão, Portugal
2   Institut Systématique Evolution Biodiversité (ISYEB), Muséum National d'Histoire Naturelle, CNRS, Sorbonne Université, EPHE, Université des Antilles, 57 rue Cuvier, CP 51, 75005 Paris, France
3   Independent Researcher, 17 Rue du Bourguet, 34230 Le Pouget, France; contact@rosalia-expertise.com
4   Association Tarnaise d'Etudes Karstiques, 52, Chemin de la Fourestole, 81990 Cambon d'Albi, France; bou.claude81@sfr.fr
5   Department of Life, Health & Environmental Sciences, University of L'Aquila, 67100 L'Aquila, Italy; dianamariapaola.galassi@univaq.it
6   Laboratoire d'Ecologie des Hydrosystèmes Naturels et Anthropisés, UMR-CNRS 5023, Université de Lyon, 43 Boulevard du 11 Novembre 1918, 69622 Villeurbanne, France; pierre.marmonier@univ-lyon1.fr
7   Ecologie, Evolution, Ecosystèmes Souterrains (E3S), Laboratoire d'Ecologie des Hydrosystèmes Naturels et Anthropisés (LEHNA), Université Lyon 1, Bât. FOREL, 43 Bd du 11 Nov.1918, 69622 Villeurbanne, France; mariejose.olivier01@gmail.com
*   Correspondence: prie.vincent@gmail.com

Abstract: The South of France is a biodiversity hotspot within Europe. Here, we present a comprehensive review of surveys conducted in the Cent Fonts aquifer, an overlooked subterranean biodiversity hotspot embedded in a region rich in stygobiotic species and threatened by climate change and water abstraction projects. Key studies, spanning from 1950 to 2006, show a progression in survey methods and results, although troglobiotic species remain poorly documented. With 43 stygobiotic species recorded, the Cent Fonts is the richest stygobiont hotspot in France. Most species are regional endemics, a quarter of which are considered vulnerable by the IUCN. The Cent Fonts also hosts several relict species and is the type locality of four species. Such a high biological value clearly deserves to be preserved. Our analysis warns of a possible decline in biodiversity, as eight of the species recorded in the 20th century were absent from the 2006 survey, suggesting potential threats of unknown origin. The capture of the Cent Font springs for water abstraction is discussed as a potential threat to this ecosystem and its unique biodiversity. Three new species of stygobiotic molluscs are described, one of which was collected in the Cent Fonts.

Keywords: stygobiont; troglobiont; conservation; karst; subterranean diversity; conservation; water abstraction

Citation: Prié, V.; Alonso, C.; Bou, C.; Galassi, D.M.P.; Marmonier, P.; Dole-Olivier, M.-J. The Cent Fonts Aquifer: An Overlooked Subterranean Biodiversity Hotspot in a Stygobiont-Rich Region. *Diversity* **2024**, *16*, 50. https://doi.org/10.3390/d16010050

Academic Editors: Sebastiano Salvidio, Tanja Pipan, David C. Culver and Louis Deharveng

Received: 7 December 2023
Revised: 31 December 2023
Accepted: 8 January 2024
Published: 12 January 2024

1. Introduction

1.1. Karst and Caves of the North-Montpellier Region

The Cent Fonts aquifer is located in the southern region of France, in the Mediterranean basin, and is part of a larger karstic system comprising the Hérault (2600 km^2), Vidourle (800 km^2), and Lez (200 km^2) river basins. These regions are already acknowledged for their abundant and remarkable subterranean biodiversity [1,2].

1.2. Description of the Cent Fonts System

The karst system supplying the Cent Fonts is located in the western part of the northern Montpellier garrigues, formed by the limestone and dolomitic massifs located between Montpellier and the Cévennes. This karstic system develops within massive dolomites and oolitic limestones of Bathonian age (Middle Jurassic). The Cent Fonts aquifer is a binary karstic system, receiving its water supply from both the rainfall on the Causse-de-la-Selle plateau and a sinkhole from the Buèges River, a tributary of the Hérault, situated more than 8 km upstream (Figure 1). The average altitude of the plateau that forms the Cent Fonts catchment area is about 300 m; the Cent Fonts springs are located at an altitude of 81 m on the right bank of the Hérault River. These springs emerge in the Bathonian dolomite, close to a fault. The system consists of about ten resurgences spread over a 300 m front, two observation points located a few meters higher, and the Cent Fonts cave, the entrance of which is situated a few meters above the observation points.

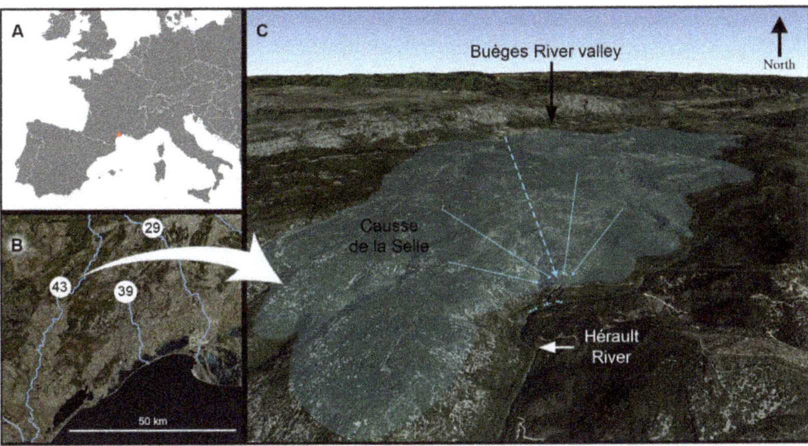

Figure 1. Location of the Cent Fonts springs. (**A**) In Western Europe. (**B**) Biodiversity hotspots of the Montpellier region, numbers refer to the number of stygobiotic species, 43 in the Cent Fonts, 39 in the Lez aquifer, and 29 in the Sauve spring. (**C**) The Causse de la Selle, aquifer of the Cent Fonts. Blue dots: springs; black line: cave topography. In bluish is the Causse-de-la-Selle plateau, which is the impluvium of the Cent Fonts system. To the north is the Buèges River, part of whose waters flow into the underground water system of the Cent Fonts (dotted blue line). Image © Google Earth.

The spring has been explored by cave divers, one of whom died in 1984. The divers reached a depth of −95 m, about 150 m from the cave entrance, and were blocked by a narrow passage.

The Cent Fonts system is the most important emergence of the Causse-de-la-Selle plateau. The land use in its catchment area consists mainly of evergreen oak forests *Quercus ilex* L., 1753 and extensive pastures. Human settlement in this area is very limited, and the presumed anthropogenic impacts are low. The Cent Fonts site falls within several protected areas (Natura 2000 site FR9101388—Gorges de l'Hérault; classified site; Grand Site de France; ZNIEFF (Natural Areas of Floristic and Faunistic Interest)). This site stands out for its remarkable vegetation associations (Salzman Pine forest); rare bird species such as the Bonelli eagle *Aquila fasciata* Vieillot, 1822 and the Cinereous vulture *Aegypius monachus* (Linnaeus, 1766); and some rare insect species, including an endemic beetle, *Cryptocephalus mayeti* Marseul, 1878. Rarely mentioned, however, is its exceptional richness in stygobiotic invertebrates.

1.3. History of Biological Studies

Following the description of the subterranean crustacean *Gallocaris inermis* (Fage, 1837) in the Gard department and its subsequent discovery in other aquifers bordering the Hérault, the aquatic fauna of the Cent Fonts massif has undergone a more extensive exploration. Initially, it was the subject of sporadic investigations utilizing rudimentary tools such as nets and baited traps [3–5], which revealed the presence of four large-sized crustacean species, *Gallocaris inermis*, *Faucheria faucheri* (Dollfus & Viré, 1900), *Sphaeromides raymondi* Dollfus, 1897, and *Niphargus virei* Chevreux, 1896, and an ostracod species, *Sphaeromicola cebennica* Rémy, 1948, a parasite of *Sphaeromides* (Table 1).

Table 1. List of the stygobiotic species recorded in the Cent Fonts system from 1950 to 2006.

Classe	Sous-Classe	Ordre	[3–5]	Rouch et al. [6]	Olivier et al. [7]	This Paper
Clitellata	Hirudinea	Arhynchobdellida			-	*Trocheta taunensis* Grosser, 2015
Gastropoda	Caenogastropoda	Littorinimorpha	-	-	*Bythinella n.* sp.	*Bythinella* sp.
			-	-	*Heraultiella exilis*	*Heraultiella exilis* (Paladilhe, 1867)
			-	-	*Islamia moquiniana*	*Islamia* cf. *moquiniana*
			-	-	*Paladilhia pleurotoma*	*Paladilhia pleurotoma* Bourguignat, 1865
			-	-	*Bythiospeum bourguignati*	*Bythiospeum bourguignati* (Paladilhe, 1866)
			-	-	*Moitessieria rolandiana*	*Moitessieria vidourlensis n.* sp.
			-	-	*Moitessieria n.* sp. 1	*Moitessieria guilhemensis* Girardi & Boeters, 2017
			-	-	*Moitessieria n.* sp. 2?	*Moitessieria* sp.
Malacostraca	Eumalacostraca	Decapoda	*Troglocaris inermis*	*Troglocaris inermis*	*Troglocaris inermis*	*Gallocaris (Troglocaris) inermis* (Fage, 1937)
		Isopoda	-	*Stenasellus buili*	*Stenasellus buili*	*Stenasellus buili* Rémy, 1949
			-	*Proasellus cavaticus cavaticus*	*Proasellus cavaticus*	*Proasellus cavaticus* (Leydig, 1871)
			-	*Microcharon doueti n.* sp.	*Microcharon doueti*	*Microcharon doueti* Coineau, 1968
			Faucheria faucheri	*Faucheria faucheri*	*Faucheria faucheri*	*Faucheria faucheri* (Dollfus & Viré, 1900)
			Sphaeromides raymondi	*Sphaeromides raymondi*	-	*Sphaeromides raymondi* Dollfus, 1897
		Amphipoda	-	-	*Niphargus laisi*	*Niphargus laisi* Schellenberg, 1936
			-	-	*Niphargus gallicus*	*Niphargus gallicus* Schellenberg, 1935
			-	-	*Niphargus kochianus*	*Niphargus kochianus* Bate, 1859
			-	-	*Niphargus pachypus*	*Niphargus pachypus* Schellenberg, 1933
			Niphargus orcinus virei	*Niphargus orcinus virei*	*Niphargus virei*	*Niphargus* cf. *virei* Chevreux, 1896 clade A
			-	*Salentinella* sp.	*Salentinella angelieri*	*Salentinella angelieri* Delamare-Deboutteville & Ruffo, 1952
			-		*Salentinella delamarei*	*Salentinella delamarei* Coineau, 1962
		Ingolfiellida	-	*Ingofiella* sp.	*Ingolfiella thibaudi*	*Ingolfiella thibaudi* Coineau, 1968
		Bathynellacea	-	-	*Clamousella* cf. *delayi*	*Gallobathynella delayi* Serban, Coineau & Delamare Deboutteville

Table 1. *Cont.*

Classe	Sous-Classe	Ordre	[3–5]	Rouch et al. [6]	Olivier et al. [7]	This Paper
Ostracoda	Podocopa	Podocopida	-	-	*Fabaeformiscandona* cf. *breuili*	*Fabaeformiscandona* cf. *breuili* (Paris, 1920)
			-	-	*Pseudocandona zschokkei*	*Marmocandona* cf. *zschokkei* (Wolf, 1920)
			-	-	*Schellencandona* cf. *simililampadis*	*Schellencandona* cf. *simililampadis* (Danielpol, 1978)
			Sphaeromicola cebennica	*Sphaeromicola cebennica*	-	*Sphaeromicola cebennica juberthiei* Danielpol, 1977
			-	-	*Candoninae* sp. 1, long, related to *Cryptocandona*	*Candoninae* sp. 1, cf. *Cryptocandona*
			-	-	*Candoninae* sp. 2, bean-shaped, related to *Pseudocandona*	*Candoninae* sp. 2, cf. *Pseudocandona*
			-	-	*Candoninae* sp. 3, triangular, related to *Pseudocandona* group *eremita*	*Candoninae* sp. 3, cf. *Pseudocandona* group *eremita*
Copepoda	Neocopepoda	Cyclopoida	-	*Acanthocyclops rhenanus*	-	*Acanthocyclops rhenanus* Kiefer, 1936
			-	*Acanthocyclops stammeri westfalicus*	*Acanthocyclops venustus westfalicus*	*Acanthocyclops venustus westfalicus* (Kiefer, 1931)
			-	-	*Graeteriella (Graeteriella)* cf. *boui*	*Graeteriella (Graeteriella) boui* Lescher-Moutoué, 1969
			-	*Graeteriella unisetiger*	-	*Graeteriella (Graeteriella) unisetigera* Graeter, 1910
			-	*Paragraeteriella* n. sp.	-	*Graeteriella (Paragraeteriella) vandeli* Lescher-Moutoué, 1969
			-	-	*Kieferiella delamarei*	*Kieferiella delamarei* (Lescher-Moutoué, 1971)
			-	*Speleocyclops* sp.	-	*Speocyclops racovitzai* Chappuis, 1923
		Harpacticoida	-	*Ceuthonectes gallicus*	*Ceuthonectes gallicus*	*Ceuthonectes gallicus* Chappuis, 1928
			-	*Elaphoidella leruthi meridionalis*	*Elaphoidella leruthi meridionalis*	*Elaphoidella leruthi meridionalis* Chappuis, 1953
			-	-	*Nitocrella omega*	*Nitocrella omega* Hertzog, 1936
			-	*Nitocrella* cf. *hirta*	*Nitocrella hirta hirta*	*Nitocrella hirta* Chappuis, 1924
			-	*Ectinosomidae* sp.	*Pseudectinosoma vandeli*	*Pseudectinosoma vandeli* (Rouch, 1969)
Arachnida	Acari	Trombidiformes	-	*Soldanellonyx chappuisi*	-	*Soldanellonyx chappuisi* Walter, 1917
Insecta	Pterygota	Coleoptera				*Laemostenus (Actenipus) oblongus balmae* (Delarouzée, 1860) [1]

[1] This is the only troglobiotic species collected so far in the Cent Fonts.

A second inventory dedicated to the Cent Fonts aquifer dates from 1967 to 1968 [6]. The system was studied under natural conditions, including all the springs and the cave. This study primarily aimed to characterize the stygobiotic fauna within the submerged zone of the karst. More methods were used and the inventory of the stygobiotic fauna was more comprehensive. Thirty-nine crustacean species were collected, including 20 stygobiotic species belonging to the orders Decapoda, Amphipoda, Isopoda, and Copepoda (Table 1). The authors concluded that the Cent Fonts aquifer was "exceptionally rich". The molluscs were not mentioned in this paper [6]. Their diversity in the northern Montpellier region was only studied later by Prié [8,9] and Girardi [10,11], but without focusing on the Cent Fonts aquifer.

In 2004, a water resource exploitation project prompted additional studies. A more thorough inventory of the stygobiotic fauna was conducted between 2005 and 2006 [7], employing extended filtrations of effluents during low-flow and flood periods, along with experimental pumping.

1.4. Threats

As mentioned above, the landscape surrounding the Cent Fonts system is relatively unaffected by human activities. However, this system is seen by the authorities as a major water resource for the entire department [12]. This human pressure on the water supply is believed to increase in the future, especially as the local climate is already dry and drought is expected to increase in the context of global warming [13].

1.5. Objectives

The aims of this paper are (i) to summarize the biospeleological studies carried out at Cent Fonts and in the surrounding caves, in particular the work of Olivier et al. [7], which has never been scientifically published; (ii) to update the taxonomy of the species present, with the description of new gastropod species; (iii) to highlight the interest in the site as a biodiversity hotspot; and (iv) to discuss the impact that the aquifer exploitation project could have on this hotspot.

2. Materials and Methods

We define the "Cent Fonts system" as the area drained by the Cent Fonts springs, extending from the Buèges River in the north to the Hérault River in the east and south, and an inactive valley to the east that, together with the rivers, outlines the boundaries of the "Causse de la Selle" (Figure 1). The stygobiotic fauna surveys were all carried out in the Cent Fonts springs, which line the right bank of the Hérault for about two hundred meters (Figure 1). The hyporheic zone of the Hérault was sampled just downstream of the springs, several meters away from the bank, using the Bou–Rouch Pump [14]. The Bou–Rouch pump allows large quantities of water to be pumped from the interstitial zone, at a depth of about 60 cm in the sandy clay alluvium. Troglobionts have not been inventoried in the Cent Fonts system. We list here species which most likely occur in the Cent Fonts, since they occupy many caves in the surroundings, outside of Figure 1.

The first surveys (1950–1951) used very simple methods such as dip nets and baited traps. Rouch et al. [6] in 1968 used a more comprehensive range of methods, including dip nets and baited traps; sight-hunting in each siphon (method only valid for large crustaceans); fine-netting (carried out in all siphons using a Bluter silk net); pumping with the Bou–Rouch pump in the Hérault River downstream of the springs; Karaman-Chappuis boreholes drilled in the siphon banks; and filtering of all the exsurgences with Bluter silk nets of various mesh sizes, left in place and lifted every week. Some water outlets were filtered almost continuously from 15 November 1967 to 23 February 1968.

The same methods were used in the years 2005 and 2006: Bou–Rouch pumping (Figure 2), surbers, and spring water filtration, and baited traps, sight hunting, plus sediment sampling for mollusc shells in the springs and in the subterranean environment (Cent Fonts cave), respectively. The latter method consists of sampling sediment and leaving it in a bucket of spring water for a few days, in a cool and dark environment (e.g., a house cellar). As oxygen becomes scarce, the animals will try to return to the surface and can be caught on the sides of the bucket with flexible forceps. After a few days, when no live snails are found, the sediment is dried out and poured into water again. The grains of sand will sink, and the empty shells will float to the surface and can be collected with a sieve.

Figure 2. Sampling the hyporheic zone of the Hérault River with the Bou–Rouch pump (© V. Prié).

All the data presented here have been deposited in the Inventaire National du Patrimoine Naturel (https://inpn.mnhn.fr (accessed on 6 November 2023)) database. The site number of the Cent Fonts is INPN 2047774. The sequences produced for the description of the new species (Appendix A) are deposited in GenBank, accession numbers PP050554 to PP050558 for COI; PP051254 to PP051258 for 16S; and PP057731 to PP057738 for 28S.

3. Results

The Lez system was considered to be the richest biodiversity hotspot for stygobiotic species in France [2]. We update here the checklist of stygobiotic species of the Lez basin given by Jourde et al. [15]: *Paladilhia umbilicata* (Locard, 1902) and *Bythiospeum articense* R. Bernasconi, 1985 are misidentifications, these species live far from the Lez system [16]; *Paladilhia subconica* Girardi, 2009 and *Moitessieria magnanae* Girardi, 2009 are considered endemic to their type locality in the Hérault basin; *Phagocata vitta* (Dugès, 1830) and *Proasellus coxalis* (Dollfus, 1892) are not stygobiotic species. Corrected in this way, the number of known stygobiotic species in the Lez system is 39. Following this study, the Cent Fonts aquifer appears as the richest system for stygobiotic taxa in France, with 43 species (Table 1). The terrestrial taxa, which are presumably not as rich as the aquatic ones, have not been studied in the Cent Fonts system itself. We present here the results of surveys carried out in neighboring caves located on the right bank of the Hérault valley, in the same geological context.

Where available, the IUCN Red List category is given for each species at global and national levels. Mollusc species were assessed at the global level in 2010 and at the regional level in 2021 (French Red List [17]). Although most species are regionally endemic, the 2010 (global) and 2021 (French) assessments sometimes differ. This is mainly due to an increased awareness of the threats to aquatic ecosystems, as human and climate change threats are increasingly documented. Most crustacean species have not been assessed at the global level, but a regional-level assessment is available [18].

3.1. Stygobionts

3.1.1. Clitellata Michaelsen, 1919; Arhynchobdellida Blanchard, 1894

- *Trocheta taunensis* Grosser, 2015 (=*T. bykowskii*)

Several populations of leeches named *T. bykowskii* have been discovered in Central and Western Europe [19]. Sket [20] was the first to suggest that *T. bykowskii* actually represents a species complex. Following Grosser [21], Lecaplain [19] considers the French populations to belong to *T. taunensis*. However, the records of *T. taunensis* in France are only from eastern France. The taxonomic status of the Cent Fonts population remains to be confirmed. The species was found in the Cent Fonts cave by F. Malard in 2002 (unpublished data).

3.1.2. Gastropoda Cuvier, 1795; Littorinimorpha Golikov & Starobogatov, 1975
Amnicolidae Tryon, 1863

- *Bythinella* sp.

A species of *Bythinella* was found in abundance in the springs of Cent Fonts (Figure 3a). It was considered a new species by Olivier et al. [7], based on the fact that it lives in a different aquifer from the regional stygobiotic *Bythinella* species described so far, i.e., *Bythinella navacellensis* Prié & Bichain 2009 endemic to the Larzac plateau (north-west of Cent Fonts) and B. *eutrepha* (Paladilhe, 1867) endemic to the Lez karst (south-west). Its identity remains unclear as no genetic data have been collected.

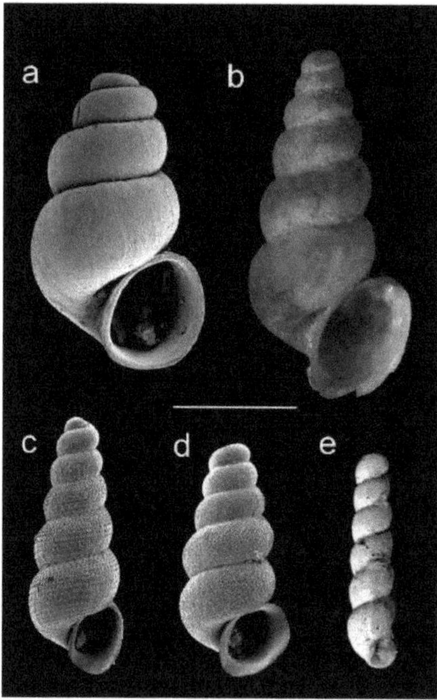

Figure 3. (**a**) *Bythinella* sp.; (**b**) *Paladilhia pleurotoma*; (**c**) *Moitessieria vidourlensis n.* sp.; (**d**) *Moitessieria guilhemensis*; (**e**) *Moitessieria* sp. nov.? or an anomalously shaped shell of *Moitessieria* sp. All these specimens were collected in the Cent Fonts sources. Scale: 1 mm.

Hydrobiidae Stimpson, 1865

- *Heraultiella exilis* (Paladilhe, 1867)

Heraultiella exilis lives in the hyporheic zone [22]. It sometimes occurs in springs but has never been found deep inside the caves. Here, it has been found in the hyporheic zone of the Hérault River, and marginally in the springs. This species is protected in France and considered Vulnerable on both the international [23] and French Red List [17].

- *Islamia* cf. *moquiniana* (Dupuy, 1851)

The genus *Islamia* is awaiting molecular revision. *Islamia moquiniana* is described from the department of Lozère (type locality "... alluvions du Lot à Mende"), far from the Cent Fonts, and the specimens collected in the Hérault basin are morphologically different from those from the department of Lozère. It is therefore likely that the population found in the Cent Fonts is part of an undescribed species.

Moitessieriidae Bourguignat, 1864

- *Paladilhia pleurotoma* Bourguignat, 1865

P. pleurotoma is restricted to a few karst areas east of the Hérault River and west of the Rhône River. It is a cave specialist and has never been collected alive in the hyporheic zone. It is not certain whether the hyporheic zone can be used by this species as a corridor, as is the case for *Bythiospeum* species. Only one shell was found in the Cent Fonts cave (Figure 3b). This shell could be allogenic (transported there by flood). This species is protected in France. It was listed as Least Concerned in the IUCN international red list in 2010 [24] but re-evaluated as Vulnerable on the French Red List in 2021 [17].

- *Bythiospeum bourguignati* (Paladilhe, 1866)

This species lives mainly in the karst on the left bank of the Hérault [16]. It is found in the hyporheic zone of the Hérault and has been marginally collected in the sediments of the springs of the Cent Fonts. It is thought to reach its westernmost distribution limit in the Cent Font, which is also the westernmost limit of the genus. This species is protected in France. It was listed as Least Concerned in the IUCN international red list in 2010 [25] but re-evaluated as Near Threatened on the French Red List in 2021 [17].

- *Moitessieria vidourlensis* n. sp. (=*Moitessieria rolandiana* Bourguignat, 1864)

Most authors consider *M. rolandiana* as a widespread species in southern France, west of the Rhône River. However, Prié [9] showed that there is a strong genetic structure within the area of occurrence of *M. rolandiana*, which reflects the structure of the hydrographic network. A description based on morphometric and molecular data is provided hereafter (Appendix A). *M. vidourlensis* n. sp. (Figure 3c) is morphologically close to *M. rolandiana* but can be distinguished by morphometric analysis. This species is protected in France under the name *Moitessieria rolandiana*.

- *Moitessieria guilhemensis* Girardi & Boeters, 2017

This species was first recognized by Prié [8] based on morphological data (shells larger and smaller than that of *M. rolandiana*, Figure 3d), but was not described as a new species, because no genetic data were available, and the morphology has proven to be misleading for stygobiont species. However, Girardi and Boeters [26] could not wait and described the species as *M. guilhemensis*. This species is protected in France under the name *Moitessieria rolandiana*.

- *Moitessieria* n. sp.?

A spectacular shell was collected at the Cent Fonts (Figure 3e), perhaps an anomalously shaped shell, perhaps something new. As this is a single shell, we prefer not to consider it as a new species, pending further data, but we do report this remarkable form.

3.1.3. Malacostraca Latreille, 1802

Decapoda Latreille, 1802

- *Gallocaris (Troglocaris) inermis* (Fage, 1937)

This is a spectacular species (Figure 4), measuring up to 2 cm long, and is one of only two species of stygobiont decapod in France, endemic to a few aquifers in the Gard and Hérault valleys where it is known from fewer than 10 localities. Interestingly, Rouch et al. [6] noted that this species only occurs in streamless waters inside the cave. Its supposed rheophobia may explain why it has never been collected in the springs, even during floods. It is listed as Near Threatened on the IUCN global Red List [27] and Vulnerable on the French Red List [18].

Figure 4. *Gallocaris inermis,* source of Sauve (Vidourle), ≈12 mm. © C. Alonso.

Isopoda Latreille, 1816

- *Stenasellus buili* Rémy, 1949

This species was described from the department of Aude, with isolated populations in the Corbières mountains and here in the Hérault valley. It is not evaluated at the international level, but it is listed as Near Threatened on the French Red List [18].

- *Proasellus cavaticus* (Leydig, 1871)

P. cavaticus is widespread in Western Europe, occurring in France along the Rhône–Rhine axis and in the Haut-Languedoc (it is also marginally present in the Atlantic basin). According to Henry [28], the population of Cent Fonts belongs to *P. cavaticus cavaticus* and is remarkable because it is the most western and the only place where *Stenasellus* and *Proasellus cavaticus* occur in syntopy. The species is considered Least Concern on the French Red List [18].

- *Microcharon doueti* Coineau, 1968

This species was discovered by Rouch by filtering the exsurgences of the Cents Fonts (type locality) and then collected in the water table of the Orb River, west of the Hérault River. It is listed as Vulnerable on the French Red List [18].

- *Faucheria faucheri* (Dollfus & Viré, 1900)

This species was originally described by Adrien Dollfus and Armand Viré in 1900 as *Cæcosphæroma faucheri* (family Sphaeromatidae), and it was reclassified by the authors in 1905 in the family Cirolanidae. Bertrand [29] lists a total of 21 localities, 10 in the upper Vidourle valley and the Hérault gorges and 11 in the eastern Corbières (Agly basin and its tributary the Verdouble). We (C.A.) add here another locality, the outlet of the Avencas cave, near Issensac, which extends the distribution of the species south to the coast. *F. faucheri* is listed as Least Concern on the French Red List [18].

- *Sphaeromides raymondi* Dollfus, 1897

S. raymondi (Figure 5) is a large species, up to 3 cm, known from a few caves in the Hérault department and the right bank drainage of the Rhône River, up to the Ardèche

River. This species is mentioned in the literature from the 1950s and by Rouch et al. [6] but was not found during the 2006 sampling. It is listed as Near Threatened on the French Red List [18].

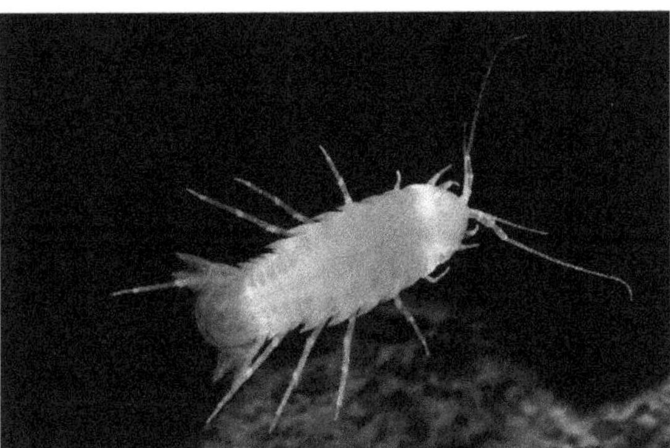

Figure 5. *Sphaeromides raymondi*, Grotte exsurgence de l'Avencas, Brissac, ≈18 mm. © J.C Queneau.

Amphipoda Latreille, 1816

- *Niphargus laisi* Schellenberg, 1936

The species is widespread in France and Germany. In France, its distribution is sporadic, from the Alsace in the north to the Rhône River aquifer near Lyon, and in the south in the Hérault basin. Its habitat is mainly represented by the hyporheic and phreatic zones. It is considered Data Deficient on the French Red List [18].

- *Niphargus gallicus* Schellenberg, 1935

This species is scarce in the southern half of France, where it lives in the porous aquifers of large alluvial floodplains (Rhône) and small streams (e.g., Triouzoune, St Angel), both in the phreatic and hyporheic zones. It has also been collected in karst areas (e.g., Prades-Le-Lez). *N. gallicus* is listed as Least Concern on the French Red List [18].

- *Niphargus kochianus* Bate, 1859

Niphargus kochianus had several subspecies, of which some are now considered as separate species. *Niphargus k. kochianus*, although frequently reported in France, is considered doubtful as it would not have a transcontinental distribution [30]. McInerney et al. [31] defined four distinct clades (A, B, C, D) based on molecular analysis. *N. kochianus* "D" would be the lineage present in France. The French form is sparsely distributed from the extreme north to the south (Pyrenees region), but is more common in the Rhône basin, Jura, and Ardèche regions. Given the large number of sites and specimens reported from France, it is difficult to provide a clear taxonomic status for the *N. kochianus* collected in the Cent Fonts system. Moreover, for this French "lineage D", the number of sites and specimens in the 2014 study [31] appears to be very low.

From an ecological point of view, it is a small species, typically interstitial, living in cool waters and stable flow conditions. In French aquifers, *N. kochianus* is often found in the upwelling zones of rivers (e.g., Rhône), or in deep alluvial and phreatic zones (e.g., wells in the Albarine valley, Jura). It is also reported in karst, where it may find conditions for an interstitial lifestyle. *N. kochianus* is listed as Least Concern on the French Red List [18].

- *Niphargus pachypus* Schellenberg, 1933

Previously described as a subspecies of *N. kochianus*, it has been raised to species level and is now recognized as a highly divergent lineage [32]. It was collected from only few sites in the Netherlands and is also reported from Belgium and Luxembourg. In France, it is widespread, with more than a hundred localities. As a small-sized species, it is typically interstitial and particularly prefers cool and hydrologically stable areas, which explains its abundance in the deep alluvia of streams and in the phreatic zone. However, it has also been collected in the karst areas, where it is probably associated with alluvial deposits. *N. pachypus* is listed as Least Concern on the French Red List [18].

- *Niphargus* cf. *virei* Chevreux, 1896 Clade A

This species is found mainly in France, but also in a few places in the Netherlands, Belgium, and Switzerland. In France, it is typically a karstic species. *Niphargus virei* has never been found in the porous aquifer, except for one specimen collected in the alluvia of the Rhône-Ardèche confluence, probably drifted from the surrounding karst. It is particularly common and abundant in the Jura and Ardèche massifs. Genetic studies have revealed the presence of three cryptic species in the French *virei* group [33]. The population of the Cent Fonts karst system belongs to the cryptic species "A", located at the extreme south of the group's geographical distribution. *N. virei* has been described from specimens collected in caves of the Jura mountains (grottes d'Arbois, Baumes-les-Messieurs, and Baumes-les-Dames). The nominal species should then belong to clade "B" of Lefebure et al. [33]. The most widespread of these cryptic species, Clade A (Figure 6), which is found from the Hérault to the Rhône and Moselle Rivers, is still awaiting formal description. This species is listed as Least Concern on the French Red List [18].

- *Salentinella angelieri* Delamare-Deboutteville & Ruffo, 1952

S. angelieri has a wide geographical distribution in Greece, Italy, and Spain. It is less common in France, where it is mainly recorded from the Rhône basin and the Hérault region. A population is also reported from Corsica. Two subspecies have been described from Croatia and Spain. In the Rhône floodplain, it is always collected in upwelling, i.e., in cool and stable interstitial water. It is listed as Near Threatened on the French Red List [18].

- *Salentinella delamarei* Coineau, 1962

This species is described from the phreatic waters of the Tech River in the department of Pyrénées-Orientales. It is reported only from France along the Rhône River, the Ardèche, and Hérault areas. Two subspecies have been described: *S. delamarei delamarei* and *S. delamarei macrocheles*. This species is listed as Least Concern on the French Red List [18].

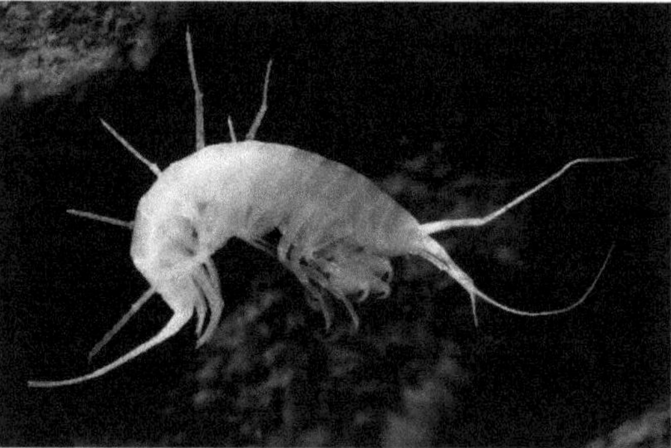

Figure 6. *Niphargus* cf. *virei*, Grotte exsurgence de l'Avencas, Brissac, ≈15 mm. © J.C Queneau.

Ingolfiellida Hansen, 1903

- *Ingolfiella thibaudi* Coineau, 1968

This species has been reported from fewer than fifteen sites, from Ruoms in the Ardèche to Tarbes and the Saint Girons area in the Pyrenees region. It has been collected in both karst and porous aquifers (hyporheic and phreatic zones). In the Cent Fonts system, several specimens have been found in the spring sediments and in the hyporheic zone of the Hérault river (Figure 7). *I. thibaudi* is listed as Least Concern on the French Red List [18].

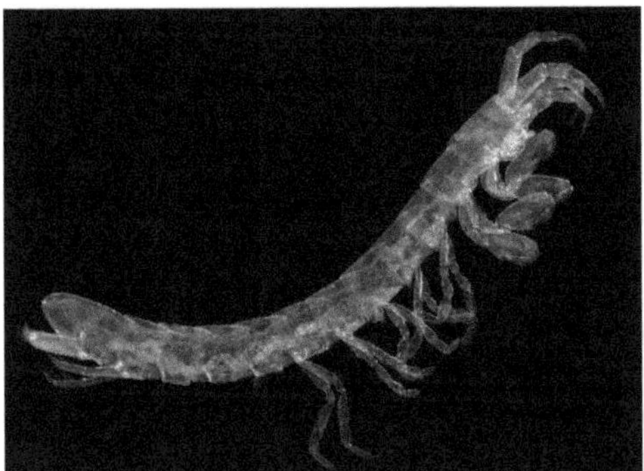

Figure 7. *Ingolfiella thibaudi*, Cent Fonts, ≈2 mm. © M.-J. Dole-Olivier.

Bathynellacea Chappuis, 1915

- *Gallobathynella* (*Clamousella*) *delayi* Serban, Coineau & Delamare Deboutteville 1971

This species was previously considered strictly endemic from the Clamouse Cave, a few kilometers downstream of the Hérault Valley, also on the right bank. The species is listed as Vulnerable on the French Red List [18].

3.1.4. Ostracoda Latreille, 1802; Podocopida Sars, 1866

- *Fabaeformiscandona* cf. *breuili* (Paris, 1920)

This species is widespread in Europe, from Poland to Spain, and certainly represents a number of subspecies or cryptic species. The taxonomic status of the Cent Fonts population needs to be clarified. On a European scale, *F. breuili* has been sampled in different habitats: wells, springs, the hyporheic zone of rivers, and, more rarely, in caves. In the Cent Fonts, *F. breuili* has only been sampled with exsurgence filtering, but not in the hyporheic zone. *F. breuili* is listed as Least Concern in the UICN French Red List [18].

- *Marmocandona* cf. *zschokkei* (Wolf, 1920)

Originally described in the genus *Candona*, it was then included in the genus *Pseudocandona*. Danielpol et al. [34] proposed the genus *Marmocandona* (whose type species is *Candona zschokkei* Wolf, 1920) for four stygobiotic species. This species is widespread in Western Europe: in Switzerland, Germany, Belgium, and France. The taxonomic status of this southern population needs to be clarified. *M. zschokkei* was often sampled in the hyporheic zone of large rivers, but also occurred in springs and wells. In the Cent Fonts, *M. zschokkei* was sampled with exsurgence filtering, but not in the hyporheic zone of the river. The species is listed as Least Concern on the UICN French Red List [18].

- *Schellencandona* cf. *simililampadis* (Danielpol, 1978)

This species was previously restricted to an artificial cave associated with the Vidourle spring at Sauve (Gard department). The taxonomic status of the population sampled in the Cent Fonts needs to be clarified. This species was sampled with exsurgence filtering. It is listed as Vulnerable on the French Red List [18].

- *Sphaeromicola cebennica juberthiei* Danielpol, 1977

This species is currently known from only two sites in the Hérault valley: the Cent Fonts and another cave a few kilometers upstream, also on the right bank of the Hérault River. It is mentioned in the literature from the 1950s and by Rouch et al. [6] but was not found during the 2006 sampling. *Sphaeromicola cebennica* is listed as Vulnerable on the French Red List [18].

- Candoninae sp. 1, 2, 3

Three other species of the subfamily Candoninae were sampled during the 2006 Cent Fonts study, but only with juveniles: a "long" form, related to the genus *Cryptocandona*; a "bean-shaped" form, related to *Pseudocandona*; and a triangular form, related to the *Pseudocandona* group *eremita*. Their taxonomic status still needs to be established by examination of adult specimens. However, although they could not be formally identified to the species level, they represent other species than those listed above.

3.1.5. Copepoda Milne Edwards, 1840
Cyclopoida Burmeister, 1834

- *Acanthocyclops rhenanus* Kiefer, 1936

This obligate groundwater cyclopoid shows a wide distribution in many groundwater habitat types of Europe. Its distribution covers several countries in central-eastern Europe, from France to Poland. The species shows no apparent habitat specialization, being recorded from almost all the groundwater habitat types. This species is mentioned by Rouch et al. [6] but was not found during the 2006 sampling. *A. rhenanus* is listed as Least Concern in France [18].

- *Acanthocyclops venustus* (*stammeri*) cf. *westfalicus* (Kiefer, 1931)

This species has an alternate representation in the current literature, and according to present knowledge, the accepted name for the subspecies *westfalicus* is *A. venustus venustus* [35]. The *venustus* group of the genus *Acanthocyclops* is in need of revision, and pending a clearer taxonomic assessment, the subspecies name *westfalicus* is provisionally maintained here. This subspecies has been recorded from Germany, Belgium, and France, and collected from phreatic habitats, the hyporheic zone of rivers, and aquifers in unconsolidated sediments. *A. venustus* is listed as Vulnerable on the French Red List [18].

- *Graeteriella boui* Lescher-Moutoué, 1974

This species is known only from France, with 11 records from both alluvial and karst aquifers, with a higher incidence in the saturated karst. It was originally described on the basis of specimens collected in the Gard department, but in the description, the author mentions the Ardèche and Hérault populations as belonging to the same species ("[The description of *Graeteriella boui* is based on individuals caught in the Gard department. Other forms collected in neighbouring departments reproduce the same characteristics; some differences, not sufficient to introduce new systematic subdivisions, are noted below]"). The population studied in the Hérault basin is that of the Cent Fonts, and Lescher-Moutoué [36] concludes the following: "The presence of *G. boui* in the Cent Fonts karstic system is all the more remarkable because two species of this genus have also been recorded in the same system: *G. unisetigera* and *G. (Paragraeteriella) vandeli* Lescher-Moutoué, 1969". The species has also been collected from the karst aquifer of the Lez River. It is listed as Vulnerable on the French Red List [18].

- *Graeteriella unisetigera* (Graeter, 1908)

This species is considered by Fiers and Ghenne [37] to be a member of the cryptozoic fauna, as it has also been found in leaf litter and in other surface habitats (e.g., mosses) in Belgium, usually with some connection to groundwater. In spite of this situation, the species has several morphological characteristics that make it a good candidate for a widespread stygobiotic species in Europe, able to live in true groundwater habitats as well as in surface ecosystems dependent on groundwater. It is mentioned by Rouch et al. [6] but was not collected again during the 2006 sampling. *G. unisetigera* is listed as Least Concern on the French Red List [18].

- *Graeteriella (Paragraeteriella) vandeli* Lescher-Moutoué, 1969

Rouch et al. [6] mention "*Paragraeteriella n. sp.*", without any further details. It was later described as *Paragraeteriella vandeli* by Lescher-Moutoué [38]. The type locality is the

Cent Fonts. It is known only from a single record from the Cent Fonts karstic system, which makes it spot endemic to this restricted area and rare in terms of abundance. At present, it has only been collected from the saturated karst. It was not found during the 2006 survey. *G. vandeli* is listed as Vulnerable on the French Red List [18].

- *Kieferiella delamarei* (Lescher-Moutoué, 1971)

This cyclopid species has exceptional stygomorphic features, such as a slender body, completely depigmented, long antennules, and long setae on the swimming legs, which make it a typical planktonic species swimming in underground karst lakes. This species is known from the Lez karst system and has also been collected from the Cent Fonts karst springs. The genus *Kieferiella* is monotypic and the only known species is from this restricted area in the south of France, making it a priority for conservation. It is listed as Vulnerable on the French Red List [18].

- *Speocyclops racovitzai* (Chappuis, 1923)

This species is mentioned by Rouch et al. 1968 as "*Speocyclops* sp. (en cours de determination)". It was not found during the 2006 sampling. *S. racovitzai* is present throughout southern France [39]. It shows a high degree of diversification in morphological microcharacteristics and is therefore divided into several subspecies with subtle morphological differences. No less important, some subspecies show overlapping distributions, raising doubts about their subspecific identity. The currently recognized subspecies need a taxonomic redefinition, but all are considered stygobionts. The nominotypical species also shows a wide distribution in the Pyrenees. It is listed as Least Concern on the French Red List [18].

Harpacticoida Sars G.O., 1903

- *Ceuthonectes gallicus* Chappuis, 1928

This species is widespread in France and always associated with groundwater habitats, both in alluvial and karst aquifers, with some preference for the latter [40]. It is endemic from France and is of Least Concern on the French Red List [18].

- *Elaphoidella leruthi meridionalis* Chappuis, 1953

The genus *Elaphoidella* is one of the most diverse harpacticoid genera in groundwater environments. In the study area, *E. leruthi meridionalis* is the only species recorded. It is known from several sites, mainly in southern France, with a clear preference for karstic groundwater, both in the saturated and unsaturated zones. *E. leruthi* is considered Data Deficient on the French Red List [18].

- *Nitocrella omega* Hertzog, 1936

The ameirid genus *Nitocrella* is considered to be of ancient direct marine origin and almost all species of this genus are known only from groundwater habitats [41]. This species collected from the Cent Fonts is rare in terms of occurrence and abundance, being known from only a few localities in France, Germany, and Hungary. It is listed as Vulnerable on the French Red List [18].

- *Nitocrella hirta* Chappuis, 1924

This species is widespread throughout Europe, with more than forty localities and collected from many groundwater habitat types. Four subspecies have been described. *N. hirta* is not evaluated on the French IUCN Red list.

- *Pseudectinosoma vandeli* (Rouch, 1969)

This minute harpacticoid was the first *Pseudectinosoma* species discovered in groundwater worldwide. The species was first mentioned by Rouch et al. in 1968 as "*Ectinosomidae* sp.". A year later, Rouch described it and placed it in the marine genus *Sigmatidium*. It was only later that Galassi et al. [42] re-analyzed the type material of the type species of the marine genus *Sigmatidium* on the occasion of the discovery of the second representative of the genus *Pseudectinosoma* in France, and they definitively placed this species in the genus *Pseudectinosoma*. The genus *Pseudectinosoma* is considered to be an ancient Tethyan relict found in the groundwater of Europe and Australia, probably the only remnant of an ancient fauna of direct marine origin. *P. vandeli* is known only from this area and has been collected in large numbers from the Cent Fonts karst system. The Cent Fonts is the type locality of the species, listed as Vulnerable on the French Red List [18].

3.1.6. Arachnida, Acari

- *Soldanellonyx chappuisi* Walter, 1917

This species is mentioned by Rouch et al. [6] but was not found (but not sought for) during the 2006 sampling.

3.2. Troglobionts

Unlike stygobionts, troglobionts have not been inventoried in the Cent Fonts system. The only troglobiont species collected in the Cent Fonts cave is the carabidae beetle *Laemostenus (Actenipus) oblongus balmae* (Delarouzée, 1860). There is currently no report of other troglobitic taxa in the Cent Fonts system itself. On the assumption that troglobionts are less drainage-dependent than stygobionts, we list below species which most likely occur in the Cent Fonts, since they occupy many caves in the surroundings.

3.2.1. Araneae

- *Palliduphantes sanctivincenti* (Simon, 1873)

This species is endemic from southern France, and it is widespread between the Pyrenees and the Alps.

3.2.2. Opiliones

- *Peltonychia clavigera* (Simon, 1872)

The genus *Peltonychia* contains the first described travunioid species. This polyphyletic genus is known from central European caves (Pyrenees, central France, and the Alps). *Peltonychia clavigera* is distributed on both slopes of the Pyrenees and in the Cevennes where it is sporadic (Figure 8A).

Figure 8. Some troglobite taxa from neighbouring karst systems. (**A**) Opiliones *Peltonychia clavigera*; (**B**) Pseudoscorpiones *Neobisium tuzetae*; (**C**) Isopoda *Trichoniscoides bonneti*; and (**D**) Diplura *Plusiocampa balsani*. © C. Alonso.

3.2.3. Pseudoscorpiones

- *Neobisium tuzetae* Vachon, 1947

N. tuzetae was described from the *Signal de la Montete* cave towards Quissac in the Gard department. This species is found in a large number of caves from the Hérault valley to the Larzac plateau (Figure 8B).

3.2.4. Isopoda

- *Trichoniscoides bonneti* Vandel, 1946

This endemic species is quite common in the caves of the limestone edge of the Cevennes, between the Hérault and Vidourle rivers (Figure 8C).

3.2.5. Diplura

- *Plusiocampa balsani* Conde, 1947

P. balsani is an endemic species found in many caves in the Massif Central. It is very common in all the caves of the Hérault valley (Figure 8D).

3.2.6. Collembola

- *Pseudosinella denisi* Gisin, 1954

This collembola is endemic from the sub-region (Gard, Ardèche, and Hérault Departments). It is widespread in caves around the Cent Fonts. It has clearly troglomorphic characteristics: eyeless, unpigmented, with elongated appendages and elongated claw.

- *Onychiurus ortus* Denis, 1935

O. ortus is endemic from the sub-region, in the departments of Hérault, Gard, and Aveyron. It is widespread in caves around the Cent Fonts. It has clearly troglomorphic characteristics: eyeless, unpigmented, and elongated claw.

3.2.7. Coleoptera

- *Laemostenus* (*Actenipus*) *oblongus balmae* (Delarouzée, 1860)

This is a widely distributed species, known from the Pyrenees to the southern and eastern edge of the Massif Central. The subspecies *balmae* is known from a few caves in the Gard and Hérault Departments, with one location in the Ardèche Department (Païolive).

3.3. Stygophilic taxa

Five stygophilic species of Cyclopids have been collected in the Cent Fonts according to Lescher-Moutoué [39]:

- *Eucyclops serrulatus* (Fisher, 1851);
- *Paracyclops fimbriatus* (Fisher, 1853);
- *Acanthocyclops vernalis* (Fisher, 1853);
- *Megacyclops viridis* (Jurine, 1820), in the hyporheic zone of the Hérault River near the Cent Fonts exurgences;
- *Diacyclops languidoides* Lilljeborg, 1901.

3.4. Troglophilic taxa and Parasites

Several other troglophilic taxa are expected to be found in the Cent Fonts system, of which the most important are listed below.

Ixodida

- *Eschatocephalus vespertilionis* (Koch, 1844) is a common bat parasite.

Araneae

- *Lessertia dentichelis* (Simon, 1884), a troglophile species, very common in the caves throughout the Hérault valley.

- *Meta bourneti* Simon, 1922, troglophile, very common in all caves in the area.
- *Meta menardi* (Latreille, 1804), troglophile, very common in all caves in the area.

Opiliones

- *Sabacon paradoxus* Simon, 1879, troglophile, found in cave entrances in France and Spain. It *is* very common in most caves in the Cévennes and in the Hérault karsts.

Julida

- *Blaniulus guttulatus* (Fabricius, 1798), troglophile, common in all caves of the region.

Isopoda

- *Oritoniscus delmasi delmasi* Vandel, 1933, endogeous and troglophile species, endemic to the southern Cévennes between the Vidourle and Lergue Rivers.
- *Phymatoniscus propinquus* (Brian, 1908), troglophile. The ocular area of this species is *generally* provided with a large single eyespot but in specimens of the Cents Fonts cave, eyes are completely invisible to external examination [43]. The species is common throughout the Cévennes in the Ardèche, Gard, and Hérault departments.

Coleoptera

- *Leptinus testaceus* P.W. Müller, 1817, is a troglophile, ectoparasite, and commensal of many *species* of micromammals. It lives mainly in subterranean mammal nests as well as in caves, on bat guano. This species is sporadic but known from many caves around the Cent Fonts system.

4. Discussion

4.1. A Biodiversity Hotspot Embedded in a Stygobiont Species-Rich Area

Only five stygobiont species were recorded from the Cent Fonts in the 1950s. Then, Rouch et al. [6] carried out a more extensive survey and found 20 stygobiont species. About 50 years later, another survey was triggered by an impact study of an important water extraction project, resulting in a total of 36 stygobionts [7]. Combining all this data, a total of 43 stygobiotic species have been identified, making the Cent Fonts system a hotspot of subterranean biodiversity in Europe (Table 1). Its stygobiont richness is higher than that of the better known Lez system (39 stygobionts), considered one of the most important biodiversity hotspots in the world [1,2]. The third-richest area of the southern Massif Central in France is the Sauve karstic system (29 stygobionts [44]), close to the Cent Fonts (Figure 1B). For the terrestrial fauna, it is expected that additional species will be found in the Cent Fonts, especially among the troglobionts known to occur in the surrounding caves (see Section 3.2,) as the Cent Fonts cave has been quickly sampled for troglobionts.

As pointed out by Rouch et al. [6], the rich fauna observed in the Cent Fonts includes groundwater genera of undoubtedly freshwater origin, such as *Elaphoidella*, *Ceuthonectes*, *Speocyclops*, and *Graeteriella* for copepods and *Gallocaris* for the decapods, and genera of no less certain marine origin such as *Microcharon* and *Sphaeromides* for isopods, *Ingolfiella* for ingolfiellids, and *Salentinella* for amphipods. Once again, the cave environment proves to be "the place of arrival of lineages of very different origins" [45]. In the stygobiotic molluscs, the origin of the family Moitessieriidae is still unclear, as all the published phylogenies have failed to anchor it in the global phylogenies of freshwater molluscs: the node linking it to the other taxa was not supported (e.g., Ref. [46]). This raises the question of the origin of this family, which could also be of marine origin.

4.2. Conservation Issues and Threats

The Cent Fons system is the second-richest biodiversity hotspot in Europe for stygobiotic species and deserves conservation measures for this reason alone. A quarter of these 43 species are considered Vulnerable by the IUCN Red List. The Cent Fonts also hosts several relict species. Furthermore, it is the type locality of four taxa: *Sphaeromicola cebennica juberthiei*, *Graeteriella vandeli*, *Microcharon doueti*, and *Pseudectinosoma vandeli*. Type localities should be preserved for future taxonomic work. Such a high biological value clearly deserves special attention.

Interestingly, eight species collected by Rouch et al. [6] (Sphaeromides raymondi, Sphaeromicola cebennica juberthiei, Acanthocyclops rhenanus, Graeteriella vandeli, Graeteriella unisetigera, Speocyclops racovitzai, Diacyclops languidoides, and Soldanellonyx chappuisi) were not collected in the 2006 inventory. Although the hydrological conditions are not documented by Rouch et al. [6], it is unlikely that the sampling conditions between November 1967 and February 1968 (4 months) were more favorable than during the whole period of the extensive survey carried out by Olivier et al. [7] between July 2005 and January 2006 (7 months), which included a major flood event. This difference in the results could be due to a lower probability of detection in 2006, but the sampling was more intensive, with water filtered for two years, a large team of experienced people both in the field and for the identification of the taxa, the use of improved collection methods, etc. So, if not the probability of detection, the absence of these species in 2006 could be due to local extirpation. It cannot be ruled out that these species are indeed in decline, but the reason for this is unknown. Although relatively well preserved in terms of land use, the Cent Fonts hydrosystem may be under unknown threats.

Water pollution from the surface is likely to be low, as this karst area has a very low human density. Climate change, which began in the early 1900s, could be a significant threat, but its effects on subterranean ecosystems are still poorly documented. However, severe droughts combined with increasing human pressure on the water resource especially in summer, with extreme fluctuations in water levels, are likely to affect subterranean ecosystems. Indeed, a short-term threat is the prospect of using this aquifer for drinking water. The Cent Fonts massif is recognized for the importance of its water supply and the quality of its water. An assessment of the volume of this resource and the possibilities for its exploitation was carried out in 2005 [12]. This study concluded that the drinking water reserve of the aquifer could not be mobilized for exploitation. However, this study is already disputed [47] and future needs may require greater resources.

4.3. Future Prospects

Troglobionts have been under-sampled, and it is likely that many more species will be found in future surveys, as described above. Intensive surveys by Rouch et al. [6] and Olivier et al. [7] have allowed the collection of many stygobiotic species, and only a few are expected to be added. However, some of the species collected in 1968 were not collected again in 2006. This may be due to local extinction and/or bias in the probability of detection. Intensive and regular surveys would give us a clearer picture of the biodiversity of the Cent Fonts and allow us to document its evolution and threats. However, these surveys require significant investment and are unlikely to be undertaken in the near-future to monitor the stygobiotic fauna.

Environmental DNA is the topical, cost-effective answer to unsatisfactory detection probabilities and the lack of taxonomic expertise. Several studies [48,49] have demonstrated its ability to detect up to 95% of aquatic organisms in surface streams, provided that optimized methods are implemented. Preliminary tests carried out in this karst with optimized methods (up to 250 L filtered, 12 PCR replicates, coverage of 300,000...) were promising, allowing the detection of most, but not all, of the gastropod and crustacean species known to occur in the area. Extensive work on sampling methods is needed to improve the detection probability. This approach deserves to be explored further and is probably the future for surveying and monitoring the fascinating stygobiotic ecosystems.

Author Contributions: Conceptualization, V.P. and M.-J.D.-O.; methodology, C.B., V.P. and M.-J.D.-O.; investigation, C.B., V.P. and M.-J.D.-O.; data curation, C.B., V.P., C.A., P.M., D.M.P.G. and M.-J.D.-O.; writing—original draft preparation, V.P.; writing—review and editing, V.P.; visualization, V.P. All authors have read and agreed to the published version of the manuscript.

Funding: The 2006 research was funded by the Bureau de Recherches Geologiques et Minières (BRGM), service EAU, unite RMD, funding number AvN1 contrat ref 03/C0275 conv-appli-brgm-cg34.

Institutional Review Board Statement: Not applicable.

Data Availability Statement: The data are available in the online open access database of the Inventaire National du Patrimoine Naturel (https://inpn.mnhn.fr/accueil/index (accessed on 6 November 2023)); INPN locality # 2047774.

Acknowledgments: We would like to thank Danielle Defaye for checking the Copepods and Ostracods. Thanks goes to Louis Deharveng and Anne Bedos for sharing the bibliography and for constructive discussions during the writing of the manuscript.

Conflicts of Interest: The authors declare no conflict of interest.

Appendix A

We present here a brief description of three new species of the genus *Moitessieria*. These new species were described by Prié [9] in a Ph.D. thesis that is not considered an official publication according to the International Code of Zoological Nomenclature (ICZN). The names used in this earlier work are therefore *nomina nuda*. The descriptions are reproduced here with the proposed new names, in line with the current trend to avoid eponyms when describing new species.

Moitessieria species are very rarely collected alive and when they are, they are difficult to preserve, because ethanol does not penetrate into the shells, hence the paucity of available sequences on Genbank. The Niku-nuki method proposed by Fukuda et al. [50] was used systematically but failed in most cases. A *Moitessieria* shell can have 7 to 8 whorls, but the animal will retract to the first three whorls when stressed. As the opening at the shell mouth is less than 1/4 mm wide, it is unlikely that the ethanol ever comes into contact with the flesh. This probably explains why, in most cases, DNA amplification fails from *Moitessieria* specimens, or only one or two genes amplify. The genetic data presented here are therefore incomplete.

The gastropod family Moitessieriidae is the only family composed entirely of stygobionts. *Moitessieria rolandiana* was considered to be a widespread species in southern France [16]. This wide distribution contrasts with that of other species in the family, which are often restricted to a small karstic area, due to the fragmentation of subterranean habitats. Prié [9] showed that *M. rolandiana* is actually composed of three cryptic species, with each occupying a distinct karstic area, which supports their reproductive isolation by geographic barriers. They can be distinguished morphometrically, and molecular data corroborate their reproductive isolation. The species delimitation is based on morphometry, molecular, and distribution data.

Appendix A.1. Material and Methods

Appendix A.1.1. Biogeography

We used drainage basins as a hypothesis for where species limits are likely to occur. Drainage basin delimitations are based on the SANDRE database [51], which describes the subsurface hydrogeological units.

Appendix A.1.2. Morphometrics

Shells unambiguously attributable to the *Moitessieria* genus were collected in the four localities: 25 shells from the Folatière spring (close to the Gourneyras cave, locality 1 in Figure A2); 33 shells from the Cabrier spring (locality 2 in Figure A2); 18 shells from the Sauve Spring (locality 3 in Figure A2); and 12 shells from the Lirou River hyporheic zone (close to the Gour Noir spring, locality 4 in Figure A2). Shells were placed on an adhesive support in a standard position, i.e., with the columellar axis standing vertically, and then digitalized with a graduated scale using a stereomicroscope connected to a digital camera. Six parameters were recorded on each picture using ImageTool 3.00 [52]: height and width of the shell, width of the last suture, width of the last whorl, and height and width of the aperture (Figure A1). These measures were log-transformed to minimize the size effect. Four ratios (H/W, H/LWW, AH/H, AH/AW) commonly used in the alpha-taxonomy of hydrobioid spring-snails, e.g., Refs. [53,54], were also calculated. Multivariate analyses were performed (Principal Component Analysis, PCA, and Linear Discriminant Analysis, LDA) to explore the distribution of these ten shell parameters using R [55].

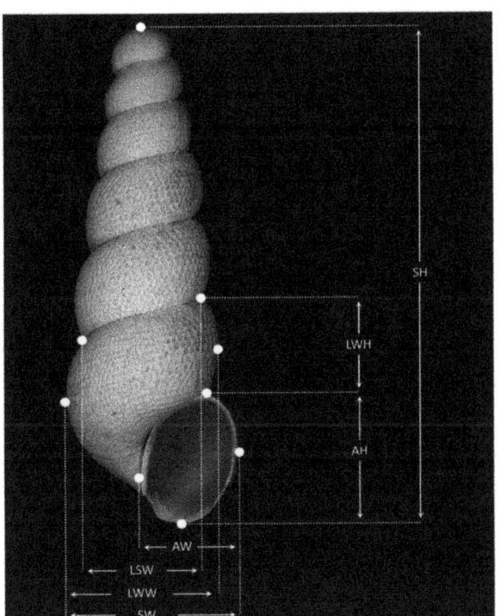

Figure A1. Measurements of a *Moitessieria* shell. SH: height of the shell; LWH: height of the last whorl; AH: height of the aperture; AW: width of the aperture; LSW: width of the suture of the last whorl; LWW: width of the last whorl; SW: width of the shell.

Appendix A.1.3. DNA Analyses

The whole specimens were used for the extractions, as their small size did not allow the animal to be taken out of the shell. DNA was extracted using the Nucleospin Tissue Kit (marketed by Macherey–Nagel), following the manufacturer's protocol. Three partial gene sequences were amplified: a fragment of the cytochrome oxidase subunit I (COI) gene (barcode fragment of Folmer et al., [56]); a fragment of the rRNA 16S gene (universal primers of Palumbi [57]); and a fragment of the rDNA 28S gene (primers C1 and D3 [58]). Extractions, amplifications, and sequencing were performed by Genoscreen (France) and Eurofins (Germany) using standard protocols. Primer pairs were newly designed when the universal primer failed to amplify the 16S gene, 16sF2: AGTCGAGCCTGCCCAGTGA and 16sR2: CAACCCTTAAAGACTTCTGCATCCTT; 16sF3: AGTCGRRCCTGCCCAGTGA and 16sR3: CAACCYTTAAAGACTTCTGCATCNTT. Sequences were automatically aligned using ClustalW multiple alignments implemented in BioEdit 7.0.5.3 [59]. The accuracy of automatic alignments was confirmed by eye. Only a few gaps, unambiguously aligned, were inferred for the 28S and 16S genes: they were conserved for the analyses. A topotype of *Paladilhia pleurotoma* Bourguignat, 1965 was included as an outgroup for both phylogenies. *Spiralix puteana* (accession numbers AF367635 and EU573992 [46]) was included as an outgroup for the mitochondrial analysis only.

Appendix A.2. Results

Appendix A.2.1. Biogeography

The region north of Montpellier in southern France is composed of distinct karst units, which have given rise to distinct faunal assemblages [9]. Not surprisingly, these distinct hydrosystems also support distinct species. The *Moitessieria* populations studied here belong to four adjacent basins. One, the Tarn basin in the west, flows into the Atlantic. The Atlantic and Mediterranean drainages are the most isolated, especially because no stygobiont gastropods were found in the upstream hydrosystems. On the Mediterranean side, the Hérault (west), the Lez (centre), and the Vidourle (east) flow. Within these main basins, different hydrogeological units can be distinguished (Figure A2). While the surface relief creates ridge lines that distinguish these catchment areas, there may or may not be underground connections between the hydrogeological units. For example, the Larzac plateau flows north to the Vis River and south to the Lergue river, but the fauna is the same on both sides, reflecting the known subterranean connections between the two drainages. The same seems to occur between the upper Hérault and Vidourle drainages.

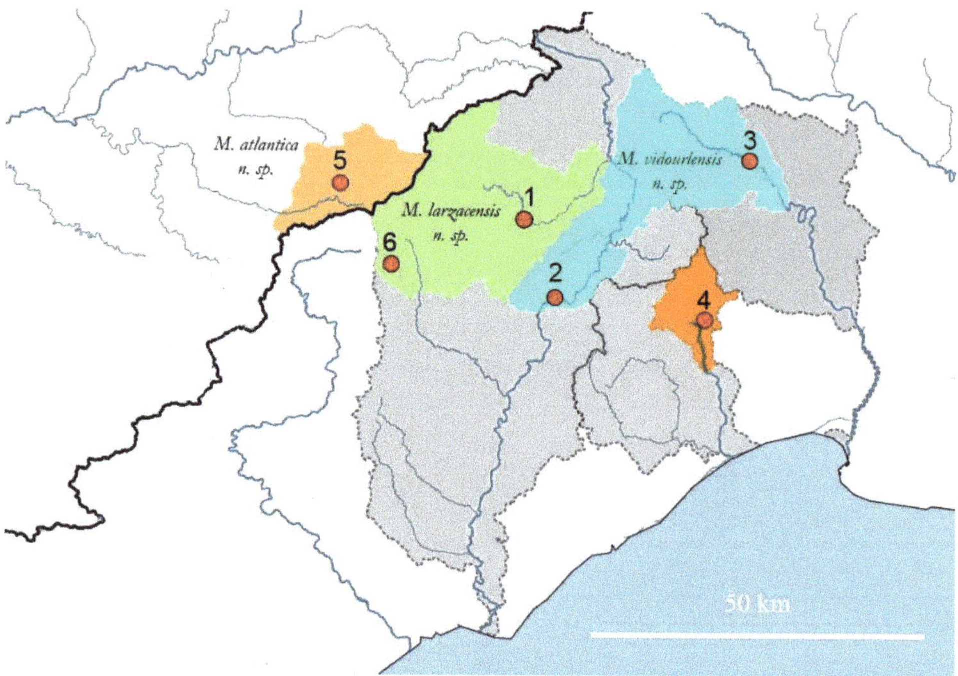

Figure A2. Biogeography of the subregion. Bold black line: separation between the Atlantic and Mediterranean drainages; thin dotted lines: separation between the major river drainages; blue lines: rivers; red dots: sampled populations, with numbers referring to the populations for which morphometric analysis was performed. The Mediterranean rivers' basins are highlighted in grey. The hydrogeological units are based on the SANDRE database. The numbers refer to the locations of the populations for which morphometric and/or molecular analyses were carried out (Figures A3 and A4).

Appendix A.2.2. Morphometry

Multivariate analyses allowed the populations from the Lez source (type locality of *Moitessieria rolandiana*), the Larzac plateau, and the upper Hérault/Vidourle to be distinguished. The populations from the upper Hérault and the upper Vidourle had the same morphology and could not be distinguished by morphometric analysis (Figure A3). Too few specimens were collected from populations 5 and 6 to be included in the morphometric analyses.

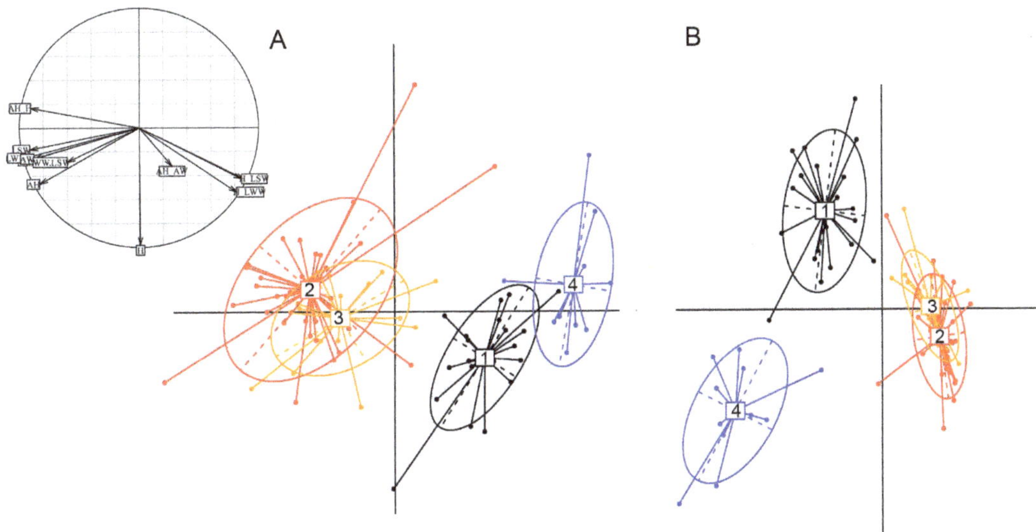

Figure A3. Multivariate analyses of the morphometrics of *Moitessieria* populations: (**A**) Principal Component Analysis; (**B**) Linear Discriminant Analysis. 1: Vis River, Larzac plateau, Hérault basin, *Moitessieria larzacensis n.* sp.; 2: Cabrier source, same system as the Cent Fonts, Hérault basin, *Moitessieria vidourlensis n.* sp.; 3: Sauve source, upper Vidourle, *Moitessieria vidourlensis n.* sp.; 4: Lez source, type locality of *Moitessieria rolandiana* (see Figure A1).

Appendix A.2.3. Genetics

Amplification was unsuccessful for several specimens or some of the genes studied, probably because *Moitessieria* species retract very deep into their shells, preventing contact between the tissues and the ethanol. The successful sequences of COI, 16S, and 28S obtained are given in Table A1. They support the biogeographic and morphometric analysis.

Table A1. Results of the tentative amplification of the three genes for the live specimens of *Moitessieria* collected. Location according to Figure A1.

Specimen	Location	COI	16S	28S
Moitessieria rolandiana	4 [1]	PP050555		
Moitessieria rolandiana	4 [1]	PP050556	PP051255	
Moitessieria larzacensis n. sp.	1	PP050554	PP051254	
Moitessieria larzacensis n. sp.	6	PP050557	PP051256	
Moitessieria atlantica n. sp. 1320	5			PP057732
Moitessieria atlantica n. sp. 1321	5		PP051257	PP057733
Moitessieria atlantica n. sp.	5			PP057734
Moitessieria atlantica n. sp.	5			PP057735
Moitessieria vidourlensis n. sp.	3			PP057731
Moitessieria vidourlensis n. sp.	2			PP057736
Moitessieria vidourlensis n. sp.	2			PP057737

[1] Type locality of Moitessieria rolandiana.

The mitochondrial genes (COI and 16S concatenated, Figure A4a) suggest that the population from the Larzac plateau is a separate species from *Moitessieria rolandiana* from the Lez drainage (type locality), as they diverge over 10% in COI, a threshold largely over interspecific divergences in related taxa [60–62]. This corresponds to a divergence of 3.8 to 4.7% in the 16S gene. No COI was available to compare the Atlantic population to *M. rolandiana*, but the 16S results can be transposed, as all mitochondrial genes share the same history.

The Atlantic population's divergence to the *M. rolandiana* type population is even higher, 6%. These results are congruent with the organization of the hydrogeological networks.

The nuclear gene (28S, Figure A4b), although it should be less variable, distinguishes the Atlantic population from the Mediterranean ones (only the upper Hérault and Vidourle basins analyzed here), but also, with a smaller divergence, the upper Hérault and upper Vidourle populations. Molecular data from more specimens are needed to determine whether the upper Vidourle and the upper Hérault populations are different species or not. A conservative attitude is adopted here and the upper Vidourle and the Cent Fonts populations are considered as belonging to the same species.

In summary, genetic data demonstrate that:

(i)   *M. larzacensis n.* sp. differs from *M. rolandiana* based on both COI and 16S.

(ii)   *M. atlantica n.* sp. differs from *M. rolandiana* and *M. larzacensis n.* sp. based on 16S.

(iii)   *M. vidourlensis n.* sp. differs from *M. atlantica n.* sp. based on 28S, but cannot be compared to *M. rolandiana* nor *M. larzacensis n.* sp. from the available molecular data. Only morphological differences, that are supported by geographical isolation, allow this species to be separated from *M. rolandiana* and *M. larzacensis n.* sp.

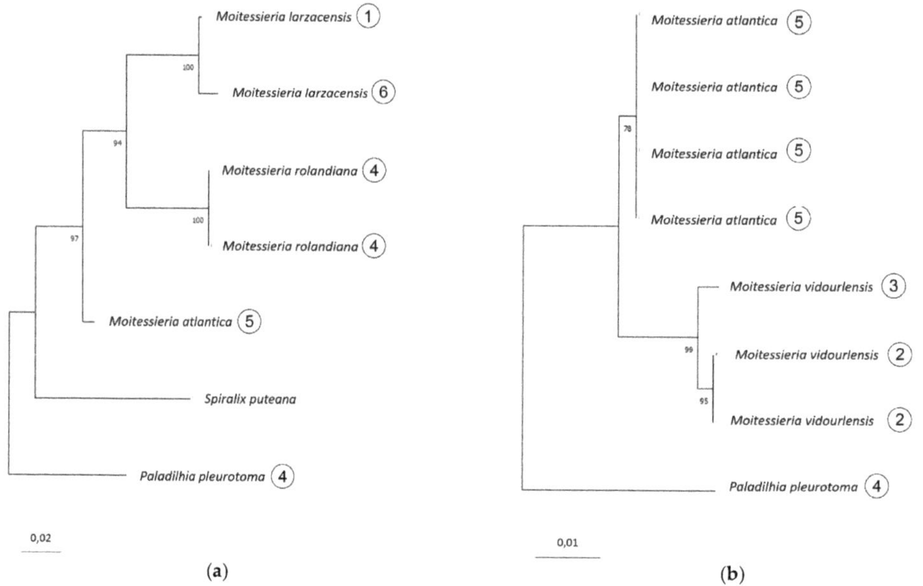

Figure A4. Phylogenetic tree of studied genes (Bayesian inference). (**a**) Concatenated COI + 16S; (**b**) 28S. Numbers correspond to localities in Figure A2.

Appendix A.2.4. Species Delimitation

Moitessieria rolandiana was considered to be a widespread species, distributed from the western tributaries of the Rhône River to the Garonne drainage [16]. Our results show that the name *Moitessieria rolandiana* should be restricted to the populations from the Lez drainage system. To the east, the adjacent Vidourle system hosts a distinct species, *M. vidourlensis n.* sp. The population sampled from the karst systems on the right bank of the Hérault River was morphologically and genetically similar to the Vidourle population and was therefore considered to belong to the same species. This distribution pattern involving two coastal river basins was unexpected. However, it is reminiscent of the distribution of some stygobiotic shrimps, *Gallocaris inermis*, *Proasellus cavaticus*, *Faucheria faucheri*, etc., a total of 12 crustacean species also known from both the Cent Fonts and Sauve (Vidourle) springs. It is therefore likely that hydrological connections exist, at least sporadically, in the complex karstic network of the upstream Hérault and Vidourle drainages. *M. larzacensis n.* sp. is likely to live in the subterranean basin of the whole Vis River, probably on both sides, as *Moitessieria* species are known to live in the hyporheic zone and can therefore easily colonize the hydrosystems of both sides of the river. The hydrosystems of the Vis drainage are isolated from the Atlantic drainages to the west. It is therefore not surprising that the population of the Atlantic drainage belongs to a distinct species, described here as *Moitessieria atlantica n.* sp., due to geographic barriers.

Appendix A.2.5. Species Turbo-Taxonomy

Moitessieria vidourlensis n. sp. Prié 2024

Nomina nuda: *Moitessieria vasseuri* (Prié 2013)

Type material: holotype IM-2000-30145; paratypes: 28 shells (IM-2000-30146), deposited at the Museum national d'Histoire naturelle in Paris (Figure A5).

Type locality: the Sauve cave, in the Sauve (Gard department) municipality; 43°56′27.2394″ N; 3°56′58.1568″ E. The live specimen was collected while scuba diving, on the ground, with forceps, a few tens of meters from the entrance of the cave.

ZooBank record: urn: lsid:zoobank.org:act:B9A3A652-DDC5-4C18-A035-CA4574F8FE8E

Etymology: This species was originally dedicated to Frank Vasseur, a highly skilled subterranean scuba diver, who collected material from deep inside caves (including *M. larzacensis n.* sp.), and ensured VP's safety while diving in the Sauve cave in search of stygobiotic snails. We prefer to avoid eponyms here and give a name that reflects the distribution of the species.

Distribution: Pending further studies, the name *Moitessieria vidourlensis* should apply to the populations of *Moitessieria* from the Vidourle drainage, and the populations from the Causse-de-la-Selle. The species' distribution probably includes part of the Hortus karstic plateau between the Hérault and Vidourle drainages. Its eastward distribution limit is unknown.

Morphological characteristics: site 3—Vidourle: shell height: 1.67 (1.47–1.98) mm; shell width: 0.75 (0.65–0.98) mm; last whorl width: 0.65 (0.60–0.73) mm; N = 31; site 2—Cent Fonts: shell height: 1.74 (1.57–2.08) mm; shell width: 0.73 (0.65–0.82) mm; last whorl width: 0.62 (0.57–0.69) mm; N = 19.

Sequences GenBank accession numbers: PP057731, PP057736, PP057737 (28S).

Moitessieria larzacensis n. sp. Prié 2024

Nomina nuda: *Moitessieria tillierae* (Prié 2013)

Type material: holotype IM-2000-30143 (Figure A6); paratypes: 19 shells (IM-2000-30144), deposited at the Museum national d'Histoire naturelle in Paris.

Type locality: The Folatière spring, exsurgence of the Folatière Cave, 43°51.871′ N, 3°31.578′ E. The live specimen used for DNA analyses was collected while scuba diving in the Gourneyras cave, 43°51′36.48 N; 3°31′38.64 E.

ZooBank record: urn:lsid:zoobank.org:act:D215581E-C50E-4707-9DA3-59B39B3FF3D1

Etymology: This species was initially dedicated to Annie Tillier, who successfully amplified the specimen collected from Gourneyras, at a time when *Moitessieria* specimen amplifications were systematically failing. We prefer to avoid eponyms here and give a name that reflects the distribution of the species.

Distribution: This species is known from molecular data only from the basins of the Vis River (north of the Larzac plateau) and the Lergue River (south of the Larzac plateau). It should therefore live in the entire hydrogeological network of the Larzac plateau. As *Moitessieria* species also inhabit the hyporheic zone, the Vis River does not represent a biogeographic barrier and the populations from the Blandas plateau are expected to belong to the same species. The distribution of *M. larzacensis n.* sp. is probably the same as that of *Bythinella navacellensis* [63], i.e., the subterranean watersheds of the Larzac and Blandas plateaus, drained to the north by the Arre, to the south by the Lergue, and in between by the Vis.

Morphological characteristics: shell height: 1.87 (1.65–2.33) mm; width: 0.63 (0.55–0.74) mm; last whorl width: 0.56 (0.50–0.67) mm; N = 20.

Sequences GenBank accession numbers: PP050554, PP050557 (COI), PP051254, PP051256 (16S).

Moitessieria atlantica n. sp. Prié 2024

Nomina nuda: *Moitessieria girardii* (Prié 2013)

Type material: holotype IM-2000-30147; paratypes: 11 shells (IM-2000-30148), deposited at the Museum national d'Histoire naturelle in Paris (Figure A7).

Type locality: The Gloriette spring, in the Sorgue drainage, municipality of Cornus (Aveyron department), 43°54′28.5114″ N; 3°10′38.0634″ E.

ZooBank record: urn:lsid:zoobank.org:act:BC712B05-70CF-4107-97B5-CBC94FE91325

Etymology: This species was initially dedicated to Henri Girardi, a famous French malacologist, author of many subterranean snails' descriptions. We prefer to avoid eponyms here and give a name that reflects the distribution of the species.

Distribution: It is known only from the type locality, but presumably present elsewhere in the Sorgue River karstic drainages. Shells from a population sampled downstream (Saint-Paul-des-Fonts) fall within the morphological range of *M. atlantica n.* sp. and could belong to the same species.

Interestingly, no subterranean snail has been collected despite important sampling in the vicinity of the limit between the Atlantic and Mediterranean watershed. *M. atlantica n.* sp. is therefore geographically isolated from the other *Moitessieria* species described here.

Morphological characteristics: shell height: 1.72 (1.58–1.93) mm; shell width: 0.68 (0.66–0.73) mm; last whorl width: 0.62 (0.60–0.67) mm; N = 9. There are no morphometric analyses for this population, as the number of adult specimens collected was too low.

Sequences GenBank accession numbers: PP051257 (16S), PP057732, PP057733, PP057734, PP057735 (28S).

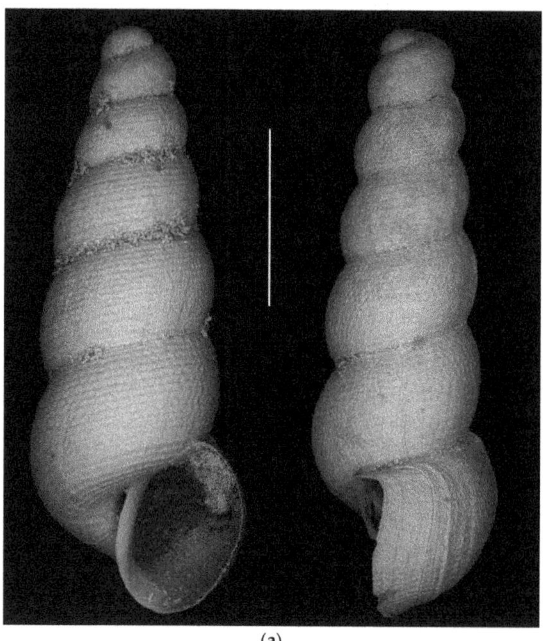
(a)

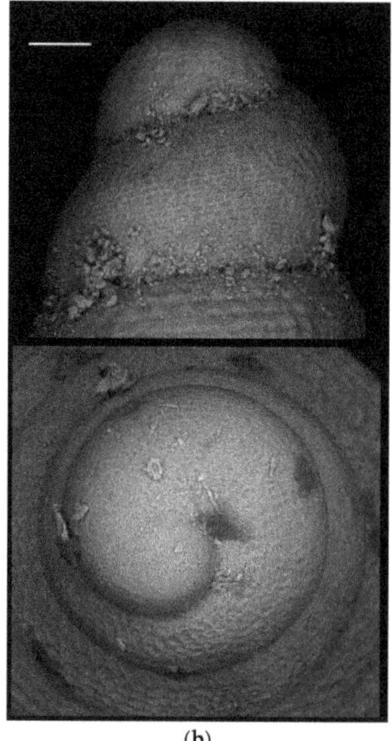
(b)

Figure A5. *Moitessieria vidourlensis*: (**a**) general view of the holotype (left) and of a paratype (right), scale = 500 μm; (**b**) details of the protoconch of the holotype (above) and of a paratype (below), scale = 50 μm.

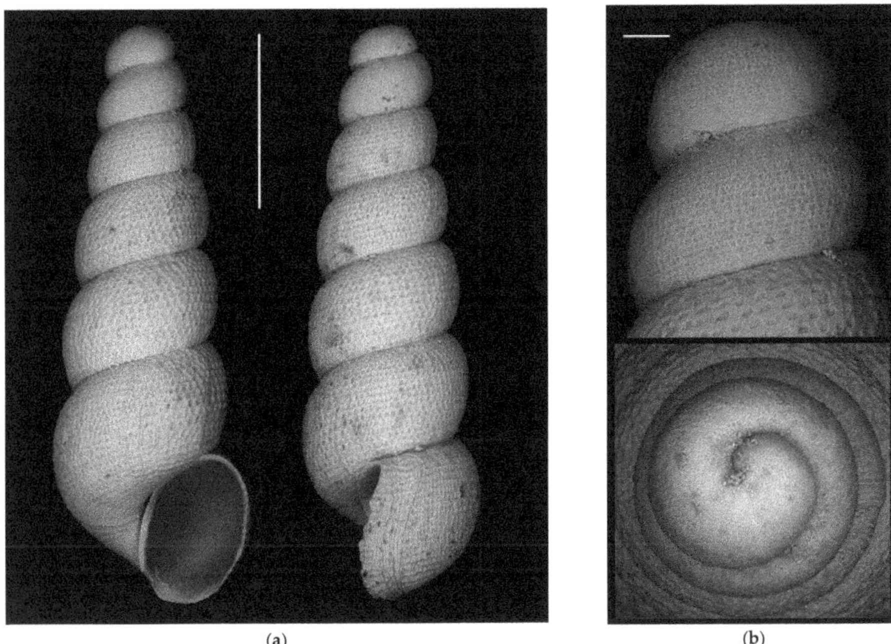

Figure A6. *Moitessieria larzacensis n.* sp.: (**a**) general view of the holotype (left) and of a paratype (right), scale = 500 μm; (**b**) details of the protoconch of the holotype (above) and of a paratype (below), scale = 50 μm.

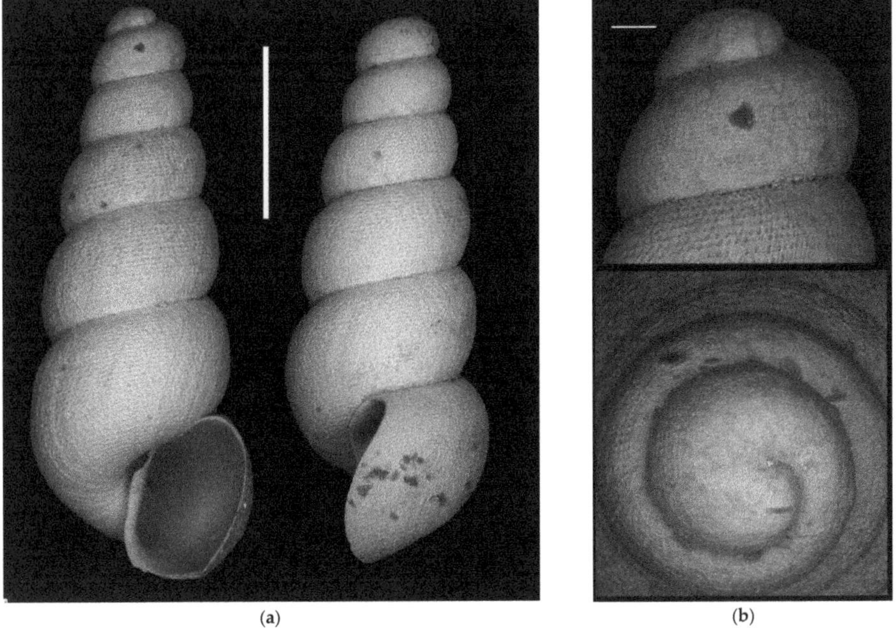

Figure A7. *Moitessieria atlantica n.* sp.: (**a**) general view of the holotype (left) and of a paratype (right), scale = 500 μm; (**b**) details of the protoconch of the holotype (above) and of a paratype (below), scale = 50 μm.

References

1. Malard, F.; Gibert, J.; Laurent, R. L'aquifère de la source du Lez: Un réservoir d'eau… et de biodiversité. *Karstologia* **1997**, *30*, 49–54. [CrossRef]
2. Culver, D.; Sket, B. Hotspots of subterranean biodiversity in caves and wells. *J. Cave Karst Stud.* **2000**, *62*, 11–17.
3. Tuzet, O.; Bonnet, A.; Bournier, E.; Du Cailar, J. Troisième contribution à la faune du Languedoc méditerranéen. *Notes Biospéologiques* **1950**, *5*, 85–95.
4. Balazuc, J.; Bonnet, A.; Bournier, E.; Du Cailar, J. Crustacés des eaux souterraines du Languedoc. Remarques sur leur répartition? *Bull. Soc. Hist. Nat. Toulouse* **1951**, *86*, 80–87.
5. Bonnet, A.; Bournier, E.; Du Cailar, J.; Quezel, P. Sur quatre crustacés aquatiques et troglobies d'une résurgence des gorges de l'Héralut. *Soc. Mér. Spéléologie Préhistoire* **1951**, *86*, 341–346.
6. Rouch, R.; Juberthie-Jupeau, L.; Juberthie, C. Recherche sur les eaux souterraines—3—Essai d'étude du peuplement de la zone noyée d'un karst. *Ann. Spéléol.* **1968**, *23*, 717–733.
7. Olivier, M.-J.; Martin, D.; Bou, C.; Prié, V. *Interprétation du Suivi Hydrobiologique de la Faune Stygobie, Réalisé sur le Système Karstique des Cent Fonts Lors du Pompage D'essai*; BRGM/RP+54865-FR; Conseil Général de l'Hérault: Montpellier, France, 2006; p. 42.
8. Prié, V. Systématique et Micro-Répartition des Mollusques Stygobies des Karsts du Nord-Montpelliérain. Master's Thesis, De l'École Pratique des Hautes Études, Paris, France, 2006.
9. Prié, V. Taxonomie et Biogéographie des Mollusques Patrimoniaux: Quelles échelles Pour la Délimitation des Taxons et des unités de Gestion? Ph.D. Thesis, Muséum National d'Histoire Naturelle, Paris, France, 2013.
10. Girardi, H. *Moitessieria wienini n.* sp. des eaux de l'Aquifère de la Montagne de la Sellette, sur la rivière Hérault, (F. 34), (Mollusca: Gastéropoda: Moitessieriidae). *Doc. Malacol.* **2001**, *2*, 3–10.
11. Girardi, H. Contribution à l'étude des gastéropodes stygobies de France. 3—*Paladilhia conica* (PALADILHE, 1867) (Gastropoda: Moitessieriidae). *Doc. Malacol.* **2003**, *4*, 89–90.
12. Ladouche, B.; Maréchal, J.C.; Dörfliger, N.; Lachassagne, P.; Lanini, S.; Le Strat, P. Pompages d'Essai sur le Système Karstique des Cent Fonts (Cne de Causse de la Selle, Hérault), Présentation et Interprétation des Données Recueillies, BRGM/RP 54426-FR, 2005. Available online: https://infoterre.brgm.fr/rapports/RP-54426-FR.pdf (accessed on 24 November 2023).
13. Tramblay, Y.; Koutroulis, A.; Samaniego, L.; Vicente-Serrano, S.M.; Volaire, F.; Boone, A.; Le Page, M.; Llasat, M.C.; Albergel, C.; Burak, S.; et al. Challenges for drought assessment in the Mediterranean region under future climate scenarios. *Earth Sci. Rev.* **2020**, *210*, 103348. [CrossRef]
14. Bou, C.; Rouch, R. Un nouveau champ de recherches sur la faune aquatique souterraine. *C. R. Acad. Sci.* **1967**, *265*, 369–370.
15. Jourde, H.; Dörfliger, N.; Maréchal, J.C.; Batiot-Guilhe, C.; Bouvier, C.; Courrioux, G.; Desprats, J.-F.; Fullgraf, T.; Ladouche, B.; Leonardi, V.; et al. Projet Gestion Multi-Usages de L'hydrosystème Karstique du Lez-Synthèse des Connaissances Récentes et Passées. BRGM/RP-60041-FR, 2011. Available online: https://brgm.hal.science/hal-03143770/document (accessed on 24 November 2023).
16. Bertrand, A. Atlas préliminaire de répartition géographique des mollusques stygobies de la faune de France (MOLLUSCA: RISSOIDEA: CAENOGASTROPODA). *Doc. Malacol.* **2004**, *2*, 82. [CrossRef]
17. UICN Comite Français; OFB; MNHN. *La Liste Rouge des Espèces Menacées en France—Chapitre Mollusques Continentaux de France Métropolitaine*; UICN: Paris, France, 2021.
18. UICN France; MNHN. *La Liste Rouge des Espèces Menacées en France—Chapitre Crustacés d'eau Douce de France Métropolitaine*; UICN: Paris, France, 2012.
19. Lecaplain, B. Sur la présence en France de *Trocheta taunensis* Grosser, 2015 (Hirudinida, Erpobdellidae). *Naturae* **2021**, *25*, 345–349. [CrossRef]
20. Sket, B. K Poznavanju Favne Pijavk (Hirudinea) v Jugoslaviji, Zur Kenntnis der Egel-Fauna (Hirudinea) Jugoslawiens. *Acad. Sci. Artium Slov. Cl. IV Hist. Nat. Med.* **1968**, *9*, 127–197.
21. Grosser, C. Differentiation of some similar species of the subfamily Trochetinae (Hirudinida: Erpobdellidae). *Ecol. Montenegrina* **2015**, *2*, 29–41. [CrossRef]
22. Prié, V. Répartition de *Heraultiella exilis* (Paladilhe, 1867) (Gastropoda, Caenogastropoda, Rissooidea). *MalaCo* **2005**, *1*, 8–9.
23. Prié, V. *Heraultiella exilis.* The IUCN Red List of Threatened Species 2011: e.T2092A9236035. 2011. Available online: https://www.iucnredlist.org/species/2092/9236035 (accessed on 24 November 2023).
24. Prié, V. *Paladilhia pleurotoma.* The IUCN Red List of Threatened Species 2010: e.T15876A5275513. 2010. Available online: https://www.iucnredlist.org/species/15876/5275513 (accessed on 24 November 2023).
25. Prié, V. *Bythiospeum bourguignati.* The IUCN Red List of Threatened Species 2010: e.T61315A12461687. 2010. Available online: https://www.iucnredlist.org/species/61315/12461687 (accessed on 24 November 2023).
26. Callot-Girardi, H.; Boeters, H.D. *Moitessieria guilhemensis*, nouvelle espèce de la résurgence du Cabrier à Saint-Guilhem-le Désert, Hérault, France. (Mollusca: Caenogastropoda: Moitessieriidae). *Avenionia* **2017**, *2*, 42–63.
27. De Grave, S. Gallocaris inermis. The IUCN Red List of Threatened Species 2013: e.T198319A2520643. Available online: https://www.iucnredlist.org/species/198319/2520643 (accessed on 1 December 2023).
28. Henry, J.-P. Contribution à l'étude du genre *Proasellus* (crustacea isopoda asellidae): Le groupe cavaticus. *Vie Milieu* **1971**, *22*, 33–77.
29. Bertrand, J.-Y. Recherches Sur l'Écologie de *Faucheria faucheri* (Crustacés, Cirolanides). Ph.D. Thesis, Université de Paris VI, Paris, France, 1974; p. 123.

30. Fiser, C.; Zagmajster, M.; Dethier, M. Overview of Niphargidae (Crustacea: Amphipoda) in Belgium: Distribution, taxonomic notes and conservation issues. *Zootaxa* **2018**, *4387*, 26. [CrossRef] [PubMed]

31. McInerney, C.E.; Maurice, L.; Robertson, A.L.; Knight, L.R.; Arnscheidt, J.; Venditti, C.; Dooley, J.; Mathers, T.C.; Matthijs, S.; Rrikkson, K.; et al. The ancient Britons: Groundwater fauna survived extreme climate change over tens of millions of years across NW Europe. *Mol. Ecol.* **2014**, *23*, 1153–1166. [CrossRef] [PubMed]

32. Stock, J.H.; Gledhill, T. The *Niphargus kochianus*—group in North-Western Europe. *Crustac. Suppl.* **1977**, *4*, 212–243.

33. Lefébure, T.; Douady, C.J.; Gouy, M.; Trontelj, P.; Briolay, J.; Gibert, J. Phylogeography of a subterranean amphipod reveals cryptic diversity and dynamic evolution in extreme environments. *Mol. Ecol.* **2006**, *15*, 1797–1806. [CrossRef]

34. Danielpol, D.L.; Namiotko, T.; Meisch, C. *Marmocandona* nov. gen. (Ostracoda, Candoninae), with comments on the contribution of stygobitic organisms to micropalaeontological studies. *Kölner Forum Geol. Paläont.* **2012**, *21*, 13–16.

35. Walter, T.C.; Boxshall, G. World of Copepods Database. *Acanthocyclops venustus venustus* (Norman & Scott T., 1906). 2023, Accessed through: World Register of Marine Species. Available online: https://www.marinespecies.org/aphia.php?p=taxdetails&id=7298 73on2023-11-08 (accessed on 11 November 2023).

36. Lescher-Moutoué, F. Recherches sur les eaux souterraines—21—Un Cyclopide nouveau du genre *Graeteriella*. *Ann. Spéléol.* **1974**, *29*, 71–76.

37. Fiers, F.; Ghenne, V. Cryptozoic copepods from Belgium: Diversity and biogeographic implications. *Belg. J. Zool.* **2000**, *130*, 11–19.

38. Lescher-Moutoué, F. Recherches sur les eaux souterraines—7—Les cyclopides de la zone noyée d'un karst. I *Graeteriella (Paragraeteriella) vandeli n. sp. Ann. Spéléol.* **1969**, *24*, 429–438.

39. Lescher-Moutoué, F. Sur la biologie et l'écologie des Copépodes Cyclopides hypogés. *Ann. Spéléol.* **1973**, *28*, 429–502 & 581–674.

40. Iannella, M.; Fiasca, B.; Di Lorenzo, T.; Biondi, M.; Di Cicco, M.; Galassi, D.M.P. Spatial distribution of stygobitic crustacean harpacticoids at the boundaries of groundwater habitat types in Europe. *Sci. Rep.* **2020**, *10*, 19043. [CrossRef]

41. Galassi, D.M.P.; Huys, R.; Reid, J.W. Diversity, ecology and evolution of groundwater copepods. *Freshw. Biol.* **2009**, *54*, 691–708. [CrossRef]

42. Galassi, D.M.P.; De Laurentiis, P.; Dole-Olivier, M.-J. Phylogeny and biogeography of the genus Pseudectinosoma, and description of *P. janineae* sp. n. (Crustacea, Copepoda, Ectinosomatidae). *Zool. Scr.* **1999**, *28*, 289–303. [CrossRef]

43. Vandel, A. *Isopodes Terrestres (Première Partie)*; Office Central de Faunistique, Fédération Française des Sociétés de Sciences Naturelles: Paris, France, 1960; p. 417.

44. Juberthie, C.; Juberthie-Jupeau, L. La réserve biologique du laboratoire souterrain du C.N.R.S. à Sauve (Gard). *Ann. Spéléol.* **1975**, *30*, 539–555.

45. Delamare Debouteville, C. *Biologie des Eaux Souterraines Littorales et Continentales*; Hermann: Paris, France, 1960; 740p.

46. Wilke, T.; Davis, G.; Falniowski, A.; Giusti, F.; Bodon, M.; Szarowska, M. Molecular systematics of Hydrobiidae (Mollusca: Gastropoda: Rissooidea): Testing monophyly and phylogenetic relationships. *Proc. Acad. Nat. Sci. USA* **2009**, *151*, 1–21. [CrossRef]

47. Machetel, P.; Yuen, D.A. Revisiting Cent-Fonts Fluviokarst Hydrological Properties with Conservative Temperature Approximation. *Hydrology* **2017**, *4*, 6. [CrossRef]

48. Valentini, A.; Taberlet, P.; Miaud, C.; Civade, R.; Herder, J.; Thomsen, P.F.; Bellemain, E.; Besnard, A.; Coissac, E.; Boyer, F.; et al. Next-generation monitoring of aquatic biodiversity using environmental DNA metabarcoding. *Mol. Ecol.* **2016**, *25*, 929–942. [CrossRef]

49. Prié, V.; Danet, A.; Valentini, A.; Lopes-Lima, M.; Taberlet, P.; Besnard, A.; Roset, N.; Gargominy, O.; Dejean, T. Conservation assessment based on large-scale monitoring of eDNA: Application to freshwater mussels. *Biol. Cons.* **2023**, *283*, 110089. [CrossRef]

50. Fukuda, H.; Haga, T.; Tatara, Y. *Niku-nuki*: A useful method for anatomical and DNA studies on shell-bearing molluscs. *Zoosymposia* **2008**, *1*, 15–38. [CrossRef]

51. SANDRE. Base de Données Sur l'Hydrographie. 2007. Available online: http://sandre.eaufrance.fr/ (accessed on 14 November 2023).

52. Wilcox, D.; Dove, B.; McDavid, D.; Greer, D. Image Tool for Windows version 3.00. 2002. UTHSCA ed. San Antonio: UTHSCA. Available online: http://ddsdx.uthscsa.edu/dig/itdesc.html (accessed on 3 July 2005).

53. Bernasconi, R. Révision du genre *Bythinella* (Moquin-Tandon, 1855) de la France du Centre-Ouest, du Midi et des Pyrénées (Gastropoda Prosobranchia Hydrobiidae: Amnicolinae Bythinellini). *Doc. Malacol.* **2000**, *1*, 126.

54. Haase, M.; Gargominy, O.; Fontaine, B. Rissoidean freshwater gastropods from the middle of the Pacific: The genus *Fluviopupa* on the Austral Islands (Caenogastropoda). *Molluscan Res.* **2005**, *25*, 145–163. [CrossRef]

55. R Core Team. *R: A Language and Environment for Statistical Computing*; R Foundation for Statistical Computing: Vienna, Austria, 2015; Available online: https://www.R-project.org/ (accessed on 3 May 2013).

56. Folmer, O.; Black, M.; Hoeh, W.; Lutz, R.; Vrijenhoek, R. DNA primers for amplification of mitochondrial cytochrome c oxidase subunit I from diverse metazoan invertebrates. *Mol. Mar. Biol. Biotechnol.* **1994**, *3*, 294–299.

57. Palumbi, S.R. Nucleic acids II: The Polymerase Chain Reaction. In *Molecular Systematics*, 2nd ed.; Hillis, D.M., Moritz, C., Mable, B.K., Eds.; Sinauer: Sunderland, MA, USA, 1996; pp. 205–247.

58. Jovelin, R.; Justine, J.-L. Phylogenetic relationships within the polyopisthocotylean monogeneans (Platyhelminthes) inferred from partial 28S rDNA sequences. *Int. J. Parasitol.* **2001**, *31*, 393–401. [CrossRef]

59. Hall, T.A. BioEdit: A user-friendly biological sequence alignment editor and analysis program for Windows 95/98/NT. *Nucleic Acids Symp.* **1999**, *41*, 95–98.

60. Hershler, R.; Liu, H.P.; Thompson, F.G. Phylogenetic relationships of North American nymphophiline gastropods based on mitochondrial DNA sequences. *Zool. Scr.* **2003**, *32*, 357–366. [CrossRef]
61. Hurt, C.R. Genetic divergence, population structure and historical demography of rare spring-snails (*Pyrgulopsis*) in the lower Colorado River basin. *Mol. Ecol.* **2004**, *13*, 1173–1187. [CrossRef] [PubMed]
62. Liu, H.P.; Hershler, R.; Clift, K. Mitochondrial DNA sequences reveal extensive cryptic diversity within a western American springsnail. *Mol. Ecol.* **2003**, *12*, 2771–2782. [CrossRef]
63. Prié, V.; Bichain, J.-M. Phylogenetic relationships and description of a new stygobite species of *Bythinella* (Mollusca, Gastropoda, Caenogastropoda, Amnicolidae) from southern France. *Zoosystema* **2009**, *31*, 987–1000. [CrossRef]

Article

Does Size Matter? Two Subterranean Biodiversity Hotspots in the Lessini Mountains in the Veneto Prealps in Northern Italy

Leonardo Latella

Museo di Storia Naturale of Verona, Lungadige Porta Vittoria, 9, 37129 Verona, Italy; leonardo.latella@comune.verona.it

Abstract: In the Lessini Mountains, the southernmost prealpine area in the Veneto region, thousands of caves are found, many of which have been extensively studied from the biological point of view. Numerous studies have been carried out on taxonomic and biogeographic aspects over the last hundred years. Two caves, in particular, have been found to be extremely rich in species adapted to life in subterranean environments. These are the Arena Cave in the Monti Lessini Veronesi and the Buso della Rana cave system in the Monti Lessini Vicentini. The two caves have extremely different development: Arena Cave is about 100 meters in length, and the Buso della Rana-Pisatela cave system is more than 37 km in length. Despite this huge difference in size, they both have the highest number of subterranean dwelling species in northern Italy (16 troglobionts and 8 stygobionts in Arena Cave, and 7 troglobionts and 11 stygobionts in the Buso della Rana-Pisatela cave system).

Keywords: caves; Arena Cave; Buso della Rana-Pisatella cave system; biospeleology; checklist; contact karst

Citation: Latella, L. Does Size Matter? Two Subterranean Biodiversity Hotspots in the Lessini Mountains in the Veneto Prealps in Northern Italy. *Diversity* **2024**, *16*, 25. https://doi.org/10.3390/d16010025

Academic Editors: Tanja Pipan, David C. Culver, Louis Deharveng and Michael Wink

Received: 6 October 2023
Revised: 14 December 2023
Accepted: 26 December 2023
Published: 30 December 2023

1. Introduction

The Lessini Mountains, located in the western part of the Veneto Prealps (southeastern Italian Alps) [1], extend over a total area of 1403 km^2 with a maximum elevation of 1865 m a.s.l. They are bordered by the Adige Valley to the west and the Leogra Valley to the east and northeast. The Val dei Ronchi separates the range from the Pasubio-Carega Group to the northwest. From the geological point of view, the Lessini Mountains are dominated by limestones from the Mesozoic and Cenozoic ages, interspersed by Cenozoic volcanic rocks and Eocenic limestone outcrops [2]. Lessinia includes the western Lessini Mountains (between the Adige and Illasi Valleys) and eastern Lessinia Mountains (between the Illasi and Leogra Valleys). The western Lessini Mountains (Dolomia Principale, Calcari grigi, Rosso Ammonitico, Scaglia Rossa, and Maiolica) mainly consist of carbonate rocks; some areas of the eastern Lessinia Mountains are primarily composed of volcanic rocks developed during the Venetian Tertiary magmatism [2].

In this mountain range, two caves have a high number of obligate subterranean species: Arena Cave (having 16 troglobionts and 8 stygobionts), which is in the Central-Western Lessini Mountains, and the Buso della Rana-Pisatella cave system (having 7 troglobionts and 11 stygobionts), in the Eastern Lessini Mountains [3,4]. The two caves are located 22 km apart as the crow flies (Figure 1).

Despite the short distance between the two caves, which are situated in the same mountain range, they show enormous differences in development, rock formations, and fauna composition, making the separate analysis and comparison of the two caves interesting. Especially evident is the huge difference in the length of the two caves. This difference in size would suggest that the larger and more diversified one has a greater richness of cave-dwelling species. To verify this hypothesis, we compare here, the number of subterranean species present in each of the two caves. However, the two caves have one important characteristic in common: both can be considered cases of contact karst caves, i.e., karst phenomena and forms influenced by the contact between two or more karstifiable rocks that differ in some characteristics, such as porosity, chemical composition and fracture density, or a karstifiable rock and a non-karstifiable rock [5].

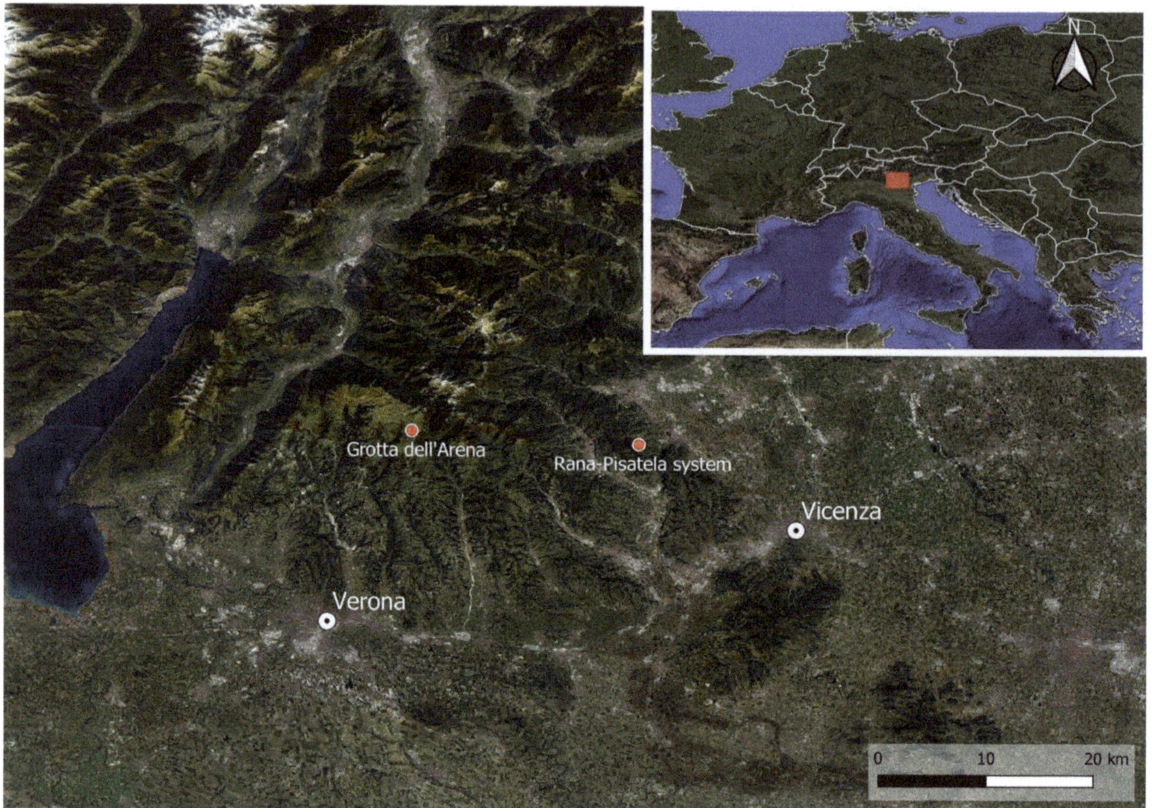

Figure 1. Location of the two studied caves in the Lessinia Mountains (the red rectangle in the small photo indicates the position in Italy).

1.1. Arena Cave (476 V/VR)

Arena Cave is registered as number 476 in the Cadastre of the Caves of Veneto Region. It is located in the Province of Verona, municipality of Bosco Chiesanuova, in the Malga Bagorno area. Its location, at 11°06′02″ E 45°39′56″ N, has an altitude of 1512 m a.s.l. (Figure 2).

The cave is 74 m long with a difference in elevation of −22 m from the entrance to the bottom. It is formed by a large chamber, roughly elliptical in plane section, with a main diameter of about 50 m. The roof coincides mostly with bedding planes. The southern part of the floor is characterized by a large, asymmetrical, funnel-shaped depression, a type of subterranean doline that developed in the collapse of debris [6] (Figure 3). The chamber is connected to the surface through some narrow passages that start from an open collapse depression located on a slope, which resembles a Roman theatre (i.e., an "Arena", hence the name of the cave).

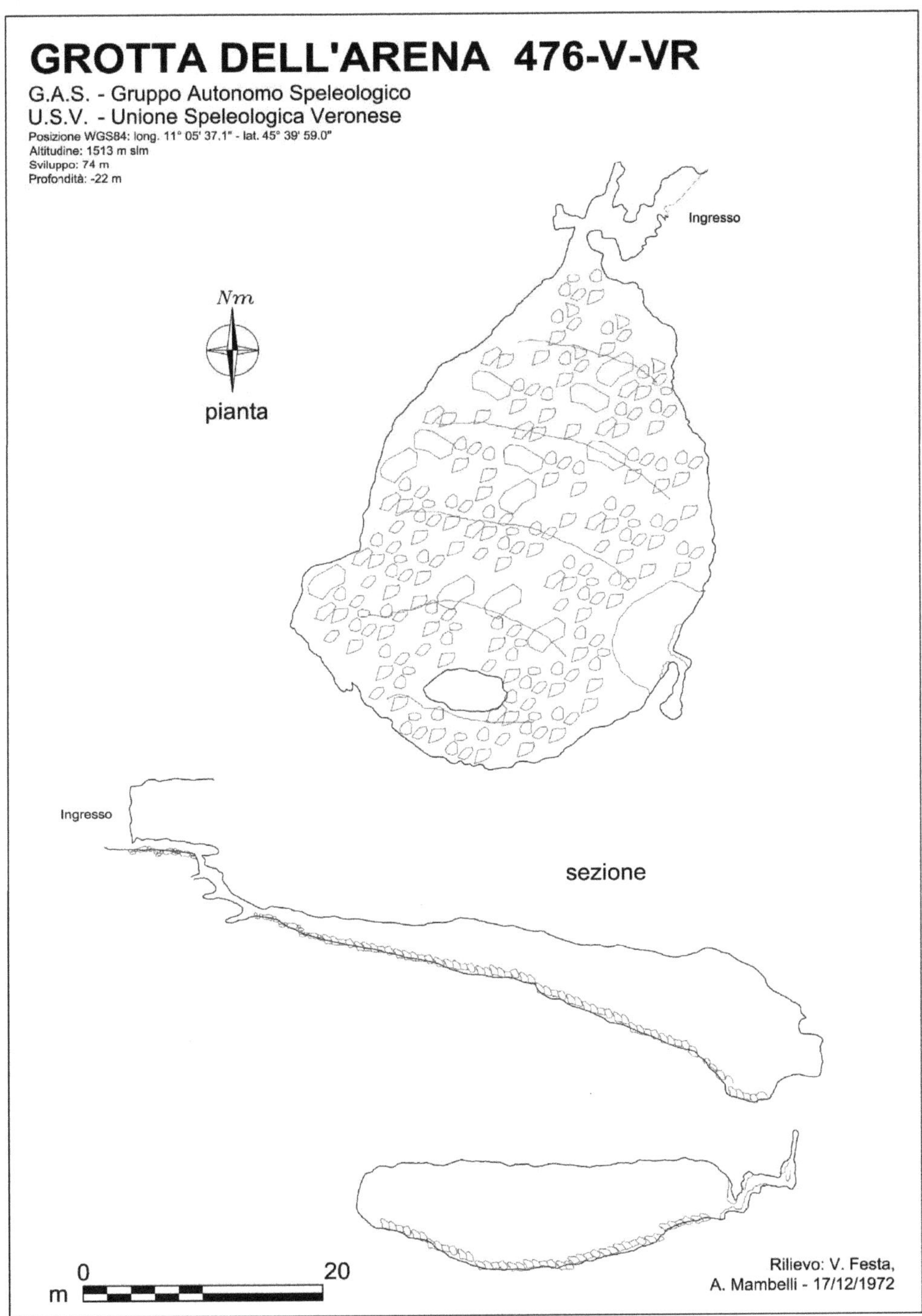

Figure 2. Cave map of Arena Cave. Sezione = cross section; pianta = plan; ingresso = entrance.

Figure 3. The large chamber of the Arena Cave (Photo: L. Latella).

From the geological point of view, the cave is an expression of a contact karst, where different limestone types are in contact both stratigraphically and along tectonic structures [5,7,8]. The limestone formations present here are "Calcari del Gruppo di San Vigilio" of lower-middle Jurassic, about 60 m in depth, both pure Oolitic and bio-sparitic/–ruditic, or reef limestones, relatively densely fractured; "Rosso Ammonitico", a condensed rock unit of middle-upper Jurassic age, about 30 m in depth, made up of nodular micritic limestone that is very resistant to erosion, crossed by widely spaced fractures; and "Biancone", a chalk-type unit, from the lower and middle Cretaceous, 100–200 m in depth, made up of whitish marly limestone that is closely stratified and densely fractured, and very sensitive to frost and atmospheric agents. Therefore, the cave develops at the stratigraphic contact between the "Calcari del Gruppo di San Vigilio" and the "Rosso Ammonitico" and is close to a fault plane, placing the two above formations in vertical contact with the "Biancone". The overlying rocks of the cave are formed by the massive beds of lower Rosso Ammonitico, whereas the inner cave is mostly developed inside the Calcari del Gruppo di San Vigilio. At the topographical surface, the line of the normal fault runs along a small valley a few meters to the east of the cave; the displacement of the fault is about 100 m.

From a hydrological point of view, the water circulates diffusely inside the dense network of discontinuities of the Biancone unit; the preferential flow is sub-parallel to the topographical surface and occurs mostly below the dry valley bottoms but is also influenced by the structural setting; vertical losses occur along the fault and fracture zones.

In contrast, the water circulation is more concentrated and mostly vertical in the Rosso Ammonitico [6].

1.2. Buso della Rana-Pisatela Cave System (40 V–VI/1707 V–VI)

Buso della Rana cave (40 V–VI) opens at an altitude of 340 m a.s.l. in the province of Vicenza, municipality of Monte di Malo. It has a length of 30,102 m and an altitudinal range of 274 m.

In 2012, the cave, which had only one entrance, was connected to the Pisatela Cave (1707 V–VI), a cavity with two entrances (Pater Noster and Pisatela), the highest of which opens at 747 m a.s.l. at a development of 7510 m. The two cavities thus give rise to the Rana-Pisatella cave system, with a development of 37,612 m and an altitude difference of 407 m between the upper (Pisatella) and lower (Rana) entrances (Figure 4).

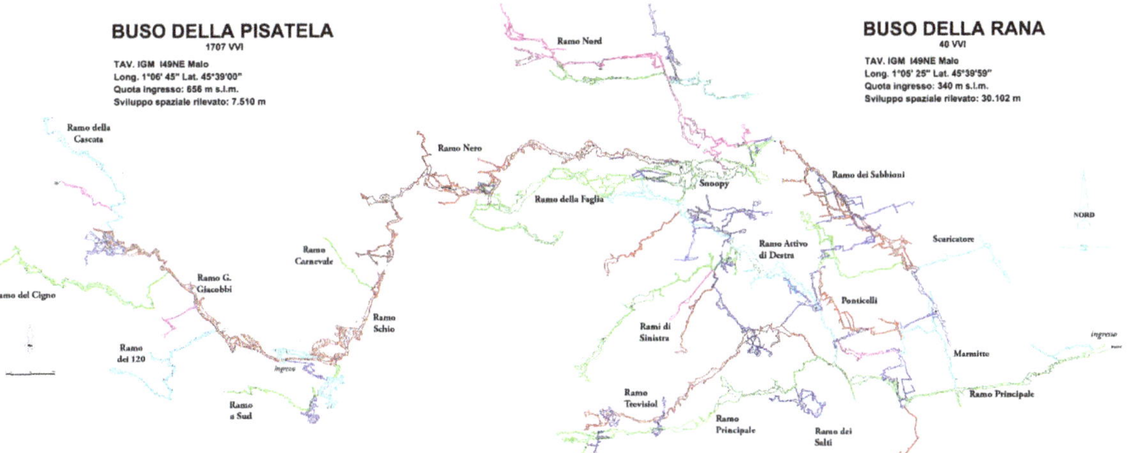

Figure 4. Plan of the Rana-Pisatela cave system (scale bar: 100 m).

The cave system is located on the Faedo-Casaron Plateau, which occupies a geographical area of 15 square kilometers in the province of Vicenza.

The area lies between the latitude of 45°36′46″ (Cornedo) to the South and 45°39′35″ (Monte di Malo) to the North and a longitude of 11°19′46″ (Monte Faedo) to the West and 11°22′22″ (Priabona) to the East. Morphologically, it is made up of limestone that gives rise to gentle and rounded surface forms typical of hills, with the valleys oriented according to the main Lessinian tectonic lines.

The Buso della Rana-Pisatela cave system, with its 37,612 m of development, is one of the longest caves in Italy. It developed in the Oligocene limestone through joint networks from the Faedo-Casaron plateau towards the less permeable basalt surface. Here, several independent and differently sized brooks descend along the gradient of the transgressive contact towards the entrance of the Buso della Rana cave, which is the main spring of the karst system forming the Rana River. Pisatela Cave is an inactive sink located inside a doline that reaches the main stream through a series of narrow meanders and shafts [9,10].

The drainage in the cave system is primarily controlled by contact with the less permeable basalt surface, basal conglomerate, and terrigenous marls of the Priabona Formation that rise above the contact in the eastern sector of the system. In some parts of the cave, the conduits evolved entirely in the Boro conglomerate, which is almost 2 m thick in the upstream part of Pisatela Cave. Here, the conglomerate covers the lower basalt contact, while in Buso della Rana, it is often absent, and the calcarenites lie directly on the basalts [9].

Cave morphologies are mostly vadose, with deep canyons and meanders and large collapse rooms at the intersection of different streams and fracture sets. Small paleophreatic conduits, evolved entirely in the calcarenite, hanging about 10–15 m above the basalt surface, are related to the phreatic primary drainage and locally ancient epiphreatic conditions. Various vadose narrow shafts reach the contact galleries from the above overlying plateau, showing condensation corrosion morphologies [9].

During periods of heavy rain, the level of the streams rises, often flooding some areas of the cave. In the terminal parts (the last few thousand meters towards the entrance), the river shows a well-structured hyporheic zone (Figures 5–9).

Figure 5. "Ramo attivo di destra", one of the hydrologically active branches in the Buso della Rana (Photo: S. Sedran).

Figure 6. The beautiful coralloid concretions in the "Ramo franchignia, Saletta broccoli" in the Buso della Rana (Photo: S. Sedran).

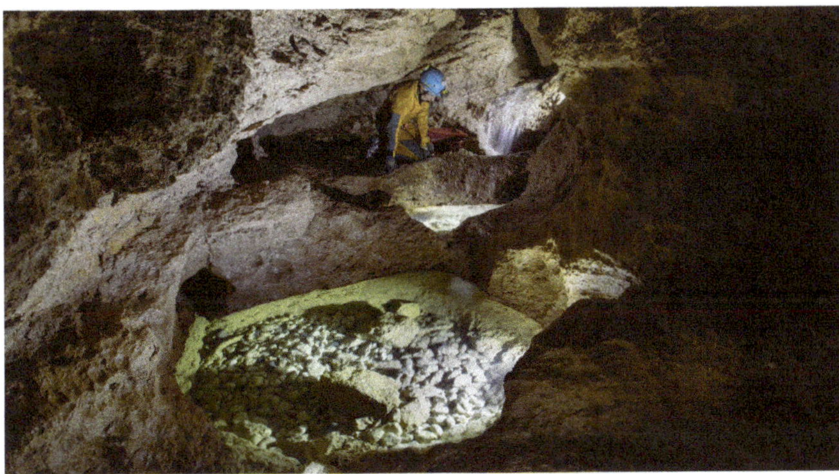

Figure 7. The left main branch of the Buso della Rana (Photo: S. Sedran).

Figure 8. The large dimensions of the "Sala dei Massi". One of the many large halls that characterize the Rana–Pisatella cave system (Photo: S. Sedran).

Figure 9. The shaft of the "Ramo del Pantano", a tributary of the Main Branch of the Buso della Rana (Photo: S. Sedran).

Since the Rana-Pisatela system has only recently been joined, much of the research (especially on aquatics) was carried out in the Buso della Rana cave, which will often be cited separately in the text.

2. Materials and Methods

2.1. Sampling and Museum Collections

Both caves have been known since the first half of the last century, and numerous biospeleological investigations have been carried out within them. The results of these surveys have largely been published, but many unstudied specimens are present in the collections of the Museo di Storia Naturale of Verona (Italy).

In recent years, research campaigns aimed to increase the knowledge of the troglobiotic and stygobiotic fauna as a whole were carried out [11–13].

Sampling of terrestrial fauna was carried out by means of pitfall traps and direct capture. The pitfall traps consisted of a plastic or glass cup, usually with an opening diameter of 10 cm and a depth of 15 cm. Each was filled with a preserving liquid (NaCl solution) in which was placed a tube containing an attracting bait of blue cheese. The traps were used on a few occasions to integrate hand-collecting and were left in place for about a month each time. In some cases, when we were unsure whether to return to the cave within a short time, the bait was placed without the trap, and fauna were collected at sight.

Aquatic fauna were sampled in both caves by direct capture, hand nets, or using a syringe to collect the water filling the pools. In the Buso della Rana, a drip funnel was placed in the main branch 600 m from the entrance [13]. A drip funnel consists of a funnel supported by a bucket that allows it to direct the dripping water into a plastic container. A 2 cm × 3 cm area on two sides of the square container was cut out and covered with a net (mesh size 60 μm) to retain the animals in the container [14,15].

The specimens sampled in the caves in the province of Verona, both in historical and recent times and not taxonomically identified, have been deposited in the 'miscellanea Biospeleologica' collection of the Museum of Verona. In the course of the study for this paper, the material collected in the two caves and not yet identified (i.e., Isopoda, Opilionida, and Collembola) was sent to specialists.

2.2. Bibliographic Research

Many papers on speleological (especially the Buso della Rana cave), geological, and biological aspects have been published from the first half of the last century. As far as biological research is concerned, in addition to the many publications devoted to the description of new species or studies and reviews of certain faunitic groups, there have been few works on the fauna as a whole of the two caves [16–19].

Therefore, we collected all publications to date and checked and updated the scientific names. This was also important so as to identify and list all the species for which the two caves represent the type locality. More than 60 publications were found in which, in various ways, the presence of animal species in one (or both) of the two (or both) caves under study is mentioned. For reasons of space and usefulness, all of them are obviously not listed in the bibliography.

3. Results

The subterranean fauna of the two caves, as a whole, consists of 46 cave-dwelling species. A total of 35 species are troglobionts or stygobionts, while 12 can be considered eutroglophiles (sensu Ruffo, 1957) [20,21]. In the Arena Cave, 24 obligate subterranean species are known, of which 16 are troglobionts and 8 are stygobionts; for the Rana-Pisatella cave system, 18 species are known, of which 7 are troglobionts and 11 are stygobionts (Table 1).

Table 1. The list of troglobiotic and stygobiotic species known in the two studied caves. Tb—troglobiont; Stb—stygobiont; 1—present; 0—absent; Tl—Type locality.

Class	Order	Family	Genus/Species/Subspecies	Status	Arena	Rana-Pisatella
Gastropoda	Ellobiida	Ellobiidae	*Zospeum globosum* Kuščer, 1928	Tb	1	1
Arachnida	Opiliones	Ischyropsalididae	*Ischyropsalis strandti* Kratochvíl, 1936	Tb	1	1
Arachnida	Pseudoscorpionida	Neobisiidae	*Neobisium (Blothrus) torrei* (Simon, 1881)	Tb	1	1
Arachnida	Pseudoscorpionida	Neobisiidae	*Balkanoroncus boldorii* (Beier, 1931)	Tb	1	0
Arachnida	Pseudoscorpionida	Chthoniidae	*Chthonius lessiniensis* Schawaller, 1982	Tb	1	0
Malacostraca	Isopoda	Trichoniscidae	*Androniscus (Dentigeroniscus) degener* Brian, 1926	Tb	1	1
Diplopoda	Chordeumatida	Craspedosomatidae	*Lessinosoma paolettii* Strasser, 1967	Tb	1 Tl	0
Diplopoda	Chordeumatida	Iulidae	*Trogloiulus boldorii* Manfredi, 1940	Tb	1	0
Collembola	Poduromorpha	Onychiuridae	*Onychiurus hauseri* Dallai, 1975	Tb	1	0
Collembola	Entomobryomorpha	Entomobryidae	*Pseudosinella concii* Gisin, 1950	Tb	1	0
Collembola	Entomobryomorpha	Entomobryidae	*Pseudosinella* sp.	Tb	1	0
Insecta	Coleoptera	Carabidae	*Italaphaenops dimaioi* Ghidini, 1964	Tb	1	0
Insecta	Coleoptera	Carabidae	*Lessynodytes pivai* Vigna Taglianti & Sciaky, 1988	Tb	1 Tl	0
Insecta	Coleoptera	Carabidae	*Orotrechus pominii* Tamanini, 1953	Tb	1	1
Insecta	Coleoptera	Carabidae	*Orotrechus vicentinus juccii* Pomini, 1940	Tb	1	0

Table 1. *Cont.*

Class	Order	Family	Genus/Species/Subspecies	Status	Arena	Rana-Pisatella
Insecta	Coleoptera	Leiodidae	*Halberria zorzii* (Ruffo, 1950)	Tb	1 Tl	0
Insecta	Coleoptera	Leiodidae	*Lessiniella trevisioli* Pavan, 1941	Tb	0	1 Tl
Insecta	Coleoptera	Leiodidae	*Neobathyscia fabianii* (Dodero, 1904)	Tb	0	1
Copepoda	Cyclopoida	Cyclopidae	*Speocyclops infernus* Kiefer, 1930	Stb	1	1
Copepoda	Harpactoida	Camptocamptidae	*Elaphoidella phreatica* (Chappuis, 1925)	Stb	0	1
Copepoda	Harpactoida	Camptocamptidae	*Elaphoidella ruffoi* Chappuis, 1953	Stb	0	1
Copepoda	Harpactoida	Camptocamptidae	*Elaphoidella* sp. A1	Stb	1	0
Copepoda	Harpactoida	Camptocamptidae	*Elaphoidella* sp. A	Stb	0	1
Copepoda	Harpactoida	Camptocamptidae	*Ceuthonectes serbicus* Chappuis, 1924	Stb	0	1
Copepoda	Harpactoida	Camptocamptidae	*Lessinocamptus insoletus* (Chappuis, 1928)	Stb	0	1 Tl
Copepoda	Harpactoida	Camptocamptidae	*Lessinocamptus pivai* Stoch, 1997	Stb	0	1 Tl
Copepoda	Harpactoida	Camptocamptidae	*Lessinocamptus caoduroi* Stoch, 1997	Stb	1 Tl	0
Copepoda	Harpactoida	Camptocamptidae	*Bryocamptus (Limocamptus) echinatus* (Mrazek, 1893)	Stb	1	0
Copepoda	Harpactoida	Camptocamptidae	*Moraria* (M.) sp. A1	Stb	1	0
Copepoda	Harpactoida	Parastenocaridiidae	*Parastenocaris ranae* Stoch, 2000	Stb	0	1 Tl
Copepoda	Harpactoida	Amaeridae	*Nitocrella psammophila* Chappuis, 1955	Stb	1	1
Malacostraca	Amphipoda	Niphargidae	*Niphargus costozzae* Schellenberg, 1935	Stb	0	1
Malacostraca	Amphipoda	Niphargidae	*Niphargus similis* Karaman & Ruffo, 1989	Stb	1	0
Malacostraca	Isopoda	Sphaeromatidae	*Monolistra (Typhlosphaeroma) bericum bericum* Fabiani, 1901	Stb	0	1
Malacostraca	Bathynellacea	Bathynellidae	*Bathynella* sp.	Stb	1	0
				Tot.	24	18

3.1. Terrestrial Fauna

Mollusca Gastropoda is represented by *Zospeum globosum* Kuščer, 1928. It is a small mollusc (rarely exceeding 2 mm in height) with a translucent or transparent shell, a body diaphanous, and totally lacking in ocular spots. It is mainly encountered on walls that are damp or wet from dripping water and often covered by silty materials [22].

Among the Opiliones, *Ischyropsalis strandi* Kratochvil, 1936 is present (Figure 10). This species is endemic to the caves in the Verona Prealps (Monte Baldo and Lessini Mountains). They are present in the two caves studied, and can be found in other caves in the Lessinia Mountains, usually above 600 meters of altitude, being a rather cryophilous species [23]. Within the Rana-Pisatella system, they are more frequently found in the higher elevations of the Grotta della Pisatela.

In regard to Pseudoscorpionida, three troglobiotic species are present in the caves: *Chthonius (Chthonius) lessiniensis* Schawaller, 1982, with Balkan affinities being a subterranean species. They show a high degree of troglomorphy, ranging from the western Venetian Prealps to the eastern Venetian Prealps [24], and are quite easy to detect under the collapsed stones at the bottom of the hall in the Arena Cave. *Neobisium (Blothrus) torrei* (E. Simon, 1881) is present in many caves in the Prealps and Alps of Veneto and Friuli Venezia Giulia regions [3]. This species is found in the Arena Cave and is the only troglobiotic pseudoscorpion currently known in the Rana-Pisatella system. *Balkanoronchus boldorii*

(Beier, 1931) is present in some caves of the Prealps of Brescia, in the Monte Baldo and the Lessini Mountains. These specimens were collected both with traps and hand-collecting in the Arena Cave. This species frequents the same habitats as *C. lessiniensis*.

Figure 10. A specimen of *Ischyropsalis strandi*: this species is present in both investigated caves (Photo: F. Rossetto).

The Terrestrial Isopoda species is represented by *Androniscus (Dentigeroniscus) degener* Brian, 1926, and is troglobiont and endemic to the Lessini Mountains in the Verona and Vicenza Provinces. It is quite common in both caves under stones in wetter areas.

To date, troglobiotic millipedes have only been found within Arena Cave; these are the Julidae *Troglojulus boldorii* Manfredi, 1940n, a species endemic to the caves of the Prealps of Lombardia, Veneto, and Trentino, and the Craspedosomatidae *Lessinosoma paoletti* Strasser, 1977, which is endemic to the Arena Cave [17].

The Troglobiotic Collembola are also known to date only from Arena Cave. The Onychiuridae species are represented by *Onychiurus hauseri* Dallai, 1975 and are endemic to the caves in the Veneto and Trentino regions, and the Entomobryidae *Pseudosinella concii* Gisin, 1950 is a species distributed in different caves in Italy and Switzerland and is quite common in the Arena Cave [3]. Some specimens belonging to the genus *Pseudosinella* Schaeffer, 1897 are not identifiable on a species level since they are very damaged; these were also sampled in the Arena Cave. However, it is possible to assert that they do not belong to *Pseudosinella concii* on the basis of the different number of labium setae.

Coleoptera are the most interesting among the terrestrial animals found in the caves considered in this study. Carabidae Trechinae, in particular, shows particularly robust adaptations for life in subterranean environments, and among them is *Italaphaenops dimaioi* Ghidini, 1964. Endemic to the Lessini Mountains in the Verona area, this species is one of the largest subterranean Trechinae in the world. The colonization of caves by this troglobiont can be traced back to an epoch preceding the Quaternary glaciations [25]. *I. dimaioi* is known from some caves of the Veronese Lessini, which develop at the contact between different types of karst formations like the Arena Cave. It is not present in the Vicenza Province, so it is not present in the Rana-Pisatella cave system.

Another extremely specialized genus of ancient pre-Quaternary origin of Trechinae is *Lessinodytes* Vigna Taglianti, 1982. This genus is distributed in the Lombardy and Veneto Prealps with three species, which are all endemic to one or a few caves and is present in the Arena Cave with the rather rare species *Lessinodytes pivai* Vigna Taglianti & Sciaky, 1988, which is endemic to that cave [26,27].

Orotrechus vicentinus juccii Pomini, 1940, is endemic to the Lessini Mountains in Verona Province and belongs to a group of species distributed in the Venetian Prealps and which, in Arena Cave, co-occurs with *Orotrechus pominii* Tamanini, 1953. *O. pomini* is the only known troglobiont Trechinae from the Rana-Pisatella system [28].

With regard to the Leiodidae Cholevinae, *Halberria zorzii* (Ruffo, 1950) is present in Arena Cave, while in the Buso della Rana, we find *Lessiniella trevisioli* Pavan, 1941 and *Neobathyscia fabianii* (Dodero, 1904) (Figure 11). The genus *Halberria* Conci & Tamanini, 1951 is present, with nine species in the caves in Eastern Veneto and Southern Trentino. In the Western Lessini Mountains, *H. zorzi* is present only in the caves that open at higher altitudes (above 1400 m a.s.l.), while in the caves at lower altitudes, the species of the genus *Neobathyscia* show vicariant distributions [17]. It is quite common in the Arena Cave [18].

The genus *Lessiniella* Pavan, 1941 is phylogenetically close to *Halberria* [29,30] and consists of two species: Lessiniella trevisoli, of which Buso della Rana is the typical locality where it is found rare in the innermost areas, and *Lessiniella berica* Piva, 1993, from the nearby Berici Mountains, which are also in the Vicenza Province [29].

To the genus *Neobathyscia* belongs nine species endemic to the Venetian Prealps, distributed between the Adige and Piave Rivers [30]. *N. fabiani* is known from several caves in the province of Vicenza that open in localities not far from the Rana-Pisatella cave system [29]. Within the system, it is rather common, especially in the branches of the Buso della Rana cave, while it seems rarer in the Pisatela cave.

Figure 11. A specimen of *Neobathyscia fabianii* from the Rana-Pisatela cave system (Photo: L. Latella).

3.2. Aquatic Fauna

In Arena Cave, six copepod species were found in a small pool fed by percolating water and a small drain, all stygobiotic. In Buso della Rana, 19 species of copepods are known, and 9 of the copepods are stygobiotic.

The Cyclopidae are represented by *Speocyclops infernus* (Kiefer, 1930), a stygobiotic species that is widespread over a broader geographical area in the epikarst and vadose zones in the eastern Alpine region and is present in both the caves under study [14,31,32]. First collected in Buso della Rana by Chappuis in the first half of the last century [33–35], S. infernus has since been found in many parts of the cave, in both small lakes and pools. It is also found in puddles inside Arena Cave, although the species attribution is not yet certain and is therefore currently reported as *S.* cf. *infernus* [3].

Among the Harpacticoida, those of the family Canthocamptidae are the most abundant stygobiotic copepods. The genus *Elaphoidella* is present in the Arena Cave and Rana-Pisatela system. *Elaphoidella* is widespread in almost all groundwater habitats in Italy, in both karstic and porous aquifers, as well as the hyporheic zones of rivers, springs, the epikarst, and the saturated karst. *Elaphoidella phreatica* (Chappuis, 1925) is widely distributed in Italy and across Europe [36]. *Elaphoidella ruffoi* Chappuis, 1953 is endemic to the Buso della Rana and is rather rare. It was first found during research in 1952 and was not found in the epikarst. Only one species of this genus was found in Arena Cave: *Elaphoidella* sp. A1 [37]. An *Elaphoidella*, different from the others, was also found in the Buso della Rana, but the scarce material made it impossible to identify at the species level (Bruno et al. 2018). *Ceuthonectes serbicus* Chappuis, 1924 was detected in recent research on epikarst fauna in the Buso della Rana cave [13].

The genus *Elaphoidella* is closely related to the genus *Lessinocamptus*. Known until a few years ago only from the vadose zone of the Lessinian caves, *Lessinocamptus* is now also known from a site in Northern Slovenia. In fact, *Lessinocamptus pivai* Stoch 1997, which was considered endemic to the Buso della Rana, was also found in the Lipnik spring complex in the Julian Alps (NW Slovenia) [38]. *Lessinocamptus caoduroi* Stoch, 1997, present in the pool in the Arena Cave, was found only in the percolating waters of the vadose zone of caves with an elevation of more than 1000 m a.s.l. in the Lessini Mountains. *Lessinocamptus insoletus* (Chappuis, 1928) was collected by Chappuis for the first time in the hypogean brook inside Buso della Rana [34]; however, further intensive sampling in the brook did not yield any specimen of *L. insoletus* and is therefore probable that the vadose zone is the main habitat of the species, from which it can be transported into the brook by percolating water [39]. The Ameiridae are represented by *Nitocrella psammophila* Chappuis, 1955 and is a stygobiotic species endemic to Italy. It is a widely distributed harpacticoid in the interstitial zone of subterranean streams in caves and the hyporheic in the Po Valley and has been reported in Apennine wells in Central Italy and caves in southern Italy. It is commonly found in the two caves under study. The Parastenocarididae are represented by *Parastenocaris ranae* Stoch, 2000, which was collected in the Buso della Rana in the large residual pools of the subterranean brook in a dry period [40].

Aquatic isopods have only been found in the Buso della Rana, where *Monolistra* (*Typhlosphaeroma*) *bericum bericum* (Fabiani, 1901) is present. It is a stygobiotic isopod endemic to the Lessini Mountains and Berici Hills (Vicenza Province).

Bathynellacea were found in the Arena Cave. Not yet identified at a species level, *Bathynella* sp. from Arena Cave was collected in a pond fed by a small water flow.

Amphipoda are present with two stygobiotic species: *Niphargus similis* G. Karaman & Ruffo, 1989 (Figure 12) in Arena Cave, and *Niphargus costozzae* Schellenberg, 1935 in the Buso della Rana cave [3].

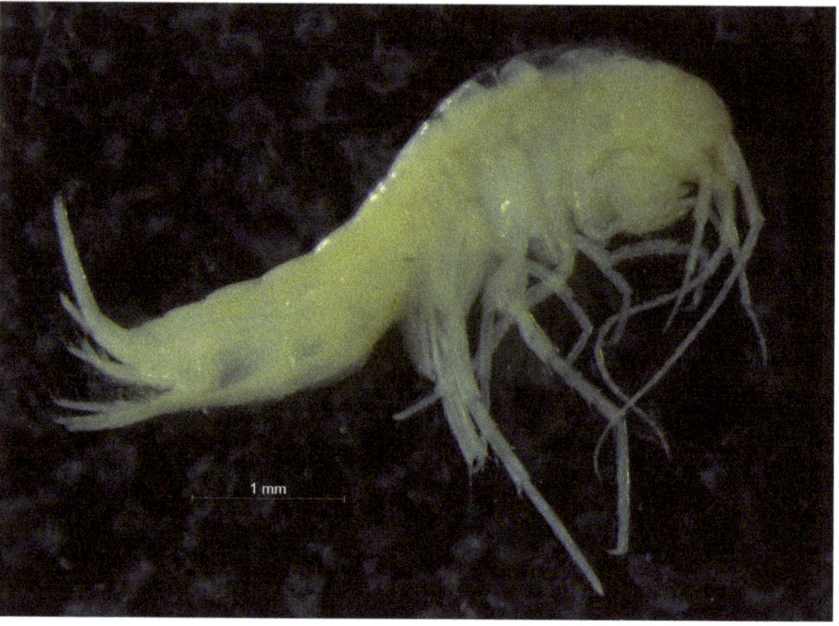

Figure 12. *Niphargus similis* from Arena Cave (Photo: L. Latella).

4. Discussion

The obligate subterranean fauna of Arena Cave and the Rana-Pisatella cave system is exceptionally rich. It comprises 35 troglobionts and stygobionts, representing 74% of the obligate subterranean fauna of the whole caves in the Lessini Mountains (more than 200 caves surveyed).

Despite its small size, Arena Cave is the richest one, with 15 troglobionts and eight stygobionts. The Rana-Pisatella system has a higher number of stygobionts (seven troglobionts and 11 stygobionts).

The geographical proximity between the two caves (22 km) would lead one to suspect a high taxonomic similarity in the fauna inhabiting them. These, on the contrary, have a very different obligate subterranean fauna in terms of terrestrial but especially aquatic species. Only seven species (five troglobionts and two stygobionts) out of thirty-five are in common for the two caves—applying the Jaccard similarity index (and expressed as a percentage similarity), the results in the similarity between the two caves is 21%.

The high richness and differences in faunal composition of the two caves can be explained by the paleogeographical events that occurred in the study area.

During the last glacial period, the Italian Alps were covered by glaciers, except at the top of the highest mountains [41]. In contrast to the Alps, Prealpine areas were only partially covered by glaciers [2,42], and glacial tongues occupied only a few deeper valleys [43]. This favored the colonization of ice-free zones by invertebrates from moist and cold habitats, like forest litter and soil, alpine grasslands, and talus areas. During interglacial periods, as glaciers retreated, populations became isolated in the highest parts of the Prealpine mountains or took refuge in cold, moist interiors of caves [5,6]. Surface populations became extinct or isolated, and there was, therefore, little or no gene flow between cave communities, boosting the evolution of the troglobiont [11]. This is known as the climatic relict hypothesis [44–47]. The effects of Quaternary glaciations also shaped the stygobiotic species distribution, as the Massif was only marginally covered by ice, and the extensive networks of fractures of the karstic system represented a refuge for stygobionts, boosting isolation and speciation [13]. In fact, the vadose zone and the epikarst of the Lessinian Massif are known to harbor a high diversity of microcrustaceans, including many endemic species due to the ancient geological age of the aquifers, high habitat fragmentation, and isolation of microhabitats, factors of which concurred to promote speciation by vicariance [13,48].

The exceptional subterranean diversity of Arena Cave and the Rana-Pisatella cave system, so different from each other in shape and development (less than 100 m for Arena Cave and about 38 km for the Rana-Pisatella system), can be explained only by their geology, where the two caves developed in different typologies of rocks, namely "contact caves" [5,6].

Contact karst is considered, in a strict sense, a karst phenomenon, where forms are influenced by the contact between a karstifiable rock and a non-karstifiable rock. In a wide sense, the karst phenomena and forms that are influenced by the contact between two karstifiable rocks differ in some of their characteristics, such as chemical composition, porosity, and fracture density [5].

These different rock characteristics create a number of different microhabitats that also influence the life and dispersion of the subterranean animals. Depending on the amount of water retained, humidity, and other factors that are not yet fully known, specimens of different species may prefer one microhabitat over another. This explains the rarity of the findings in Arena Cave of the trechine *Lessynodites pivai*, a species that most probably do not frequently inhabit the cave proper but rather lives in the wetter interstices of the Rosso Ammonitico formations. The same can be said for *Italaphaenops dimaioi*; ongoing studies by the lab of the Museum of Verona show that *I. dimaioi* was sampled almost exclusively in the caves in contact with Rosso Ammonitico rock in the Lessini Mountains.

The same microhabitat characteristics are probably the reason for the abundance of copepods and the presence of *Bathynella* sp., present in waters that flow from the environments in the small pool inside the cave. It is in similar conditions that, collecting water from the epikarst, we found *Ceuthonectes serbicus* for the first time in Buso della Rana [13].

The smaller cave is, therefore, the richest in diversity. As is often said, size does not always matter.

Funding: This research received no external funding.

Institutional Review Board Statement: Not applicable.

Data Availability Statement: Data are contained within the article.

Acknowledgments: I would like to thank Dave Culver (Washington, DC, USA), Louis Deharveng (Paris, France), Valeria Lencioni (Trento, Italy), and Francesco Sauro (Verona, Italy) for their helpful suggestions in the elaboration of the text; Ivan Petri (Trento, Italy), Barbara Valle (Milan, Italy), Fabio Stoch (Brussels, Belgium) for their taxonomic identification or suggestions; Arianna Spada (Verona, Italy) for the graphic elaboration of some figures; Sandro Sedran (Dolo, Italy) and Filippo Rossetto (Verona, Italy) for allowing the use of photographs of the Buso della Rana and *Ischyropsalis strandii*. I am also grateful to all the speleologists who shared the cave surveys with me, especially Giorgio Annichini (Verona, Italy), Tarcisio Battagini (Verona, Italy), and Andrea Pasotto (Verona, Italy)—and a special thanks to Dave Culver for revising the English text and a thanks to the four anonymous reviewers.

Conflicts of Interest: The author declares no conflict of interest.

References

1. Marazzi, S. *Atlante Orografico Delle Alpi—SOIUSA*; Priuli and Verlucca: Scarmagno, Italy, 2005.
2. Sauro, U. *Il Paesaggio Degli Alti Lessini: Studio Geomorfologico*; Museo Civico di Storia Naturale di Verona: Verona, Italy, 1973; Volume 6, 161p.
3. Caoduro, G.; Ruffo, S. La Grotta dell'Arena, un biotopo di eccezionale interesse negli alti Lessini. In *La Lessinia Ieri Oggi Domani: Quaderno Culturale*; Editrice La Grafica: Lavagno, Italy, 1998; pp. 39–44.
4. Culver, D.C.; Deharveng, L.; Bedos, A.; Lewis, J.J.; Madden, M.; Reddell, J.R.; Sket, B.; Trontelj, P.; White, D. The mid-latitude biodiversity ridge in terrestrial cave fauna. *Ecography* **2006**, *29*, 120–128. [CrossRef]
5. Sauro, U. Aspects of contact karst in the Venetian Fore-Alps. *Acta Carsologica* **2001**, *30*, 89–102.
6. Latella, L.; Sauro, U. Aspects of the Evolution of an Important Geo-Ecosystem in the Lessinian Mountain (Venetian Prealps, Italy). *Acta Carsologica* **2007**, *36*, 69–75. [CrossRef]
7. Pasa, A. Carsismo ed idrografia carsica del Gruppo del Monte Baldo e dei Lessini Veronesi. *CNR Cent. Studi Geogr. Fis. Ric. Sulla Morfol. Idrogr. Carsismo* **1954**, *5*, 1–150.
8. Sauro, U. Aspetti dell'evoluzione carsica legata a particolari condizioni litologiche e tettoniche negli Alti Lessini. *Boll. Soc. Geol. Ital.* **1974**, *93*, 945–969.
9. Tisato, N.; Sauro, F.; Bernasconi, S.M.; Bruijn, R.H.; De Waele, J. Hypogenic contribution to speleogenesis in a predominant epigenic karst system: A case study from the Venetian Alps, Italy. *Geomorphology* **2012**, *151*, 156–163. [CrossRef]
10. Sistema carsico Rana-Pisatella. Available online: www.busodellarana.it (accessed on 1 August 2023).
11. Latella, L.; Verdari, N.; Gobbi, M. Distribution of terrestrial cave-dwelling arthropods in two adjacent Prealpine Italian areas with different glacial histories. *Zool. Stud.* **2012**, *51*, 1113–1121.
12. Avesani, D.; Latella, L. Spatio-temporal distribution of the genus Chionea (Diptera, Limoniidae) in the Buso del Valon ice cave and other caves in the Lessini Mountains (Northern Italy). *Boll. Mus. Civ. St. Nat. Verona* **2016**, *40*, 11–16.
13. Bruno, M.C.; Cottarelli, V.; Grasso, R.; Latella, L.; Zaupa, S.; Spena, M.T. Epikarst crustaceans from some Italian caves: Endemisms and spatial scales. *Biogeographia* **2018**, *33*, 1–18.
14. Pipan, T. *Epikarst—A Promising Habitat*; ZRC Publishing: Ljubljana, Slovenia, 2005; pp. 1–101.
15. Pipan, T.; Christman, M.C.; Culver, D.C. Dynamics of epikarst communities: Microgeographic pattern and environmental determinants of epikarst copepods in Organ Cave, West Virginia. *Am. Midl. Nat.* **2006**, *156*, 75–87. [CrossRef]
16. Allegranzi, A.; Bartolomei, G.; Broglio, A.; Pasa, A.; Rigobello, A.; Ruffo, S. Il Buso della Rana (40 V-VI). *Ras. Speleol. Ital.* **1960**, *12*, 99–163.
17. Caoduro, G.; Osella, G.; Ruffo, S. La fauna cavernicola della regione veronese. *Memorie del Museo Civico di Storia Naturale di Verona. Sez. Biol.* **1994**, *6*, 1–144.
18. Latella, L.; Sauro, U. Note di Storia Naturale del sottosuolo dei Monti Lessini e del suo popolamento. *Quaderno Culturale Lessinia Ieri Oggi Domani* **2006**, *34*, 57–64.
19. Latella, L. Il contributo del Museo Civico di Storia Naturale di Verona allo sviluppo della biospeleologia. *Studi Trentini Di Sci. Nat. Acta Biol.* **2004**, *81*, 15–22.
20. Ruffo, S. Le attuali conoscenze sulla fauna cavernicola della Regione Pugliese. *Mem. Biogeogr. Adriat.* **1957**, *3*, 1–143.
21. Sket, B. Can we agree on an ecological classification of subterranean animals? *J. Nat. Hist.* **2008**, *42*, 1549–1563. [CrossRef]
22. Pezzoli, E. Il genere Zospeum Bourguignat, 1856 in Italia (Gastropoda Polmonata Basommatophora). Censimento delle stazioni ad oggi segnalate. *Nat. Brescia.* **1992**, *27*, 123–169.
23. Petri, I.; Ballarin, F.; Latella, L. Seasonal abundance and spatio-temporal distribution of the troglophylic harvestman Ischyropsalis ravasinii (Arachnida, Opiliones, Ischyropsalididae) in the Buso del Valon ice cave, Eastern Italian Prealps. *Subterr. Biol.* **2022**, *42*, 151–164. [CrossRef]

24. Gardini, G. The species of the Chthonius heterodactylus group (Arachnida, Pseudoscorpiones, Chthoniidae) from the eastern Alps and the Carpathians. *Zootaxa* **2014**, *3887*, 101–137. [CrossRef]
25. Casale, A.; Vigna Taglianti, A. Note su Italaphaenops dimaioi Ghidini (Coleoptera, Carabidae). *Boll. del Mus. Civ. Stor. Nat. di Verona* **1976**, *2*, 293–314.
26. Sciaky, R.; Vigna Taglianti, A. The genus Lessinodytes (Coleoptera, Carabidae, Trechinae). *Mém. Biospéol.* **1990**, *17*, 169–173.
27. Monguzzi, R. Nuovi Dati Per La Conoscenza Del Genere Lessinodytes Vigna Taglianti, 1982 (Coleoptera Carabidae Trechinae). *Nat. Brescia.* **1993**, *29*, 179–183.
28. Faille, A.; Casale, A.; Balke, M.; Ribera, I. A molecular phylogeny of Alpine subterranean Trechini (Coleoptera: Carabidae). *BMC Evol. Biol.* **2013**, *13*, 248. [CrossRef] [PubMed]
29. Latella, L. The Subterranean Cholevinae of Italy. In *Cave Biodiversity: Speciation and Diversity of Subterranean Fauna*; Johns Hopkins University Press: Baltimore, MD, USA, 2022; p. 164.
30. Giachino, P.M.; Vailati, D. I Cholevidae delle Alpi e Prealpi italiane: Inventario, analisi faunistica e origine del popolamento del settore compreso fra i corsi dei fiumi Ticino e Tagliamento (Coleoptera). *Biogeogr. J. Integr. Biogeogr.* **2005**, *26*, 229–378. [CrossRef]
31. Deharveng, L.; Stoch, F.; Gibert, J.; Bedos, A.; Galassi, D.; Zagmajster, M.; Brancelj, A.; Camacho, A.; Fiers, F.; Martin, P.; et al. Groundwater biodiversity in Europe. *Freshw. Biol.* **2009**, *54*, 709–726. [CrossRef]
32. Papi, F.; Pipan, T. Ecological studies of an epikarst community in Snežna jama na planini Arto an ice cave in north central Slovenia. *Acta Carsologica* **2011**, *40*, 3. [CrossRef]
33. Chappuis, P.A. Nouveaux Copépodes cavernicoles des genres Cyclops et Canthocamptus (Note préliminaire). *Bul. Soc. De Stiint. Din Cluj* **1923**, *1*, 584–590.
34. Chappuis, P.A. Nouveaux Copépodes cavernicoles. *Bul. Soc. De Stiint. Din Cluj* **1928**, *2*, 20–34.
35. Chappuis, P.A. Biospeologica LIX. Copepodes (premiere serie), avec l'énumeration de tous les copepodes cavernicoles connus en 1931. *Arch. De Zool. Expérimentale Et Générale* **1933**, *76*, 1–56.
36. Galassi, D.M.P. Groundwater copepods: Diversity patterns over ecological and evolutionary scales. *Hydrobiologia* **2001**, *453*, 227–253. [CrossRef]
37. Ruffo, S.; Stoch, F. *Checklist and Distribution of the Italian Fauna. Memorie del Museo Civico di Storia Naturale di Verona, II Serie, Sezione Scienze Della Vita 17*; Ministero dell'Ambiente e della Tutela del Territorio e del Mare: Roma, Italy, 2006; with CD-ROM.
38. Mori, N.; Brancelj, A. Differences in aquatic microcrustacean assemblages between temporary and perennial springs of an alpine karstic aquifer. *Int. J. Speleol.* **2013**, *42*, 3–9. [CrossRef]
39. Stoch, F. A new genus and two new species of Canthocamptidae (Copepoda, Harpacticoida) from caves in northern Italy. *Hydrobiologia* **1997**, *350*, 49–61. [CrossRef]
40. Stoch, F. New and little known Parastenocaris (Copepoda, Harpacticoida, Parastenocarididae) from cave waters in Northeastern Italy. *Boll. Del Mus. Civ. Di Stor. Nat. Di Verona* **2000**, *24*, 195–206.
41. Hughes, P.D. Quaternary glacial history of the Mediterranean mountains. *Progr. Phys. Geogr.* **2006**, *30*, 334–364. [CrossRef]
42. Bassetti, M.; Borsato, A. Evoluzione geomorfologica e vegetazionale della bassa valle dell'Adige dall'ultimo massimo glaciale: Sintesi delle conoscenze e riferimenti ad aree limitrofe. *Studi Trentini Sci. Nat. Acta Geol.* **2007**, *82*, 31–42.
43. Sommaruga, M.; Zorzin, R. Evidences of Morphologies and Glacial Deposits into High Valle del Chiampo (Northern Italy, Vicenza Province): First Results. *Boll. Del Mus. Civ. Di Stor. Nat. Di Verona Geol. Paleontol. Preist.* **2018**, *42*, 43–71.
44. Sbordoni, V. Advances in speciation of cave animals. In *Mechanisms of Speciation*; Barigozzi, C., Ed.; Liss: New York, NY, USA, 1982; pp. 219–224.
45. Humphreys, W.F. Relict fauna and their derivation. In *Ecosystems of the World*; Wilkens, H., Culver, D.C., Humphreys, W.F., Eds.; Subterranean Ecosystems; Elsevier: Amsterdam, The Netherlands, 2000; Volume 30, pp. 417–432.
46. Assmann, T.; Casale, A.; Drees, C.; Habel, J.C.; Matern, A.; Schuldt, A. Review: The dark side of relict species biology: Cave animals as ancient lineages. In *Relict Species: Phylogeography and Conservation Biology*; Assmann, T., Habel, J.C., Eds.; Springer: Berlin/Heidelberg, Germany, 2010; pp. 91–103.
47. Hampe, A.; Jump, A.S. Climate relicts: Past, present, future. *Annu. Rev. Ecol. Evol. Syst.* **2011**, *42*, 313–333. [CrossRef]
48. Galassi, D.M.; Stoch, F.; Fiasca, B.; Di Lorenzo, T.; Gattone, E. Groundwater biodiversity patterns in the Lessinian Massif of northern Italy. *Freshw. Biol.* **2009**, *54*, 830–847. [CrossRef]

Article

Feihu Dong, a New Hotspot Cave of Subterranean Biodiversity from China

Sunbin Huang [1,2], Mingzhi Zhao [1], Xiaozhu Luo [1], Anne Bedos [3], Yong Wang [4], Marc Chocat [5], Mingyi Tian [1] and Weixin Liu [1,*]

1   Department of Entomology, College of Plant Protection, South China Agricultural University, 483 Wushan Road, Guangzhou 510642, China; huangsunbin@163.com (S.H.); zhaomzhai@gmail.com (M.Z.); xiaozhu.luo@scau.edu.cn (X.L.); mytian@scau.edu.cn (M.T.)
2   Henry Fok School of Biology and Agriculture, Shaoguan University, 288 Daxue Road, Shaoguan 512005, China
3   Institut de Systématique, Évolution, Biodiversité (ISYEB), UMR7205, CNRS, Muséum National d'Histoire Naturelle, Sorbonne Université, EPHE, 45 rue Buffon, 75005 Paris, France; bedosanne@yahoo.fr
4   Xiangxi Cave Expedition, 27 Jianxin Road, Jishou 416000, China; wangyong19711018@163.com
5   SHAG Caving Association, 24 rue des Roses, 25000 Besançon, France; mchocat@aol.com
*   Correspondence: da2000wei@163.com

Abstract: China is a country with abundant karst landscapes, but research on cave biodiversity is still limited. Currently, only Ganxiao Dong, located in Huanjiang, Guangxi, has been reported as a hotspot for cave biodiversity. Many of the world's most troglomorphic species in the major groups of cave animals have been recently discovered in China, making the existence of many more hotspots in the country likely. Feihu Dong, one of these potential hotspot caves, has been systematically investigated to complement a preliminary species list of 1995, leading to the discovery of 62 species of animals from the cave. Among them, 27 are considered troglobionts or stygobionts, 26 are considered troglophiles or stygophiles, and nine are classified as trogloxenes or stygoxenes. Research on the cave biodiversity of Feihu Dong has demonstrated that it currently holds the highest number of known cave animal species in China. Among the most remarkable features of this fauna is the co-occurrence of five species of cave-obligate beetles, all modified for cave life. The biological survey was limited to a small part of the cave. Several habitats (like guano) have not been investigated so far, and several important cave groups have been insufficiently or not sampled (like Ostracoda). Meanwhile, the system increases in length with each new caving expedition. Further discoveries of cave organisms in Feihu Dong are therefore expected. As Feihu Dong and Ganxiao Dong are the only caves in China that have been extensively studied for a large range of organisms, and as they are located in karstic areas that are similar in richness to other regions of southern China, it can be confidently assumed that several other caves of high biodiversity will be discovered in the coming years.

Keywords: South China Karst; Hunan; Wulongshan; cave fauna; stygobionts; troglobionts; diversity; checklist; conservation

Citation: Huang, S.; Zhao, M.; Luo, X.; Bedos, A.; Wang, Y.; Chocat, M.; Tian, M.; Liu, W. Feihu Dong, a New Hotspot Cave of Subterranean Biodiversity from China. *Diversity* **2023**, *15*, 902. https://doi.org/10.3390/d15080902

Academic Editor: Michael Wink

Received: 30 June 2023
Revised: 26 July 2023
Accepted: 26 July 2023
Published: 30 July 2023

1. Introduction

China has the largest karst area in the world, covering 3.4 million km^2 of soluble rock, including an exposed carbonate rock area of 910,000 km^2 [1–3]. The South China Karst, one of the world's largest and most geomorphologically diverse wet tropical-subtropical karst landscapes, stretches from the Qinling Mountains in the north to the Guangxi Basin in the south and from the Hengduan Mountains in the west to the Luoxiao Mountains in the east. This karst crosses the three-step terrain in China from west to east, with an elevation of 110–2100 m and an area of 550,000 km^2 [4]. It contains seven exceptional karst landscape clusters that have been designated World Heritage Sites [5].

China not only presents stunning karst landforms but also abundant caves and an exceptional cave biodiversity [6]. It is estimated that there are more than 500,000 caves in China [2]. Currently, the longest explored cave system is Shuanghe Dong in Suiyang County, Guizhou Province, with a reported length of 257.4 km in 2021 [7], which has since been extended to 400.8 km with connected caves increasing from 64 to 105 (Qian Z., pers. comm.). The cave with the largest chamber volume is Miaoting, located in Getuhe, Guizhou Province, with a volume of 10.78 million cubic meters [8]. China is also home to many beautiful and huge caves and tiankengs (the term for giant dolines), such as Zhijin Dong in Guizhou, Shui Dong of Benxi, Liaoning, Shihua Dong in Beijing, or the tiankengs of Dashiwei in Guangxi and Xiaozhai in Chongqing [9,10].

In recent years, the descriptions of subterranean species new to science has considerably increased [11–17]. To date, only Ganxiao Dong, situated at the junction of the Mulun Karst in Guangxi and the Maolan Karst in Guizhou, has been reported as a regional hotspot of cave biodiversity, with 26 species of cave invertebrates reported, including 20 species of troglobionts and six species of troglophiles [18]. However, our investigation indicates that numerous other areas in China also have a high number of cave species, such as Huoyan Karst in northwestern most Hunan, Hanzhong Karst in southern Shaanxi, Du'an Karst in Guangxi, and Wulong Karst in Chongqing, among others [19].

Wulongshan National Geopark is located in Longshan County, Xiangxi Tujia and Miao Autonomous Prefecture, Hunan Province, at the junction of Hunan and Hubei Provinces and Chongqing Municipality. The most famous scenic spots in Wulongshan Park are in the Huoyan Karst, where 212 caves have been recorded [20], including Feihu Dong, Wulong Dong, Shihua Dong, Feng Dong, and Lianyu Dong. Among them, Feihu Dong is the most spectacular. From 1993 to 2002, cave explorers and biologists from the Sino-French joint expedition team and from other countries such as Slovenia, Belgium, and Japan came to Feihu Dong for several scientific expeditions [21–23]. The cave is today nearly 20 km long and has not been fully explored (Figure 1). Underground rivers, lakes, boulders, and side passages have been reported in Feihu Dong [22,24]. The cave is inhabited by abundant and diversified subterranean fauna that includes a number of troglobiotic/stygobiotic and endemic species [25–27], such as *Triplophysa xiangxiensis* (Yang, Yuan and Liao, 1986); *Caridina longshan* Cai and Ng, 2018; *Toshiaphaenops ovicollis* Ueno, 1999; *Angustopila huoyani* Jochum, Slapnik and Páll-Gergely, 2014 [23,28–30].

After the joint expeditions of 1993–2002, there has not been any further systematic cave survey or biological sampling in Feihu Dong. Recently, in February 2023, an exploration project of Feihu Dong was resumed under the leadership of the Caves Committee of the Geological Society of China. The project was mainly executed by the Xiangxi Cave Exploration Team, which used paperless cave surveying and 3D laser scanning technology to re-explore and investigate Feihu Dong. In April 2023, the team have completed the re-survey of approximately 4 km from the main entrance and the newly discovered approximately 2.2 km of cave passage during the last three explorations (Wang Y., pers. comm.). At the same time, the South China Agricultural University (SCAU) biocaving team conducted a week-long systematic survey of cave fauna in Feihu Dong in February 2023 and discovered additional cave animals.

The purpose of this study is to provide an updated list of animals living in Feihu Dong, to draw attention to the scientific importance of these species and their fascinating habitats, and to contribute to the subterranean biodiversity in China.

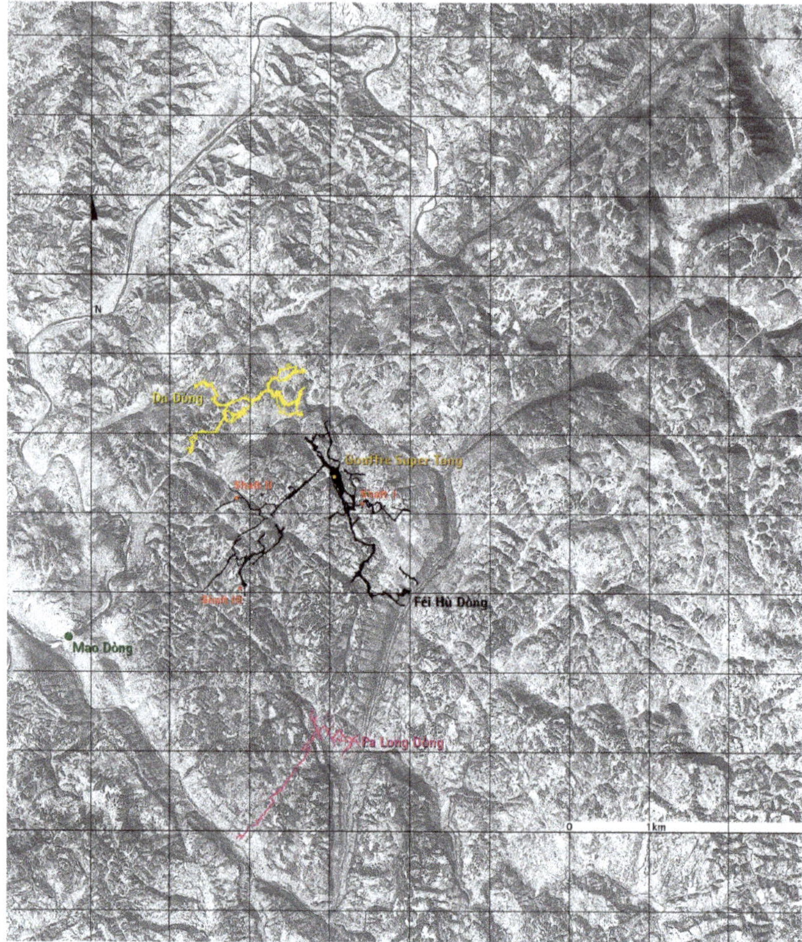

Figure 1. Overview of the geographical location of Feihu Dong and other caves in Huoyan Karst: base map and data from SHAG Caving Association, except the location data of three potential shafts from the Xiangxi Cave Expedition.

2. Materials and Methods

2.1. Research Site

Feihu Dong (飞虎洞, "Flying Tiger Cave" in Chinese) (Figure 1) is located in the Grand Canyon of Wulong Mountain, in Longshan County, Xiangxi Tujia and Miao Autonomous Prefecture, northeastern Hunan Province.

Feihu Dong is a complex cave system (Figure 2) with three entrances (Figures 3 and 4A–D), one of which is a −320 m deep shaft (Figure 3), long galleries (Figure 4E) adorned with speleothems (Figure 4F,G), large chambers, subterranean lakes and rivers. The length of the explored and surveyed cave passages is about 20 km in total. The main entrance of the cave, Feihu Dong (coordinates: 29°12′28.4″ N 109°18′16.4″ E), is at an altitude of around 360 m a.s.l. A three-kilometer-long gallery from the entrance is connected to a large chamber of 26,400 m² in surface. A large shaft named "Gouffre Super Tong", with a waterfall inside, opens on the karst surface and leads 320 m deeper to the northern part of this chamber (Figure 3). In addition, three potential shafts that may be connected to Feihu Dong are marked on the map (Figure 1) (Wang Y., pers. comm.). Some location names within Feihu Dong are translated between French, English, and Chinese in Table 1.

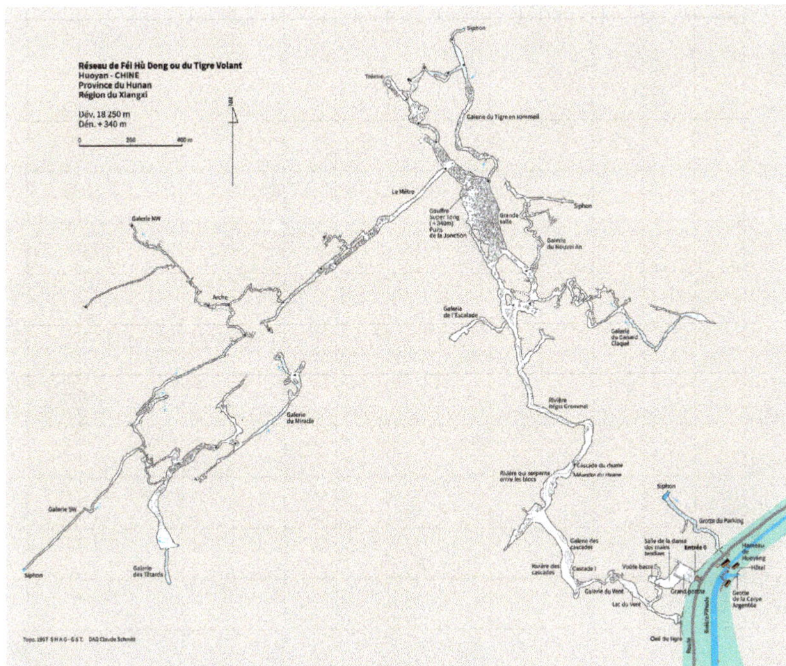

Figure 2. Cave map of Feihu Dong: "Topographie de la Grotte du Tigre Volant (Feihu Dong)" drawn by SHAG Caving Association.

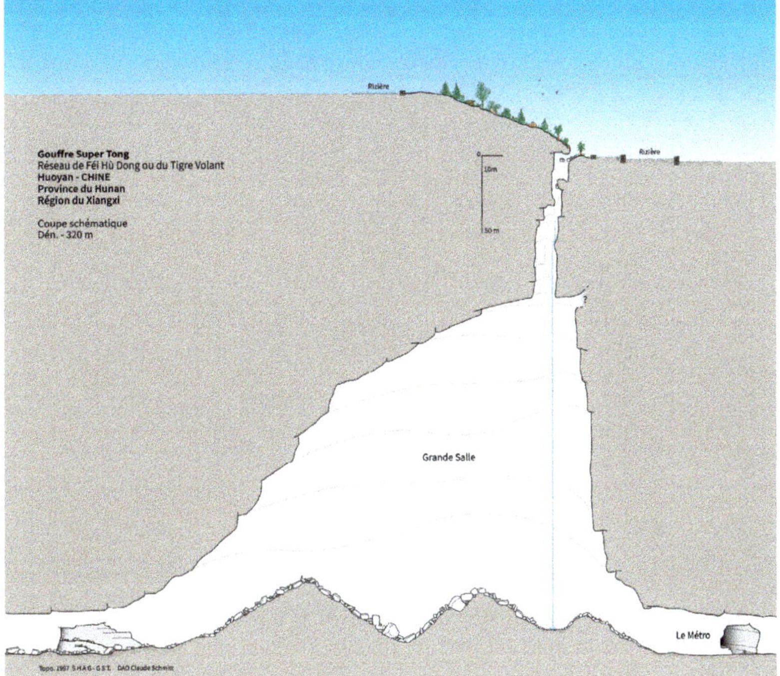

Figure 3. Cave map of the shaft "Gouffre Super Tong", explored and mapped by the SHAG Caving Association.

Figure 4. Entrances and habitats of Feihu Dong: (**A,B**) the main entrance, "Large Porch Entrance"; (**C,D**) the entrance of "Tiger Eye"; (**E**) Waterfall Gallery; (**F,G**) speleothems.

Table 1. Names of some locations in Feihu Dong.

French	English	Chinese
Grand porche	Large Porch Entrance	洞口
Oeil du tigre	Tiger Eye	虎眼
Salle de la danse des mains tendues	Room of the Dance of Outstretched Hands	摆手厅
Voûte basse	Low Vault	低厅
Lac du vent	Wind Lake	浴心池
Galerie du vent	Wind Gallery	风谷
Cascade I	Waterfall No. I	洞内瀑布1号
Galerie des cascades	Waterfall Gallery	瀑布大道
Rivière des cascades	Waterfall River	瀑布小河
Rivière qui serpente entre les blocs	River that meanders between the Blocks	石中河
Méandre du rhume	Flu Meander	盲鱼峡
Cascade du rhume	Flu Waterfall	望天瀑布
Rivière Régis Gremmel	Régis Gremmel River	子午谷河
Gouffre Super Tong	Super Tong Shaft	超级竖井

2.2. Sampling

To update the list of cave fauna in Feihu Dong, we reviewed the available literature and conducted explorations from February to April 2023, led by the South China Agricultural University Biocaving Team and Xiangxi Cave Expedition, to upgrade the first biological exploration carried out in August 1995 by Louis Deharveng and Anne Bedos from the Muséum National d'Histoire Naturelle (MNHN Paris) during the Xiangxi 95 expedition. We sampled different habitats within the cave, and we used a combination of direct sampling, baited traps, and litter extraction methods to collect small invertebrates. Direct sampling was performed by hand or using an aspirator, and baited traps were set up using bananas and fish gut as attractants for terrestrial and aquatic animals. Litter extraction was performed using a sifter to separate soil-dwelling animals from organic matter. All specimens were kept in 75% ethanol for morphological studies and identification or 95% ethanol for DNA sequencing. Photos of the cave animals were taken by a Canon EOS 6D camera (Tokyo, Japan) with a Sigma 50 mm F2.8 EX DG Macro lens (Kanagawa, Japan) and an adapted Meike MK-14 ext E-TTL macro flash (Hongkong, China). They were then processed using Photoshop CC 2019 (San Jose, CA, USA).

2.3. Terminology

Ecological classification of cave animals and troglomorphy are defined as in Huang et al., 2021 [18].

3. Results

3.1. Fauna Composition of Feihu Dong

In total, 62 species of animals have been discovered in Feihu Dong. In the current state of our knowledge, 27 of these species are considered troglobionts (23 species) or stygobionts (four species) taxa, 26 as troglophiles (24 species) or stygophiles (two species), and the remaining nine as trogloxenes (six species) or stygoxenes (three species) (Table 2).

3.2. Notes on Animals Found in Feihu Dong

Firstly, it has to be stressed that the taxonomic coverage of our species list is biased. The group that contributes the most to aquatic diversity, Microcrustacea, has not been sampled.

Aside from this gap, a total of 62 animal species occurs in Feihu Dong, and Insecta are the most abundant group with 22 recorded representatives. Many of them are classified as troglobionts as their occurrence is limited to cave habitats, and they frequently exhibit troglomorphic characters. The major and interesting groups are listed and discussed below.

Table 2. Species list of cave animals found in cave Feihu Dong. Column status: Tb = troglobiont; Tp = troglophile; Tx = trogloxene; Sb = stygobiont; Sp = stygophile; Sx = stygoxene; * = known endemic of the Huoyan karst or of Feihu Dong; ? = uncertain ecological status or species under study; SCAU = South China Agricultural University.

No.	Taxon	Taxonomic Classification	Status	Source
01	cf. *Dugesia japonica* Ichikawa and Kawakatsu, 1964	Turbellaria: Tricladida: Dugesiidae	Sx ?	[25]
02	*Angustopila huoyani* Jochum, Slapnik and Páll-Gergely, 2014	Gastropoda: Stylommatophora: Hypselostomatidae	Tb *	[25,30]
03	*Synprosphyma* cf. *lyra* (Gredler,1887)	Gastropoda: Stylommatophora: Clausiliidae	Tp	SCAU
04	Subulinidae sp.	Gastropoda: Stylommatophora: Subulinidae	Tp ?	SCAU
05	Helixarionoidea sp.	Gastropoda: Stylommatophora: Helixarionoidea	Tp ?	SCAU
06	Oligochaeta sp.	Oligochaeta	Tx ?	SCAU
07	*Papiliocoelotes guitangensis* Zhao and Li, 2016	Arachnida: Araneae: Agelenidae	Tp ?*	[31], SCAU
08	Agelenidae sp.	Arachnida: Araneae: Agelenidae	Tb	SCAU
09	Linyphiidae sp.	Arachnida: Araneae: Linyphiidae	Tp ?	SCAU
10	*Belisana* sp.	Arachnida: Araneae: Pholcidae	Tb	SCAU
11	Pholcidae sp.	Arachnida: Araneae: Pholcidae	Tb	SCAU
12	*Telema wunderlichi* Song and Zhu, 1994	Arachnida: Araneae: Telemidae	Tb *	SCAU
13	Rhagidiidae sp.	Arachnida: Acari: Rhagidiidae	Tb ?	[25], SCAU
14	Spinturnicidae sp.	Arachnida: Acari: Spinturnicidae	Tp ?	SCAU
15	Acari sp. 1	Arachnida: Acari	Tp ?	SCAU
16	Acari sp. 2	Arachnida: Acari	Tp ?	SCAU
17	Laniatores sp.	Arachnida: Opilionida: Laniatores	Tb	[26]
18	*Schenkeliobunum* cf. *wuxi* Lu, Wang and Zhang, 2022	Arachnida: Opilionida: Sclerosomatidae	Tp ?	SCAU
19	*Glyphiulus deharvengi* Golovatch, Geoffroy, Mauriès and Van Den Spiegel, 2006	Diplopoda: Spirostreptida: Cambalopsidae	Tb *	[25,32], SCAU
20	*Epanerchodus* sp.	Diplopoda: Polydesmida: Polydesmidae	Tb *	SCAU
21	*Eutrichodesmus sketi* Golovatch, Geoffroy, Mauriès and Van Den Spiegel, 2015	Diplopoda: Polydesmida: Haplodesmidae	Tb *	[25,33], SCAU
22	cf. *Lithobius* (*Monotarsobius*) sp.	Chilopoda: Lithobiomorpha: Lithobiidae	Tb	SCAU
23	Geophilidae sp.	Chilopoda: Geophilomorpha: Geophilidae	Tp	SCAU
24	*Caridina longshan* Cai and Ng, 2018	Malacostraca: Decapoda: Atyidae	Sb *	[25,29], SCAU
25	*Gammarus* sp.	Malacostraca: Amphipoda: Gammaridae	Sb	[25], SCAU
26	*Trogloniscus* sp.	Malacostraca: Isopoda: Styloniscidae	Tb	SCAU
27	Lernaeidae sp.	Copepoda: Cyclopoida: Lernaeidae	Sb ?	Zhou J.J.
28	*Coecobrya* sp.	Collembola: Entomobryomorpha: Entomobryidae	Tb	[25], SCAU
29	*Tomocerus* sp.	Collembola: Entomobryomorpha: Tomoceridae	Tb	[25], SCAU
30	*Vitronura* sp.	Collembola: Poduromorpha: Neanuridae	Tx	[25]
31	Campodeidae sp.	Entognatha: Diplura: Campodeidae	Tb	SCAU
32	*Toshiaphaenops ovicollis* Ueno, 1999	Insecta: Coleoptera: Carabidae	Tb *	[23], SCAU
33	*Huoyanodytes tujiaphilus* Tian and Huang, 2016	Insecta: Coleoptera: Carabidae	Tb *	SCAU
34	*Cathaiaphaenops* (*Cathaiaphaenops*) *delprati* Deuve, 1996	Insecta: Coleoptera: Carabidae	Tb *	[21,25], SCAU
35	*Sinotroglodytes bedosae* Deuve, 1996	Insecta: Coleoptera: Carabidae	Tb *	[21,25], SCAU

Table 2. *Cont.*

No.	Taxon	Taxonomic Classification	Status	Source
36	*Zopherobatrus tianmingyii* Yin and Li, 2015	Insecta: Coleoptera: Staphylinidae	Tb	SCAU
37	*Nipponobythus* sp.	Insecta: Coleoptera: Staphylinidae	Tp ?	SCAU
38	*Quedius feihuensis* Smetana, 1999	Insecta: Coleoptera: Staphylinidae	Tx ?*	[34], SCAU
39	*Pseudeurostus hilleri* (Reitter, 1877)	Insecta: Coleoptera: Ptinidae	Tx	SCAU
40	*Mycetina* sp.	Insecta: Coleoptera: Endomychidae	Tx	SCAU
41	*Tachycines* (*Gymnaeta*) *omninocaecus* (Gorochov, Rampini and Di Russo, 2006)	Insecta: Orthoptera: Rhaphidophoridae	Tb *	[25,35,36], SCAU
42	*Eutachycines crenatus* (Gorochov, Di Russo and Rampini, 2006)	Insecta: Orthoptera: Rhaphidophoridae	Tb *	[12,25,35], SCAU
43	*Tachycines* (*Gymnaeta*) *solidus* (Gorochov, Rampini and Di Russo, 2006)	Insecta: Orthoptera: Rhaphidophoridae	Tp ?	[25,35,36], SCAU
44	Ischnopsyllidae sp.	Insecta: Siphonaptera: Ischnopsyllidae	Tb ?	SCAU
45	*Sarasaeschna* sp.	Insecta: Odonata: Aeshnidae	Sp ?	SCAU
46	Perlidae sp.	Insecta: Plecoptera: Perlidae	Sp ?	SCAU
47	Trichoptera sp.	Insecta: Trichoptera	Sx ?	SCAU
48	Triphosa sp.	Insecta: Lepidoptera: Geometridae	Tp	SCAU
49	Tineidae sp.	Insecta: Lepidoptera: Tineidae	Tp ?	SCAU
50	Anisolabididae sp.	Insecta: Dermaptera: Anisolabididae	Tp ?	SCAU
51	Culicidae sp.	Insecta: Diptera: Culicidae	Tp ?	SCAU
52	Limoniidae sp.	Insecta: Diptera: Limoniidae	Tp ?	SCAU
53	Psychodinae sp.	Insecta: Diptera: Psychodidae	Tp ?	SCAU
54	*Oreolalax rhodostigmatus* Hu and Fei, 1979	Amphibia: Anura: Pelobatidae	Tp	[25], SCAU
55	*Rana* sp.	Amphibia: Anura: Ranidae	Tx ?	SCAU
56	*Triplophysa xiangxiensis* (Yang, Yuan and Liao, 1986)	Actinopterygii: Cypriniformes: Nemacheilidae	Sb *	[25,28], SCAU
57	*Misgurnus anguillicaudatus* (Cantor, 1842)	Actinopterygii: Cypriniformes: Cobitidae	Sx ?	SCAU
58	*Myotis chinensis* (Tomes, 1857)	Mammalia: Chiroptera: Vespertilionidae	Tp	SCAU
59	*Myotis altarium* Thomas,1911	Mammalia: Chiroptera: Vespertilionidae	Tp	[37], SCAU
60	*Rhinolophus pearsonii* Horsfield, 1851	Mammalia: Chiroptera: Rhinolophidae	Tp	SCAU
61	*Rhinolophus pusillus* Temminck, 1834	Mammalia: Chiroptera: Rhinolophidae	Tp	SCAU
62	*Rhinolophus sinicus* Andersen, 1905	Mammalia: Chiroptera: Rhinolophidae	Tp	SCAU

3.2.1. Mollusca

Mollusca in Feihu Dong comprise four species. Unidentified Subulinidae (Figure 5A), a very frequent troglophile of tropical caves, and Helixarionoidea (Figure 5B) shells were located inside the Tiger Eye. *Synprosphyma* cf. *lyra* (Figure 5C), found in the Low Vault and Flu Waterfall, appears to be the most common species. This species may enter the cave through the underground water system since the specimens near the waterfall are buried in litter, which arrives from outside. Another minute gastropod, *Angustopila huoyani*, discovered in the entrance corridor [30], was not encountered during the 2023 survey.

3.2.2. Arachnida

Six species of spiders are found in various sections of Feihu Dong. *Telema wunderlichi* (Figure 6A), a *Belisana* species (Figure 6B), and an Agelenidae species (Figure 6D) were mostly observed in a moist microhabitat in Tiger Eye, hiding under scattered rubble. A Linyphiidae species (Figure 6C) and *Papiliocoelotes guitangensis* (Figure 6E) were found on the ground of the Low Vault. The unidentified Pholcidae species was collected far inside the cave, after the large chamber; it has extremely long legs and unpigmented, reduced eyes. We can confidently say that *Telema wunderlichi* is a trogobiont because of its eyelessness. In addition, we assume that *Belisana*, as well as the unidentified Agelenidae, which have reduced pigmentation in the body and eyes, are also troglobionts. *Papiliocoelotes guitangensis*, though only known from caves, has no adaptive characters related to cave life [31] and is here assumed to be a troglophile.

Figure 5. Animals found in Feihu Dong: (**A**) Subulinidae sp.; (**B**) Helixarionoidea sp.; (**C**) *Synprosphyma* cf. *lyra* (Gredler,1887); (**D**) Oligochaeta sp.

The Opilionida *Schenkeliobunum*, probably *S. wuxi* Lu, Wang and Zhang, 2022 (Figure 7A), discovered in subtropical forest in Chongqing [38], was documented on the ground of Tiger Eye. It is likely a widely distributed species, inhabiting different humid habitats. A single specimen of an unidentified, blind, and highly modified Laniatores was found in the large chamber [26]. It seems to be similar to the cave restricted Opilionida of Southeast Asia, but these usually have eyes.

We collected one species of Rhagidiidae on the ground (Figure 7B). In addition, three ectoparasitic mites, one Spinturnicidae (Figure 7C) and two other unidentified Acari (Figure 7D,E), were found perched on the wings of bats (*Rhinolophus pusillus* Temminck, 1834).

3.2.3. Diplopoda

Millipedes are among the most common large invertebrates in Feihu Dong. *Glyphiulus deharvengi* (Figure 8A) and *Eutrichodesmus sketi* (Figure 8B) are widespread in the cave. The presence of these species was expected, as both genera are very frequently found and highly diversified in South China caves [11,14,39]. *Eutrichodesmus sketi* is eyeless. *Glyphiulus deharvengi* has eyes but is unpigmented and is likely a troglobiont. An undescribed *Epanerchodus* species (Figure 8C) has a scattered distribution inside the cave. This *Epanerchodus* exhibits an unusual morphological polymorphism related to its spatial distribution in the cave (Figure 8C,D). However, the different forms recognized were confirmed to be conspecific according to both genital features and preliminary barcoding results.

Figure 6. Animals found in Feihu Dong: (**A**) *Telema wunderlichi* Song and Zhu, 1994; (**B**) *Belisana* sp.; (**C**) Linyphiidae sp.; (**D**) Agelenidae sp.; (**E**) *Papiliocoelotes guitangensis* Zhao and Li, 2016.

Figure 7. Animals found in Feihu Dong: (**A**) *Schenkeliobunum* cf. *wuxi* Lu, Wang and Zhang, 2022; (**B**) Rhagidiidae sp.; (**C**) Spinturnicidae sp.; (**D**) Acari sp. 1; (**E**) Acari sp. 2.

Figure 8. Animals found in Feihu Dong: (**A**) *Glyphiulus deharvengi* Golovatch, Geoffroy, Mauriès and Van Den Spiegel, 2007; (**B**) *Eutrichodesmus sketi* Golovatch, Geoffroy, Mauriès and Van Den Spiegel, 2015; (**C**) *Epanerchodus* sp. from Régis Gremmel River; (**D**) *Epanerchodus* sp. from Tiger Eye.

3.2.4. Chilopoda

Chilopoda in Feihu Dong comprise two species, viz. *Lithobius* (*Monotarsobius*) sp. (Figure 9A) and a Geophilidae species (Figure 9B). The depigmented *Lithobius* is probably a troglobiont and was found near the Tiger Eye and at a deeper site, Flu Waterfall. In southern China caves, Lithobiidae are rare, with only two *Australobius* species reported from Guizhou and Guangxi [19,40]. The Geophilidae species is probably a troglophile, found only near the Tiger Eye.

3.2.5. Crustacea

Crustaceans are represented by three moderately troglomorphic species, viz., *Caridina longshan* (Figure 10A), *Gammarus* sp. (Figure 10B), and *Trogloniscus* sp. (Figure 10C). All of them lack either pigment or eyes. *Gammarus* are common in most of the aquatic microhabitats inside Feihu Dong. *Caridina longshan* was collected in Flu Meander, and *Trogloniscus* sp. was collected in the Régis Gremmel River (Figure 4F).

Figure 9. Animals found in Feihu Dong: (**A**) *Lithobius* (*Monotarsobius*) sp.; (**B**) Geophilidae sp.

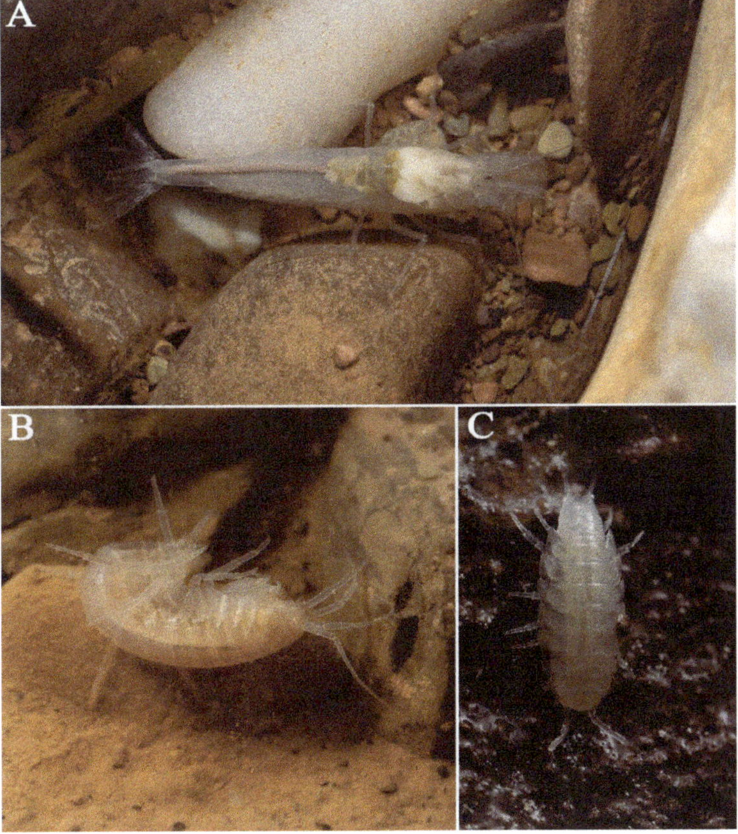

Figure 10. Animals found in Feihu Dong: (**A**) *Caridina longshan* Cai and Ng, 2018; (**B**) *Gammarus* sp.; (**C**) *Trogloniscus* sp.

In addition, a species of lernaeid copepod had been observed on the blind fish *Triplophysa xiangxiensis* (Zhou J.J., pers. comm.).

3.2.6. Non-Insect Hexapods

A Campodeidae species (Figure 11A) inhabits the dried and sandy habitats of the Régis Gremmel River. It occurs individually and keeps crawling all the time. The combination of its long antennae, cerci, and large size suggests that the species might be a troglobiont. The knowledge of the eight species of Chinese cave-dwelling Campodeidae was summarized by Sendra et al. in 2021 [41]. Remarkable troglomorphic features are obvious in most species, and the Feihu Dong Campodeidae is another highly troglomorphic species of southern China.

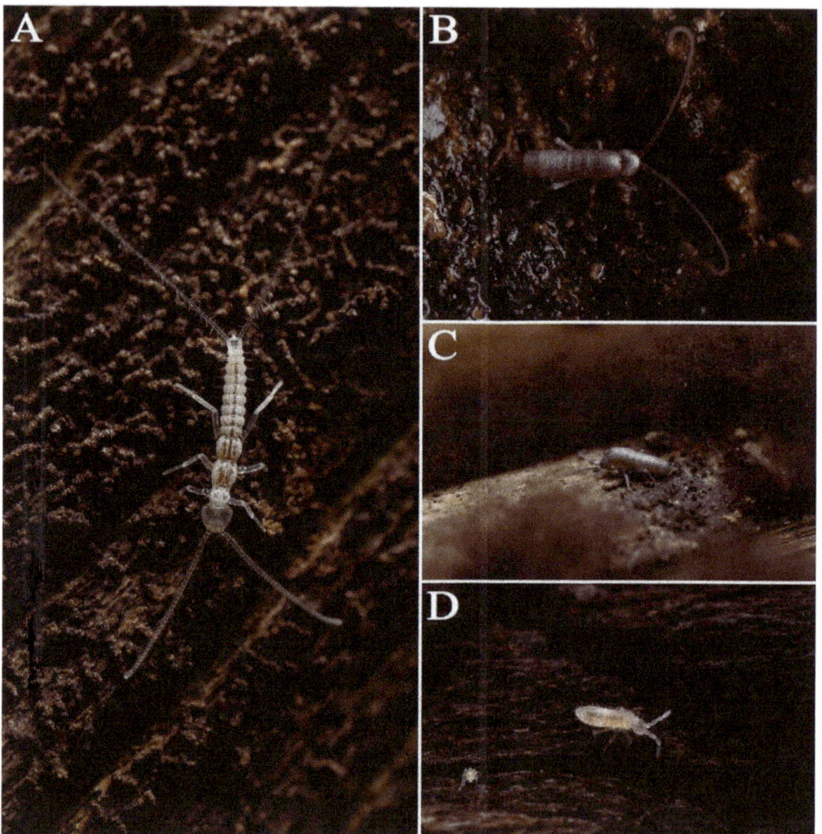

Figure 11. Animals found in Feihu Dong: (**A**) Campodeidae sp.; (**B,C**) *Tomocerus* sp.; (**D**) *Coecobrya* sp.

Springtails were found on decayed wood and leaves that were brought into the cave by the water flow or human activity. We found species of *Tomocerus* (Figure 11B,C) and *Coecobrya* (Figure 11D) in several collecting points that were associated with predators, e.g., ground beetles and pselaphine beetles. These two springtail genera are highly diversified in southern China's caves.

3.2.7. Insecta

Coleoptera

Nine species of beetles were collected in Feihu Dong, including four Carabidae, three Staphylinidae, one Ptinidae, and one Endomychidae. The troglomorphic carabids *Toshi-*

aphaenops ovicollis (Figure 12A), *Huoyanodytes tujiaphilus* (Figure 12B), *Cathaiaphaenops* (*Cathaiaphaenops*) *delprati* (Figure 12C), and *Sinotroglodytes bedosae* are widespread in the cave system. During the survey in 2023, the former three species were spotted. In addition, a larva of *C. delprati* (Figure 12D) was captured among an adult population located in Flu Waterfall. Three elytra of ground beetles were uncovered under the compacted sand on the Régis Gremmel River during our survey, which can be attributed to *C. delprati* (two pieces) (Figure 12E) and *T. ovicollis* (one piece) (Figure 12F), respectively. We assume that blind ground beetles are abundant during the rainy season, when strong water flow is carrying lots of resources inside the cave.

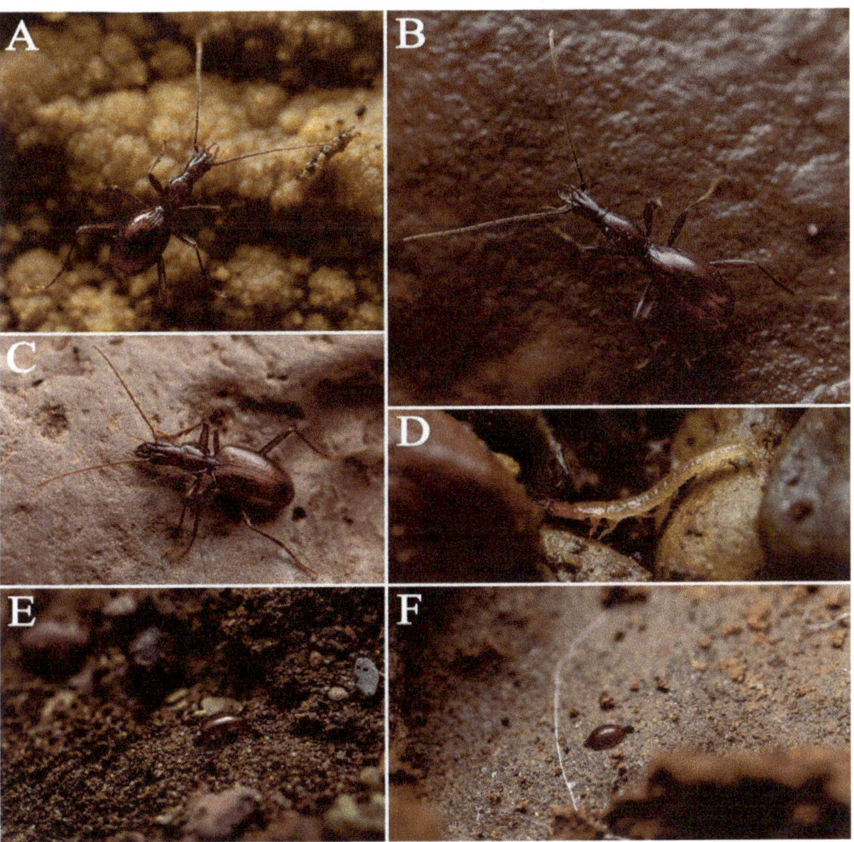

Figure 12. Animals found in Feihu Dong: (**A**) *Toshiaphaenops ovicollis* Ueno, 1999; (**B**) *Huoyanodytes tujiaphilus* Tian and Huang, 2016; (**C–E**) adult, larva, and elytra of *Cathaiaphaenops* (*Cathaiaphaenops*) *delprati* Deuve, 1996; (**F**) elytra of *Toshiaphaenops ovicollis* Ueno, 1999.

For staphylinids, the troglomorphic *Zopherobatrus tianmingyii* (Figure 13A), which is also reported from a limestone cave in Guizhou [42], was found near Tiger Eye, accompanied with a lot of springtails, while a possibly troglophile *Nipponobythus* (Figure 13B) species was found alone near the Waterfall No. I. The genus *Zopherobatrus* contains three species, all troglobionts with reduced eyes, previously known from Guizhou, Chongqing, and Sichuan [19]. *Quedius feihuensis* Smetana, 1999 was collected together with *Cathaiaphaenops*. In spite of being only known from Feihu Dong, the species has no troglomorphic character and is likely a trogloxene [34].

Figure 13. Animals found in Feihu Dong: (**A**) *Zopherobatrus tianmingyii* Yin and Li, 2015; (**B**) *Nipponobythus* sp.; (**C**) *Pseudeurostus hilleri* (Reitter, 1877); (**D**) *Mycetina* sp.

In addition, two other beetles are trogloxenes, *viz.* *Pseudeurostus hilleri* (Figure 13C) and a *Mycetina* species (Figure 13D). They were attracted by baits (banana and fish gut), and we settled in Low Vault.

Orthoptera

Three species of the family Rhaphidophoridae have been spotted in Feihu Dong. *Tachycines* (*Gymnaeta*) *solidus* (Figure 14B), restricted to the Room of the Dance of Outstretched Hands, is probably a troglophile due to its normal-sized eyes and dark, striped body. *Eutachycines crenatus* (Figure 14C) occupies a deeper section of the cave, from Tiger Eye to Wind Gallery, including Low Vault. It has a depigmented body, medium-sized eyes, and a strongly crenate abdomen, which demonstrate its link to cavernicolous life. The third species, *Tachycines* (*Gymnaeta*) *omninocaecus* (Figure 14A), was found along the Régis Gremmel River, far distant from the two aforementioned species. *Tachycines omninocaecus* is a highly troglomorphic cricket on account of its eyelessness and pale body. The degree of troglomorphy seems related to the spatial distribution of this species within the Feihu Dong system when compared to that of the two other cave crickets.

Figure 14. Animals found in Feihu Dong: (**A**) *Tachycines* (*Gymnaeta*) *omninocaecus* (Gorochov, Rampini and Di Russo, 2006); (**B**) *Tachycines* (*Gymnaeta*) *solidus* (Gorochov, Rampini and Di Russo, 2006); (**C**) *Eutachycines crenatus* (Gorochov, Rampini and Di Russo, 2006).

Siphonaptera

A peculiar flea (Figure 15A), which is non-jumpable and belongs to the family Ischnopsyllidae, was observed by the second author (Zhao M.Z.) (Figure 15B). The species has a rather elongated body and legs. Based on our observations, it seems it crawls slowly and uses its forelegs to detect the environment. Congeneric specimens were also obtained from caves in Guizhou Province and convincingly support their parasitism on bats. Further studies regarding its taxonomy and biology are currently being conducted.

Figure 15. Animals found in Feihu Dong: (**A,B**) Ischnopsyllidae sp.

Aquatic insects

The aquatic insect fauna in Feihu Dong is represented by three species belonging to Odonata, Plecoptera, and Trichoptera. A *Sarasaeschna* species of the dragonfly family Aeshnidae was found in the Waterfall River. Four final instar (Figure 16A) and one probably penultimate instar (Figure 16B) nymphs were found in the shallow water. Unlike the *Sarasaeschna* species in Ganxiao Cave [18], troglomorphic characters are absent in the final instar nymph of the species found in Feihu Dong. However, we noticed that the younger nymph has a depigmented body and partially developed eyes. The troglophilic status of this species is under investigation. In addition, two mature Trichoptera larvae (Figure 16C) were hidden under the stones of the Régis Gremmel River. We had also uncovered final stage nymphs (Figure 16D) of a Perlidae (Plecoptera) in shallow water of Flu Meander. The discovery of these aquatic insects inside Feihu Dong reveals the complexity of the subterranean water system. The most likely is that a connection exists between the subterranean water and a sinking stream from the surface that remains to be spotted.

Figure 16. Animals found in Feihu Dong: (**A**,**B**) ultimate and penultimate instar nymphs of *Sarasaeschna* sp.; (**C**) Trichoptera sp.; (**D**) Perlidae sp.

Other insects

Triphosa species (Figure 17A), one of the common troglophilic moths in China, were frequently seen on the walls from the entrance to deep sections (*ca.* 0–4 km inside the cave Feihu Dong), sometimes infected by fungi (Figure 17B). Additionally, Tineidae (Lepidoptera) (Figure 17C) and Anisolabididae (Dermaptera) (Figure 17G), as well as Psychodidae (Figure 17D), Culicidae (Figure 17E) and Limoniidae (Figure 17F) (all Diptera), are each represented by one species. Except for the Limoniidae, others are spotted as single individuals. These five species are difficult to assign to an ecological category. However, tineid and psychodid are very often linked to guano in caves, while the three other Diptera belong to families that are among the dominant troglophiles of temperate cave entrances.

Figure 17. Animals found in Feihu Dong: (**A,B**) Triphosa sp.; (**C**) Tineidae sp.; (**D**) Psychodinae sp.; (**E**) Culicidae sp.; (**F**) Limoniidae sp.; (**G**) Anisolabididae sp.

3.2.8. Vertebrates

Vertebrates encompass two fish, two frogs, and five bats. The only stygobiotic vertebrate is *Triplophysa xiangxiensis* (Figure 18A), a completely blind fish. The genus comprises 102 species in China, and all the identified cave-dwelling species of the genus were reported from China. More than one fourth of them are typical cavefish, with eyes and pigmentation reduced or completely lost. They are restricted to the karst regions of Yunnan, Hunan, Guizhou, Guangxi, and Chongqing [43–45].

Figure 18. Animals found in Feihu Dong: (**A**) *Triplophysa xiangxiensis* (Yang, Yuan and Liao, 1986); (**B**) adult *Oreolalax rhodostigmatus* Hu and Fei, 1979; (**C**) *Rana* sp.; (**D**) larva and adult *Oreolalax rhodostigmatus* Hu and Fei, 1979.

Triplophysa xiangxiensis coexists with tadpoles of *Oreolalax rhodostigmatus* (Figure 18B,D), a troglophilic frog, and with the degenerated eyes of the shrimp *Caridina longshan*. Moreover, two trogloxenes, viz., the frog *Rana* sp. (Figure 18C) and the fish *Misgurnus anguillicaudatus*, were discovered accidentally. During winter, bats are scarcely seen along our explored sections inside the cave, yet they accommodate five species (three *Rhinolophus* and two *Myotis*) (Figure 19). Due to the roof of the cave being quite high in most sections, we may have missed some hibernating bats. The bats we observed do not form colonies during the winter season. Their guano is rather scattered in the cave, providing an important source of nourishment for cave-dwelling invertebrates.

3.2.9. Other Animals

Besides what we discussed above, the fauna of Feihu Dong also includes a flatworm and an earthworm. The flatworm may be assigned to *Dugesia japonica*, but the photo is not available [25]. A trogloxene or stygoxene earthworm (Figure 5D) was found in Flu Waterfall, where a lot of organic matter accumulates from outside.

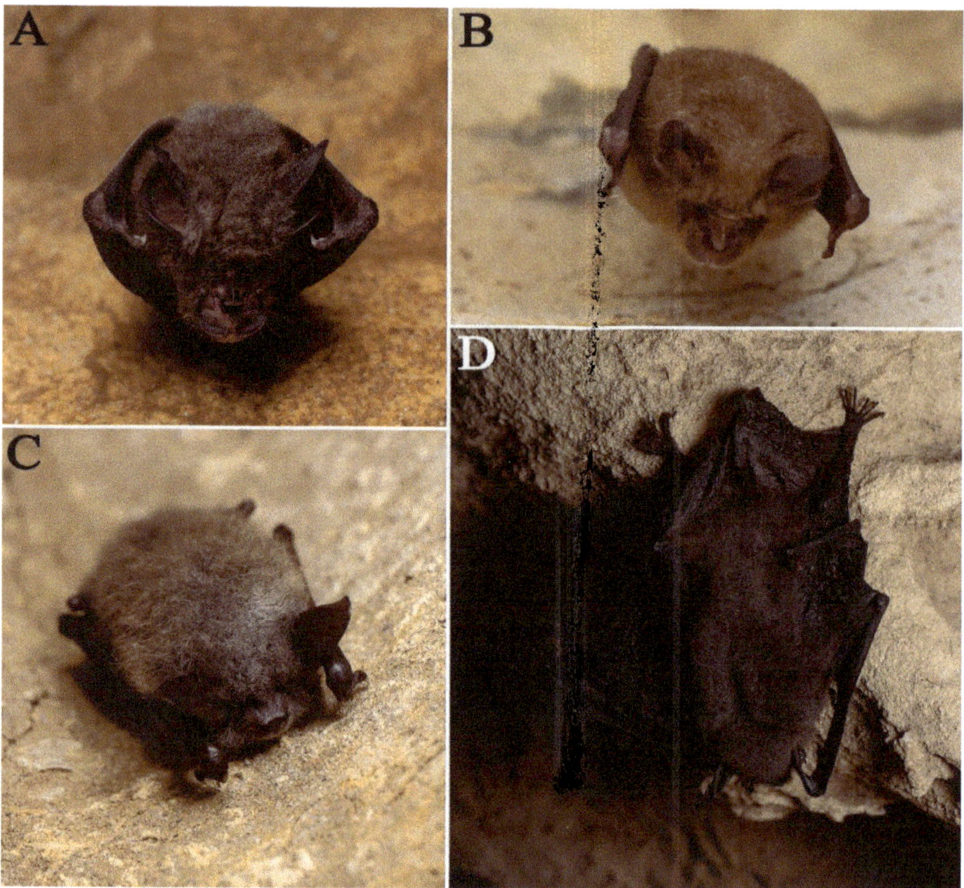

Figure 19. Animals found in Feihu Dong: (**A**) *Rhinolophus pearsonii* Horsfield, 1851; (**B**) *Rhinolophus pusillus* Temminck, 1834; (**C**) *Myotis altarium* Thomas, 1911; (**D**) *Myotis chinensis* (Tomes, 1857).

4. Discussion

Explorations into the subterranean biodiversity of Feihu Dong revealed an abundance of cave-dwelling fauna, exceeding initial expectations. Here we discuss the implications of these findings for underscoring the significance of Feihu Dong as a site warranting further research and conservation efforts.

4.1. Biodiversity Features of Feihu Dong

The subterranean ecosystem of Feihu Dong has been found to be highly diverse and unique, with a total of 27 troglobiotic species recognized in this study. This is a significantly higher number compared to other known caves in South China, such as Ganxiao Dong (20 troglobionts) [18]. Among the different taxonomic groups of subterranean fauna, insects were found to be the most diverse, with eight troglobiotic species identified, including five beetles.

Regarding endemism, 15 of the cited species are restricted to the surrounding karst area. All but one of the 13 troglobiotic (11 species) or stygobiotic (two species) species identified at the species level are endemic to Feihu Dong or to the karst area of Huoyan, except *Zopherobatrus tianmingyii*, which is also reported from Guizhou [42].

The troglobionts (23 species) and stygobionts (four species) found in Feihu Dong represent a diverse range of taxa that have reached different levels of troglomorphy. Though generally less morphologically modified than species of cave communities in southern Guizhou and northern Guangxi, some species of Feihu Dong are highly troglomorphic, such as the blind ground beetle *Huoyanodytes tujiaphilus* or the blind fish *Triplophysa xiangxiensis*. Interestingly, several of the large stygobionts *Gammarus* sp., *Caridina longshan*, *Triplophysa xiangxiensis*, and the tadpoles of the frog *Oreolalax rhodostigmatus* may occur in high numbers (such as *G.* sp. in Waterfall River) and may even co-occur (such as the tadpoles of *O. rhodostigmatus*, *T. xiangxiensis*, and *C. longshan* in Flu Meander).

4.2. Discussion on Some Animals Found in Cave Feihu Dong

Millipedes are commonly seen in caves, where they typically rest on cave walls or attach themselves to decayed wood. In our 2023 investigations, we did not observe large numbers of millipedes in Feihu Dong. This may be due to a lack of food, as bats are known to hibernate during the winter and water flow is often reduced.

Reduced eyes *Caridina* (with cornea pigmentation ranging from totally absent to a small black spot) and blind *Gammarus* are known by many narrowly distributed species from several caves in south China [46–48], mostly described in recent years. Their presence in Feihu Dong is in line with this distribution. *Trogloniscus* sp. belongs to an oligospecific genus of more or less amphibious species known from Guangxi, Guizhou, and Guangdong [19,49]. Its presence in Feihu Dong extends significantly to the north of the distribution of the genus.

Regarding the blind carabid beetle, *Cimmeritodes* (*Cimmeritodes*) *huangi* Deuve, 1996, reported from the cave Baiyan Dong in Huoyan Karst [21]. It is less than 1 km between Baiyan Dong and Feihu Dong in straight-line distance and the beetle is very likely to occur in Feihu Dong.

Some of the unidentified troglobiotic or stygobiotic species may also be endemic to the area, e.g., the springtails *Coecobrya* sp. and *Tomocerus* sp., the spiders *Belisana* sp. and Agelenidae sp., the Chilopoda *Lithobius* (*Monotarsobius*) sp., the Gammaridae *Gammarus* sp., the woodlice *Trogloniscus* sp., and highlighting the need for further investigation and taxonomic research in this region.

4.3. Threats and Conservation

One of the threats to the biodiversity of Feihu Dong is the potential impact of tourism. With the increasing popularity of the cave as a tourist destination, excessive tourism development and overcrowding have the potential to negatively affect the cave's ecosystem. It is important to note that the impact of tourism on the cave biodiversity is not yet significant, as its development has been limited to the entrance area, which is, like in all touristic caves, severely disturbed, while kilometers of very large undisturbed galleries exist beyond this area. However, to prevent future negative impacts on biodiversity, it would be useful to maintain the majority of passages, chambers, and cave floors in their original state. For this purpose, several measures could be taken. Firstly, limiting the extent of touristic passages and the number of visitors permitted in the cave at any time. Secondly, monitoring cave biodiversity and conducting regular scientific research as a background to sound conservation measures. Thirdly, encouraging a moderate development that would preserve most cave passages. Fourthly, keep the cave entrance open and not subject to drastic human impacts, to allow the easy passage of bats in and out of the cave. It is less the problem of the entrance itself, which is very large, than of human activity at this entrance, which should remain reasonable in terms of light and noise. It may be beneficial to consider limiting festivals or infrastructural works within the cave from the main entrance to the "Room of the Dance of Outstretched Hands" to further reduce human impact on the cave's ecosystem. Fifthly, it is crucial to provide adequate protection for the unique and best documented cave species found in Feihu Dong, the blind fish *Triplophysa xiangxiensis*, which had been assessed for The IUCN Red List of Threatened Species in 1996 and was listed as Vulnerable under criteria D2 [50]. It is also classified as Category II in the list of National Key Protected Wild Animals in China [51]. Moreover, two species of Carabidae of Feihu Dong

(*Cathaiaphaenops delprati, Huoyanodytes tujiaphilus*), which are the best known invertebrates of the cave, have been recently assigned to the IUCN category "Data Deficient" [52,53], highlighting the fact that further investigations on the Feihu Dong invertebrate fauna are needed to understand the magnitude and extent of the local cave biodiversity. The current and fast development of tourism and associated potential disturbance in the area should be carefully followed during the coming years, in completing for aquatic microcrustacea the baseline inventory proposed here, in assessing the vulnerability of the other troglobionts of the cave system, and in using some of them to monitor changes in biodiversity. That would provide the background needed to implement appropriate conservation measures if they become necessary.

4.4. Limitations and Prospects

The subterranean biodiversity of Feihu Dong is far from being fully explored and understood. Although several surveys have been conducted, there are still limitations and prospects that need to be addressed to have a comprehensive understanding of the subterranean biodiversity of Feihu Dong.

One limitation is that the survey area is still relatively limited, and most deeper areas have yet to be explored. Regarding the sampling effort in the aquatic environment, we conducted very few investigations concerned with the baited traps and litter extraction in the water. The aquatic fauna of the Feihu Dong system remains poorly known, and further surveys are needed to fully understand the diversity and ecology of this unique ecosystem. Another gap to fill is that many species have not been identified yet, making their ecological traits difficult to assess. We have not been able to determine the relationship between some of the trogloxenes or stygoxenes and the cave environment, given our poor knowledge of cave fauna in the region. The group to which these species belong has also sometimes been reported from caves. For example, the staphylinid beetle *Quedius feihuensis* was only found in Feihu Dong and is considered here as a trogloxene based on morphology, but other species of the same genus and of similar morphology are also reported from caves [54,55]. This suggests that it may be benefiting from the cave environment in some way. So, we took into account trogloxenes and stygoxenes in this paper. We also face the challenge of finding taxonomists who can deal with the selected material, which is also a classical issue in countries where cave fauna remains under-investigated. Overcoming these limitations would require more extensive and in-depth surveys across the Feihu Dong system, and in the surrounding karst.

There is therefore a significant potential for further research on the subterranean biodiversity of Feihu Dong. The prospects are still far-reaching, as many small passages and connections have not been explored. Systematic surveys of the cave fauna will be carried out in parallel with the four-year exploration project of Chinese cavers, which is expected to shed more light on the subterranean biodiversity of Feihu Dong.

Moreover, the newly discovered passages of Feihu Dong may also potentially connect to other caves, such as Tujiamei Dong (Chushui Dong or Parking Cave "Grotte du Parking"), which could increase the richness of species in the area. Tujiamei Dong, which has been partly surveyed, is very likely linked to Feihu Dong as an outlet for water (Huang, S.B., pers. comm.). It has some species in common with Feihu Dong: the terrestrial species of four cave ground beetles, *Toshiaphaenops ovicollis*, *Huoyanodytes tujiaphilus*, *Cathaiaphaenops delprati* and *Sinotroglodytes bedosae* [56], and millipede *Glyphiulus deharvengi*; as well as a large number of aquatic species, including shrimps *Caridina longshan*, blind fish *Triplophysa xiangxiensis*, and mud fish *Cobitidae* sp. (Figure 20E). But it has other species that are absent from Feihu Dong: the millipede *Epanerchodus tujiaphilus* Liu and Golovatch, 2018 (Figure 20B) [57], a widespread troglophilic scutiger *Thereuopoda clunifera* (Wood, 1862) (Figure 20A), a spider *Nesticella huomachongensis* Lin, Ballarin and Li, 2016 (Figure 20C), and a nymph of damselfly Synlestidae sp. (Figure 20D).

Moreover, further caves, Da Dong, Panlong Dong (Pa Long Dong), and Mao Dong, might also potentially be connected to Feihu Dong (Figure 1), as well as three potential shafts (Wang Y., pers. comm.) that may act as sources of food through the dripping or sinking waters (Figure 3). The three shafts are positioned above the cave system of Feihu Dong, but their connection with Feihu Dong would require further exploration. Numerous other caves that exist around Huoyan (in Longshan, Rongshun, and Sangzhi regions) have been less intensively, or not at all surveyed biologically. The rare available records that have been published indicate that they probably possess levels of species richness similar to those of Feihu Dong.

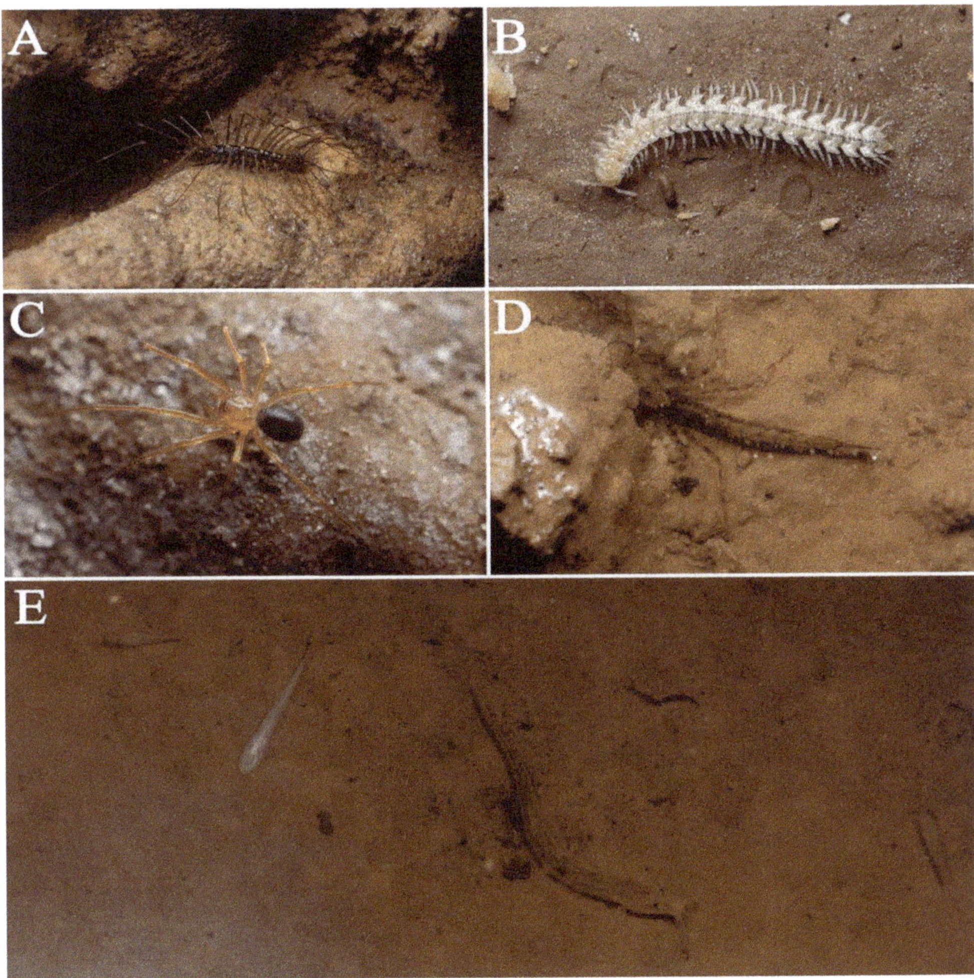

Figure 20. Animals found in Tujiamei Dong: (**A**) *Thereuopoda clunifera* (Wood, 1862); (**B**) *Epanerchodus tujiaphilus* Liu and Golovatch, 2018; (**C**) *Nesticella huomachongensis* Lin, Ballarin and Li, 2016; (**D**) Synlestidae sp.; (**E**) shrimps *Caridina longshan* Cai and Ng, 2018, blind fish *Triplophysa xiangxiensis* (Yang, Yuan and Liao, 1986); and mud fish Cobitidae sp.

Although we know the significance of water for cave fauna, our understanding of the hydrogeology of Feihu Dong is still limited. No validated surface streams were addressed, except for the shaft 'Gouffre Super Tong' and three potential shafts (Figure 1) that may act as inlets for rainwater during the rainy season. The relationship between the cave's water system and the external water system is not well understood and requires further research.

In conclusion, the study of Feihu Dong has exceeded expectations in terms of species richness, and it is likely that many other caves in China have a similar level of subterranean biodiversity. A further significant increase in species richness in South China's caves and karsts can be expected in the near future. Even if research on cave fauna in China still lags behind that of some western countries [58–60], the results obtained on Feihu Dong are demonstrating the potential for South China to become a world hotspot for cave biodiversity. This is indeed stimulating for further investigations, even if there is still a long way to go to fully uncover the diversity of cave fauna in the huge karsts of South China. It is demanding to expand the scope and intensity of cave fauna surveys, and to explore and document the intricate subterranean habitats of caves and karsts in China through collaborative work.

Author Contributions: Conceptualization, M.T., W.L. and S.H.; methodology, S.H., M.Z. and W.L.; software, S.H., M.Z. and W.L.; validation, M.T., A.B., X.L., Y.W., M.C. and W.L.; formal analysis, S.H., M.Z. and W.L.; investigation, S.H., M.Z., A.B., M.C. and Y.W.; resources, Y.W., M.C., A.B. and S.H.; data curation, S.H. and M.Z.; writing—original draft preparation, S.H., M.Z., X.L. and W.L.; writing—review and editing, S.H., M.Z., A.B., M.T., X.L., M.C. and W.L.; visualization, S.H. and M.Z.; supervision, M.T. and W.L.; project administration, M.T. and W.L.; funding acquisition, M.T. and W.L. All authors have read and agreed to the published version of the manuscript.

Funding: This research received no external funding.

Institutional Review Board Statement: Not applicable.

Data Availability Statement: Not applicable.

Acknowledgments: We would like to express our sincere gratitude to various organizations, individuals, and teams who have contributed to the success of our research project on the subterranean biodiversity of Feihu Dong. We extend our appreciation to the Speleology Geological Professional Committee of the Geological Society of China, Xiangxi UNESCO Geopark, Wulongshan National Geopark, and Wulongshan Canyon Scenic Area for their support and cooperation throughout the project. We would also like to thank Yuanhai Zhang of the Institute of Karst Geology, Chinese Academy of Geological Sciences, for arranging the biocaving in Feihu Dong, which was instrumental in our research efforts. Special thanks go to Louis Deharveng from the Muséum National d'Histoire Naturelle for his efforts to inform the foundation about the biodiversity in Feihu Dong, Jian Zhou (member of the team Xiangxi Cave Expedition) for sharing valuable information and data about Feihu Dong, and Zhi Qian of the Guizhou Speleology Association for providing information about Shuanghe Dong. We are also grateful to the following individuals who assisted us in identifying various species in Feihu Dong: Jiajun Zhou of Zhejiang Forest Survey, Planning and Design Company and Xingliang Wang of Guizhou Normal University and for their expertise in identifying bats; Qidi Zhu of Hebei University for her help with crickets; Ziwei Yin of Shanghai Normal University for his assistance with Staphylinidae; Jisheng Wang of Dali University for his expertise in Siphonaptera; Yejie Lin of the Institute of Zoology, Chinese Academy of Sciences for identifying spiders; Feng Lu of Shenzhen University for his help with Harvestmen; and Hengjie Huang of South China Agricultural University for assisting with Scutigeridae. Additionally, we thank Lu Qiu of Mianyang Normal University for his help with mollusks. We would like to express our sincere gratitude to the three anonymous reviewers for their valuable feedback and suggestions, which have helped us to improve the quality of our work. We would also like to thank Georgi Angelov Geshev (Imperial College London, United Kingdom) and Sergei I. Golovatch (Russian Academy of Sciences, Moscow, Russia) for their assistance in checking and validating the English of our manuscript. Finally, we extend our thanks to other people from the South China Agricultural University Biocaving Team (Haomin Yin, Xinhui Wang, Mingruo Tang, Zijun Ma, Xinyang Jia, Yi Zhao) and the team Xiangxi Cave Expedition (烧鸟, 天悟, 末路, 三皮, 华华, 小爱, 虫子, 三二, 马儿, 笑笑, 水瓶, 畅畅, 鸡腿, 伙头军, 燕子, 不二, 余老师, 小郭耳朵, 浆糊, 烧成灰) for their assistance with caving and fieldwork in Feihu Dong. Thanks also to the Speleo-Club de Paris (France) for the support of the biological team during the 1995 speleological expedition, and to the SHAG Caving Association of Besançon (France) for the final cave topographies, conducted after the 1997 speleological expedition, with the support of the authorities from Hunan (the Institute of Geology of Hunan Province and the Government of Longshan County).

Conflicts of Interest: The authors declare no conflict of interest.

References

1. Che, Y.T.; Yu, J.Z. *Karst of China*; Science Press: Beijing, China, 1985; pp. 1–230.
2. Chen, W.H. An outline of speleology research progress. *Geol. Rev.* **2006**, *52*, 783–792.
3. Song, X.; Gao, Y.; Green, S.M.; Dungait, J.A.J.; Peng, T.; Quine, T.A.; Xiong, B.; Wen, X.; He, N. Nitrogen loss from karst area in China in recent 50 years: An in-situ simulated rainfall experiment's assessment. *Ecol. Evol.* **2017**, *7*, 10131–10142. [CrossRef] [PubMed]
4. Xiong, K.N. Karst and World Natural Heritage in Southern China. In Proceedings of the Southwest Region Conference of Mountain Environment and Ecological Civilization Construction of the Chinese Geographical Society Annual Academic Conference, Kunming, Yunnan, 27 April 2013.
5. Williams, P.W. Karst in UNESCO World Heritage Sites. In *Karst Management*; van Beynen, P.E., Ed.; Springer: Dordrecht, The Netherlands, 2011; pp. 459–480. ISBN 978-94-007-1206-5/978-94-007-1207-2. [CrossRef]
6. Latella, L. Biodiversity: China. In *Encyclopedia of Caves*; Elsevier: Amsterdam, The Netherlands, 2019; pp. 127–135. ISBN 978-0-12-814124-3. [CrossRef]
7. Luo, S.W.; He, W.; Zhang, Y.H.; Deng, Y.D.; Wu, K.H.; Zhou, W.L.; Zhang, H.Z.; Luo, S.Q. Influence of Shuanghe Cave development on the changes of surface water system in Suiyang County, Guizhou, China. *Carsologica Sin.* **2021**, *40*, 920–931. [CrossRef]
8. Zhou, Y.J. Exploring the underground world: World's largest cave chamer by volume–Ziyun Miao Room Chamber. *New Era Ecol. Civiliz.* **2018**, *2*, 41–46.
9. Wang, X.Y. Longest karstic caves in China. *Carsologica Sin.* **1988**, *7*, 384.
10. Zhu, X.W.; Chen, W. Tiankengs in the karst of China. *Speleogenesis Evol. Karst Aquifers* **2006**, *4*, 1–18.
11. Liu, W.; Wynne, J.J. Cave millipede diversity with the description of six new species from Guangxi, China. *Subterr. Biol.* **2019**, *30*, 57–94. [CrossRef]
12. Zhu, Q.; Chen, H.; Shi, F. Remarks on the genus *Tachycines* Adelung, 1902 (Orthoptera: Rhaphidophoridae: Aemodogryllinae) with description of eight new species from caves in southern China. *Zootaxa* **2020**, *4809*, 71–94. [CrossRef]
13. Zhu, W.; Yao, Z.; Zheng, G.; Li, S. Six new species of the spider genus *Belisana* Thorell, 1898 (Araneae: Pholcidae) from southern China. *Zootaxa* **2020**, *4810*, 175–197. [CrossRef]
14. Zhao, Y.; Guo, W.-R.; Golovatch, S.I.; Liu, W.-X. Revision of the javanicus species group of the millipede genus *Glyphiulus* Gervais, 1847, with descriptions of five new species from China (Diplopoda, Spirostreptida, Cambalopsidae). *ZooKeys* **2022**, *1108*, 89–118. [CrossRef]
15. Zhou, X.; Yang, W. Ten new species of genus *Tachycines* (Orthoptera, Rhaphidophoridae, Aemodogryllinae) from karst caves in Guizhou, China. *ZooKeys* **2022**, *1109*, 115–140. [CrossRef]
16. Tian, M.Y.; Huang, S.B.; Jia, X.Y. A contribution to cavernicolous beetle diversity of South China Karst: Eight new genera and fourteen new species (Coleoptera: Carabidae: Trechini). *Zootaxa* **2023**, *5243*, 1–66. [CrossRef]
17. Hou, Y.; Feng, Z.; Zhang, F. Diversity of cave-dwelling pseudoscorpions from Guizhou in China, with the description of twenty-four new species of the genus *Tyrannochthonius* (Pseudoscorpiones, Chthoniidae). *Zootaxa* **2023**, *5062*, 1–158. [CrossRef]
18. Huang, S.B.; Wei, G.F.; Wang, H.S.; Liu, W.X.; Bedos, A.; Deharveng, L.; Tian, M.Y. Ganxiao Dong: A hotspot of cave biodiversity in northern Guangxi, China. *Diversity* **2021**, *13*, 355. [CrossRef]
19. Tian, M.Y.; Liu, W.X.; Huang, S.B.; Luo, X.Z. *Biospeleology*; Geological Publishing House: Beijing, China, 2023; pp. 1–248.
20. Li, M.Q. Afforestation design for Wulong Mountain Sights. *J. Nanjing Agric. Technol. Coll.* **1999**, *15*, 52–54.
21. Deuve, T. Descriptions de trois Trechinae anophtalmes cavernicoles dans un karst du Hunan, Chine (Coleoptera, Trechidae). *Rev. Française D'entomologie N.S.* **1996**, *18*, 41–48.
22. *Xiangxi 95, Expédition Franco-Chinoise, Report*; Delprat, B. (Ed.) La Salamandre: La Ciotat, France, 1998.
23. Uéno, S.I. New genera and species of aphaenopsoid trechines (Coleoptera, Trechinae) from South-Central China. *Elytra* **1999**, *27*, 617–633.
24. Yao, Y.R. Genetic Diversity and Physiology of the Cavefish *Triplophysa xiangxiensis*. Ph.D Dissertation, Huazhong Agricultural University, Wuhan, China, 2012.
25. Deharveng, L.; Bedos, A. La faune des grottes de Huoyan. In *Xiangxi 95, Expédition Franco-Chinoise, Report*; Delprat, B., Ed.; La Salamandre: La Ciotat, France, 1998; pp. 183–185.
26. Deharveng, L.; Bedos, A. The cave fauna of southeast Asia. Origin, evolution and ecology. In *Ecosystems of the World Vol 30 Subterranean Ecosystems*; Wilkens, H., Culver, D.C., Humphreys, W.F., Eds.; Elsevier: Amsterdam, The Netherlands, 2000.
27. Deharveng, L.; Bedos, A. Diversity patterns in the tropics. In *Encyclopedia of Caves*; Elsevier: Amsterdam, The Netherlands, 2012; pp. 238–250.
28. Yang, G.R.; Yuan, F.X.; Liao, R.M. A new blind Cobitidae fish from the subteranean water in Xiangxi, China. *J. Huazhong Agric. Univ.* **1986**, *5*, 219–223.
29. Cai, Y.; Ng, P.K.L. Freshwater Shrimps from Karst Caves of Southern China, with Descriptions of Seven New Species and the Identity of *Typhlocaridina linyunensis* Li and Luo, 2001 (Crustacea: Decapoda: Caridea). *Zool. Stud.* **2018**, *57*, e27.
30. Jochum, A.; Slapnik, R.; Kampschulte, M.; Martels, G.; Heneka, M.; Pall-Gergely, B. A review of the microgastropod genus *Systenostoma* Bavay & Dautzenberg, 1908 and a new subterranean species from China (Gastropoda, Pulmonata, Hypselostomatidae). *ZooKeys* **2014**, *410*, 23–40. [CrossRef]

31. Zhao, Z.; Li, S. *Papiliocoelotes* gen. N., a new genus of Coelotinae (Araneae, Agelenidae) spiders from the Wuling Mountains, China. *ZooKeys* **2016**, *585*, 33–50. [CrossRef]
32. Golovatch, S.I.; Geoffroy, J.-J.; Mauries, J.-P.; Van Den Spiegel, D. Review of the millipede genus *Glyphiulus* Gervais, 1847, with descriptions of new species from Southeast Asia (Diplopoda, Spirostreptida, Cambalopsidae). Part 1: The granulatus-group. *Zoosystema* **2007**, *29*, 7–49.
33. Golovatch, S.I. Cave Diplopoda of southern China with reference to millipede diversity in Southeast Asia. *ZooKeys* **2015**, *510*, 79–94. [CrossRef]
34. Smetana, A. Contributions to the knowledge of the *Quediina* (Coleoptera, Staphylinidae, Staphylinini) of China. Part 15. Genus Strouhalium Scheerpeltz, 1962. Section 3. Genus Quedius Stephens, 1829. Subgenus Microsaurus Dejean, 1833. Section 9. *Elytra* **1999**, *27*, 519–534.
35. Gorochov, A.V.; Rampini, M.; Di Russo, C. New species of the genus *Diestrammena* (Orthoptera: Rhaphidophoridae: Aemodogryllinae) from caves of China. *Russ. Entomol. J.* **2006**, *15*, 355–360.
36. Zhu, Q.D.; Shi, F.M. Description of four new species of the subgenus *Tachycines* (*Gymnaeta*) Adelung, 1902 (Orthoptera: Rhaphidophoridae) from caves in China and additional notes on some previously known species. *Eur. J. Taxon.* **2021**, *764*, 1–17. [CrossRef]
37. Fu, D.F.; Zhang, Y.X.; Jiang, X.; Liu, Z.X.; Yan, Z.J.; Yang, W.W.; Zeng, W.X. Distribution Record of *Myotis altarium* Thomas in Hunan Province of China. *J. Jishou Univ. Nat. Sci. Ed.* **2010**, *31*, 106–108.
38. Lu, F.; Wang, L.Y.; Zhang, C. *Schenkeliobunum wuxi* sp. nov., a new harvestmen species from Chongqing, China (Opiliones: Sclerosomatidae). *Acta Arachnol. Sin.* **2022**, *31*, 75–80.
39. Golovatch, S.I.; Geoffroy, J.-J.; Mauriès, J.-P.; Van Den Spiegel, D. Review of the millipede genus *Eutrichodesmus* Silvestri, 1910, in China, with descriptions of new cavernicolous species (Diplopoda, Polydesmida, Haplodesmidae). *ZooKeys* **2015**, *505*, 1–34. [CrossRef]
40. Li, Q.; Pei, S.; Guo, X.; Ma, H.; Chen, H. *Australobius Tracheoperspicuus* sp. N., the first subterranean species of centipede from southern China (Lithobiomorpha, Lithobiidae). *ZooKeys* **2018**, *795*, 83–91. [CrossRef]
41. Sendra, A.; Komerički, A.; Lips, J.; Luan, Y.; Selfa, J.; Jiménez-Valverde, A. Asian cave-adapted diplurans, with the description of two new genera and four new species (Arthropoda, Hexapoda, Entognatha). *Eur. J. Taxon.* **2021**, *731*, 1–46. [CrossRef]
42. Yin, Z.-W.; Li, L.-Z. *Zopherobatrus* gen. n. (Coleoptera: Staphylinidae: Pselaphinae), a new troglobitic batrisine from southwestern China. *Zootaxa* **2015**, *3985*, 291. [CrossRef] [PubMed]
43. Ma, L.; Zhao, Y.; Yang, J. Cavefish of China. In *Encyclopedia of Caves*; Elsevier: Amsterdam, The Netherlands, 2019; pp. 237–254. ISBN 978-0-12-814124-3. [CrossRef]
44. Lu, Z.-M.; Li, X.-J.; Lü, W.-J.; Huang, J.-Q.; Xu, T.-K.; Huang, G.; Qian, F.-Q.; Yang, P.; Chen, S.-G.; Mao, W.-N.; et al. *Triplophysa xuanweiensis* sp. Nov., a new blind loach species from a cave in China (Teleostei: Cypriniformes: Nemacheilidae). *Zool. Res.* **2022**, *43*, 221–224. [CrossRef] [PubMed]
45. Liu, F.; Zeng, Z.-X.; Gong, Z. Two new hypogean species of *Triplophysa* (Cypriniformes: Nemacheilidae) from the River Yangtze drainage in Guizhou, China. *J. Vertebr. Biol.* **2022**, *71*, 22062.1–22062.14. [CrossRef]
46. Xu, D.-J.; Li, D.-X.; Zheng, X.-Z.; Guo, Z.-L. *Caridina sinanensis*, a new species of stygobiotic atyid shrimp (Decapoda, Caridea, Atyidae) from a karst cave in the Guizhou Province, southwestern China. *ZooKeys* **2020**, *1008*, 17–35. [CrossRef]
47. Guo, G.-C.; Chen, Q.-H.; Chen, W.-J.; Cai, C.-H.; Guo, Z.-L. *Caridina stellata*, a new species of atyid shrimp (Decapoda, Caridea, Atyidae) with the male description of *Caridina cavernicola* Liang & Zhou, 1993 from Guangxi, China. *ZooKeys* **2022**, *1104*, 177–201. [CrossRef]
48. Hou, Z.; Zhao, S.; Li, S. Seven new freshwater species of *Gammarus* from southern China (Crustacea, Amphipoda, Gammaridae). *ZooKeys* **2018**, *749*, 1–79. [CrossRef]
49. Taiti, S.; Xue, Z. The cavernicolous genus *Trogloniscus* nomen novum, with descriptions of four new species from southern China (Crustacea, Oniscidea, Styloniscidae). *Trop. Zool.* **2012**, *25*, 183–209. [CrossRef]
50. Kottelat, M. The IUCN Red List of Threatened Species 1996: E.T22202A9364418. In *Triplophysa xiangxensis*; IUCN: Gland, Switzerland, 1996. [CrossRef]
51. Zhao, Y.; Zhang, Y.Y.; Zhao, Y.H. Assessment of endangerment category on Chinese cavefish: A case study of two national protected fish species. *Carsologica Sin.* **2021**, *40*, 1032–1037.
52. Deharveng, L.; Huang, S.; Tian, M. The IUCN Red List of Threatened Species 2020: E.T120544161A120547523. In *Huoyanodytes tujiaphilus*; IUCN: Gland, Switzerland, 2019. [CrossRef]
53. Deharveng, L.; Huang, S.; Tian, M. The IUCN Red List of Threatened Species 2020: E.T120504866A120504946. In *Cathaiaphaenops delprati*; IUCN: Gland, Switzerland, 2019. [CrossRef]
54. Viana, A.C.M.; Ferreira, R.L. A new troglobitic species of *Allochthonius* (subgenus *Urochthonius*) (Pseudoscorpiones, Pseudotyrannochthoniidae) from Japan. *Subterr. Biol.* **2021**, *37*, 43–55. [CrossRef]
55. Brunke, A.J.; Hansen, A.K.; Salnitska, M.; Kypke, J.L.; Predeus, A.V.; Escalona, H.; Chapados, J.T.; Eyres, J.; Richter, R.; Smetana, A.; et al. The limits of Quediini at last (Staphylinidae: Staphylininae): A rove beetle mega-radiation resolved by comprehensive sampling and anchored phylogenomics. *Syst. Entomol.* **2021**, *46*, 396–421. [CrossRef]
56. Tian, M.Y.; Huang, S.B.; Wang, X.H.; Tang, M.R. Contributions to the knowledge of subterranean trechine beetles in southern China's karsts: Five new genera (Insecta, Coleoptera, Carabidae, Trechinae). *ZooKeys* **2016**, *564*, 121–156. [CrossRef]

57. Liu, W.; Golovatch, S. The millipede genus *Epanerchodus* Attems, 1901 in continental China, with descriptions of seven new cavernicolous species (Diplopoda, Polydesmida, Polydesmidae). *Zootaxa* **2018**, *4459*, 053–084. [CrossRef]
58. Zagmajster, M.; Polak, S.; Fišer, C. Postojna-Planina Cave System in Slovenia, a Hotspot of Subterranean Biodiversity and a Cradle of Speleobiology. *Diversity* **2021**, *13*, 271. [CrossRef]
59. Polak, S.; Pipan, T. The Subterranean Fauna of Križna Jama, Slovenia. *Diversity* **2021**, *13*, 210. [CrossRef]
60. Iliffe, T.M.; Calderón-Gutiérrez, F. Bermuda's Walsingham Caves: A global hotspot for anchialine stygobionts. *Diversity* **2021**, *13*, 352. [CrossRef]

diversity

Article

Tham Chiang Dao: A Hotspot of Subterranean Biodiversity in Northern Thailand

Louis Deharveng [1], Martin Ellis [2], Anne Bedos [1] and Sopark Jantarit [3,*]

[1] Institut de Systématique, Evolution, Biodiversité (ISYEB)—UMR 7205 CNRS, Museum National d'Histoire Naturelle, Sorbonne Université, 45 rue Buffon, 75005 Paris, France; dehar.louis@wanadoo.fr (L.D.); bedosanne@yahoo.fr (A.B.)

[2] 212, m8, Wat Pha, Lom Sak 67110, Phetchabun, Thailand; thailandcaves@gmail.com

[3] Excellence Center for Biodiversity of Peninsular Thailand, Faculty of Science, Prince of Songkla University, Hat Yai 90110, Songkhla, Thailand

* Correspondence: fugthong_dajj@yahoo.com

Abstract: The Doi Chiang Dao massif, which became a UNESCO Biosphere Reserve in 2021, is the highest karst mountain in Thailand. Tham Chiang Dao cave is located at the foot of this massif and is among the best-known caves in Thailand, having been visited since prehistoric times, and being a sacred place for the local Shan and Thai people. The cave consists of five main interconnected passages with a total length of 5342 m which ranks it as the 11th longest cave in Thailand. Tham Chiang Dao is the best studied cave in Thailand with a long series of explorations, investigations and zoological collecting. Here, we summarize the 110 years of biological exploration and investigation devoted to this cave. A total of 149 taxa have been recognized in Tham Chiang Dao, of which 61 have been identified to species level. The cave is the type locality for 14 species. The obligate subterranean fauna includes 37 species, of which 33 are troglobionts and 4 are stygobionts. Conservation issues are addressed in the discussion. This work is intended to provide a reference for the knowledge of cave fauna of the Chiang Dao Wildlife Sanctuary and a tool for its management by the local cave management committee, the National Cave Management Policy Committee, and the Department of Mineral Resources. It also documents the biological importance of Tham Chiang Dao in the Doi Chiang Dao UNESCO Biosphere Reserve.

Keywords: biosphere reserve; cave fauna; karst; troglobionts; stygobionts

Citation: Deharveng, L.; Ellis, M.; Bedos, A.; Jantarit, S. Tham Chiang Dao: A Hotspot of Subterranean Biodiversity in Northern Thailand. *Diversity* **2023**, *15*, 1076. https://doi.org/10.3390/d15101076

Academic Editor: Ana Sofia Reboleira

Received: 14 August 2023
Revised: 17 September 2023
Accepted: 18 September 2023
Published: 11 October 2023

1. Introduction

Doi Chiang Dao mountain in Chiang Mai province, northern Thailand, is the highest karst mountain in Thailand (2195 m asl.) and is connected to other karst massifs, forming the Daen Lao mountain range. It is the third highest peak in the country after Doi Inthanon (2565 m asl.) and Doi Pha Hom Pok (2285 m asl.). The Doi Chiang Dao massif is formed by the Doi Chiang Dao Limestone which consists of mainly pale gray, massive limestone with occasional dark colored and moderately bedded limestones, particularly in the lowermost part of the massif, with frequent dolomitic levels. This limestone is essentially free from siliciclastic materials throughout the thick succession. The total thickness is at least 1000 m in total. Fossil foraminifers show that the Doi Chiang Dao Limestone ranges from the Visean (Mississippian/Early Carboniferous) to the Changhsingian (Late Permian), a period of about 90 Ma [1–7], Figure 1B. It rests on a basal pillow basalt of Tournaisian–Visean age [6,7]. Doi Chiang Dao was originally an oceanic sea mount in the Paleotethys Ocean and developed as carbonates capped the sea mount. These carbonates were later structurally incorporated within a closed remnant sea of the Paleotethys Ocean [6,7].

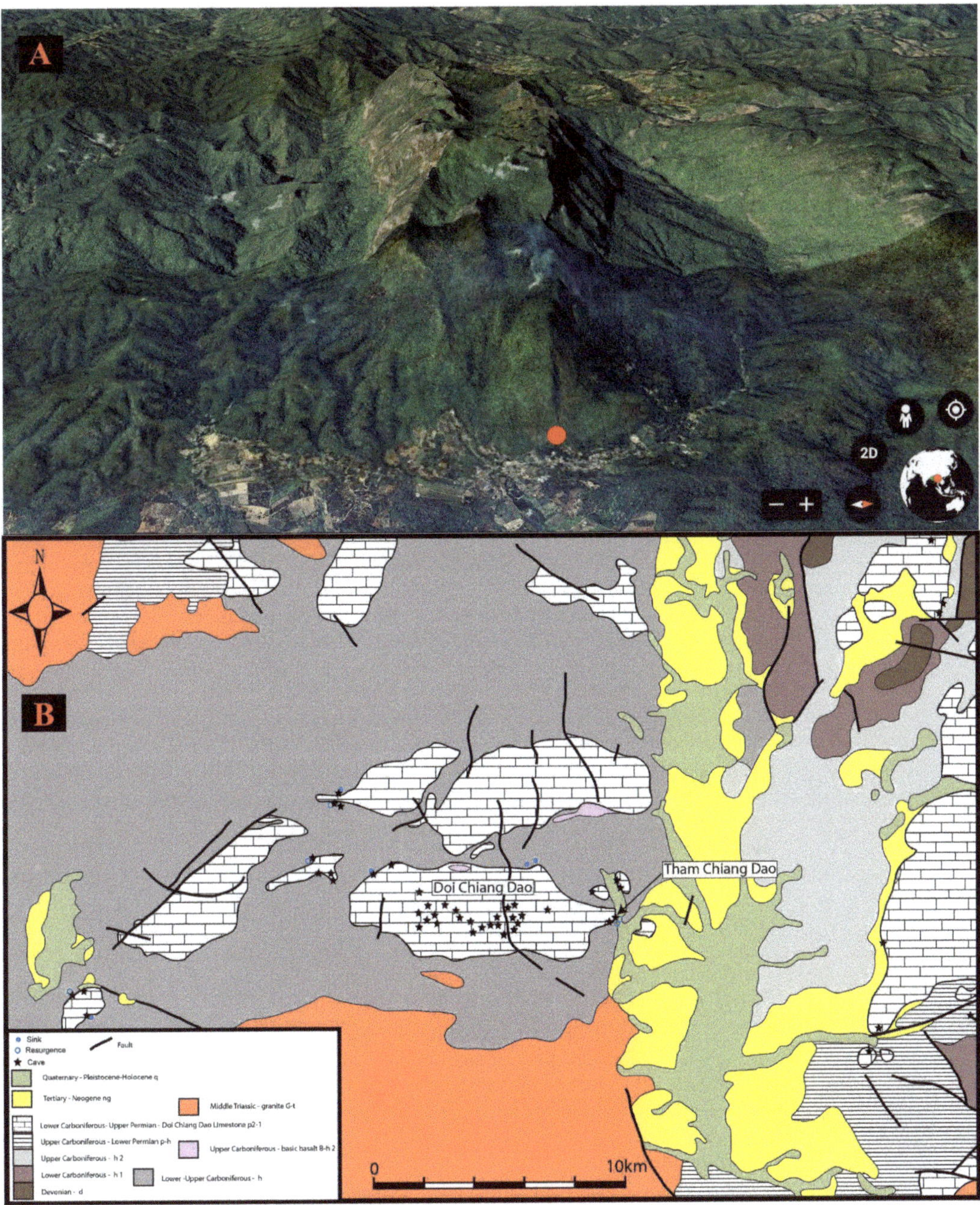

Figure 1. (**A**) Aerial view of Doi Chiang Dao. Red dot indicates cave entrance at the base of the mountain (from Google Earth Pro); (**B**) Geological map of Doi Chiang Dao and Tham Chiang Dao.

Doi Chiang Dao is a protected area as part of the Chiang Dao Wildlife Sanctuary which is managed by the Department of National Parks, Wildlife and Plant Conservation (DNP). In 2021 it was recognized as a UNESCO Biosphere Reserve, the fifth one in Thailand, with an area of 85,909 ha. It is the only region in the country to be covered with a sub-alpine ecosystem (with flora similar to the Himalayas and the southern part of China) and is home to an abundance of rare, endangered, and vulnerable species of plants and animals along with a constellation of tribal peoples. At least 821 plant species and 697 vertebrate animal species are recognized from Doi Chiang Dao with many uncounted invertebrate species [8,9]

In the large limestone massif of Doi Chiang Dao at least 40 caves and shafts have been documented [5], and there are many other unknown and unexplored caves. Among these caves, Tham Chiang Dao is the largest and most famous, being a popular tourist attraction, and it is the best-known cave in northern Thailand. The cave is located at the base of Doi Chiang Dao (the entrance is at 460 m asl.). It has been known for over 1000 years and has a long history of speleological exploration and investigation, with most caving expeditions to the region having visited it. Tham Chiang Dao was the first cave to be speleologically explored in northern Thailand when 2.1 km of high-grade mapping was done in 1972 by Windecker and his team [10]. The cave was also mapped by Deharveng and Gouze in 1980 [11] and was mapped again in 1983 by the American Thailand Karst Hydrologic Project expedition (unpublished). The Association Pyrénéenne de Spéléologie (APS) from France carried out the most detailed exploration and survey in 1985 when 5.1 km was mapped [12]. The most recent, and most complete, mapping has been done by Chiang Mai Rock Climbing Adventures in 2021 (unpublished data).

The cave fauna of Tham Chiang Dao is among the best studied and surveyed of all the Thailand caves. The first biological collecting for bats was done in 1913, mosquitoes were studied in 1969, and in the 1970s, several speleobiologists visited the cave and made limited fauna collections. Since the 1980s, more thorough collections have been made by several expeditions conducted by both national and international organizations (see details in Table 1). This is because the cave is a very popular tourist attraction, is easily accessible, has impressive natural cave formations, has subterranean habitats with both terrestrial and aquatic ecosystems within a complex of interconnected passages, and is of high biological interest.

Tham Chiang Dao is among the eleven pilot caves of Thailand designated in 2019 by the National Cave Management Policy Committee (NCMPC) to be studied as references to set up policies and guidelines for cave management. The goal is to increase public awareness and to support operations beneficial to cave natural resources, maintenance, conservation, rehabilitation, and environment-friendly tourist attractions. These pilot schemes are undertaken by the Department of Mineral Resources (DMR). Hence, the present work will not only document the first hotspot of subterranean biodiversity in Thailand, but also serve as a primary database on Tham Chiang Dao for the NCMPC and DMR development objectives.

2. Materials and Methods

2.1. A Historical Overview of Tham Chiang Dao

Tham Chiang Dao has probably been known for several thousand years as there is archeological evidence that Chiang Dao town, less than 5 km away, has been an important settlement since prehistoric times [13]. In the nearby cave of Tham Bia (1 km away) prehistoric evidence such as pottery, stone tools (polished stone axes), and human and animal skeletal fragments have been found. It is assumed that these items are from the Neolithic period, 3500 to 4500 years ago [14]. Tham Chiang Dao is a sacred place for the local Shan and Thai people and is used for important religious rituals. The oldest religious objects found in the cave are a Buddha image and a 200 kg bronze bell which was made in 1615, indicating that Tham Chiang Dao has been an important religious site for many centuries. Under a skylight near the entrance, which is known as Plong Jaeng, the Shan built several Buddha images and shrines in 1635. The earliest published record of a visit to Tham Chiang Dao by a foreigner is by the American missionary Daniel McGilvary in June or July 1876 [15]. In the 1880s, the abbot of the temple blasted a new horizontal entrance to the cave which is still in use today. Prior to this, the only entrance was through the skylight at Plong Jaeng, which involved a risky 10 m vertical descent on bamboo ladders. Since then, the cave and temple have been restored, developed with shrines, statues, and Buddha images inside the cave, and nowadays it is a major tourist attraction in the region.

2.2. A Brief History of Cave Fauna Investigation

The cave fauna of Tham Chiang Dao has been of scientific interest for over a century, since the first collection of bats was conducted in 1913 by Thomas Harold Lyle, who was the British consul in Nan. Subsequently, many visits have been made for biological collecting as presented in Table 1.

Table 1. A historical overview of cave fauna investigation and study in Tham Chiang Dao.

Date	Researchers	Institution	Biological Survey	Notes	Reference
January 1913	T. H. Lyle	British consul, Nan	Bats		[16]
25 June 1914	N. Gyldenstolpe	Swedish Zoological Expeditions to Siam	Biological survey	No bats seen in the cave, but there were large deposits of guano	[17]
March–June 1937		Harvard Asiatic Primate Expedition, USA	Bats		[18]
19 January 1958	T. Umesao and K. Yoshikawa	Osaka City University, Japan	General cave fauna collecting		[19]
18 July 1958	B. Degerbøl Hansen	Zoological Museum, University of Copenhagen, Denmark	General cave fauna collecting		[20]
1967	F. Stone and R. Montgomery	Cornell University, USA	General cave fauna collecting		[21]
11 and 19 December 1969	B. A. Harrison and K. Mongkolpanya	SEATO Laboratory, Bangkok	Mosquitos		[22]
1968–1971	C. Boutin	Faculté des Sciences de Phnom Penh	Diptera		[23]
27 December 1972	F. Stone	Bishop Museum, Honolulu, USA	Invertebrates		[24]
May 1974	J. Sedlacek	Bishop Museum, Honolulu, USA	General cave fauna collecting		[25]
15 February 1975	P. Strinati	Switzerland	General cave fauna collecting		[26]
December 1980–January 1981	L. Deharveng and A. Gouze	Université Paul Sabatier, Toulouse, France	General cave fauna collecting	Cave exploration and survey	[11]
1980–1987	M. Kottelat	Laboratoire d'Ichthyologie, Delémont, Switzerland	Fish		[27]
July 1981	F. Stone	Bishop Museum, Honolulu, USA	Invertebrates		[28]
14 and 16 August 1981	F. Stone	Bishop Museum, Honolulu, USA	Invertebrates		[29]
24 December 1983	R. Hemperly	Thailand Karst Hydrologic Project, USA	Bats	Cave exploration and survey	[30]
10 June 1984 and November 1984	P. Beron and S. Andreev	National Museum of Natural History, Bulgaria	General cave fauna collecting		[31]
July 1985	L. Deharveng, P. Leclerc, A. Bedos, J.-P. Besson et al.	Association Pyrénéenne de Spéléologie, France	General cave fauna collecting	Cave exploration and survey	[32]

Table 1. *Cont.*

Date	Researchers	Institution	Biological Survey	Notes	Reference
5 and 31 July 1986	F. Stone	Bishop Museum, Honolulu, USA	General cave fauna collecting		[24,28]
10 January 1989	J. Trautner and K. Geigenmüller	Staatliches Museum für Naturkunde, Stuttgart, Germany	General cave fauna collecting		[33]
6 March 1989	M. Anderson and H. Read	Natural History Museum of Denmark	Spiders		[34]
2007–2010	S. Watiroyram	Nakhon Phanom University, Nakhon Phanom	Copepods		[35]
2010	L. Chintapitasakul and colleagues	National Institute of Animal Health, Bangkok	Bat viruses		[36]
24, 25 and 28 June 2014	P. Jaeger, S. Li, E. Shaw and E. Grall	Senckenberg Museum, Frankfurt am Main, Germany	Spiders		[37]
25 October 2015	Animal Systematics Research Unit	Chulalongkorn University, Bangkok	Molluscs		[38]
10 March 2019	S. Jantarit	Prince of Songkla University, Hat Yai	Collembola		[39]
8–11 January 2023	S. Jantarit, R. Promdam, P. Pitaktunsakul, N. Boonkanpai, B. Noipracha, Y. Tokiri, C. Siripornpibul, W. Jaitrong, T. Jeenthong, K. Thongsri	DMR/Kanchanaburi Rajabhat University	General cave fauna collecting	First field visit	[14]
9–11 June 2023	S. Jantarit, R. Promdam, P. Pitaktunsakul, N. Boonkanpai, B. Noipracha, Y. Tokiri, C. Siripornpibul, W. Jaitrong, T. Jeenthong, K. Thongsri	DMR/Kanchanaburi Rajabhat University	General cave fauna collecting	Second field visit	[14]

2.3. Cave System

Tham Chiang Dao is located in Ban Tham subdistrict, Chiang Dao district, Chiang Mai province in northern Thailand (19.3942° N 098.9277° E). The peak of the Doi Chiang Dao karst mountain has an elevation of 2195 m asl., but the cave is situated at the base of the mountain with the main entrance at 460 m asl. (Figures 1 and 2). This entrance is on the grounds of a Buddhist temple (Wat Tham Chiang Dao) which is built in the Lanna style. Covered steps lead up to the gated entrance from a man-made pond of crystal-clear water, fed by the streams resurging from the cave, which is home to numerous fish. The cave extends sub-horizontally directly into the mountain and has a total length of 5342 m, updated [5,12], which ranks it as the 11th longest cave in Thailand and the 6th longest cave in northern Thailand, [5] and Figure 2. A short distance inside the entrance, the cave splits into two branches which head north and south. Each branch has an active phreatic system, and these hydrological systems are not connected until the resurgence. No water tracing has been done, but the northern branch is thought to be fed by sinks 3.5 km to the north-west (700 m asl.), while the source of the water in the southern branch is unknown. Each branch has a network of seasonally flooded and dry passages at different levels above the phreatic system. The cave is divided into five main passages (Figure 2):

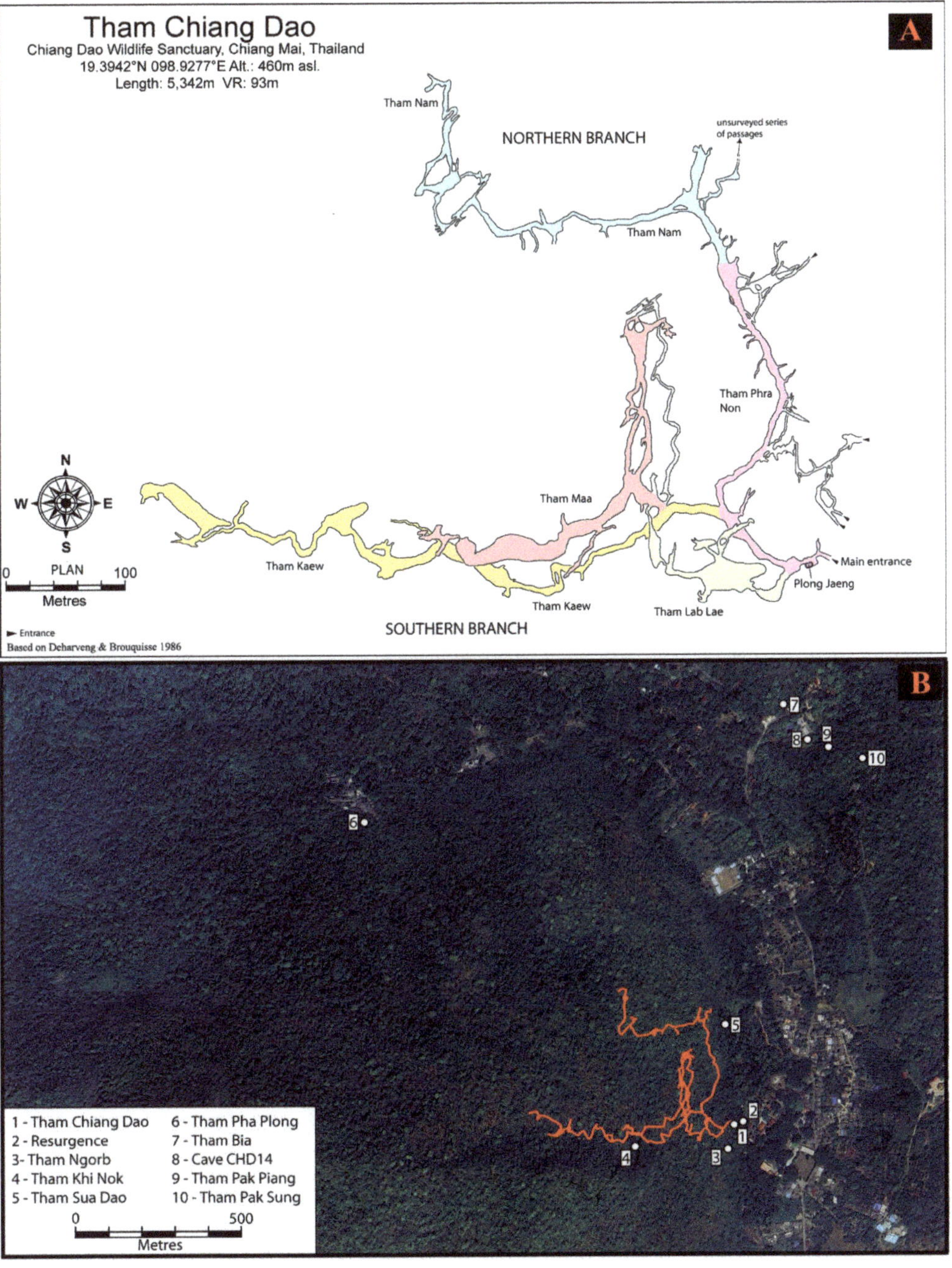

Figure 2. (**A**) Map of Tham Chiang Dao system, modified from Deharveng and Brouquisse (1986); (**B**) Tham Chiang Dao system with nearby cave entrances overlaid on Doi Chiang Dao (from Google Earth Pro).

(1) Tham Phra Non (Sleeping Buddha Cave) in the northern branch is the main tourist cave for self-guided tours with a concrete path, bridges, and electric lighting throughout this horizontal passage. Further into the cave, most of the passage floor is fine sand and the passages are flooded to a depth of 1 m to 2 m during the wet season. This passage is decorated by several natural cave formations as well as many historical statues, shrines, and Buddha images, including a Reclining Buddha built in 1913 which is located at the end of the tourist section. The length of Tham Phra Non is 450 m.

(2) Tham Nam (Water Cave) is the continuation of Tham Phra Non. This passage is without electric lighting and is not developed for tourism. Its length is about 1000 m and it has numerous speleothems throughout. To the north of the main Tham Nam passage is a series of dry passages extending for over 600 m which are infrequently visited as the entrance to this section is an obscure low crawl (these passages are not on the 1985 survey by the APS). Towards the end of Tham Nam are sump pools into the underlying phreatic system. In the wet season, these passages become active and the water backs up to near the start of Tham Phra Non. The floor of Tham Nam is either sand or thick mud and it is home to a variety of cave fauna, both terrestrial and aquatic.

(3) Tham Lab Lae (Secret Cave) and (4) Tham Maa (Horse Cave) are in the southern branch and are a series of dry upper levels branching off from Tham Phra Non near Plong Jaeng, with a total length of 1500 m. These two sections form a longer guided tour, without a path or electric lighting, through passages that are larger and better decorated than Tham Phra Non. Towards the end of Tham Maa, holes in the floor connect with Tham Kaew.

(5) Tham Kaew (Crystal Cave) in the southern branch is at the same level as Tham Phra Non, but it is associated with a separate stream system. This passage has not been developed for tourism and has a length of 900 m. Tham Kaew remains in a more natural condition than the tourist parts of the cave and supports a diversified cave fauna. Similar to Tham Nam, this section of the cave floods seasonally and has thick clay and sand deposits and has sump windows into the underlying phreatic system.

2.4. Checklist and Sampling of Cave Fauna

A checklist of the cave fauna of Tham Chiang Dao has been compiled from the available taxonomic, biological, and speleological literature published until July 2023. The checklist of cave fauna in Table 2 only includes the taxa identified to species. Taxa identified as morphospecies (sp., spp.), referring to a named species (*cf.*) and those of unidentified/undetermined species (i.e., *Gen.* sp. *Gen.* spp.), as well as those which are only identified to a higher taxonomic level, are excluded from the list. However, for the obligate cave species listed in Table 3, the morphospecies, *cf*, and those which are only identified at a higher taxonomic level are counted as troglobionts or stygobionts.

The subterranean fauna that had been reported in the previous studies was re-investigated during January and June 2023 as part of a joint Department of Mineral Resources/Kanchanaburi Rajabhat University biodiversity project with the senior author (SJ) as part of the team. The subterranean fauna (troglobiotic species) was searched for carefully in almost all the passages in both aquatic and terrestrial habitats and were collected by hand, with an entomological aspirator and a net for aquatic fauna, as well as in situ photographed with an Olympus Tough 4 or 6 camera.

3. Results and Discussion

3.1. Diversity of Cave Fauna in Tham Chiang Dao

Overall, a total of 149 taxa have been recognized from Tham Chiang Dao. Most of the collected specimens (88 taxa, 59%) are unstudied or are only identified at a high taxonomic level, while 61 have been identified to species level (Table 2). Tham Chiang Dao is the type locality for 14 species with 13 of the species being endemic to the cave (Table 2). Of these 61 known species, 21 are troglobionts/stygobionts, 23 are troglophiles/stygophiles, and 17 are trogloxenes (Table 2). Among the 149 taxa there are 37 troglobionts/stygobionts

(Table 3), 54 troglophiles/stygophiles, 27 trogloxenes, and 29 species with an unknown ecological category (not listed).

Tham Chiang Dao today has the richest cave fauna in Thailand reported so far. Other caves in the country which have been well-studied include Tham Le Stegodon in Satun province with 126 documented taxa [40], Tham Khao Chang Hai in Trang province with 102 taxa [41,42], Tham Phu Pha Phet and Tham Loko in Phatthalung province with 94 and 79 taxa, respectively [42], and Tham Thalu and Tham U-Rai Thong, Satun Province, with 85 and 66 taxa, respectively [40]. The high value of alpha diversity in Tham Chiang Dao reflects, however, primarily the zoological collecting effort, as the cave has been sampled for a long time and its fauna studied by several specialists (Table 1). These numbers are underestimates, as many mites, spiders, springtails, crustaceans and insects have not been worked up beyond family or genus level and several are expected to be new to science [14,43].

The 61 named species of Tham Chiang Dao (including 21 troglobionts/stygobionts) represent a steep increase from the 47 previously known in December 2020 (including 19 troglobionts/stygobionts) [44]. Despite numerous samplings, covering various kinds of microhabitats, large sections of Tham Chiang Dao remain unexplored (e.g., passages with high levels of carbon dioxide, permanently flooded sections, and vertical passages) and several groups are clearly undersampled (e.g., Copepoda, Insecta). More species, including troglobionts/stygobionts, may therefore be expected to be found in the cave.

In Southeast Asia many caves have been zoologically investigated reasonably thoroughly. In Indonesia, Ngalau Surat, Sumatra, had 74 species (of which 20 were troglo/stygobionts); Batu Lubang, Halmahera, had 72 species (of which 16 were troglo/stygobionts [43]; and Towakkalak and Saripa System, Sulawesi, had 93 species (of which 28 were troglo/stygobionts) [45]. The Batu Caves of Malaysia is the best studied cave system in Southeast Asia with 314 taxa with 183 identified to species (type locality for 63 species) [46]. The high species richness of the Batu Caves is the result of intensive samplings and studies since the end of the 19th century and almost all groups of animals have been diagnosed at a species level. However, only 50 troglo/stygobionts are known from this cave (accounting for only 14.6% in the total fauna of a cave), a relatively low number compared to the caves cited above (Batu Lubang = 22%, Tham Chiang Dao = 25%, Ngalau Surat = 27% and Towakkalak = 30%), which indicates an artifact of collecting bias, in that more common surface species have been identified from the comparatively smaller and shallower Batu Caves system, and further suggesting that sampling effort alone may be a poor predictor of cave-obligate species richness even in a climatically homogeneous region.

Table 2. List of known species from Tham Chiang Dao, Chiang Mai, Thailand; TB: troglobiont, TP: troglophile, TX: trogloxene, SB: stygobiont, SP: stygophile, TL: type locality; *: type locality and only recorded locality; SMF: Senckenberg Museum, Frankfurt am Main, Germany.

Phylum	Class	Order	Family	No.	Species	Reference(s)	Status
Mollusca	Gastropoda	Architaenioglossa	Pupinidae	1	*Pupina artata* Benson, 1856	[38]	TP
		Stylommatophora	Achatinidae	2	*Allopeas gracile* (Hutton, 1834)	[14]	TP
Annelida	Clitellata	Haplotaxida	Haplotaxidae	3	*Heterochaetella glandularis* (Yamaguchi, 1953)	[47,48]	SB
Arthropoda	Arachnida	Opiliones	Assamiidae	4	*Bandona palpalis* Roewer, 1927	[14,29]	TP
				5	*Neopygoplus siamensis* Suzuki, 1985	[20]	TP
		Pseudoscorpiones	Chernetidae	6	*Megachernes trautneri* Schawaller, 1994 *	[33]	TP, TL
		Palpigradi	Eukoeneniidae	7	*Eukoenenia thais* Condé, 1988 *	[49]	TB, TL

Table 2. *Cont.*

Phylum	Class	Order	Family	No.	Species	Reference(s)	Status
		Araneae	Clubionidae	8	*Systaria lannops* Jäger, 2018	[37]	TB
			Psilodercidae	9	*Althepus tibiatus* Deeleman-Reinhold, 1985 *	[24]	TB, TL
			Ochyroceratidae	10	*Theotima minutissima* (Petrunkevitch, 1929)	[24]	TP
			Sparassidae	11	*Heteropoda venatoria* Linnaeus, 1767	[14]	TP
				12	*Sinopoda ruam* Grall & Jäger, 2020 *	[34]	TB, TL
			Nesticidae	13	*Nesticella beccus* Grall & Jäger, 2016	[34]	TP
				14	*Nesticella mogera* (Yaginuma, 1972)	[43]	TP
			Theridiidae	15	*Nesticodes rufipes* (Lucas, 1846)	[43]	TP
			Gnaphosidae	16	*Micythus anopsis* Deeleman-Reinhold, 2001 *	[50]	TB, TL
			Liocranidae	17	*Jacaena schwendingeri* (Deeleman-Reinhold, 2001)	Unpublished record. Specimen in SMF	TX
	Chilopoda	Scolopendromorpha	Scolopendridae	18	*Scolopendra dehaani* Brandt, 1840	[51]	TP
	Diplopoda	Polydesmida	Paradoxosomatidae	19	*Tylopus perarmatus* Hoffman, 1973	[14,52]	TX
			Haplodesmidae	20	*Eutrichodesmus gremialis* (Hoffman, 1982) *	[14,26]	TB, TL
	Maxilliopoda	Cyclopoida	Cyclopidae	21	*Tropocyclops prasinus* (Fischer, 1860)	[43]	SP
		Harpacticoida	Canthocamptidae	22	*Elaphoidella namnaoensis* Brancelj, Watiroyram & Sanoamuang, 2010	[53]	SB
				23	*Epactophanes richardi* Mrázek, 1893	[53]	SP
	Malacostraca	Bathynellacea	Parabathynellidae	24	*Siambathynella janineana* Camacho & Leclerc, 2022 *	[54]	SB
		Isopoda	Oniscidae	25	*Exalloniscus beroni* Taiti & Ferrara, 1988 *	[14,31]	TB, TL
		Decapoda	Palaemonidae	26	*Macrobrachium yui* Holthuis, 1950	[14]	SP
	Collembola	Entomobryomorpha	Isotomidae	27	*Folsomides parvulus* Stach, 1922	[14,43]	TB
				28	*Folsomina onychiurina* Denis, 1931	[43]	TP
			Paronellidae	29	*Salina pulchella* Goto, 1955	[19,55]	TX
				30	*Troglopedetes fredstonei* Deharveng 1988 *	[14,56]	TB, TL
				31	*Troglopedetes leclerci* Deharveng, 1990 *	[28]	TB, TL
			Entomobryoidae	32	*Pseudosinella chiangdaoensis* Deharveng, 1990 *	[14,28,55]	TB, TL
				33	*Coecobrya guanophila* Deharveng, 1990 *	[28]	TB, TL
				34	*Coecobrya similis* Deharveng, 1990	[14,28,55]	TB
		Poduromorpha	Hypogastruridae	35	*Acherontiella colotlipana* Palacios-Vargas & Thibaud, 1985	[14,57]	TB
		Symphypleona	Arrhopalitidae	36	*Arrhopalites anulifer* Nayrolles, 1990	[58]	TP

Table 2. *Cont.*

Phylum	Class	Order	Family	No.	Species	Reference(s)	Status
	Insecta	Coleoptera	Carabidae	37	*Arrhopalites chiangdaoensis* Nayrolles, 1990 *	[15,58]	TB, TL
				38	*Itamus castaneus* Schmidt-Goebel, 1846	[14]	TP
			Staphylinidae	39	*Bironium troglophilum* Löbl, 1990	[25]	TB
		Lepidoptera	Tineidae	40	*Crypsithyris spelaea* Meyrick, 1908	[43]	TP
				41	*Tinea antricola* Meyrick, 1924	[14,43]	TB
				42	*Wegneria cerodelta* (Meyrick, 1911)	[43]	TP
		Pscoptera	Liposcelididae	43	*Liposcelis bostrychophilus* Badonnel, 1931	[14,43]	TP
				44	*Liposcelis entomophilus* Enderlein, 1907	[43]	TP
			Psyllipsocidae	45	*Psocathropos lachlani* Ribaga, 1899	[43]	TP
		Diptera	Culicidae	46	*Culex harrisoni* Sirivanakorn, 1977 *	[22]	TB, TL
		Hymenoptera	Formicidae	47	*Carebara diversa* (Jerdon, 1851)	[14]	TX
				48	*Anoplolepis gracilipes* Smith, 1857	[14]	TX
Chordata	Actinopterygii	Cypriniformes	Cyprinidae	49	*Neolissochilus stracheyi* (Day, 1871)	[14]	TX
	Reptilia	Squamata	Colubridae	50	*Elaphe taeniura* (Cope 1861)	[14]	TP
	Mammalia	Chiroptera	Soricidae	51	*Suncus murinus* (Linnaeus, 1766)	[59]	TX
			Hipposideridae	52	*Aselliseus stoliczkanus* Dobson, 1871	[14,30]	TX
				53	*Hipposideros armiger* (Hodgson, 1835)	[14,16,59]	TX
				54	*Hipposideros diadema* (Geoffroy, 1813)	[60]	TX
				55	*Hipposideros lylei* Thomas, 1913	[14,16,59]	TX, TL
			Pteropodidae	56	*Eonycteris spelaea* (Dobson, 1871)	[60]	TX
				57	*Macroglossus sobrinus* Andersen, 1911	[60]	TX
				58	*Rousettus leschenaulti* (Desmarest, 1820)	[60]	TX
			Rhinolophidae	59	*Rhinolophus pusillus lakkhanae* Yoshiyuki, 1990	[14,59]	TX
			Vespertilionidae	60	*Ia io* Thomas 1902	[18]	TX
				61	*Pipistrellus paterculus* (Thomas, 1915)	[59]	TX

Note: *Nesticella mogera* was originally described as *Howaia mogera* and *Psocathropos lachlani* was originally described as *Psocathropos microps*.

3.2. The Subterranean Fauna of Tham Chiang Dao

The obligate cave fauna of Tham Chiang Dao belongs to 3 phyla, 8 classes, 23 orders, 33 families, 36 genera, and 37 species, of which 33 are troglobionts and 4 are stygobionts (Table 3). The best represented class is Arachnida (12 species), followed by Insecta (10 species) and Collembola (6 species) (Table 3). The Araneae are the most diversified order with five species, followed by Entomobryomorpha with four species. Troglobiotic species are much more numerous than stygobiotic species, as in other Thai caves studied so far. In contrast, temperate caves often have more stygobionts than troglobionts [43,61–67]. This difference is clearly linked to different sampling efforts in terrestrial versus aquatic habitats and the real pattern remains unknown for tropical caves.

3.2.1. Terrestrial Fauna

(1)   Gastropoda

A single troglobiotic microsnail *Acmella* sp. has been recently discovered in Tham Chiang Dao [14]. It was mainly found in the cave hygropetric where thin biofilms of bacteria and fungi are probably the main food source for this minute snail. Specimens were found in Tham Lab Lae, Tham Maa, and Tham Nam (Figure 3).

Figure 3. Gastropoda. A troglobiotic microsnail *Acmella* sp., photo by R. Promdam with permission.

(2)   Acari

A white, long-legged mite has been collected in oligotrophic habitats (on the surface of standing rock pools and on the mud floor) which is probably a Leeuwenhoekiidae, similar to those encountered in many caves of Southeast Asia (Figure 4).

(3)   Araneae

Five troglobiotic spiders from five different families have been reported from this cave: *Systaria lannops*, *Micythus anopsis*, *Spermophora* sp., *Althepus tibiatus*, and *Sinopoda ruam*. *Systaria lannops* were collected in the dark zone by P. Jäger, S. Li, and E. Grall in June 2014 and are also known from two other caves in Chiang Mai: Tham Tab Tao (the type locality) 35 km to the NE and Tham Pha Daeng 25 km ENE of Tham Chiang Dao [37]. *Micythus anopsis* was collected in July 1985 by L. Deharveng [50]. Deharveng and Bedos [43] listed a blind *Scotophaeus* sp. (Gnaphosidae) in their table of terrestrial cave fauna, which probably refers to this specimen. A blind unidentified *Spermophora* species was collected in Tham Chiang Dao by the APS (Deeleman-Reinhold identification). *Althepus tibiatus* was collected from the dark zone by F.D. Stone in December 1972 [24], with further specimens collected in July 1985 by L. Deharveng and in July 1986 by F.D. Stone. The only other known locality of this species is Tham Pha Daeng 2 which is 25 km to the ENE [24]. *Sinopoda ruam* is only known from Tham Chiang Dao. It was collected by M. Anderson and H. Read in March 1989 and by P. Jäger, E. Shaw, S. Li, and E. Grall in June 2014 [34]. *Heteropoda* sp. and *Sinopoda ruam* distributions in the cave narrowly overlap, which is rare for Sparassidae spiders [68].

Figure 4. Acari. A troglobiotic Leeuwenhoekiidae, photos by S. Jantarit.

(4)   Opiliones

The single troglobiotic species of Opiliones recorded from the cave is an unidentified microphthalmic and troglomorphic *Paratakaoia* sp. [43]. The troglophilic species *Bandona palpalis* Roewer, 1927 is abundant in Tham Chiang Dao [29].

(5)   Palpigradi

Two micro-whipscorpions have been reported from Tham Chiang Dao: *Eukoenenia thais* and *Eukoenenia* cf. *lyrifer*. *Eukoenenia thais* is a troglobiotic species that was collected by L. Deharveng and A. Gouze in December 1980 and in July 1985 in Tham Maa [49,69]. *Eukoenenia* cf. *lyrifer* was collected by P. Leclerc as an adult female, on the wall of Tham Kaew in July 1985. Despite the proximity of the place of collection, it is not possible to relate this specimen to *E. thais*, which is larger and exhibits significant differences in morphology. As for *E. lyrifer* from Tham Ku Kaeo in Chiang Rai province [69], *Eukoenenia* cf. *lyrifer* seems to be intermediate between the *euedaphic E. siamensis* and the troglomorphic *E. thais*.

(6)   Pseudoscorpion

A blind troglobiotic species of *Tyrannochthonius* is recorded by Deharveng and Bedos [43] and DMR [14]. In addition, two troglophilic pseudoscorpion species are reported from Tham Chiang Dao (Figure 5). *Megachernes trautneri* Schawaller, 1994, was collected by J. Trautner and K. Geigenmüller in January 1989 and has also been found in surface habitats on other mountains in Chiang Mai province [48]. *Megachernes* cf. *grandis* (Beier, 1930) was listed by Deharveng and Bedos [43] as an unidentified guanophilic pseudoscorpion that was referred to *M. grandis*, but it is probably *M. trautneri* or another species of the genus as several species of *Megachernes* are sometimes present in caves in Afghanistan, China, Japan, and Turkmenistan, some being associated with guano [70].

Figure 5. Pseudoscorpiones. A troglobiotic *Tyrannochthonius* sp., photos by S. Jantarit.

(7)   Schizomida

A single species of Hubbardiidae, presumably troglobiotic, is recorded from guano by Deharveng and Bedos and DMR [14,43] under the name *Schizomus* sp.

(8)   Diplopoda

Two species of troglobiotic millipedes are found in Tham Chiang Dao. *Eutrichodesmus gremialis* is a small, blind, pale species that was found mainly in Tham Nam and Tham Keaw on the cave walls and cave floor in oligotrophic habitats. Another micropolydesmoid millipede is an undescribed species of the family Opisotretidae which is rarer than *E. gremialis* and sometimes co-occurs with it (Figure 6). Only two species of the species-rich genus *Eutrichodesmus* are described from Thai caves and cave Opisotretidae were unknown from Thailand [44].

Figure 6. Diplopoda. (**Left**) *Eutrichodesmus gremialis* (Hoffman, 1982); undescribed species of Opisotretidae sp. (**Right**), photos by S. Jantarit.

(9) Isopoda

Three troglobiotic species of isopods have been found in Tham Chiang Dao: *Exalloniscus beroni*, *Cubaris* sp., and a blind Philosciidae (Figure 7). The first species is found throughout the caves, except in Tham Phra Non. It is a colorless and blind species that was collected mainly on the cave mud floor and sometimes in scattered bat feces. *Cubaris* sp. is more abundant with large colonies that gather on the cave walls and floor throughout the cave except in Tham Phra Non. Both genera have cave species in several regions of Southeast Asia. The Philosciidae is blind, but its ecological status is uncertain, as another blind Philosciidae has been found in the soil on Doi Chiang Dao.

Figure 7. Isopoda: Oniscidea. (**Left**) Philosciidae sp.; *Cubaris* sp. (**Right**), photos by S. Jantarit.

(10) Collembola

Springtails are often numerically dominant in Thai caves. Tham Chiang Dao is amongst the richest caves in the tropics for its collembolan fauna with 17 species, including six troglobionts [43]. This is the highest number of species found in a Thai cave, as the highest richness in other caves of the country does not exceed 10 species per cave, with an average of 3–5 species per cave [71]. Collembola are well represented in tropical caves across Southeast Asia. For example, 14 species are listed from Batu Caves in Malaysia, including 2–3 troglobionts [46], 12 species from Batu Lubang in Halmahera including 5 troglobionts, 22 from Ngalau Surat in Sumatra including 5 troglobionts [43], and 24 from the Towakkalak System in Sulawesi including 6 troglobionts [45]. The Tham Chiang Dao springtail fauna is, therefore, in line with other Southeast Asian caves. It is also the type locality for five species which are endemic to the cave. *Coecobrya guanophila* is a white, blind, and guanobiotic springtail only known from Tham Chiang Dao. There are several records of this endemic species from the Tham Kaew part of the cave, where it is abundant in humid guano deposits, collected by P. Leclerc, F. D. Stone, and L. Deharveng in December 1980, July 1981, and July 1985 [28,72]. *Coecobrya* has many cave species in Southeast Asia. *Pseudosinella chiangdaoensis* is a white, eyeless, slightly troglomorphic springtail that is only known from Tham Chiang Dao. Specimens were caught in July 1985 by P. Leclerc and L. Deharveng in Tham Maa [28]. The genus is very diversified in temperate caves, but rare in tropical caves, and *P. chiangdaoensis* is the only cave *Pseudosinella* of continental Southeast Asia. *Troglopedetes fredstonei* is a troglomorphic species with no eyes, no pigment, long appendages, large body size, and slender claws. It was collected by the APS in July 1988 and by F.D. Stone, and was found on humid mud banks with scattered bat guano in the lower levels of Tham Kaew and Tham Nam. It was not found in the upper level Tham Maa or outside the cave [56]. *Troglopedetes leclerci* was collected in December 1980 and July 1985 by L. Deharveng and P. Leclerc on the walls of Tham Kaew and in Tham Maa [28]. *Acherontiella colotlipana* (Palacios-Vargas and Thibaud, 1985) is a troglobiotic and guanobiotic springtail that was originally found in guano in a Mexican cave. As the species seems to be well characterized morphologically, we provisionally assume that this disjunct distribution reflects sampling gaps in the cave guano habitats of tropical caves. Such a wide distribution among cave guano species is known in several other species of Collembola, such as *Xenylla yucatana*. Four specimens were collected in Tham Chiang Dao by the APS from guano and soil [57]. *Arrhopalites chiangdaoensis* is a pale, troglobiotic collembola that is only known from Tham Chiang Dao. It was collected by L. Deharveng in December 1980 and July 1985 in Tham Nam/Tham Phra Non, Tham Maa, and Tham Kaew. This species was shown to be polyphagous after the dissection of its gut, which was found to contain clay or mycelia mixed with clay, and sometimes fragments of collembola or pieces of scale, probably from Tineoidea (Lepidoptera) which are abundant in the cave [58].

(11) Diplura

A single species of unidentified Japygidae was reported by Deharveng and Bedos [43] as slightly troglomorphic. Although no Japygidae have been described from Thai caves, they can be found in caves throughout the country.

(12) Blattodea

Two troglomorphic Nocticolidae are present in Tham Chiang Dao: *Helmablatta* sp. and *Spelaeoblatta* sp. Nocticolidae are widespread in Southeast Asian caves, but few species have been described. The genus *Helmablatta*, originally characterized by extremely modified upstanding tergal glands [73], was only known by a single species from a Vietnamese cave. The presence of a species of *Helmablatta* in Tham Chiang Dao is an interesting discovery. Nocticolidae are very common on muddy cave floors with scattered guano in Tham Nam and Tham Keaw. Another guanobiotic cockroach, *Blattella* cf. *cavernicola* (Shelford, 1907), is rather common in the dry upper parts of Tham Lab Lae and Tham Maa, especially on guano deposits and under the mats near the statues and Buddha images (Figure 8).

Figure 8. Blattodea. Troglomorphic cave cockroaches: (**above left**) *Spelaeoblatta* sp. and *Helmablatta* sp. (**above right**); photos by S. Jantarit and guanobiotic cockroach *Blattella* cf. *cavernicola* (**below**), photos by T. Jeenthong with permission.

(13) Orthoptera

At least one species of cricket found in Tham Chiang Dao is troglobiotic, the ant cricket *Myrmecophilus* sp. which is reported for the first time in a Thai cave [14]. This ant cricket is rare and found on the mud floor in Tham Nam. Two additional species of cave cricket are also recognized from this cave, *Rhaphidophora* sp. and *Paradiestrammena* sp. (Figure 9). They are abundant in almost all the cave passages, except in Tham Phra Non which is a main tourist passage and has electric lighting. Cave crickets are often troglophiles which leave the cave at night for feeding. We here omit them in the list of cave-obligate species, though further studies on their ecology may change their status.

Figure 9. Orthoptera. (**A**) A troglomorphic ant-cricket *Myrmecophilus* sp. (photo by T. Jeenthong with permission); two troglophilic crickets (**B**) *Rhaphidophora* sp. and *Paradiestrammena* sp. (**C**), photos by S. Jantarit.

(14) Hymenoptera

An interesting species of ant was found in the cave throughout the undisturbed passages, especially in Tham Lab Lae, Tham Maa, and Tham Keaw. It is a *Brachyponera* sp. which exhibits a reduction of eyes and unusually long appendages for the genus (Figure 10). This ant species is currently being formally described. Its colonies are established in rock cracks or muddy soil and sometimes under stones. This ant is rather common. It appears to be omnivorous and can hunt small invertebrates found in cave environments. If confirmed, it would be the second cave-ant of Southeast Asia, after *Leptogenys khammouanensis* Roncin & Deharveng, 2003 from a cave in Laos.

Figure 10. (**Left**) undescribed species of staphylinid beetle (Oxytelinae); a possible troglobiotic ant species *Brachyponera* sp. (**right**), photos by S. Jantarit.

(15) Coleoptera

At least two subterranean beetles have been reported from Tham Chiang Dao: *Bironium troglophilum* Löbl, 1990 and an undescribed species of staphylinid beetle (Oxytelinae) (Figure 10). *Bironium troglophilum* was collected by J. Sedlacek from Tham Chiang Dao. The type locality is Tham Hued in Mae Hong Son and the beetle is also known from another small cave in Mae Hong Son. Although it has only been recorded from caves, *B. troglophilum* has fully developed wings and does not exhibit any morphological adaptation [25]. Löbl does not give a date for the specimen collected in Tham Chiang Dao, but there is circumstantial evidence that this was in May 1974. The undescribed species of the staphylinid beetle (Oxytelinae) was found on the passage wall in Tham Lab Lae by the DMR in 2023.

(16) Diptera

Non-glowing larvae of a fungus gnat, *Chetoneura* sp., have been found in the cave. This predatory larva builds sticky threads to catch flying insects by hanging them down from the ceilings of the cave passages (Figure 11). The species is rather common throughout the cave, especially in wet habitats and/or near water pools. Its adult stage is still unknown, but there is a report of an epigean species, *Chetoneura oligoradiata*, from the Doi Chiang Dao nature trail [74]. We here place this fungus gnat as a possible troglobiotic species.

The troglobiotic mosquito *Culex harrisoni* Sirivanakorn, 1977 was reported from Tham Chiang Dao (Table 1), breeding in two rock pools of 38–45 cm in diameter and 8.5–10.0 cm in depth, located 300–400 m inside the cave. Most adult specimens came from rearing the larvae and only a few were collected on the wall of the cave near the breeding site. The adult biology is unknown [22]. This mosquito has also been found in Tham Borichinda in the Doi Inthanon National Park, Chiang Mai.

(17) Lepidoptera

Tineid moths are abundant on guano deposits in Tham Chiang Dao with three species identified: *Crypsithyris spelaea* Meyrick, 1908, *Tinea antricola*, and *Wegneria cerodelta* (Meyrick, 1911). Only *T. antricola* (Figure 11) is considered a troglobiont, the two other species being troglophilic. *Tinea antricola* was collected by the APS [43]. The larvae feed on guano and the species is common in caves in Southern Asia.

Figure 11. Non-glowing sticky worm, *Chetoneura* sp. (**A**) its sticky threads and (**B**) its larva, photos by R. Promdam with permission.

3.2.2. Aquatic Fauna

Only five species are stygobiotic, though many aquatic taxa were sampled and described from this cave.

(1) Nematoda

A species of the genus *Tobrilus* sp. was collected from the pool at the end of Tham Nam. Its ecological assignation is not possible.

(2) Annelida

The stygobiotic species, *Heterochaetella glandularis* (Yamaguchi, 1953) was reported in the pools at the end of Tham Nam and Tham Kaew [47,48]. An unidentified Enchytraeidae from the same section of the cave might be stygobiotic as well [47].

(3) Harpacticoida

A single stygobiotic copepod species *Elaphoidella namnaoensis* was found in Tham Chiang Dao in 2007–2011 by S. Watiroyram as part of a study into the cave Harpacticoida of northern Thailand. The samples were taken from individual pools on the floor of the caves, which were filled exclusively by percolation water. *E. namnaoensis* is rather common in the caves of northern and central Thailand, in both the unsaturated and saturated zones [53]. In addition, three stygophilic copepod species are also reported from water pools in this cave: *Tropocyclops prasinus* (Fischer, 1860), *Elaphoidella* cf. *grandidieri* (Guerne & Richard, 1893) [43], and *Epactophanes richardi* Mrázek, 1893 [35].

(4) Bathynellacea

The micro-stygobiotic species *Siambathynella janineana* was collected by the APS in July 1985 from muddy pools in Tham Maa, where hundreds of specimens were found, and one specimen from a sump in Tham Nam. The species was also found outside the cave in the resurgence pool and in the hyporheic of the stream at −40 cm, about 25 m downstream of the resurgence [47,54].

The number of stygobiotic species recorded from Tham Chiang Dao is rather small. Ostracoda and Cyclopoidea have been collected, but remain unidentified [47]. Stygobiotic amphipods, decapods, and fish are known from tropical subterranean habitats, but have not been found in Tham Chiang Dao. Blind fish and shrimps have long been mentioned by local people to exist in Tham Chiang Dao, but attempts to find them have failed so far [14,27]. The shrimp *Macrobrachium yui* Holthuis, 1950, which is present in the permanent pools of Tham Nam, does not show adaptations to cave life [14]. Several specimens of the Cyprinidae fish *Neolissochilus stracheyi* (Day, 1871) were observed, but the species does not show any sign of cave adaption and is considered as a stygoxene.

3.2.3. Other Fauna

Surprisingly, the long-legged centipede (*Thereuopoda longicornis* (Fabricius, 1793)) and bent-toed geckos (*Cyrtodactylus* sp.) are not reported even though Tham Chiang Dao has long been zoologically investigated. These taxa are very common and widespread in the caves of Thailand [44]. No amphibians nor birds have been reported from Tham Chiang Dao, while only a single species of snake, the common and widespread cave racer *Elaphe taeniura*, has been found recently [14]. There are also no reports of rodents, especially *Rattus tanezumi* Temminck, 1844 and *Leopoldamys nielli* (Marshall, 1976), which are common visitors in Thai caves. However, footprints were seen on the floor of many passages suggesting that rodents may visit the cave.

Bats are common in Tham Chiang Dao, which is the type locality of *Hipposideros lylei*. Tham Chiang Dao is among the best caves in the region for bats, supporting large colonies and at least 10 species of bats (Table 1). All of them roost in the habitats where there is less impact from tourist visits or in the chambers where electric lights are absent. Many colonies exist even in the deep parts of the cave, near the end of the passages (>500 m from the entrance), suggesting that there are several small openings through which bats can enter and leave the cave.

Table 3. List of obligate cave species present in Tham Chiang Dao, Chiang Mai, Thailand.

#	TB/SB	Species	Taxonomic Classification	Notes	Reference(s)
1	SB	*Heterochaetella glandularis* (Yamaguchi, 1953)	Clitellata: Haplotaxida: Haplotaxidae	(TM)	[47]
2	SB?	Undetermined sp.	Clitellata: Enchytraeida: Enchytraeidae		[47]
3	TB	*Acmella* sp.	Gastropoda: Caenogastropoda: Assimineidae	TM?	[14]
4	TB	Undetermined sp.	Arachnida: Acari: Leeuwenhoekiidae (?)	TM	[14,43]
5	TB	*Systaria lannops* Jäger, 2018	Arachnida: Araneae: Clubionidae		[37]
6	TB	*Micythus anopsis* Deeleman-Reinhold, 2001	Arachnida: Araneae: Gnaphosidae	* TM	[50]
7	TB	*Spermophora* sp.	Arachnida: Araneae: Pholcidae	TM	[43]
8	TB	*Althepus tibiatus* Deeleman-Reinhold, 1985	Arachnida: Araneae: Psilodercidae	TL	[24,75]
9	TB	*Sinopoda ruam* Grall & Jäger, 2020	Arachnida: Araneae: Sparassidae	*	[34]
10	TB	*Paratakaoia* sp.	Arachnida: Opiliones: Epedanidae	TM	[43]
11	TB	*Eukoenenia thais* Condé, 1988	Arachnida: Palpigradi: Eukoeneniidae	* TM	[41,69]
12	TB	*Eukoenenia* sp. (*E*. cf. *lyrifer* Condé, 1992)	Arachnida: Palpigradi: Eukoeneniidae		[69]
13	TB	*Tyrannochthonius* sp.	Arachnida: Pseudoscorpiones: Chthoniidae	(TM)	[14,43]

Table 3. *Cont.*

#	TB/SB	Species	Taxonomic Classification	Notes	Reference(s)
14	TB?	Undetermined sp.	Arachnida: Schizomida: Hubbardiidae	G	[43]
15	TB	*Eutrichodesmus gremialis* Hoffman, 1982	Diplopoda: Polydesmida: Haplodesmidae	*	[26,76]
16	TB	Undetermined sp.	Diplopoda: Polydesmida: Opisotretidae		[14]
17	SB	*Elaphoidella namnaoensis* Brancelj, Watiroyram & Sanoamuang, 2010	Maxillopoda: Harpacticoida: Canthocamptidae		[47,53]
18	SB	*Siambathynella janineana* Camacho & Leclerc, 2022	Malacostraca: Bathynellacea: Parabathynellidae	*	[47,54]
19	TB	*Cubaris* sp.	Malacostraca: Isopoda: Armadillidae	(TM) G	[14,43]
20	TB	*Exalloniscus beroni* Taiti & Ferrara, 1988	Malacostraca: Isopoda: Oniscidae	* (TM)	[31]
21	TB?	Undetermined sp.	Malacostraca: Isopoda: Philosciidae	(TM)	[43]
22	TB	*Coecobrya guanophila* Deharveng, 1990	Collembola: Entomobryomorpha: Entomobryidae	* G	[28]
23	TB	*Pseudosinella chiangdaoensis* Deharveng, 1990	Collembola: Entomobryomorpha: Entomobryidae	* (TM)	[28]
24	TB	*Troglopedetes fredstonei* Deharveng 1988	Collembola: Entomobryomorpha: Paronellidae	* TM	[56]
25	TB	*Troglopedetes leclerci* Deharveng, 1990	Collembola: Entomobryomorpha: Paronellidae	* G	[28]
26	TB	*Acherontiella colotlipana* Palacios-Vargas & Thibaud, 1985	Collembola: Poduromorpha: Hypogastruridae	G	[57]
27	TB	*Arrhopalites chiangdaoensis* Nayrolles, 1990	Collembola: Symphypleona: Arrhopalitidae	*	[58]
28	TB?	Undetermined sp.	Insecta: Diplura: Japygidae	(TM)	[43]
29	TB	*Helmablatta* sp.	Insecta: Blattodea: Nocticolidae	TM	[14]
30	TB	*Spelaeoblatta* sp.	Insecta: Blattodea: Nocticolidae	TM	[14]
31	TB	*Myrmecophilus* sp.	Insecta: Orthoptera: Myrmecophilidae	TM	[14]
32	TB?	*Brachyponera* sp.	Insecta: Hymenoptera: Formicidae		[14]
33	TB	*Bironium troglophilum* Löbl, 1990	Insecta: Coleoptera: Scaphidiidae		[25]
34	TB?	Undetermined sp.	Insecta: Coleoptera: Staphylinidae: Oxytelinae	(TM)	[14,43]
35	TB	*Tinea antricola* Meyrick, 1924	Insecta: Lepidoptera: Tineidae	G	[43]
36	TB?	*Culex harrisoni* Sirivanakorn, 1977	Insecta: Diptera: Culicidae	TL	[22]
37	TB?	*Chetoneura* sp.	Insecta: Diptera: Keroplatidae		[14,23]

TB: troglobiont; TB?: probable troglobiont; SB: stygobionts; SB?: probable stygobiont; TL: type locality; *: type locality and only recorded locality; TM: troglomorphic; (TM): slightly troglomorphic; G: guanobiont or guanophile.

4. Cave Management and Conservation

Tham Chiang Dao is situated in a protected area under the Chiang Dao Wildlife Sanctuary, managed by the DNP, where all the fauna is protected by laws and regulations. In practice, the entrance to the cave is located in a Buddhist monastery and it is a very popular tourist attraction which is managed by a local cave management committee. There are two tours: (1) self-guided through electrically lit horizontal passages (Tham Phra Non) and (2) a longer guided tour through unlit passages with the guide using a kerosene storm lantern (Tham Lab Lae and Tham Maa).

Tham Phra Non is the main religious tourism attraction and contains lots of shrines, statues, images, and other sights of interest. Permanent infrastructure such as concrete paths, bridges, CCTV, and a 4G mobile telephone network has been built. Electric lights are all along the tourist cave passage for illumination, decoration, and the safety and comfort of visitors. The passage has been illuminated for many years and today the electric lights are switched on for at least 8 to 9 consecutive hours every day, which directly stimulates the growth of lampenflora, especially algae, mosses, and ferns ([77,78] and Figure 12A–C).

The proliferation of lampenflora has considerable impacts on cave formations and the cave environment as it creates habitats for various external opportunistic species that may compete with or prey on the original obligate cave species [79], though hard data are still very scarce. Lampenflora in Tham Chiang Dao supports the colonization of invasive species such as the yellow crazy ant *Anoplolepis gracilipes*. This ant species is one of the worst invasive alien species in the world and is today widespread in the tropics and subtropics. It can affect the population dynamics of obligate subterranean species, being rather aggressive and having been reported to prey on and attack mollusks, arachnids, myriapods, isopods, insects, and earthworms [80]. The species is, however, limited to the most disturbed areas or entrance zone in caves and preserving passages in their natural state should largely limit its impact. In any case, it is highly recommended that the lampenflora in Tham Chiang Dao is controlled or cleaned by non-chemical agents, that lights which do not heat the cave and with a low emission in the wavelengths that are not absorbed for growth by the lampenflora are installed, and that lights are switched off when visitors are absent by using automatic light sensors.

Figure 12. The proliferation of lampenflora (**A–C**) and little towers by piling up stones in Tham Chiang Dao (**D**), photos by P. Chananin with permission.

Tham Lab Lae and Tham Maa are frequently visited by tourists as these two interconnected passages are more adventurous with various kinds of cave formations through unlit passages. Local guides prefer to follow the tradition of using a kerosene storm lantern to illuminate the cave. This has caused serious problems not only to the cave ecosystem and its biodiversity, but also to the health of the guides and visitors. It has long been known that using kerosene is smelly and irritating to the eyes, skin, and respiratory system [81]. When used for lighting, the kerosene lanterns emit toxic and carcinogenic gases, such as carbon monoxide, nitric oxides, and sulfur dioxide, and fine particulates [82]. It has been shown [83] that these lamps emit significant amounts of black carbon, 20 times more than previously thought, which directly affects the beauty of the cave formations, prevents the accumulation of calcite, and contributes to microclimate pollution. At least 70 people are working part-time or full-time as guides in the cave, mostly women. Replacing kerosene lamps with LED lamps is, therefore, recommended not only for the environment, but also for the health and welfare of the local guides and tourists.

In Tham Phra Non it has become common in the last 10 years for tourists to construct, for good luck, little towers by piling up stones (Figure 12D). Aside from creating unsightly artificial eyesores, this activity also poses a threat to the cave fauna as moving the stones disturbs their habitat. Tourists should be advised not to construct these piles and existing towers should be removed so that future visitors are not inspired to make their own.

The carbon dioxide in Tham Chiang Dao was measured in the wet season in July 1985 [84] and June 2023 [14] and in the dry season in January 2023 [14] (Figure 13). In the wet season, CO_2 reached the highest concentration (2.9%) at the end of the northern branch of Tham Nam near the water, a high level (1.3–2.2%) in Tham Kaew, and had the lowest concentration (0.1–0.5%) near the entrance. In the dry season, CO_2 levels were much lower in all passages, with the maximum level at the western end of Tham Kaew (0.46%). At the beginning of the wet season (June), the minimal levels of CO_2 were higher than in the dry season (January) and lower than later in the wet season (July) [14] (Figure 13). It is noteworthy that the cave sections which had the highest CO_2 level in the wet season seemed to be richer in troglobionts, in support of Howarth and Stone's observations of a positive impact of CO_2 on biodiversity in an Australian cave [85]. These parts of the caves should, therefore, be closed to tourist visits in order to keep habitats in their original state, aside from the fact that high peaks of CO_2 in the wet season may be uncomfortable or dangerous for visitors.

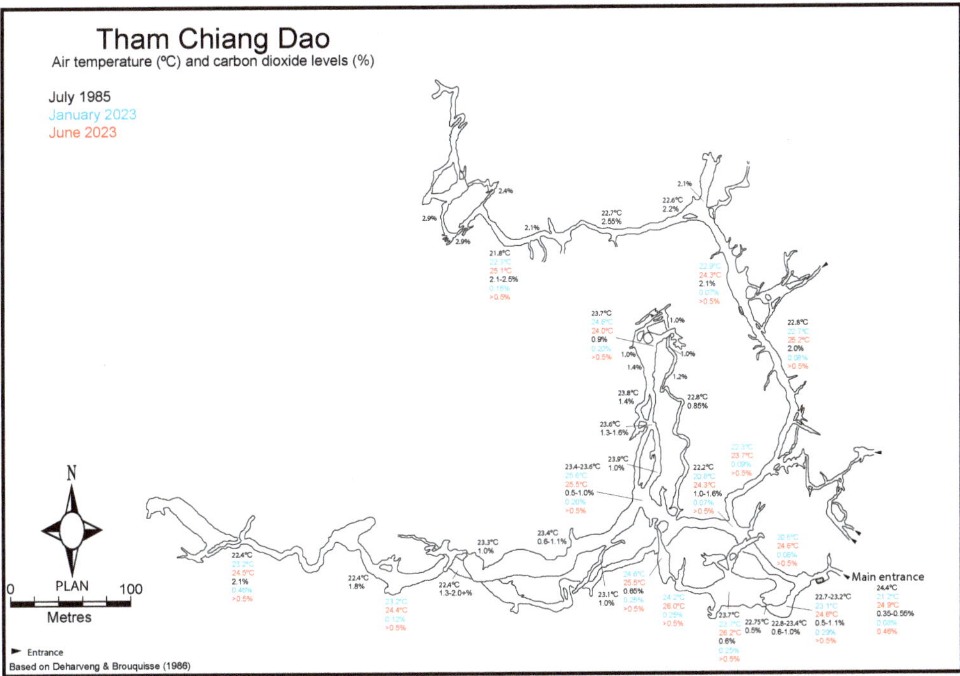

Figure 13. Temperature and carbon dioxide in Tham Chiang Dao. Black indicates measurements done in July 1985 [84], blue, those done in January 2023 [14], and red, those done in June 2023 [14].

The detailed zoological record extending back more than 40 years shows some indications of changes in the fauna, including the possible extirpation of some species. This needs to be investigated in more detail before any conclusions can be drawn. Deharveng and Bedos [43] tabulated 92 taxa from Tham Chiang Dao while only 50 are listed in the recent survey of the DMR [14]. However, the former dataset was carried out over a much longer period than the later dataset, and the comparison is not conclusive. The slow, but continuous, increase in tourist frequentation, habitat disturbance, installation of infrastructure in the tourist section (concrete path, bridge, electric lights), as well as the use of kerosene storm lanterns, may directly and indirectly drive changes in cave animal population dynamics, as well as favoring the spread of invasive species. This is supported by the observation that the tourist passages with electric lighting contain a smaller number of cave-obligate species and more alien species than the natural passages [14]. Although Doi Chiang Dao became a UNESCO Biosphere Reserve in 2021, its cave fauna appears to have played no part in the designation. The present paper fills this gap and shows the biological importance of the Tham Chiang Dao cave fauna, especially its endemic species, in this Biosphere Reserve. It might also serve as a basic reference for the bodies in charge of the management of the Chiang Dao Wildlife Sanctuary, i.e., the local cave management committee, the NCMPC, and the DMR, and in a larger scope will be a tool for conservation purposes in the future.

Author Contributions: Conceptualization, L.D. and S.J.; Formal analysis, S.J. and M.E.; Investigation and data curation by all authors; Writing—original draft, L.D. and S.J.; Writing—review and editing by all authors. All authors have read and agreed to the published version of the manuscript.

Funding: This research received no external funding.

Institutional Review Board Statement: Not applicable.

Data Availability Statement: Not applicable.

Acknowledgments: We would like to thank Piyaporn Pitaktunsakul, Weeyawat Jaitrong, Banchobporn Noipracha, Yuppayao Tokeeree, Norarat Boonkanpai, Chaiporn Siripornpibul, Tadsanai Jeenthong, Rueangrit Promdam, Phinit Chananin, Areeruk Nilsai, Mattrakan Jitpalo, Katthaleeya Surakhamhaeng, and Kanchana Jantapaso for offering valuable help and knowledge in the field and providing some photos for S.J. We are thankful to the Division of Biological Science (Biology), Faculty of Science, and Princess Maha Chakri Sirindhorn Natural History Museum, PSU for providing facilities and support.

Conflicts of Interest: The authors declare no conflict of interest.

References

1. Baum, F.; Hahn, L. *Geological Map of Northern Thailand 1:250,000-Sheet 2 Chiang Rai*; German Geological Mission in Thailand/Department of Mineral Resources, Federal Institute of Geosciences and Natural Resources: Hannover, Germany, 1976.
2. Baum, F.; Hahn, L. *Geological Map of Northern Thailand 1:250,000-Sheet 3 Phayao*; German Geological Mission in Thailand/Department of Mineral Resources, Federal Institute of Geosciences and Natural Resources: Hannover, Germany, 1977.
3. Baum, F.; Hahn, L. *Geological Map of Northern Thailand 1:250,000-Sheet 4 Chiang Dao*; German Geological Mission in Thailand/Department of Mineral Resources, Federal Institute of Geosciences and Natural Resources: Hannover, Germany, 1979.
4. Miyahigashi, A.; Ueno, K.; Charoentitirat, T. Late Permian (Lopingian) foraminifers from the Doi Chiang Dao limestone in the Inthanon zone of Northern Thailand. *Acta Geosci. Sin.* **2009**, *30*, 40–43.
5. Ellis, M. (Shepton Mallet Caving Club, Somerset, UK). Thailand Cave Database. 2023; Unpublished database.
6. Ridd, M.F.; Barber, A.J.; Crow, M.J. (Eds.) *The Geology of Thailand*; Geological Society of London: London, UK, 2011; pp. 1–312.
7. Ueno, K.; Charoentitirat, T.; Sera, Y.; Miyahigashi, A.; Suwanprasert, J.; Sordsud, A.; Boonlue, H.; Pananto, S. The Doi Chiang Dao Limestone: Paleo-Tethyan mid-oceanic carbonates in the Inthanon Zone of North Thailand. In Proceedings of the International Symposia on Geoscience Resources and Environments of Asian Terranes (GREAT 2008) 4th IGCP 516 and 5th APSEG, Bangkok, Thailand, 24–26 November 2008; pp. 42–48.
8. Putiyanan, S.; Maxwell, J.F. Survey and herbarium specimens of medicinal vascular flora of Doi Chiang Dao. *CMU. J. Nat. Sci.* **2007**, *6*, 159.
9. Chiang Dao Wildlife Sanctuary. Significant Wild Animal. 2023. Available online: https://www.chiangdao-biosphere.com/ (accessed on 5 August 2023).
10. Windecker, R.C.; Mangkrontong, N.; Siriwitayakorn, S.; Chiraprapoosak, S. Map of Chiang Dao Cave. Tripod-mounted compass and tape survey 1972–73. *Nat. Hist. Bull. Siam. Soc.* **1975**, *26*, 1–10.
11. Deharveng, L.; Gouze, A. *Expédition en Thailande-Rapport Speleologique*; Privately Circulated Report; Université Paul Sabatier: Toulouse, France, 1983; pp. 1–14.

12. Deharveng, L.; Brouquisse, F. *Le massif du Doï Chiang Dao. Les karsts de l'est de Chiang Mai. Expédition Thaï-Maros 85, Rapport Spéléologique et Scientifique*; Association Pyrénéenne de Spéléologie: Toulouse, France, 1986; pp. 23–30.

13. Sidisunthorn, P.; Gardner, S.; Smart, D. *Caves of Northern Thailand*; River Books: Bangkok, Thailand, 2006; pp. 1–392.

14. Department of Mineral Resources, Ministry of Natural Resources and Environment, Thailand. *Biodiversity Assessment of Tham Chiang Dao System, Chiang Mai Province*; Kanchanaburi Rajabhat University and Parties: Kanchanaburi, Thailand, 2023; pp. 1–152.

15. McGilvary, D. *A Half Century among the Siamese and the Lao, an Autobiography*; Fleming H. Revell Company: New York, NY, USA, 1912; pp. 1–435.

16. Thomas, O. Some new Ferae from Asia and Africa, Annals and Magazine of Natural History Series 9 Vol. 12 pp. 88–89, reprinted as a new species of bat from Siam. *Nat. Hist. Bull. Siam. Soc.* **1913**, *1*, 49–50.

17. Gyldenstolpe, N. Zoological results of the Swedish Zoological Expeditions to Siam 1911–1912 & 1914–1915. V. Mammals II. *Kungl. Svenska Vetenskapsakad. Handl.* **1916**, *57*, 3–59.

18. Allen, G.M.; Coolidge, H.J., Jr. Mammal and bird collections of the Asiatic Primate Expedition. Mammals. *Bull. Mus. Comp. Zool.* **1940**, *87*, 131–166.

19. Yosii, R. On some Collembola from Thailand. *Nat. Life SE Asia* **1961**, *1*, 171–201.

20. Suzuki, S. A synopsis of the Opiliones of Thailand (Arachnida) I. Cyphophthalmi and Laniatores. *Steenstrupia* **1985**, *11*, 69–110.

21. Stone, F.D. Caving in South-east Asia. *New York State Grotto Dripstone* **1967**, *5*, 3–5.

22. Sirivanakarn, S. Redescription of four Oriental species of *Culex* (*Culiciomyia*) and the description of a new species from Thailand (Dipteria: Culicidae). *Mosq. Syst.* **1977**, *9*, 93–111.

23. Boutin, C. Observations biospelologiques en Asie du Sud-Est. *Ann. Fac. Sc. Phnom Penh* **1971**, *4*, 168–185.

24. Deeleman-Reinhold, C.L. New *Althepus* Species from Sarawak, Sumatra and Thailand (Arachnida: Araneae: Ochyroceratidae). *Sarawak Mus. J. SMJ* **1985**, *34*, 115–123.

25. Löbl, I. Review of the Scaphidiidae (Coleoptera) of Thailand. *Rev. Suisse Zool.* **1990**, *97*, 505–621. [CrossRef]

26. Hoffman, R.L. A new genus and species of Doratodesmid millipede from Thailand. *Arch. Sci. Genève* **1982**, *35*, 87–93.

27. Kottelat, M. Two species of cavefishes from Northern Thailand in the genera *Nemacheilus* and *Homaloptera* (Osteichthyes: Homalopteridae). *Rec. Aust. Mus.* **1988**, *40*, 225–230. [CrossRef]

28. Deharveng, L. Fauna of Thai Caves II: New Entomobryoides Collembola from Chiang Dao Cave, Thailand. *Occas. Pap. Bernice P. Bishop Mus.* **1990**, *30*, 279–287.

29. Suzuki, S.; Stone, F.D. The fauna of Thai caves I: Three phalangides from Thailand (Arachnida). *Occas. Pap. Bernice P. Bishop Mus.* **1986**, *26*, 123–127.

30. Blood, B.R.; McFarlane, D.A. Notes on some bats from northern Thailand with comments on the subgeneric status of *Myotis altarium*. *Z. Säugetierkd.* **1988**, *53*, 276–280.

31. Taiti, S.; Ferrara, F. Revision of the genus *Exalloniscus* Stebbing, 1911 (Crustacea: Isopoda: Oniscidea). *Zool. J. Linn. Soc.* **1988**, *94*, 339–377. [CrossRef]

32. Association Pyrénéenne de Spéléologie. *Expédition Thai-Maros 85: Rapport Spéléologique et Scientifique*; Association Pyrénéenne de Spéléologie: Toulouse, France, 1986; pp. 1–215.

33. Schawaller, W. Pseudoskorpione aus Thailand (Arachnidae: Pseudoscorpiones). *Rev. Suisse Zool.* **1994**, *101*, 725–759. [CrossRef]

34. Grall, E.; Jäger, P. Forty-seven new species of *Sinopoda* from Asia with a considerable extension of the distribution range to the South and description of a new species group (Sparassidae: Heteropodinae). *Zootaxa* **2020**, *4797*, 1–101. [CrossRef] [PubMed]

35. Watiroyram, S. Species diversity and distribution of freshwater Copepoda in caves in Northern Thailand. Ph.D. Thesis, Khon Kaen University, Khon Kaen, Thailand, 2012; pp. 1–186.

36. Chintapitasakul, L.; Choengern, N.; Bumrungsri, S.; Chalamat, M.; Molee, L.; Parchariyanon, S.; Ratanamungklanon, S.; Kongsuwan, K. RT-PCR survey of emerging Paramyxoviruses in cave-dwelling bats. *Thai J. Vet. Med.* **2012**, *42*, 29–36. [CrossRef]

37. Jäger, P. On the genus *Systaria* (Araneae: Clubionidae) in Southeast Asia: New species from caves and forests. *Zootaxa* **2018**, *4504*, 524–544. [CrossRef] [PubMed]

38. Jirapatrasilp, P.; Sutcharit, C.; Panha, S. Annotated checklist of the operculated land snails from Thailand (Mollusca, Gastropoda, Caenogastropoda): The family Pupinidae, with descriptions of several new species and subspecies, and notes on classification of *Pupina* Vignard, 1829 and *Pupinella* Gray, 1850 from mainland Southeast Asia. *ZooKeys* **2022**, *1119*, 1–115.

39. Jantarit, S. *Cavernicolus Springtails (Hexapoda: Collembola) in Northern and Western Part of Thailand*; Thailand Research Fund: Bangkok, Thailand, 2021; pp. 1–151.

40. Soisook, P.; Jantarit, S.; Rodcharoen, E.; Sa-ardrit, P.; Samoh, A.; Promdam, R.; Mitpuangchon, N.; Pimsai, A.; Wattanasen, S.; Watcharakul, S.; et al. *Management of Biodiversity and Indigenous Knowledge of the Cave Ecosystem in Satun UNESCO Global Geopark for Sustainable Tourism*; National Science and Technology Development Agency: Khlong Luang, Thailand, 2020; pp. 1–380.

41. Department of Mineral Resources, Ministry of Natural Resources and Environment, Thailand. *Biodiversity Assessment of Tham Khao Chang Hai, Na Muen Si, Na Yong, Trang Province*; Kanchanaburi Rajabhat University and Parties: Kanchanaburi, Thailand, 2022; pp. 1–180.

42. Jantarit, S.; Thongtip, U.; Boonrotpong, S.; Soisook, P.; Sa-ardrit, P.; Promdam, R.; Mitpuangchon, N.; Pimsai, A. *An Interdisciplinary Approach to Evaluate Geological Structures, Physical Factors and Biodiversity of Cave Ecosystem: A Case Study in Tourist Caves of Satun, Phatthalung and Trang Province*; Biodiversity-Based Economy Development Office (Public Organization) (BEDO): Bangkok, Thailand, 2020; pp. 1–200.

43. Deharveng, L.; Bedos, A. The cave fauna of Southeast Asia. Origin, evolution and ecology. In *Ecosystems of the World Vol. 30: Subterranean Ecosystems*; Wilkens, H., Culver, D.C., Humphreys, W.F., Eds.; Elsevier: Amsterdam, The Netherlands, 2000; pp. 603–632.

44. Jantarit, S.; Ellis, M. *The Cave Fauna of Thailand*; Prince of Songkla University Press: Songkhla, Thailand, 2023; pp. 1–396.

45. Deharveng, L.; Rahmadi, C.; Suhardjono, Y.R.; Bedos, A. The Towakkalak System, a hotspot of subterranean biodiversity in Sulawesi, Indonesia. *Diversity* **2021**, *13*, 392. [CrossRef]

46. Moseley, M.; Lim, T.W.; Lim, T.T. Fauna reported from Batu Caves, Selangor, Malaysia: Annotated checklist and bibliography. *Cave Karst Sci.* **2012**, *39*, 77–92.

47. Gibert, J. Le système karstique du Doi Chiang Dao (Thailande). Peuplements aquatiques souterrains, repartition, relations entre le milieu karstique et le sous-ecoulement de l'exutoire. In *Expédition Thaï-Maros 86, Rapport Spéléologique et Scientifique*; Association Pyrénéenne de Spéléologie: Toulouse, France, 1987; pp. 117–128.

48. Giani, N.; Bouguenec, V. *La Faune Aquatique de Thailande Généralités et Catalogue. Expeditions de l'APS en Asie du Sud-Est: Travaux Scientifiques–1*; Association Pyrénéenne de Spéléologie: Toulouse, France, 1988; pp. 29–38.

49. Condé, B. Nouveaux Palpigrades de Trieste, de Slovenié, de Malte, du Paraguay, de Thaïlande et de Bornéo. *Rev. Suisse Zool.* **1988**, *95*, 723–750. [CrossRef]

50. Deeleman-Reinhold, C.L. *Forest spiders of South East Asia: With a revision of the sac and ground spiders (Araneae: Clubionidae, Corinnidae, Liocranidae, Gnaphosidae, Prodidomidae and Trochanterrudae)*; Brill Academic Publishers: Leiden, The Netherlands, 2001; pp. 1–591.

51. Siriwut, W.; Edgecombe, G.D.; Sutcharit, C.; Tongkerd, P.; Panha, S. A taxonomic review of the centipede genus *Scolopendra* Linnaeus, 1758 (Scolopendromorpha, Scolopendridae) in mainland Southeast Asia, with description of a new species from Laos. *ZooKeys* **2016**, *590*, 1–124.

52. Likhitrakarn, N.; Golovatch, S.I.; Prateepasen, R.; Panha, S. Review of the genus *Tylopus* Jeekel, 1968, with descriptions of five new species from Thailand (Diplopoda, Polydesmida, Paradoxosomatidae). *ZooKeys* **2010**, *72*, 23–68.

53. Watiroyram, S.; Brancelj, A.; Sanoamuang, L. Two new stygobiotic species of *Elaphoidella* (Crustacea: Copepoda: Harpacticoida) with comments on geographical distribution and ecology of harpacticoids from caves in Thailand. *Zootaxa* **2015**, *3919*, 81–99. [CrossRef]

54. Camacho, A.I.; Leclerc, P. A new species of the genus *Siambathynella* Camacho, Watiroyram & Brancelj, 2011 (Crustacea, Bathynellacea, Parabathynellidae) from a Thai cave. *Subterr. Biol.* **2022**, *44*, 139–152.

55. Jantarit, S.; Bedos, A.; Deharveng, L. An annotated checklist of the Collambolan fauna of Thailand. *Zootaxa* **2016**, *4169*, 301–360. [CrossRef] [PubMed]

56. Deharveng, L. A new troglomorphic Collembola from Thailand: *Troglopedetes fredstonei*, n. sp. (Collembola: Paronellidae). *Occas. Pap. Bernice P Bishop Mus.* **1988**, *28*, 95–98.

57. Thibaud, J. Révision du genre *Acherontiella* Absolon, 1913 (Insecta, Collembola). *Bull. Mus. Natl.* **1990**, *12*, 401–414. [CrossRef]

58. Nayrolles, P. Fauna of Thai caves III: Two new cavernicolous species of *Arrhopalites* from Thailand (Insecta: Collembola). *Occas. Pap. Bernice P Bishop Mus.* **1990**, *30*, 288–293.

59. Roguin, L. *Expéditions Thai-Maros 85 et Thai 87-Mammiferes. Expeditions de l'APS en Asie du Sud-est: Travaux Scientifiques–1*; Association Pyrénéenne de Spéléologie: Toulouse, France, 1988; pp. 47–52.

60. Yenbutra, S.; Felton, H. Bat species and their distribution in Thailand according to the collections in TISTR and SMF. In *Contributions to the Knowledge of the Bats of Thailand*; Felton, H., Ed.; Courier Forschungsinstitut Senckenberg: Frankfurt, Germany, 1986; Volume 87, pp. 9–45.

61. Culver, D.C.; Sket, B. Hotspots of subterranean biodiversity in caves and wells. *J. Caves Karst Stud.* **2000**, *62*, 11–17.

62. Camacho, A.I.; Puch, C. Ojo Guareña: A hotspot of subterranean biodiversity in Spain. *Diversity* **2021**, *13*, 199. [CrossRef]

63. Brad, T.; Iepure, S.; Sarbu, S.M. The chemoautotrophically based Movile cave groundwater ecosystem, a hotspot of subterranean biodiversity. *Diversity* **2021**, *13*, 128. [CrossRef]

64. Polak, S.; Pipan, T. The subterranean fauna of Križna Jama, Slovenia. *Diversity* **2021**, *13*, 210. [CrossRef]

65. Hutchins, B.T.; Gibson, J.R.; Diaz, P.H.; Schwartz, B.F. Stygobiont diversity in the San Marcos artesian well and Edwards aquifer groundwater ecosystem, Texas, USA. *Diversity* **2021**, *13*, 234. [CrossRef]

66. Zagmajster, M.; Polak, S.; Fišer, C. Postojna-Planina cave system in Slovenia, a hotspot of subterranean biodiversity and a cradle of Speleobiology. *Diversity* **2021**, *13*, 271. [CrossRef]

67. Faille, A.; Deharveng, L. The Coume Ouarnède system, a hotspot of subterranean biodiversity in Pyrenees (France). *Diversity* **2021**, *13*, 419. [CrossRef]

68. Jäger, P. The second true troglobiont *Heteropoda* species from a limestone cave system in Palawan, Philippines (Araneae: Sparassidae: Heteropodinae). *Arachnology* **2018**, *17*, 427–431. [CrossRef]

69. Condé, B. Palpigrades cavernicoles et endogés de Thaïlande et de Célèbes (1ere Note). *Rev. Suisse Zool.* **1992**, *99*, 665–672. [CrossRef]

70. Harvey, M.S.; Ratnaweera, P.B.; Udagama, P.; Wijesinghe, M.T. A new species of the pseudoscorpion genus *Megachernes* (Pseudoscorpiones: Chernetidae) associated with a threatened Sri Lankan rainforest rodent, with a review of host associations of *Megachernes*. *J. Nat. Hist.* **2012**, *46*, 2519–2535. [CrossRef]

71. Jantarit, S. *Biodiversity of Collembola in Subterranean Habitat of Thailand*; Thailand Research Fund: Bangkok, Thailand, 2022; pp. 1–136.
72. Zhang, F.; Deharveng, L.; Chen, J. New species and rediagnosis of *Coecobrya* (Collembola: Entomobryidae), with a key to the species of the genus. *J. Nat. Hist.* **2009**, *43*, 2597–2615. [CrossRef]
73. Vidlička, Ľ.; Vršanský, P.; Kúdelová, T.; Kúdela, M.; Deharveng, L.; Hain, M. New genus and species of cavernicolous cockroach (Blattaria, Nocticolidae) from Vietnam. *Zootaxa* **2017**, *4232*, 361–375. [CrossRef]
74. Ševčík, J. *Pseudochetoneura* gen. nov., a peculiar new genus from Ecuador, with notes on *Chetoneura* (Diptera: Keroplatidae). *Acta Entomol. Mus. Natl. Pragae.* **2012**, *52*, 281–288.
75. Deeleman-Reinhold, C.L. The Ochyroceratidae of the Indo-Pacific Region (Araneae). *Raffles Bull. Zool. Suppl.* **1995**, *2*, 1–103.
76. Golovatch, S.I.; Geoffroy, J.-J.; Mauriès, J.-P.; VandenSpiegel, D. Review of the millipede genus *Eutrichodesmus* Silvestri, 1910 (Diplopoda, Polydesmida, Halpodesmidae), with descriptions of new species. *ZooKeys* **2009**, *12*, 1–46. [CrossRef]
77. Castello, M. Species diversity of Bryophytes and ferns of lampenflora in Grotta Gigante (NE Italy). *Acta Carsologica* **2014**, *43*, 185–193. [CrossRef]
78. Mulec, J.; Kosi, G. Lampenflora algae and methods of growth control. *J. Cave Karst Stud.* **2009**, *71*, 109–115.
79. Kurniawan, I.D.; Rahmadi, C.; Ardi, T.E.; Nasrullah, R.; Willyanto, M.I.; Setiabudi, A. The impact of lampenflora on cave-dwelling arthropods in Gunungsewu karst, Java, Indonesia. *Biosaintifika* **2018**, *10*, 275–283. [CrossRef]
80. O'Dowd, D.J. Crazy Ant Attack. *Wingspan* **1999**, *9*, 7.
81. Health Protection Agency; Chilcott, R.P. Compendium of Chemical Hazards: Kerosene (Fuel Oil). Available online: https://silo.tips/download/health-protection-agency (accessed on 12 August 2023).
82. Lam, N.L.; Smith, K.R.; Gauthier, A.; Bates, M.N. Kerosene: A review of household uses and their hazards in low- and middle-income countries. *J. Toxicol. Environ. Health B Crit. Rev.* **2012**, *15*, 396–432. [CrossRef]
83. Tedsen, E. *Black Carbon Emissions from Kerosene Lamps*; Ecologic Institute: Berlin, Germany, 2013; pp. 1–45.
84. Deharveng, L.; Bedos, A. Gaz carbonique. In *Expédition Thaï-Maros 85 Rapport Spéléologique et Scientifique*; Association Pyrénéenne de Spéléologie: Toulouse, France, 1986; pp. 144–152.
85. Howarth, F.G.; Stone, F.D. Elevated carbon dioxide levels in Bayliss Cave, Australia: Implications for the evolution of obligate cave species. *Pac. Sci.* **1990**, *44*, 207–218.

Article

A Hotspot of Subterranean Biodiversity on the Brink: Mo So Cave and the Hon Chong Karst of Vietnam

Louis Deharveng [1,*], Cong Kiet Le [2], Anne Bedos [1], Mark L. I. Judson [1], Cong Man Le [2], Marko Lukić [3,4], Hong Truong Luu [5,6], Ngoc Sam Ly [6,7], Tran Quoc Trung Nguyen [5], Quang Tam Truong [7] and Jaap Vermeulen [8]

[1] Institut de Systématique, Évolution, Biodiversité (ISYEB), UMR 7205 CNRS, Muséum National d'Histoire Naturelle, Sorbonne Université, 57 rue Cuvier, 75005 Paris, France; bedosanne@yahoo.fr (A.B.); judson@mnhn.fr (M.L.I.J.)

[2] Department of Ecology-Evolutionary Biology, Faculty of Biology-Biotechnology, University of Science, Vietnam National University Ho Chi Minh City, 227 Nguyen Van Cu Street, District 5, Ho Chi Minh City 749000, Vietnam; lecongkiet@gmail.com (C.K.L.); lecongman2202@gmail.com (C.M.L.)

[3] Croatian Biospeleological Society, Rooseveltov trg 6, HR-10000 Zagreb, Croatia; lukic.mar@gmail.com

[4] Department of Molecular Biology, Ruđer Bošković Institute, Bijenička cesta 54, HR-10000 Zagreb, Croatia

[5] Southern Institute of Ecology, Institute of Applied Materials Science, Vietnam Academy of Science and Technology, 1D, TL 29 Street, Thanh Loc Ward, District 12, Ho Chi Minh City 729110, Vietnam; hongtruongluu@gmail.com (H.T.L.); ntqtrung@gmail.com (T.Q.T.N.)

[6] Vietnam Academy of Science and Technology, Graduate University of Science and Technology, 18 Hoang Quoc Viet, Cau Giay District, Ha Noi 122000, Vietnam; lysamitb@gmail.com

[7] Institute of Tropical Biology, Vietnam Academy of Science and Technology, 85 Tran Quoc Toan Street, Vo Thi Sau Ward, District 3, Ho Chi Minh City 722700, Vietnam; truongtam58@yahoo.com

[8] JK Art and Science, Lauwerbes 8, 2318 AT Leiden, The Netherlands; jk.artandscience@gmail.com

* Correspondence: dehar.louis@wanadoo.fr

Citation: Deharveng, L.; Le, C.K.; Bedos, A.; Judson, M.L.I.; Le, C.M.; Lukić, M.; Luu, H.T.; Ly, N.S.; Nguyen, T.Q.T.; Truong, Q.T.; et al. A Hotspot of Subterranean Biodiversity on the Brink: Mo So Cave and the Hon Chong Karst of Vietnam. *Diversity* 2023, *15*, 1058. https://doi.org/10.3390/d15101058

Academic Editor: Michel Baguette

Received: 17 August 2023
Revised: 21 September 2023
Accepted: 25 September 2023
Published: 2 October 2023

Abstract: The southern part of the Mekong Delta Limestones of Vietnam (MDL-HC or Hon Chong karst) comprises numerous small limestone hills. It is a hotspot of biodiversity for soil and cave invertebrates. Here, we synthesize the results of biological surveys carried out in Hang Mo So, the richest MDL-HC cave for troglobionts, and in surrounding karsts. Methodologies for the ecological characterization of species are discussed, with emphasis on parallel sampling (external soil plus cave). Hang Mo So has 27 troglobionts, including many still undescribed. An additional 40 cave-obligate species are known from other caves of MDL-HC. Among them, several are expected to be found in Hang Mo So. Most troglobionts of MDL-HC are endemic. Several relictual taxa without close relatives in Southeast Asia occur in Hang Mo So and in MDL-HC, reflecting an ancient origin of the fauna. The reasons for this richness are uncertain, but the cause of its current destruction—quarrying—is all too evident. Most of the original 4 km^2 of the MDL-HC karst has been destroyed or soon will be, ultimately leaving only 1.6 km^2 unquarried. Endemic species linked to karst habitats are, therefore, under clear threat of extinction. The Hon Chong karst (MDL-HC) was listed among the ten most endangered karsts on the planet 25 years ago. Today it would probably top the list.

Keywords: invertebrates; Mekong Delta; karst; species richness; stygobionts; troglobionts; quarrying; sampling; caves; species extinction

1. Introduction

Vietnam has many large and famous karsts in its central and northern regions, in particular at Ke Bang and Ha Long Bay. However, further south (below 17° N), limestones are limited to small, sparse isolated outcrops in two regions: Da Nang where a few small limestone towers are located in the town itself, and a larger region straddling the border between Vietnam and Cambodia (the Mekong Delta Limestones or MDL). The latter is composed of numerous small limestone hills of Permian age, extending from the southwestern part of the Mekong Delta to southern Cambodia. Along 70 km, from Hon Chong hill in the southeast to Laang hill in the west, limestone outcrops emerge above the Mekong plain. Many are grouped in loose clusters of small, karstified hills that rarely exceed a square kilometer in surface area. Those located southeast of the Giang Thanh River in Vietnam—collectively referred to as Hon Chong karst in the literature and as MDL-HC here—have highly dissected surfaces, deep dolines and numerous cave passages, remnants of larger cave systems.

Long caves with underground rivers are generally those richest in biodiversity [1], such as Mammoth Cave in USA [2], the Postojna-Planina Cave System in Slovenia [3], Ojo Guarena in Spain [4], Feihu Dong in China [5], the Towakkalak system in Sulawesi [6] or Agua Clara in Brazil [7]. The MDL caves are quite different: they are short and shallow, without active hydrological circulation, and are supplied with large amounts of nutrients (roots, guano, debris) due to surface proximity. The best-known and richest cave of the MDL karst is Hang Mo So (Hang = cave), in the Nui Bai Voi hill of MDL-HC. It is the focus of this work.

1.1. Karst and Caves of MDL

The MDL karst is spatially isolated from the other karst areas of the region: the Battambang karst in Cambodia is 250 km to the northwest; the small tower karst of Danang in Vietnam [8] is 700 km to the north; and the huge karst of Ke Bang-Khammouane is over 750 km to the north. No significant active system is known in the MDL karst, due to the small size of the limestone hills, their isolation, and the low level (0–2 m a.s.l.) of the alluvial plain in which they are situated. Nevertheless, numerous small, hydrologically unconnected caves are present. Many are of archeological, aesthetic, or historical interest, such as the old Khmer temples and monasteries in several caves of the Cambodian part of the karst [9], the cave-temple at Chua Hang, or the Mo So Cave in MDL-HC, which served as an important logistics base of Vietnamese liberation troops during the Vietnam war, with an arsenal and an army hospital. Shell impacts and bomb craters can still be seen, and munitions were still numerous in the underground passages 25 years ago [10]. Even today, ordnance is sometimes found in and around MDL caves.

The MDL karst is constituted of loose clusters of small limestone outcrops intermixed with non-limestone hills. The northwest part of MDL in Cambodia (MDL-C) includes two groups of hills: the Kampot group and the Tuk Meas group. The Ha Tien cluster straddles the border between Cambodia and Vietnam (Area 1 of [8]). Southeast of Ha Tien, in Vietnam, the Hon Chong karst (MDL-HC) is formed of four clusters (Figure 1): the Kien Luong group, between the Giang Thanh River and the Ba Hon canal (Area 2 of [8]); the northern Hon Chong group, between the Ba Hon canal and Nui Binh Tri (Area 3 of [8]); the southern Hon Chong group, south of Nui Binh Tri (Area 4 of [8]); and the minute, isolated karst of Hon Nghe island, 15 km southwest of Nui Hon Chong.

The MDL-HC karst landscape ranges from gentle hills to spectacular pinnacle formations on steep slopes and extremely rugged, impassable karst terrain (Figure 2). Caves are numerous, but all are short and hydrologically inactive [8]. They can be assigned to three categories:

(1) 'Tidal caves'. These are smooth-surfaced, horizontal, shallow caves at sea level, devoid of speleothems, associated with horizontal undercutting of the hill circumference. Undercuttings (notches) are often visible at two or three levels. They may have been caused by acids released from the sulphidic mangrove mud acting as a dissolving agent, since mangrove woodland probably surrounded most limestone hills [11]. They sometimes develop in deeper hill-foot caves (Figure 2C,D).

(2) 'Hill-foot caves'. These resemble "tidal caves" in their smooth-surfaced walls, horizontal development and lack of speleothems, but they penetrate deep into the hill (Figure 2E). They are frequent in tropical tower karsts on alluvial plains [12]. They are hydrologically inactive or filled with slow-moving water connected to surface swamp waters in the MDL karst.

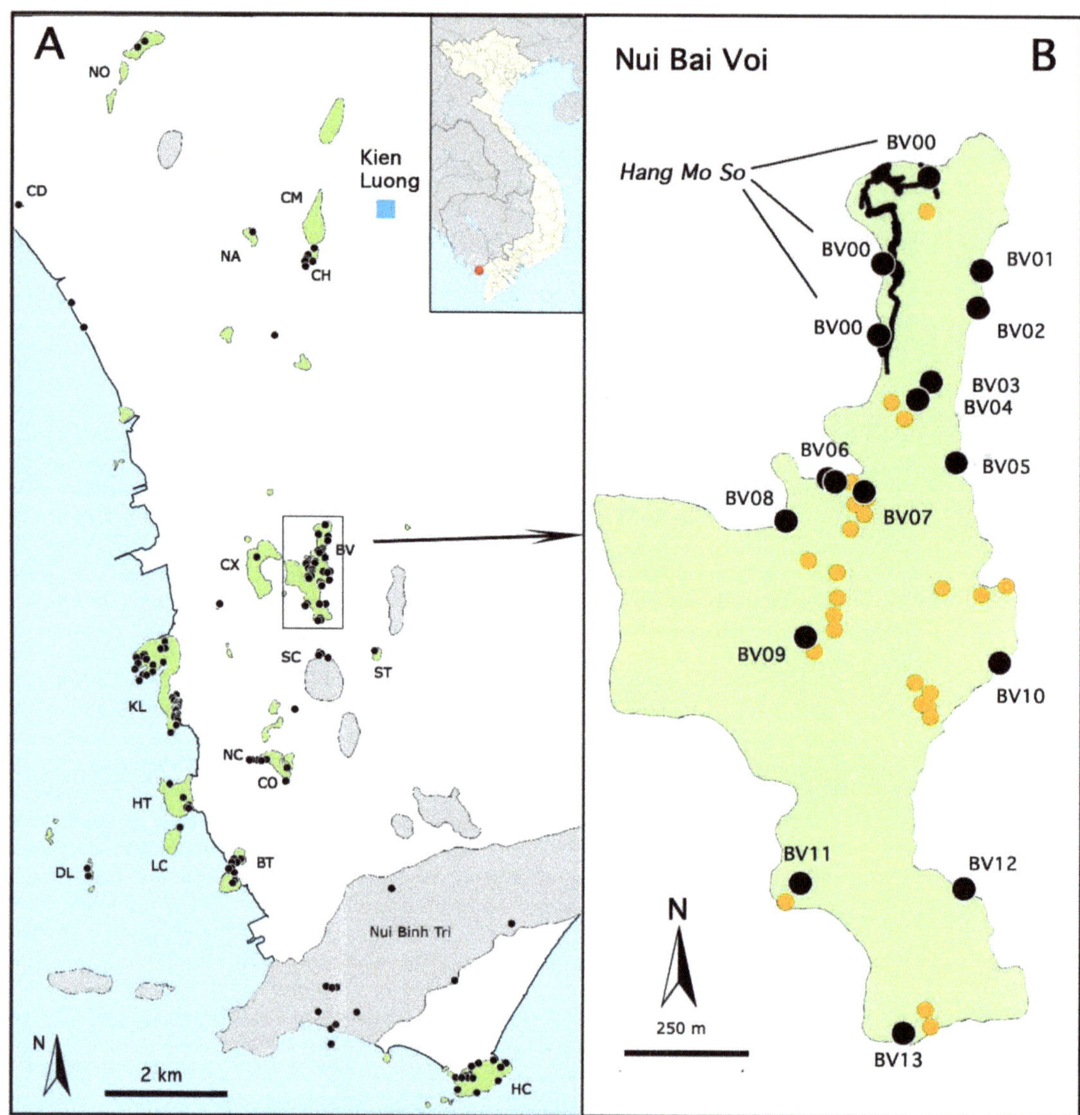

Figure 1. The MDL-HC karst. (**A**) Hills, islands and samples. Red dot on the inserted map, location of the MDLS-HC karst in Vietnam; blue square, the town of Kien Luong (Kien Giang Province); green, limestone hills; grey, non-limestone hills. Hill name abbreviations (Nui and Phnom mean hill, Hon means island, Hang means cave): BT, Nui Ba Tai; BV, Nui Bai Voi; CD, Nui Ca Danh; CH, Nui Chau Hang; CM, Nui Com; CO, Nui Hang Cay Ot; CX, Nui Cay Xoai; DL, Hon Da Lua; HC, Nui Hon Chong; HT, Nui Hang Tien; KL, Nui Khoe La; LC, Hon Lo Coc; LV, Nui Lo Voi; NA, Nui Nai; NC, Nui Coc; NO, Nui Ong; NT, Nui Trau; SC, Nui Son Cha; ST, Phnom Sray Toch. (**B**) Caves and samples in Nui Bai Voi. Black lines, main galleries of Hang Mo So; black circles, main entrances of sampled caves; orange circles, sites of soil and litter samples outside caves; one spot represents 1 to 20 samples. List of caves sampled in Nui Ba Voi: BV00, Hang Mo So, with its 3 main entrances; BV01, Guano Lake Cave; BV02, Roots Cave; BV03, Cliff Cave; BV04, Pass Cave; BV05, Hang Hei Truong; BV06, Old Man Caves; BV07, Feaellidae Cave; BV08, Hang Tai; BV09, French Man Cave; BV10, Cows Cave; BV11, Bai Voi SW cave 1; BV12, Hang Phat Man; BV13, Bai Voi SW cave 2.

Figure 2. Karstic landscapes of Hon Chong hills: (**A**) sharp karstic relief along the coast of Nui Hon Chong; (**B**) southeast of Nui Bai Voi, access to the karst hill through marshes (ph. ML); (**C**) notches at Nui Hang Cay Ot; (**D**) a subvertical slope of Hon Da Lua with basal notches and specialized vegetation with *Cycas* on rocks; (**E**) network of galleries at entrance of Hang Mo So. Photos: Louis Deharveng (**A**,**C–E**), Marko Lukić (**B**).

(3) 'Phreatic or vadose caves'? These are formed by the dissolving action of rainwater seeping through rock fissures, or by flowing freshwater. They often have an uneven, tilted floor, and speleothems are common. In the MDL karst, phreatic caves are remnants of old cave systems dissected by surface erosion of the hills.

Eighteen caves of MDL-HC have been mapped [8,13], several others have been explored, and many remain to be discovered, especially shafts in almost impassable terrain. From 1993 to 2014, biologists and cavers extensively sampled many caves, particularly Hang Mo So (also called Hang Moi Chau in [8] and Grotte-hôpital in [13]). Hang Mo So, with 1 km of passages, is the longest cave of MDL (Figure 3).

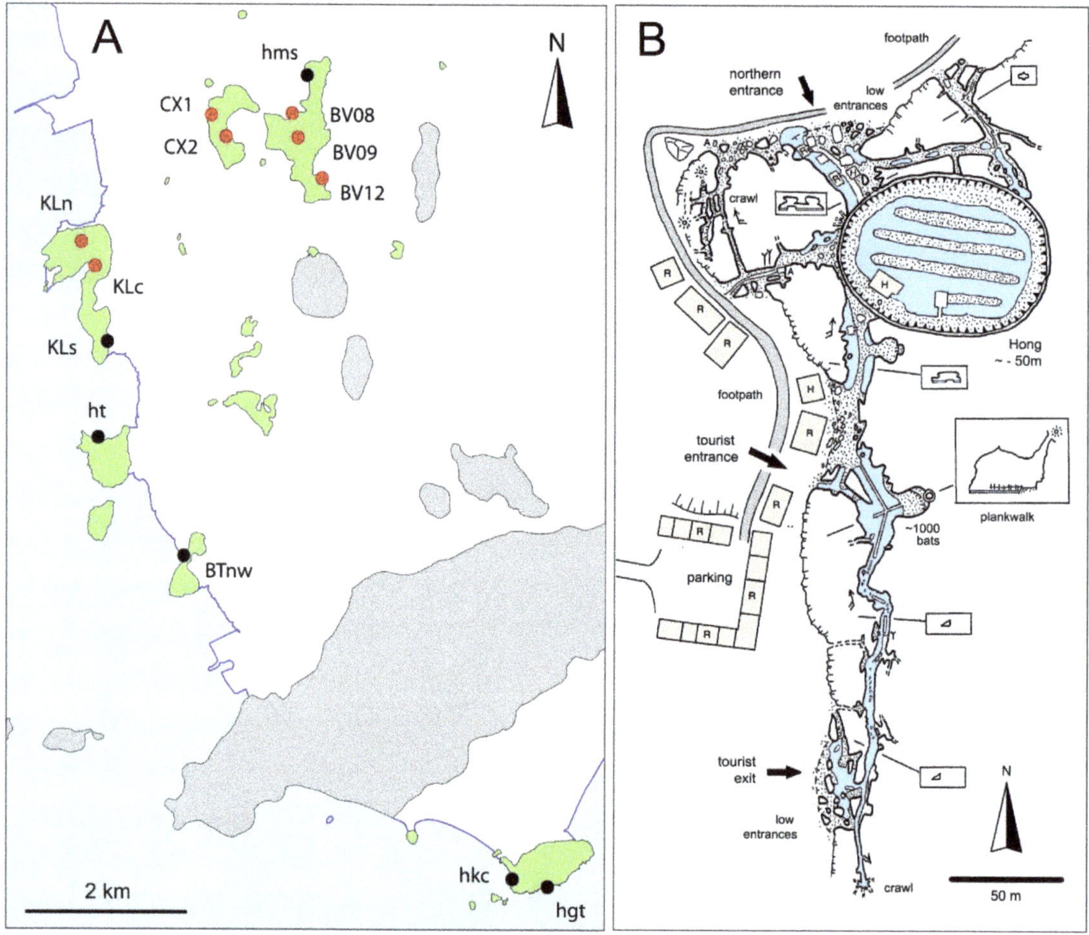

Figure 3. (**A**) Location of the MDL-HC caves cited in the text. Blue line, sea coastline; green, limestone hills; grey, non-limestone hills; black dots, richest caves in unquarried hills or parts of hills not planned to be quarried; red dots, caves of interest in hills quarried or planned to be quarried. BTnw, cave northwest of Nui Ba Tai; BV08, Hang Tai; BV09, French Man Cave; BV12, cave south of Nui Bai Voi; CX1, CX2, Nui Cay Xoai caves 1 and 2; hgt, Hang Gieng Tien; hkc, Hang Kim Cuong; hms, Hang Mo So; ht, Hang Tien; KLn, KLc, KLs, caves north, center and south of Nui Khoe La. (**B**) Map of Hang Mo So after Laumanns [8], modified; cave length 1003 m; survey dates, 6 and 8 July 2001. Blue, water; A, altar; H, house; R, restaurant.

Two other significant caves of MDL-HC penetrate almost perpendicularly into the hill, suggesting that larger systems existed in the past. These are Hang Gieng Tien in Nui Hon Chong, the southernmost hill of MDL-HC [8,13], and Hang Tien, a cave that runs through the northwestern arm of Nui Hang Tien (Figure 3A). Both contain a large array of oligotrophic habitats, unusual in the MDL karst, and host a rich fauna.

1.2. Description of Hang Mo So

Hang Mo So, the largest cave of Nui Bai Voi, extends 1003 m into the limestone. It was mapped and described in 1995 [13] and later, in more detail, in 2011 [8] (Figure 3B). Nui Bai Voi was formerly surrounded by mangroves and semi-natural wetlands, but these have been largely converted into commercial ponds. The main cement plant of MDL-HC is situated along the western side of Nui Bai Voi and most of the hill is currently being quarried (see Section 5.3). Hang Mo So is located at the northern tip of Nui Bai Voi, which is supposed to be left untouched (Figure 1B). Its main passages are horizontal and accessible to tourists. These include a large entrance hall, with a ceiling window and a lake, followed by a spacious (often over 3 × 3 m) phreatic gallery that leads to the southern entrances. This gallery often shows a keyhole cross-section, with its narrow lower cutting, more or less filled with brackish water or sediment. Brackish water is also present, at least seasonally, in most parts of the cave, flowing very slowly from south to north. The main hall is connected to a large central depression used for farming and fish ponds (annotated "hong" on the cave map). Several much smaller passages branch on each side of the main gallery, at the same level (Figure 2E). At the southernmost part of the cave, a low crawl has been followed for 30 m and might continue further. Calcite formations are small and rare throughout the cave.

Habitats in the cave are quite diverse. A colony of fruit bats that roosts in a large chimney near the entrance produces a big amount of guano. Impressive hanging root bundles are present near the entrance and further along the main gallery (Figure 4A). The limestone floor is covered with soil in the main hall, but often becomes exposed in the galleries, which are locally very damp, especially in short blind passages and in the southernmost low crawl. These passages have only sparse deposits of organic debris, approaching oligotrophic conditions. It is likely that most habitats at floor level are seasonally washed during flooding, with an unknown impact on the fauna. Human disturbance and pollution are obvious in most parts of the cave. Bags of concrete and sand are placed on the ground, a wooden walkway has been constructed along the entrance lake, and hanging roots have been cut (Figure 4). There are accumulations of rubbish in some recesses and numerous spots for religious offerings, with many incense sticks left along the main gallery [8].

1.3. History of Biological Studies

In 1970–1974 [14,15], Le Cong Kiet recognized two formations that characterize the unique vegetation of the MDL-HC karst:

(1)   Cremnophyte and xerophilous shrub formations, dominated by *Cycas clivicola* subsp. *lutea*, *Euphorbia antiquorum*, and *Dracaena cambodiana*, which develop on patchy soil accumulated between exposed rocks and cliffs.

(2)   Mesophilic tree formations, found in small patches on shallow soil, near the base of the hills and on low slopes. These may develop into a semi-deciduous forest dominated by *Tetrameles nudiflora*, *Diospyros crumenata*, *Sterculia foetida*, and *Ficus* spp. The most significant example was on Nui Com near Kien Luong, but the hill was totally erased by quarrying shortly after the publication of Le Cong Kiet's study. Roots of *Ficus* spp. host rich subterranean invertebrate communities in Hang Mo So and in several other caves of MDL-HC.

The MDL-HC flora is not species-rich, but several micro-endemic phanerogams that are strictly linked to limestone habitats have been described from there [16–18], providing additional evidence for the interest of this karst.

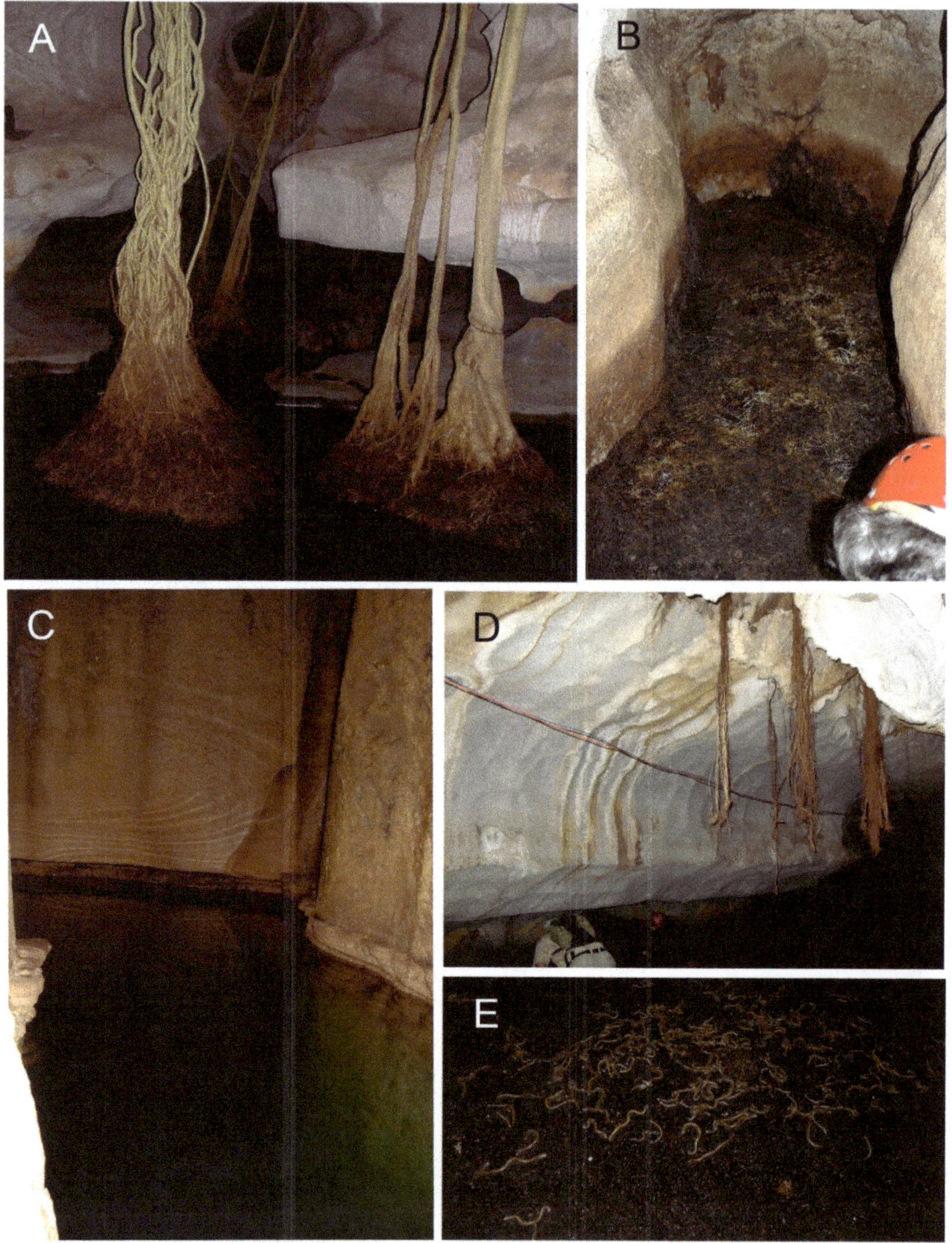

Figure 4. Cave habitats: (**A**) hanging roots in Hang Mo So; (**B**) root-mat in cave BV08 of Nui Bai Voi; (**C**) deep freshwater lake in Hang Gieng Tien (Nui Hon Chong); (**D**) hanging roots cut in the touristic gallery of Hang Mo So in 2008; (**E**) swarming of *Trachyjulus singularis* on guano in the cave of Hon Lo Coc. Photos: Louis Deharveng (**A,C,D**), Anne Bedos (**B**), Marko Lukić (**E**).

Research on the cave and soil fauna in MDL-HC started 20 years after the pioneering work of Le Cong Kiet, with two Franco-Vietnamese biological surveys in 1993 and 1994. These evaluated the invertebrate biodiversity value of selected caves and limestone hills, where drilling had just started as a prelude to quarrying. They discovered several cave endemics from various zoological groups. The first described species were two troglobionts: the springtail *Lepidonella lecongkieti* Deharveng & Bedos, 1995 and the beetle *Eustra honchongensis* Deuve, 1996. These results provided the impetus for a succession of biological excursions up to 2014, which yielded an exceptionally rich cave, soil and limestone associated fauna, including new endemic species and supra-specific taxa, especially among snails and soil beetles.

In 2001, Deharveng et al. [19] listed nine troglobionts for MDL-HC, seven of which were undescribed. At the same time, Boutin [20] reported four troglobionts for MDL-C, none of which were shared with MDL-HC. By 2009, the troglobiotic invertebrate richness had increased to 30 species for MDL-HC, while remaining unchanged for MDL-C [21]. Some additional troglophilic or guanobiotic taxa were later recorded by Steiner [22] from several caves of the MDL karst.

1.4. Threats and Focus

We focus on the MDL-HC karst for two reasons. Firstly, it is currently the best-known part of MDL. Many micro-endemic and relictual invertebrates have been discovered in these hills, whereas the MDL-C karst remains insufficiently investigated. Secondly, the threats to this karst are more acute than for any other karst in the world [21,23]. Quarrying on a massive scale since the 1990s has erased entire hills, critically endangering a greater number of species than in any other subterranean system, with the possible exception of the karstic groundwater fauna of the Dutch Caribbean island of Curaçao [24]. Several spots of high aesthetic and biological value (Nui Com, Nui Cay Xoai, and the natural cirque at the center of Nui Bai Voi) have been irreversibly lost to limestone exploitation, despite the international efforts led and coordinated by Tony Whitten [25] to curb the appetites of mining companies [26]. At present, many of the largest limestone hills have either been erased (Nui Trau, Nui Com, Nui Cay Xoai) or are undergoing destruction (most of Nui Bai Voi, Nui Khoe La, Nui Hang Cay Ot), thus increasing the risk of extinction for micro-endemic species. In addition, one remarkable landmark in the flat and densely populated Mekong Delta—Hang Mo So—is being degraded by an increasing flux of tourists.

In most parts of MDL-HC, the severity of conservation threats is obvious at a glance. We therefore analyse the cave fauna of Hang Mo So (the focus of the present paper) in the context of the hill in which it is situated, namely Nui Bai Voi (Figure 1B), and the wider Hon Chong cluster of hills (Figure 1A). This extended coverage provides an idea of the number of troglobionts that have not yet been collected and of the proportion of site endemics in the fauna, several of which are on the brink of extinction.

2. Materials and Methods

2.1. Assessment of the Ecological Status of Species

Many publications define, or refine, the concepts of troglophiles, stygophiles, troglobionts and stygobionts. However, the grounds on which the ecological status of a species is assessed are less frequently addressed (but see [27,28]), despite the importance of understanding the degree of dependence on the subterranean environment when evaluating cave biodiversity.

Caves in the MDL-HC karst are usually short and shallow, with frequent terrestrial and aquatic connections to the exterior, allowing constant inputs of nutrients. This results in cave communities dominated by troglophiles and tramp species, making the ecological category assessment of individual species hazardous. We have therefore adopted four approaches:

(1)   Morphological inference, based on the presence or absence of troglomorphic traits. This approach is straightforward but raises two caveats. Firstly, depigmentation and

eye reduction are frequently considered troglomorphic, but they actually occur in most deep-soil species as well. Thus, a cave arthropod that has these two characteristics can only be qualified as troglomorphic if additional traits are present that have been established as cave-dependent [29], such as appendage elongation or larger body size. It is these additional traits that make the difference between troglomorphic species, which are almost always linked to subterranean life, and euedaphomorphic species, which dominate in the deep soil but are also present in caves. Secondly, troglomorphy is clearly linked to cave-restricted life, whereas euedaphomorphy can be linked to either cave or soil life. Since cave invertebrates are more often euedaphomorphic than troglomorphic, especially in lowland caves of the humid tropics like MDL-HC, morphological inference alone will not work for them.

(2) Parallel-sampling inference, based on the occurrence of species outside subterranean habitats. Many species found in caves are described as troglobionts in the literature, even though they do not exhibit typical troglomorphic traits, or only show euedaphomorphic traits similar to those of many deep-soil species. The absence of a species outside caves can be a good indicator in such cases. This information is often available in the literature for well-investigated regions, but not for the tropics, where it is thus necessary to gather data both inside and outside caves in comparable microhabitats. Extensive parallel sampling in the MDL-HC karst, with 270 cave samples and 674 non-cave samples (including 322 in mineral soil), allows a reasonably reliable ecological status assessment. Because the strength of such inference depends on sampling effort and species frequency, it will be less reliable for species with low population densities or patchy distributions.

(3) Taxonomic inference, based on the ecological status of related taxa. Certain groups are particularly prone to diversify in subterranean habitats, even though the underlying biological mechanisms are not well understood [30–32]. This may cast suspicion on the putative troglobiont status of a species when it belongs to a group that is otherwise not known for having cave-obligate species.

(4) Barcoding inference, based on levels of genetic divergence between populations. Cryptic diversity poses problems for the recognition of species using morphological characters alone. Molecular barcoding often reveals lineages with identical morphologies that show divergence levels as high as those encountered between traditional morphospecies, as has been demonstrated for several Collembola [33]. Moreover, such cryptic lineages may differ in their degree of dependence on cave habitats [34]. Molecular sequencing can, therefore, provide greater accuracy in the delimitation and ecological characterization of species. Another advantage is that it can allow otherwise unidentifiable larval forms to be correlated with their adult stages, especially in insects.

Here, we classify as a troglobiont (TB) or stygobiont (STB) a species (or morphospecies) that has been collected exclusively in caves and exhibits unambiguous troglomorphic traits, or that has euedaphomorphic traits (reduced eyes or/and depigmentation) without being known from soil. We consider as a putative troglobiont ("TB?") or stygobiont ("STB?") a species (or morphospecies) that shows neither euedaphomorphic nor troglomorphic traits, but has been collected exclusively in caves (e.g., guano-associated species). In the latter case, the assessment is less reliable when the species is rare.

We classify as a troglophile or a stygophile a species (or morphospecies) that has been collected in both caves and non-cave habitats in significant numbers, or is numerous in subterranean communities even if much less so than in the outside (as is the case for several tramp species), or is mostly found in the twilight zone. Troglophiles in Hon Chong caves are either eutroglophiles or subtroglophiles [29]. It is noteworthy that parietal communities in the MDL-HC caves and in the tropics include both eutroglophiles and non-seasonal subtroglophiles, whereas seasonal subtroglophiles largely dominate in temperate caves.

2.2. Sampling

2.2.1. Sampled Habitats

Though limited in extent, the MDL-HC karst displays a wide range of surface habitats due to its rugged relief. Adjacent microhabitats may differ greatly in soil thickness, inclination, limestone denudation, drainage and vegetation. Within caves, we sampled soil, roots (root-mats and hanging roots), habitats rich in nutrients (guano piles, scattered bat feces, organic debris from outside, water rich in organic matter and human-generated debris), nutrient-poor (oligotrophic) habitats (clay, speleothem surface, lakes, puddles, dripping water and endogenic streamlets), screes and boulders, and habitats at the subterranean/epigean interface (entrance zone, walls, springs and soil/limestone bedrock interface) (Figure 4). Non-limestone hills and the alluvial plain were also sampled for comparison. Terrestrial habitats were extensively sampled, but aquatic habitats only occasionally.

Soils and roots, which are often neglected in biological surveys, received special attention. In the caves of MDL-HC, soils are often rich in organic matter and in fauna, due to the massive input of material from the outside, through karst windows or water movement across the swamp areas of the alluvial plain, and the frequent occurrence of bats. Roots of *Ficus* (12 species in the area [15]) are common in subterranean passages, due to the shallowness of the caves and the deeply dissected karst. These occur as isolated roots or in dense aggregates of finer, hanging roots (Figure 4A,B,D). Less frequently, rootlets form a dense carpet on the floor of some passages that are probably flooded during the rainy season. Cave roots may host a rich and characteristic fauna in both groundwater [35] and terrestrial habitats [36–38], as was the case in Hang Mo So and several MDL-HC caves (see below).

2.2.2. Parallel Sampling and Techniques

For the reasons explained in (Section 2.1), we systematically carried out parallel sampling during surveys, pairing inside cave and external sampling. Samples taken outside caves outnumbered those from inside caves, in order to cover the higher diversity of surface microhabitats and offset patchy distributions.

Sampling techniques used for the aquatic fauna were limited to netting and filtration of water from puddles or lakes. The terrestrial fauna was much more extensively sampled by means of various techniques both inside and outside caves, combining techniques traditionally used in soil and subterranean arthropod surveys [39,40] with other more rarely employed methods (5 and 6 below):

(1) Collection by sight (timed or not timed) using a fine brush or pooter in all visited caves
(2) Bulk extraction of arthropods on Berlese funnels from litter and soil cores of standardized volumes, associated with larger unstandardized samples for rare species
(3) Sieving litter and debris for arthropods and gastropods
(4) Baiting and pitfall trapping for arthropods active on the ground—this being the only technique that produces significant numbers of invertebrates in oligotrophic cave habitats
(5) Beating of hanging roots in caves
(6) Mineral soil washing to collect deep-soil arthropods by flotation
(7) Flotation of litter and debris for snails

3. Results

The cave-obligate species or morphospecies of Hang Mo So and other MDL-HC caves (Figures 1 and 3) are listed in Table 1. Here, we comment on them, explaining the rationale for their ecological status. We also place them in their wider taxonomic and ecological contexts, and provide basic information on the main troglophilic species in the area. Authorship of stygobiotic and troglobiotic species is given in Table 1 and not repeated in the text.

Table 1. List of stygobionts and troglobionts of MDL-HC, with emphasis on Hang Mo So. Columns: Ecol, ecological category of species (STB, stygobiont; TB, troglobiont; "?", when uncertain); RL, published IUCN category for red-listed species (CR, critically endangered; EN, endangered; NT, near threatened; VU, vulnerable); End: single hill endemic; HMS, number of samples that contained the species in Hang Mo So; NBV, caves of Nui Bai Voi (number of samples that contained the species in parentheses), cave numbers as in Figure 1B; MDL-HC: hills of MDL-HC (number of samples that contained the species in parentheses), hill abbreviations as in Figure 1A.

Taxon	Species	Ecol	RL	End	HMS	NBV	MDL-HC
Gastropoda: Pomatiopsidae							
	Pseudoiglica sp.	STB		x	0		HC (2)
Actinotrichida: Leeuwenhoekiidae							
	gen. sp.	TB			1	bv00 (1), bv10 (1), bv12 (2)	BV (4), HT (1), KL (1)
Anactinotrichida: Opilioacaridae							
	Siamacarus sp.	TB			0	bv12 (1)	BV (1), HC (2), KL (2), LC (1), NA (1)
Amblypygi: Charinidae							
	Weygoldtia sp.	TB?			3	bv00 (3), bv12 (2), bv13 (1)	BT (1), BV (6), HC (8), KL (1), LC (1), NA (1)
Araneae: Ctenidae							
	gen.sp. 1	TB		x	0		KL (1)
	gen. sp. 2	TB			5	bv00 (5), bv08 (1), bv11 (1), bv12 (1), bv13 (1)	BT (1), BV (9), CH (1), HC (1), HT (3), KL (3), NA (1), NO (1)
Araneae: Halonoproctidae							
	Latouchia schwendingeri Decae, 2019	TB		+	0		HT (1)
Araneae: Ochyroceratidae							
	gen. sp. 1	TB?		x	0		BT (1)
	gen. sp. 2	TB			0		BT (1), HC (3)
	gen. sp. 3	TB?			0	bv12 (2)	BT (2), BV (2), KL (2), NO (1)
Araneae: Oonopidae							
	gen. sp. 1	TB?		x	0	bv12 (1)	BV (1)
	gen. sp. 2	TB			3	bv00 (3), bv08 (1)	BV (4), KL (1)
	gen. sp. 3	TB?			0	bv13 (1)	BT (1), BV (1)
Araneae: Pholcidae							
	gen. sp. 1	TB?		x	0		NA (1)
	gen. sp. 2	TB			0	bv13 (1)	BT (2), BV (1), HC (1), LC (1), NA (1)
	gen. sp. 3	TB?			0		HC (1), NA (1)
	gen. sp. 4	TB?			1	bv00 (1), bv08 (1)	BV (2), HC (1), KL (1)
Araneae: Telemidae							
	gen. sp. 1	TB?		x	0		HC (1)
	gen. sp. 2	TB?		x	1	bv00 (1)	BV (1)
	gen. sp. 3	TB?		x	0	bv13 (1)	BV (1)
	gen. sp. 4	TB		x	0		KL (2)
	gen. sp. 5	TB?		x	0	bv13 (1)	BV (1)
Araneae: Tetrablemmidae							
	gen. sp.	TB?		x	1	bv00 (1)	BV (1)
Opiliones: Epedanidae							
	gen. sp. 1	TB			4	bv00 (4), bv09 (1)	BT (1), BV (5)
	gen. sp. 2	TB		x	0		HC (1)
Opiliones: family undet.							
	gen. sp.	TB?			0		BT (1)
Pseudoscorpiones: Chthoniidae							
	Lagynochthonius fragilis Judson, 2007	TB		x	5	bv00 (5)	BV (5)

Table 1. *Cont.*

Taxon	Species	Ecol	RL	End	HMS	NBV	MDL-HC
Diplopoda: Haplodesmidae							
	gen. sp.	TB?		x	1	bv00 (1)	BV (1)
Diplopoda: Pyrgodesmidae							
	gen. sp.	TB			0		HC (1), HT (1)
Diplopoda: Trichopolydesmidae							
	gen. sp.	TB?			0		HT (1)
Diplopoda: Siphonophoridae				x			
	gen. sp.	TB?		x	1	bv00 (1)	BV (1)
Diplopoda: Cambalopsidae							
	Glyphiulus sp.	TB?			0		BT (3), NC (1)
	Trachyjulus singularis Attems, 1938	TB?			14	bv00 (14), bv01 (3), bv05 (2), bv08 (3), bv09 (1), bv10 (1), bv11 (1), bv12 (3), bv13 (1)	BV (29), CH (2), HT (4), KL (6), LC (3), NA (3)
Diplopoda: Stemmiulidae							
	Eostemmiulus caecus Mauriès, Golovatch & Geoffroy, 2010	TB	CR	x	1	bv00 (1)	BV (1)
Amphipoda: Bogidiellidae							
	gen. sp.	STB		x	0		BT (1)
Isopoda Oniscidea: family undet.							
	gen. sp.	TB?		x	0		BT (1)
Isopoda Oniscidea: Armadillidae							
	Sumatrillo sp.	TB	VU		0		HC (1), HT (2)
Isopoda Oniscidea: Philosciidae							
	Burmoniscus sp.	TB	EN	x	3	bv00 (3), bv12 (2)	BV (5)
	gen. sp.	TB?		x	0		LC (1)
Isopoda Asellota: Stenasellidae							
	Stenasellus sp.	STB		x	0		HC (1)
Collembola: Hypogastruridae							
	Acherontiella sp.	TB?		x	0		NO (1)
	Ceratophysella sp.	TB	CR	x	0		HC (1)
Collembola: Tullbergiidae							
	gen. sp.	TB			0		BT (1), HT (1), LC (1), NC (1)
Collembola: Entomobryidae							
	Acrocyrtus (cf.) sp.	TB	VU	x	0		KL (1)
	Ascocyrtus sp.	TB?		x	1	bv00 (1)	BV (1)
	Coecobrya sp.	TB		x	1	bv00 (1), bv11 (1)	BV (2)
	Lepidocyrtinae gen. sp.	TB			0		BT (1), LC (1)
	Lepidosinella sp.	TB			2	bv00 (2)	BV (2), KL (1)
Collembola: Isotomidae							
	Folsomides anops Deharveng, Bedos & Lukić, 2020	TB	VU	x	0		BT (2)
	Folsomides whitteni Deharveng, Bedos & Lukić, 2020	TB		x	2	bv00 (2), bv08 (3)	BV (5)
Collembola: Paronellidae							
	Lepidonella lecongkieti Deharveng & & Bedos, 1995	TB	NT		7	bv00 (7), bv01 (2), bv08 (3), bv12 (2), bv13 (3)	BT (3), BV (17), CH (1), HC (6), HT (8), KL (9), NA (1), NO (2)
	Lepidonella sp.	TB?		x	1	bv00 (1)	BV (1)

Table 1. *Cont.*

Taxon	Species	Ecol	RL	End	HMS	NBV	MDL-HC
Collembola: Neelidae							
	Spinaethorax adamantis Schneider & Deharveng 2017	TB		x	0		HC (4)
	Spinaethorax sp. 1	TB		x	0	bv01 (1)	BV (1)
	Spinaethorax sp. 2	TB?		x	0		NC (1)
Diplura: Japygidae							
	gen. sp.	TB?		x	0	bv01 (1), bv12 (1)	BV (2)
Zygentoma: Family ind.							
	gen. sp.	TB			0	bv13 (1)	BV (1), HC (1)
Zygentoma: Nicoletiidae							
	gen. sp. 1	TB?		x	0		BT (1)
	gen. sp. 2	TB		x	1	bv00 (1)	BV (1)
Blattodea: Nocticolidae							
	Spelaeoblatta sp.	TB			4	bv00 (4), bv01 (1), bv08 (1), bv12 (1), bv13 (1)	BV (8), CH (1), KL (3), NA (1)
Coleoptera: Carabidae							
	Eustra honchongensis Deuve, 1996	TB	EN	x	4	bv00 (4), bv01 (1), bv12 (1), bv13 (1)	BV (7)
Coleoptera: Curculionidae							
	gen. sp.	TB			0		BT (1), KL (1)
Coleoptera: Tenebrionidae							
	Harvengia vietnamita Ferrer, 2004	TB	EN		0		BT (2), CO (1), HC (5), KL (1)
	Pseudochillus honchongensis Schawaller & Faille, 2023	TB		x	1	bv00 (1)	BV (1)
Heteroptera: Reduviidae: Harpactorinae							
	gen. sp.	TB?			0	bv12 (1)	BV (1), HC (1), NA (1), NC (1)
Heteroptera: Schizopteridae							
	gen. sp.	TB?		x	0		BT (1)
Homoptera: Cixiidae							
	gen. sp. 1	TB			2	bv00 (2), bv07 (1), bv08 (1), bv11 (1), bv12 (1), bv13 (2)	BT (1), BV (8), CO (1), KL (2)
	gen. sp. 2	TB		x	0		KL (2)
Homoptera: Delphacidae							
	gen. sp.	TB		x	1	bv00 (1), bv08 (2)	BV (3)
Homoptera: Kinnaridae: Kinnarini							
	gen. sp.	TB?		x	1	bv00 (1)	BV (1)
Total				**43**	**27**	**38**	**70**

3.1. Gastropoda (Snails)

The land snail fauna (including several introduced species) of the MDL-HC karst is an 'island fauna' [41] with relatively few species (only 87 recorded so far). Many species that are common and widespread in Indochina are notably absent. However, the rate of endemism (to MDL or parts of MDL) is staggering: 52 species or 60% of the fauna, with several groups still in need of investigation. Snail surveys in the MDL karst were aimed at collecting shells rather than the entire organisms. The resulting collection provides an adequate overview of the fauna, but it is biased against slugs and semi-slugs and does not allow the observation of adaptations to cave life. None of the collected species can be qualified as a troglobiont, but several are considered troglophiles here because they are also found in deep-soil samples [42], and one is a stygobiont.

Several genera found in soils and caves of MDL are of special interest. *Macrochlamys psyche* Vermeulen, Luu, Theary & Anker, 2019 (Ariophantidae) (Figure 5A) is endemic to Nui Bai Voi and nearby Nui Khoe La, and is frequent in Hang Mo So. The genus *Speleocyclotus* Vermeulen, Luu, Theary & Anker, 2019 (Cyclophoridae) was established to accommodate four endemic MDL species. Among these are *S. macrocoryphe* Vermeulen, Luu, Theary & Anker, 2019 (Figure 5B), from soil and nutrient-poor areas in a cave in Nui Hon Chong, and *S. microcoryphe* Vermeulen, Luu, Theary & Anker, 2019, from Hang Mo So and deep soil deposits in Nui Bai Voi, and three localities in Cambodia. Elsewhere, the genus occurs in the Malaysian Peninsula. The troglophilic genus *Notharinia* Vermeulen, Phung & Truong, 2007 (Diplommatinidae) is represented in MDL by 10 tiny (ca 2 mm long) species with partly overlapping ranges. Surprisingly, none have been reported yet from the most intensely sampled hill, Nui Bai Voi, although nearby Nui Khoe La is home to two site-endemic [43] species. Elsewhere, the genus is only known from two species, one from northern Laos and one from Sarawak. Finally, the cave Hang Gieng Tien in Nui Hon Chong has, at its end, a basin of a few cubic meters (Figure 4C), which is the only known locality of a site-endemic hydrobioid freshwater snail, probably of the genus *Pseudoiglica* Grego, 2018.

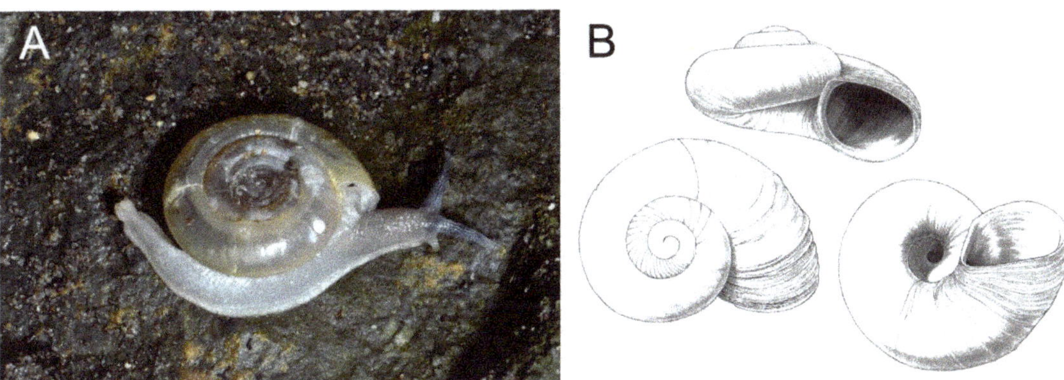

Figure 5. Gastropoda: (**A**) *Macrochlamys psyche*; (**B**) *Speleocyclotus macrocoryphe*. Photo and drawings: Jaap Vermeulen.

3.2. Arachnida

- Anactinotrichida (parasitiform mites)

An opilioacarid of the genus *Siamacarus* Leclerc, 1989 is remarkable for its extremely elongated legs (Figure 6A). It has only been collected in caves, mostly under loose rocks. It was not found in Hang Mo So, but it occurred in cave BV12 of Nui Bai Voi and in six other MDL-HC caves. Otherwise, *Siamacarus* is only known from two cave species in southern Thailand [44].

- Actinotrichida (acariform mites)

Very few cave mites have been identified from Southeast Asia [45]. The numerous small mites associated with guano have not been analyzed in the present study, but a large obligate cave species collected in Hang Mo So belongs to Leeuwenhoekiidae (Figure 6B), a family of Trombidioidea that occurs frequently in caves of Southeast Asia.

Caecothrombium deharvengi Makol & Gabrys, 2005 is only known from deep soil in the twilight zone near an entrance of Hang Tien in Nui Hang Tien. It is one of the three species of the monogeneric subfamily Caecothrombiinae (Eutrombidiidae). The species is small (body length about 1 mm), eyeless, and clearly euedaphomorphic, but it is not possible to reliably determine its ecological status because most mites in our samples remain unidentified.

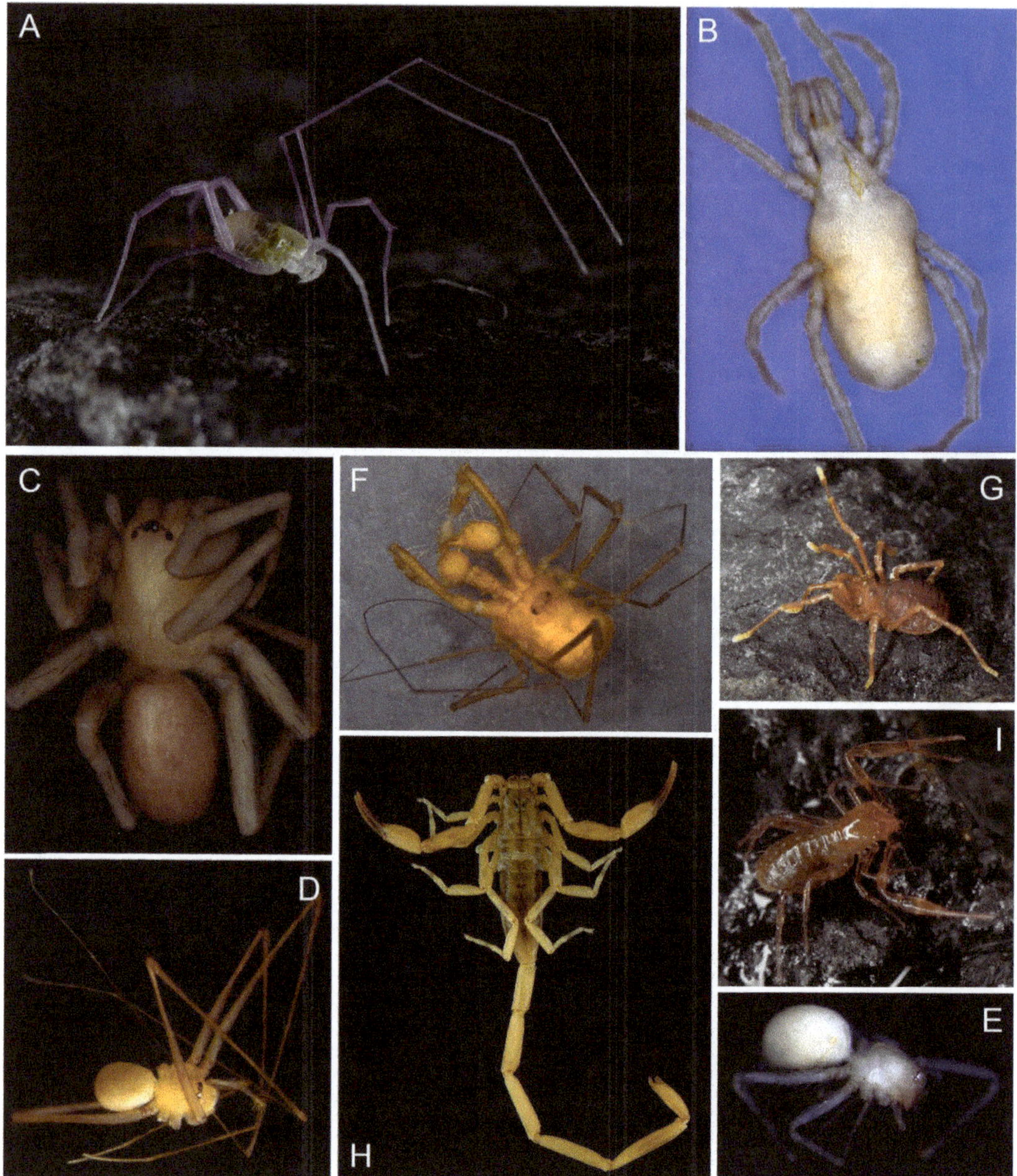

Figure 6. Arachnida: (**A**) *Siamacarus* sp. from a cave of Nui Nai; (**B**) *Leeuwenhoekiidae* sp. from Hang Tien (Nui Hang Tien); (**C**) *Ctenidae* sp. 2 from cave BV08 of Nui Bai Voi; (**D**) *Telemidae* sp. 5 from cave BV13 of Nui Bai Voi; (**E**) *Ochyroceratidae* sp. 2 from Hang Kim Cuong (Nui Hon Chong); (**F**) *Epedanidae* sp. 1 from cave BV09 of Nui Bai Voi; (**G**) *Gnomulus bedoharvengorum* from cave BV12 of Nui Bai Voi; (**H**) *Isometrus (Reddyanus) deharvengi* from Hang Gieng Tien (Nui Hon Chong); (**I**) *Lagynochthonius fragilis* from Hang Mo So. Photos: Louis Deharveng (**B–F**), Marko Lukić (**A,G,I**), Elise-Anne Leguin from the Museum national d'Histoire naturelle, Paris (**H**).

- Amblypygi (whip spiders)

Three species of Amblypygi are present in MDL, but only one, *Weygoldtia* sp. (probably the same as "cf. *Sarax*" in [22] and *Stygophrynus* sp. in [19]), has been found in MDL-HC, including Hang Mo So. Elsewhere, three species of *Weygoldtia* are known from surface habitats and under stones in Southeast Asia [46–48]. However, Zhu et al. [49] found another species in Hainan, China, in karst crevices—a habitat that has not been sampled intensively in MDL-HC. Therefore, we consider the Hon Chong *Weygoldtia* a questionable troglobiont, even though it is currently only known from caves.

- Araneae (spiders)

With the possible exception of mites, spiders are the most diverse terrestrial invertebrates in the caves of MDL-HC. Many of the species collected are troglophiles or trogloxenes, and several are probably troglobionts (Figure 6C–E). They have yet to be studied taxonomically. Interpreting a spider as a troglobiont or troglophile can be difficult, even in temperate lineages [50]. In MDL-HC, the diversity and euedaphomorphic morphology of the many small species from soil or caves makes the task particularly complicated, even with parallel sampling. A number of these species, which are rather similar in appearance, are not included in our list since they were mostly obtained from soil samples and are often immature.

Anapidae. One species, probably *Pseudanapis paroculus* (Simon, 1899), was collected twice in Hang Mo So. It is also present in the nearby cave of Son Cha. In the Kien Luong karst, it was found outside caves on Nui Chau Hang. It is known from soils elsewhere in SE Asia [51].

Ctenidae. A rather large species, possessing eyes but devoid of pigment, is the most common troglobiotic spider in the caves of MDL-HC. It was not found outside caves. A similar, but eyeless, species was found only in the southern cave of Nui Khoe La.

Mygalomorphae. Three troglophilic mygalomorphs have been recorded in caves of MDL-HC: a species of Ctenizidae, found in six caves of MDL-HC, including Hang Mo So, and in a single litter sample on Nui Ba Tai; *Latouchia schwendingeri* (Halonoproctidae), a small species known only from the type locality, Hang Tien in Nui Hang Tien [52]; and a relatively large species found in soil on Nui Bai Voi and Nui Hon Chong, and in caves at Nui Ba Tai and Nui Chau Hang.

Ochyroceratidae and Telemidae occur rather frequently in the Hon Chong caves and soils. They are small (less than 2 mm), with reduced pigmentation and eyes. Six morphospecies belong to Telemidae: five are rare troglobionts (including one in Hang Mo So) and the sixth is frequent in caves of MDL-HC, but also present in the soil. Three cave morphospecies belong to Ochyroceratidae, of which one, from some Nui Ba Tai and Nui Hon Chong caves, probably belongs to the speciose genus *Speocera*, known from caves and soil in Southeast Asia [53].

Oonopidae. Five morphospecies of this family are restricted to caves of MDL-HC. That from BV12 in Nui Bai Voi has a reduced number of unpigmented eyes. The other two, from Bai Voi Hill, have well-developed eyes; one, with a small dorsal plate on the abdomen, occurs in three caves, including Hang Mo So. Surprisingly, the only blind morphospecies of Oonopidae encountered in MDL-HC was collected more frequently from soil than from caves.

Pholcidae. Five morphospecies of Pholcidae occur in the MDL-HC caves. All have eyes, but their bodies are, at most, feebly pigmented. One, with pale patches of pigment on the abdomen, was found in several caves but never outside; three others are potential troglobionts, but, because of the rarity of adult specimens in our samples, we cannot now confirm their ecological status.

Sparassidae. Steiner [22] recorded *Heteropoda* sp. in only one cave of MDL-HC while citing it as present in several caves in the nearby mountain of Da Dung, close to Ha Tien. We confirm this unusual rarity of huntsman spiders, which are otherwise very common in most Southeast Asian caves.

Tetrablemmidae. Few cave species were known in the 1980s [54], but several have since been described, particularly from southern China [55]. A single species occurs in Hang Mo So and seems to be limited to this cave. Elsewhere in MDL-HC, two widespread eyed morphospecies are troglophiles. Another blind species, found in two caves and in the outside soil, is also present in MDL-C.

- Opiliones (harvestmen)

Harvestmen (Laniatores) are rare but diverse in caves of MDL-HC (Figure 6F,G). *Gnomulus bedoharvengorum* Schwendinger & Martens, 2006 (Sandakanidae) is a robust troglophile from Hang Mo So, Hang Tien, and the soils of Nui Bai Voi and Nui Hon Chong [56]. Later, it was also collected on Nui Khoe La. A long-legged, unpigmented but eyed Epedanidae, resembling morphospecies from several caves of Southeast Asia, was found as isolated individuals in oligotrophic habitats in two caves of Nui Bai Voi and one cave of Nui Ba Tai, on walls and on the floor. In the latter cave, it co-occurs with another troglobiont harvestman, unidentified to family. In Hang Gieng Tien cave (Nui Hon Chong), a cave-adapted Epedanidae and a less modified Tithaeidae occur. Interestingly, a small, unpigmented Petrobunidae with reduced eyes is common in soil and litter, but rare inside caves (only in Nui Ba Tai).

- Palpigradi

Palpigradi are common in soils, but were only collected in six MDL-HC caves, including Hang Mo So. Two genera occur in soil: *Eukoenenia* (Börner, 1901) and *Prokoenenia* Börner, 1901 [21]. The cave specimens, not identified to genus, are provisionally classified as troglophiles. The two caves in Southeast Asia where Palpigradi have been studied, Tham Chiang Dao in Thailand and Towakkalak in Sulawesi, each hosted two different species [6,57–59]. Cave species can, therefore, be expected in our unidentified MDL-HC material.

- Pseudoscorpiones

One species, the chthoniid *Lagynochthonius fragilis* (Figure 6I), is a troglobiont restricted to Hang Mo So [60]. An undescribed species of *Cryptocheiridium* (Chamberlin, 1931) (Cheiridiidae) has been found in deep soil at the entrance of cave BV03 on Nui Bai Voi. Two Feaellidae are known in the MDL karstic hills: *Cybella deharvengi* Judson, 2017 from Nui Bai Voi (in the small cavity BV07) and near Hang Mo So (in soil), and *Cybella bedosae* Judson, 2017 from a shallow cave of MDL-C. The subfamily Cybellinae was erected for the new genus *Cybella* and *Protofeaella* Henderickx, 2016, a fossil genus from Burmese amber [61]. Later, two additional *Cybella* species were described from caves in Malaysia [62]. These records are of biogeographical interest because Feaellidae were previously unknown from Southeast Asia. More troglobionts and troglophiles might be found among the numerous Chernetidae, Chthoniidae and Ideoroncidae in our material from MDL.

- Schizomida

Schizomida of the family Hubbardiidae are not uncommon in the MDL-HC caves and soil. They are absent in Hang Mo So, but were collected on several occasions in cave BV13 of the same hill (Nui Bai Voi). They are also present in several caves of Nui Ba Tai, Nui Khoe La and Nui Hon Chong. Schizomida are under-studied in Southeast Asia, but their high diversification in Australasia [63] might prefigure a similar richness in Southeast Asia caves.

- Scorpiones

Isometrus (*Reddyanus*) *deharvengi* Lourenço & Duhem, 2010 (Figure 6H) is an elegant buthid scorpion [64,65], collected as isolated specimens in Nui Bai Voi caves (Hang Mo So and BV07), Nui Khoe La (one cave) and Nui Hon Chong (one cave). We list it as troglophile because it also occurred in a deep crevice of Nui Cay Ot.

3.3. Diplopoda (Millipedes)

Compared with central Indochina or China [66,67], the MDL-HC caves, including Hang Mo So, are poor in millipedes. Typically, one troglobiotic Cambalopsidae and one troglophilic Haplodesmidae occur in a cave. The tramp species *Orthomorpha coarctata* (de Saussure, 1860) is locally present in caves and soils.

As elsewhere in Southeast Asia, Cambalopsidae occur in dense monospecific populations on guano (Figure 4E) in many MDL caves and more rarely in soils. They include four troglobionts (*Trachyjulus singularis*, *Plusioglyphiulus boutini* Mauriès, 1970, *Plusioglyphiulus biserratus* Likhitrakarn et al., 2020 and *Glyphiulus* sp.) and two epigean species (*Plusioglyphiulus khmer* Likhitrakarn et al., 2020 and an undescribed species from Nui Ong). *Plusioglyphiulus boutini*, *P. biserratus* and *P. khmer* are only known from MDL-C [68]. *Trachyjulus singularis* (Figure 7A) is the only Cambalopsidae in Hang Mo So. It also occurs in several other MDL-HC caves. The undescribed *Glyphiulus* sp. is limited to Nui Ba Tai and Nui Coc caves. All Cambalopsidae species show mutually exclusive distributions. For instance, *Glyphiulus* sp. replaces *T. singularis* on the hills mentioned, notwithstanding the short distance between the distribution areas of the two species, their high abundance where present, and the relatively large range of *T. singularis* (southeast Vietnam and southeastern Thailand). In MDL-HC and in Thailand, *T. singularis* occurs exclusively in caves, but the population at its type locality ("Pulau Condor" = Con Son Island), 270 km to the southeast, is epigean [69]. The MDL-HC populations are here considered as a troglobiont lineage of *T. singularis* of uncertain taxonomic status, as is also suggested by unpublished barcode studies. Interestingly, no Cambalopsidae have ever been found in Nui Hon Chong caves or soils. This is quite unusual in southeast Asian karsts and may reflect a long isolation.

Eostemmiulus caecus, the only species of the genus, is known from Hang Mo So (Figure 7B). Geographically, it is isolated from other Stemmiulida; the nearest species occur in Sri Lanka, 2600 km to the west, and in Maluku (Indonesia), 2800 km to the southeast [70]. Morphologically, *E. caecus* is blind—a trait only shared with the non-cave African *Stemmiulus oculiscaptus* Demange & Mauriès, 1975 among Stemmiulida. *Eostemmiulus caecus* was only found in a small oligotrophic recess of Hang Mo So, but we assume that it persists in the other narrow under-sampled southern passages of the cave.

Among Haplodesmidae, *Eutrichodesmus griseus*, described from soil and rather widespread in MDL-HC, is more frequent inside than outside caves in our samples. Its closest relative, *E. cambodiensis* Srisonchai & Panha, 2020, endemic to the Tuk Meas karst, occurs outside caves [71]. Another Haplodesmidae unique to Hang Mo So might be a troglobiont. The related family Pyrgodesmidae is absent from Hang Mo So but represented by a rare cave species in oligotrophic habitats of Hang Tien (Nui Hang Tien) and Hang Kim Cuong (Nui Hon Chong). A slightly troglomorphic Trichopolydesmidae has also been found in oligotrophic Hang Tien habitats, while epigean representatives of this family are frequent and diversified in soil and litter of MDL [72].

Hang Mo So hosts at least one Siphonophorida with slightly elongated antennae, unknown elsewhere, which could, therefore, be a troglobiont.

3.4. Malacostracea: Isopoda

- Oniscidea (woodlice) (Figure 7C–E)

The oniscid fauna of MDL-HC is characterized by (i) the absence of Trichoniscidae and Styloniscidae and the dominance of Armadillidae and Philosciidae, and (ii) diverse and abundant troglophilic species. Troglobionts include four rare species.

Most remarkable is an undescribed *Sumatrillo* Taiti, Paoli & Ferrara, 1998, white and eyeless, from oligotrophic habitats of two caves (Hang Tien in Nui Hang Tien and Hang Gieng Tien in Nui Hon Chong). Another species of unidentified family, also white and eyeless, is a putative troglobiont in a cave of Nui Ba Tai. The third troglobiont is a *Burmoniscus* Collinge, 1914 with an unmistakable colour pattern, only found in hanging roots of Hang Mo So and in BV12. The fourth is a small Philosciidae with reduced eyes and pigment from a cave in Hon Lo Coc.

Figure 7. Diplopoda and Crustacea: (**A**) *Trachyjulus singularis* from a cave of Nui Nai; (**B**) *Eostemmiulus caecus* from Hang Mo So; (**C**) *Sumatrillo* sp. from Hang Gieng Tien (Nui Hon Chong); (**D**) *Troglodillo* sp. from a cave entrance in Phnom Sray Toch; (**E**) *Burmoniscus* sp. from Hang Mo So; (**F**) *Stenasellus* sp. from Hang Gieng Tien (Nui Hon Chong). Photos: Louis Deharveng (**B,D,E**), Marko Lukić (**A,C,F**).

Troglophiles are much more abundant than troglobionts in the MDL-HC caves. Two Philosciidae, a *Burmoniscus* and *Pseudotyphloscia alba* (Dollfus, 1898), are often swarming over hanging roots at Hang Mo So. In and around guano piles, Philosciidae are replaced by various species of Armadillidae, Platyarthridae and Trachelipodidae, which sometimes occur in large colonies. Several have a wide distribution, such as *Cubaris murina* Brandt, 1833, a very common species in Hang Mo So. *Nagurus pallidipennis* (Dollfus, 1898), only recorded from caves in MDL-HC, is reported from epigean habitats in southern Asia [73]. A large *Troglodillo* Jackson, 1937, showing an attractive colour pattern, is often found in large numbers at entrances of several MDL-HC caves and less frequently in litter and soil. The soil inside the caves of MDL-HC often hosts a small blind species of *Hybodillo* Taiti, Paoli & Ferrara, 1998, which occurs rarely in deep soils outside caves. It is absent from Hang Mo So.

- Asellota (Figure 7F)

Stenasellus cambodianus Boutin & Magniez, 1985 was described from a cave of the Tuk Meas karst in MDL-C. We discovered another large species in the freshwater lake of Hang Gieng Tien (Nui Hon Chong) (Figure 4C) at the extreme south of MDL-HC. The genus likely occurs in between these localities, but it has not been found in Hang Mo So, where accessible waters are slightly brackish.

3.5. Collembola (Springtails)

Springtails form an abundant and diverse component of cave faunas worldwide [31]. In MDL-HC, 15 troglobiont or probable troglobiont species occur. Troglophilic species also abound, encompassing both regional species and pantropical "tramp" species. The most common springtails of eutrophic MDL caves are tramps, such as *Xenylla yucatana* Mills, 1938, a bisexual, pigmented and oculate Hypogastruridae, and five species of Isotomidae, all parthenogenetic, pantropical, and with reduced eyes and pigment. They also colonize a wide range of surface habitats in the tropics, especially in disturbed areas.

- Poduromorpha

Hypogastruridae. Two troglobionts occur in MDL-HC. In Hang Kim Cuong (Nui Hon Chong), we found an undescribed relictual species of the large Holarctic genus *Ceratophysella* Börner, 1932, which has very few representatives in southern Asia, none of which was cave-restricted so far. The second is a putative troglobiont from a cave in Nui Ong. Two troglophilic Hypogastruridae are also present in MDL-HC caves. *Willemia* cf. *buddenbrocki*, a species of euedaphomorphic facies, was found in several caves and in deep soil outside. The tramp species *Xenylla yucatana* is often the dominant springtail in guano and is also common in soils.

Tullbergiidae. A new genus occurs in the soils of several hills. One species so far is cave-restricted.

- Entomobryomorpha (Figure 8A,B)

Isotomidae. This large family is abundant and diversified in surface habitats. Soil cores taken inside caves of MDL-HC produced the first troglobiotic Isotomidae for Southeast Asia, *Folsomides anops* and *F. whitteni*. Both are rare and exhibit euedaphomorphic characters: minute size and regression of the pigment, eyes and furca. Other Isotomidae that are common in Hang Mo So and in many MDL-HC caves are pantropical tramp species. They are often the most abundant springtails in surface soils and subterranean habitats: *Folsomides centralis* (Denis, 1931), *F. parvulus* Stach, 1922, *Folsomina onychiurina* Denis, 1931, *Isotomiella nummulifer* Deharveng & Oliveira, 1990 and *I. symetrimucronata* Najt & Thibaud, 1987.

Entomobryidae: Entomobryinae. *Coecobrya* Yosii, 1956 is represented by several eyeless or microphthalmic morphospecies in soils and caves of MDL [21]. The only described species in MDL, *C. tukmeas* Zhang, Deharveng, & Chen, 2009, occurs in a cave at Tuk Meas. Among the blind morphospecies, one with slightly elongated antennae, found in two caves of Nui Bai Voi, is listed as a troglobiont here.

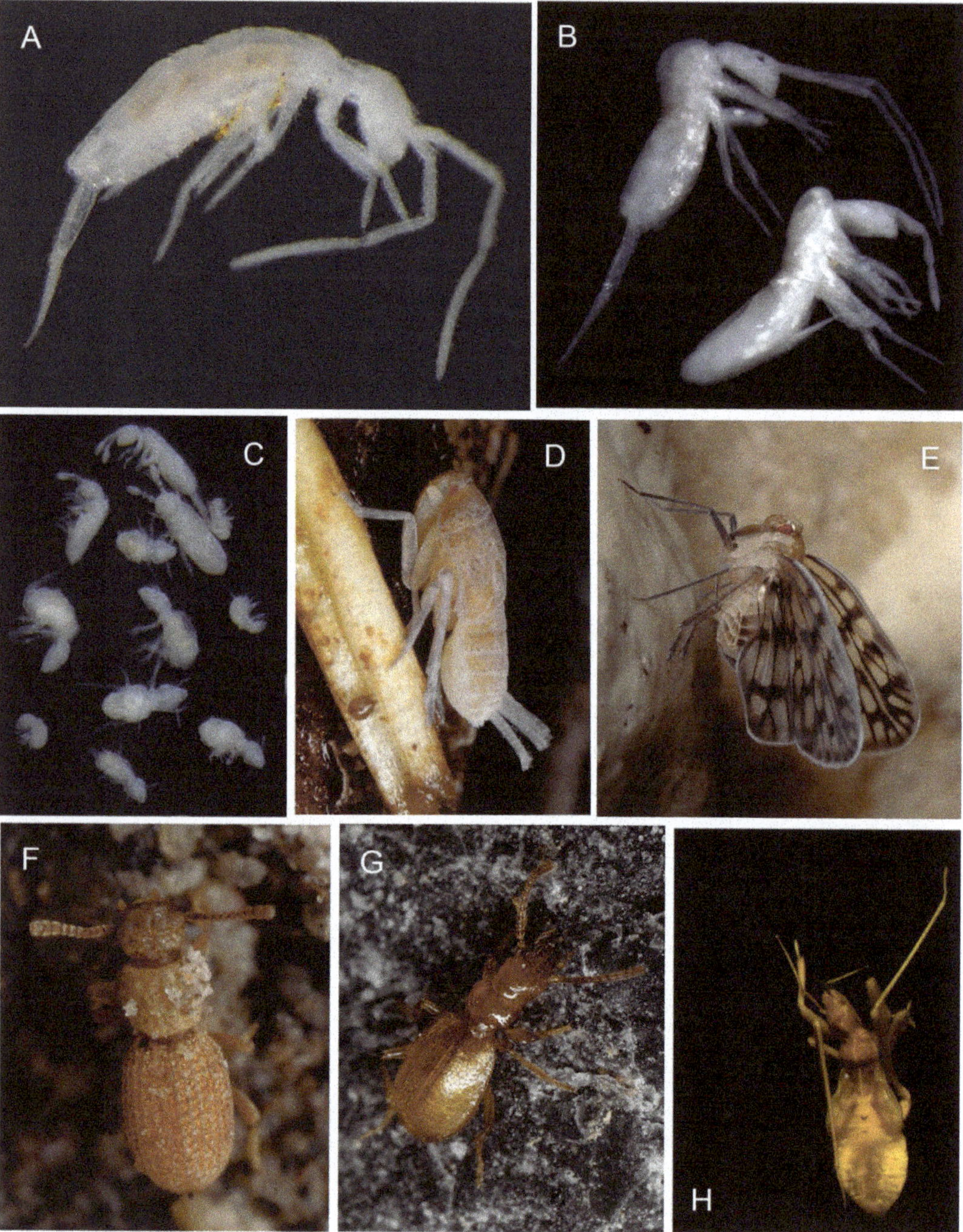

Figure 8. Hexapoda: (**A**) *Lepidosinella* sp. from Hang Mo So; (**B**) *Lepidonella lecongkieti* and *Acrocyrtus* (cf.) sp. from cave KLs of Nui Khoe La; (**C**) *Spinaethorax adamantis* and two troglophilic species, *Megalothorax laevis* and a blind *Rambutsinella* from Hang Kim Cuong (Nui Hon Chong); (**D**) larva of *Cixiidae* sp. 1 from cave BTnw of Nui Ba Tai; (**E**) *Borysthenes* sp. from a cave of Nui Khoe La; (**F**) *Harvengia vietnamita* from a cave of Nui Hon Chong; (**G**) *Eustra honchongensis* from cave BV12 of Nui Bai Voi; (**H**) Harpactorinae sp. from a cave of Nui Coc. Photos: Louis Deharveng (**A–C,H**), Marko Lukić (**D–G**).

Entomobryidae: Lepidocyrtinae. This subfamily is rich in troglobionts in temperate regions, but not in the tropics. Two MDL-HC species are listed as troglobionts: one eyeless species of an undescribed genus resembling *Acrocyrtus* (Yosii, 1959) occurs in the southern cave of Nui Khoe La; the second is an oculate species of uncertain generic assignment that lives in cave guano of two hills (Nui Ba Tai and Hon Lo Coc). A species of the genus *Ascocyrtus* Yosii, 1963 is abundant in hanging roots of Hang Mo So and might be troglobiotic. Another MDL-HC *Ascocyrtus*, also common in hanging roots of Hang Mo So, is found occasionally in humid litter on Nui Bai Voi and nearby hills, and can be considered a troglophile. In Southeast Asian soils, the genus *Rambutsinella* Deharveng & Bedos, 1996 is among the dominant springtails [74]. The species have a reduced number of eyes and are usually faintly pigmented. The most common in MDL-HC is *R. honchongensis* Deharveng & Bedos, 1996, from soils of Nui Ba Tai and Nui Bai Voi. In line with its euedaphomorphic morphology, it is frequent in deep soils but also occurs in litter and caves. Another undescribed, white, blind *Rambutsinella* is common in caves and deep soils, but restricted to Nui Hon Chong. *Pseudosinella* is the Collembolan genus with the highest number of troglobionts in temperate regions. It includes a few cave-obligate species in Southeast Asia (Thailand [75], Halmahera [76], Sulawesi [77]). The single *Pseudosinella* that occurs in the MDL-HC caves is a small, white, eyeless species which is one of the commonest troglophilic springtails.

Entomobryidae: Willowsiinae. The most remarkable springtail of Hang Mo So belongs to the genus *Lepidosinella* Handschin, 1920, previously known from a single termitophilous species, *L. armata* Handschin, 1920, in East Java [78]. *Lepidosinella* sp. occurs in Hang Mo So as rare, isolated specimens, not associated with termites or ants. It is a troglomorphic species, white and eyeless like *L. armata*, with significantly elongated appendages. *Hawinella* Bellinger & Christiansen, 1974 (with two species from Hawaii) and *Lepidosinella* are the only genera of the large subfamily Willowsiinae that have colonized subterranean habitats.

Paronellidae. This mainly tropical family includes epigean, soil, and many cave species. *Lepidonella*, the only genus present in the MDL-HC caves, includes numerous undescribed species from soils and caves in Vietnam, from Hon Chong to Ha Long Bay, in peninsular Malaysia and on Sumatra. Most cave species are clearly troglomorphic, showing reduced eyes and depigmentation, and having longer appendages and a larger size than epigean species. *Lepidonella lecongkieti*, the first troglobiotic invertebrate to be described from MDL-HC [79], and *L. doveri* (Carpenter, 1933) from the Batu Caves of Malaysia are the only troglomorphic species described in the genus [80]. *Lepidonella lecongkieti* is widespread in oligotrophic habitats of the MDL-HC caves. Another putative troglobiont, *Lepidonella* sp., is more similar to surface species and has only been found in Hang Mo So.

Cyphoderidae. Widespread across the world, this family includes many eyeless and unpigmented myrmecophilous or termitophilous species. Species resembling the temperate, strictly myrmecophilous *Cyphoderus albinus* Nicolet, 1842 form a major component of cave communities in Hang Mo So and other MDL-HC caves. As in Thailand [81], they are not associated with ants or termites in caves, and are rarely present in soils.

- Symphypleona

Sminthuridae. In most tropical regions, *Pararrhopalites* Bonet & Tellez, 1947 replaces the temperate Arrhopalitidae genera *Arrhopalites* Börner, 1906 and *Pygmarrhopalites* Vargovitsh, 2009. All three are diversified in caves and soils, and generally have reduced pigment and eyes. An unpigmented and microphthalmic *Pararrhopalites* species is present in soils and several caves of MDL-HC, including Hang Mo So.

- Neelipleona (Figure 8C)

Neelidae. The genera *Neelus* (Folsom, 1896), *Megalothorax* (Willem, 1900) and *Spinaethorax* Papáč & Palacios-Vargas, 2016 are present in several MDL-HC caves. The latter two occur in Hang Mo So. *Spinaethorax*, known from two cave-restricted species in Mexico [82], was unexpectedly discovered in the MDL-HC caves. *Spinaethorax adamantis* was described from Hang Kim Cuong (Nui Hon Chong) [83], and other undescribed and possibly troglobiotic species occur on Nui Bai Voi and Nui Coc (Table 1), as well as in Thailand (Surat Thani Province; Deharveng and Bedos, personal unpublished data). *Megalothorax*, which includes the smallest species of Collembola, is very common in all cave habitats and in soils. Several species occur in caves of MDL-HC, including Hang Mo So, but their taxonomic status needs to be updated following the new standards of Schneider [84].

3.6. Diplura

Campodeidae are frequent in caves and soils of MDL-HC. *Lepidocampa* cf. *weberi*, widespread in Southeast Asia, occurs in Hang Mo So and in soils of Nui Bai Voi, Nui Ba Tai and Nui Khoe La. Japygidae are not infrequent in the MDL-HC soils and caves. Large individuals with elongated antennae are possible troglobionts in Nui Bai Voi caves.

3.7. Zygentoma

Zygentoma are extremely rare in caves of Southeast Asia. Three potentially troglobiotic species were found as isolated specimens in a few MDL-HC caves, including a Nicoletiidae at Hang Mo So.

3.8. Pterygota

- Blattodea (cockroaches)

Nocticolidae are among the most common troglobionts of Southeast Asia [29] but remain largely unstudied [85]. An unpigmented and microphthalmic *Spelaeoblatta* Bolivar, 1897 occurs in oligotrophic habitats of several caves of MDL-HC (especially in Nui Bai Voi, including Hang Mo So). The genus has four species in Thailand and Myanmar, and is probably present elsewhere in caves of Indochina. Other cockroaches of the MDL-HC caves are troglophiles living at cave entrances or in guano. Larger cockroaches, such as *Pycnoscelus* spp. or *Periplaneta* spp., which often swarm in hot and disturbed caves elsewhere in Southern Asia, are exceptional in MDL-HC [19,22].

- Orthoptera (crickets)

Rhaphidophoridae are very common in caves of MDL-HC [21,22], as elsewhere in Southeast Asian caves. They have never been recorded outside. However, given their pigmentation, large eyes, and the biology of other species of the family, which can leave caves at night to feed outside, they are probably subtroglophiles, rather than troglobionts as assumed in [21]. The MDL-HC species seems to be undescribed.

- Heteroptera (Figure 8H)

Large Harpactorinae nymphs (Reduviidae), yellowish and with slightly reduced eyes, occur in some MDL-HC caves [19]. Similar species of uncertain taxonomic status and unknown biology are present in caves of China, Southeast Asia and Australia [29,86]. A Schizopteridae from a cave of Nui Ba Tai, which has reduced eyes, reduced pigment and slightly elongated appendages, is a possible troglobiont. Species of this family are otherwise frequent in MDL-HC litter.

- Homoptera (Figure 8D,E)

Root-sucking bugs are frequent in many MDL-HC caves, in hanging roots, or in the ground root-mat of cave BV08 in Nui Bai Voi. They belong to several genera of the families Kinnaridae (rare), Cixiidae (probably two species, one frequent) and Delphacidae (one species). One Cixiidae, only known from the southern cave of Nui Khoe La, is eyeless; the other species is frequent as nymphs and has reduced eyes. Adults of *Borysthenes* sp. also occur in hanging roots and may be the adult phase of these nymphs. We consider them probable troglobionts, by analogy with similarly eyed species from Hawaii [87] and Sulawesi [88]. The mealybug *Ripersiella ficaria* (Williams, 2004) was described from hanging roots in Hang Mo So. It is probably this species that is sometimes found in dense populations in root habitats of other caves of MDL-HC. Since no mealybug is known to be cave-restricted, we provisionally consider this species to be a troglophile.

- Coleoptera (beetles) (Figure 8F,G)

In lowland tropical caves, troglobionts are exceptional among beetles [29,89]. The only troglobiotic carabid of MDL-HC is *Eustra honchongensis*, restricted to oligotrophic habitats in four caves of Nui Bai Voi, including Hang Mo So [90]. Being unpigmented, with very reduced eyes and slightly elongated appendages, it resembles cave and soil species of the same genus from northern Thailand [90]. Tenebrionidae are much more frequent than Carabidae in the MDL-HC caves. A remarkable guanobiont-troglobiont, *Harvengia vietnamita*, the sole species of its genus, was discovered in four hills close to Nui Bai Voi, but never in Nui Bai Voi itself. It forms large populations on relatively dry guano. Placed in a new subtribe, Harvengina [91], it was at that time the only blind species known in the tribe Stenosini. Another Stenosini, *Pseudochillus honchongensis*, occurs in guano in Hang Mo So. Although dark-coloured and lacking troglomorphic traits, it has never been found outside this cave. A morphologically similar species, *Dichillus kuschstaberi* Kaszab, 1980, is known from a guano cave in Thailand [92]. We assume the Hang Mo So species is a guanobiont-troglobiont. A very rare, blind Curculionidae is only known from two specimens collected in caves of two MDL-HC hills, and in a cave of Phnom Ang in MDL-C. A second curculionid is a troglophilic Entiminae, common in Hang Mo So and several other MDL-HC caves, and occasionally found in soil samples. It is characteristic of root invertebrate communities. Other beetles of various families are encountered in cave guano or debris in MDL-HC, such as Histeridae and Ptiliidae, but none are troglobionts. Dark and eyed Ptiliidae are frequent in caves, whereas blind and pale species are well represented in soil, but absent from caves.

- Lepidoptera (moths)

Adult moths are regular components of cave-wall assemblages in temperate regions. MDL-HC cave-wall assemblages generally do not include moths, but larvae are abundant in two high-energy habitats: guano (Tineidae) and roots (Erebidae). In terms of biomass, Tineidae caterpillars, in their characteristic cases, are often the most important invertebrates of guano. Except for one species with reduced eyes in the Philippines, authors usually qualify cave-dwelling Tineidae as troglophiles [93]. The MDL-HC caves host several species of tineids, unstudied so far, that we provisionally consider troglophiles.

Caterpillars of the genus *Schrankia* (Erebidae), of which one species is probably a troglobiont, are numerous in the invertebrate root communities of Hawaiian caves [94]. Similarly, erebid caterpillars of an undetermined genus were abundant in subterranean root habitats of MDL-HC, particularly in the hanging roots that once flourished in Hang Mo So. Interestingly, erebid moths have rarely been reported from cave roots in other tropical regions.

4. Discussion

4.1. Species Richness

The richness of Hang Mo So in troglobionts is comparable to that of the Towakkalak system in Indonesia (27 versus 26 species), but it does not include a single stygobiont, as opposed to the 10 in Towakkalak [6]. This similarity in the number of troglobionts is surprising since Hang Mo So is a shallow cave, rich in organic debris and bats, which lacks active karstic water circulation, whereas Towakkalak is large and deep, with a wider array of habitats and is active hydrogeologically. Both caves share faunal characteristics typical of tropical lowland terrestrial caves [29], such as the prevalence of arachnids over insects, high diversity of guano species, rarity of troglobiotic beetles, and a low number of troglomorphic species. Cave species richness is often linked to the regional stock of epigean (especially edaphic) species, which may be potential colonizers [95]. This only partly holds true for the MDL-HC cave fauna, which is less rich than could be expected from the exceptional diversity of deep-soil invertebrates, while at the same time being richer than might be expected from the low biodiversity observed so far in other external habitats [21].

How complete is the current species list for Hang Mo So? The high proportion of troglobionts collected only once (14 out of 27) suggests that other rare species remain to be discovered. From a nested perspective (cave > hill > karst), it appears that, of the 38 troglobionts present in the caves of Nui Bai Voi, 11 have not yet been found in Hang Mo So (Table 1). With all caves of the hill being close to each other (Figure 1B) and likely interconnected by fissures, several of these 'missing' species might yet be found in Hang Mo So.

At the MDL-HC level, 29 cave-obligate species are absent in Nui Bai Voi. It is unlikely that many of these will be found in Hang Mo So, because species dispersion between MDL-HC hills is more difficult than between caves of Nui Bai Voi. Nevertheless, the configuration of the MDL-HC karst makes the exchanges of cave-restricted fauna between hills conceivable, provided that they are not too distant from one another. The potential dispersal vector is water, which sometimes floods large parts of the plain and flows slowly in various directions during the rainy season. The limestone hills near Nui Bai Voi also share the same limestone stratum underground, so it is possible that they once had, or even still have, subterranean connections. This is not the case for Nui Hon Chong because it is separated from this cluster of hills by the sandstone hill of Nui Binh Tri (Figure 1A), which might explain why its cave fauna is rather different.

Another limitation of the Hang Mo So species list is the uncertain ecological status of several species (those marked '?' in Table 1), despite the parallel-sampling strategy employed and the extensive sampling that was carried out in MDL-HC. However, the same is true for most surveys carried out in under-sampled karsts of the world.

4.2. Endemism

Endemism rates are very high among karst-dependent invertebrates of MDL-HC, i.e., snails [42], soil arthropods and cave invertebrates [21], as in many tropical karst areas. This endemism is particularly narrow, with single-hill endemics on most surveyed hills of MDL-HC (Table 1). A number of generic or suprageneric taxa have been erected, reflecting an exceptional level of phyletic and geographic isolation among this fauna, probably caused by a long isolation of the MDL karst—these can be provisionally qualified as relicts.

Deep-soil species include several endemics at the supra-specific level, notably among mites [96], beetles [91,97,98] and springtails [21]. Snail endemism rates in MDL-HC are also high, while several species widespread in Indochina are absent [42]. Most MDL-HC troglobionts are endemic, many to a single cave (Table 1). Among these are two taxonomically isolated genera: the millipede *Eostemmiulus* from Hang Mo So, basal to the order Stemmiulida [99], and the tenebrionid *Harvengia*, type genus of the subtribe Harvengina [91]; the pseudoscorpion genus *Cybella*, type of the subfamily Cybellinae, was described from two troglophilic endemics of the MDL karsts [61]. Altogether, the levels of endemism and relictuality among the MDL-HC fauna are currently unmatched in other karst areas of Asia.

Other less studied caves of MDL-HC are known to host some remarkable endemic troglobionts, so they may ultimately prove to have levels of biodiversity similar to that of Hang Mo So. Examples are the southern cave of Nui Khoe La, Hang Tien in Nui Hang Tien, Hang Gieng Tien in Nui Hon Chong, and a cave NW of Nui Ba Tai (Figure 3A). The same might also have been true of several MDL-HC caves now destroyed by quarrying, such as BV08 (Hang Tai) and BV09 in Nui Bai Voi (Figure 1B), a large cave of central Nui Khoe La, believed to have been the longest of the MDL-HC karst, a cave in northern Nui Khoe La, and two caves of Nui Cay Xoai (Figure 3A).

4.3. Gaps

Our sampling largely neglected the aquatic fauna, leaving a major gap in the data set (Table 1). Marine or anchialine habitats were only sampled once, in Hang Tien (Nui Hang Tien), but are presumably developed along the kilometers of highly karstified coasts of MDL-HC (Figure 2A,D). No stygobiont has been found in Hang Mo So itself. This might be due to the scarcity of phreatic waters, since the water in the cave (Figure 3B) is directly connected to the slightly brackish surface waters of the alluvial plain. The few freshwater stygobionts in our list mostly come from Hang Gieng Tien (Nui Hon Chong), the only cave with deep karstic water (Figure 4C). Besides *Pseudoiglica* (Gastropoda) and *Stenasellus* (Isopoda), these include Copepoda and Ostracoda. Two unstudied freshwater samples from caves BV08 and BV09 of Nui Bai Voi, taken before their destruction by quarrying, contained Gastropoda, Hydracari, Copepoda and Ostracoda. The presence of stygobionts is therefore highly probable in other MDL-HC caves.

Among terrestrial groups, sampling has been much more extensive, but highly diverse cave-dwellers collected in large numbers, such as spiders, mites and woodlice, have been only identified as morphospecies.

Geographical gaps, in addition to those mentioned for aquatic fauna, are due to access restrictions to some hills being quarried, or managed by the military. Hills that are inaccessible due to rugged terrain are likely to be rich in unexplored shafts, deep crevices and caves.

4.4. Causes of Species Richness

Overall, the richness in troglobionts of MDL-HC is high for a tropical karst by current standards [1]. Several factors have been advanced to explain the similarly high richness of the Towakkalak system, a hotspot of subterranean diversity in Southeast Asia [100]. These were seasonal climate, proximity to the sea, intensive sampling effort, and location at the foot of mountains, all of which, except for the last, apply to MDL-HC. Other well-sampled caves in Southeast Asia are Tham Chiang Dao (Thailand [76]), the Batu Caves (Western Malaysia [101]), the Clearwater Cave System in Mulu (Sarawak [86]), and Niah Cave (Sarawak [102]). Tham Chiang Dao's richness in troglobionts is comparable to that of Towakkalak or Hang Mo So. These three caves are under a seasonal tropical climate (category Am or Aw of Köppen-Geiger [103]). Interestingly, the three other caves, in Western Malaysia and Sarawak, are under a permanent non-seasonal tropical climate (category Af); they are noticeably less rich than the caves listed above.

Differences in species richness are often best explained by unequal sampling effort [104]. The main drivers of diversity are climate and geological history, but their impact may only become apparent after standardized sampling. Quaternary sea-level fluctuations, linked to glaciation periods, are considered to have had a strong influence on faunal diversification in Southeast Asia, through habitat fragmentation [105]. For instance, the sea was 120 m below its current level 21,000 BP [106], and the Mekong Delta extended more than 220 km southwards into the South China Sea 8000 BP [107,108]. Slight faunal differences between adjacent hills could be the result of past isolation by higher sea levels. Differences at higher taxonomic levels probably result from more ancient events that are not yet decipherable in the absence of phylogenies and information on past changes in the karst environment.

5. Conservation Issues

In 1997, the Karst Water Institute (USA) listed Hon Chong among the 10 most endangered karsts worldwide based on threats to its fauna [109]. Since then, existing quarries have been expanded and new ones opened, leaving unaffected an ever smaller surface of the original karst and severely impacting the surroundings (Figure 9). Of the original karst surface of about 4.01 km^2, 1.71 km^2 had already been quarried by 2018, and this is set to rise to 2.41 km^2 within a few years, leaving, at best, only 1.60 km^2 unquarried. In terms of the number of hills, 13 had been erased by 2018, a further two will disappear in the coming years, and the two largest will be erased over most of their surface (Figure 10). Hang Mo So and a small part of Nui Bai Voi around it will be spared, along with a few other hills, provided that the present quarrying plans are not expanded again.

Figure 9. Evolution of limestone quarrying in the MDL-HC karst from 1985 to 2018. Year 1985: before quarrying; abbreviations of hills as in Figure 1A. Photos Google Earth.

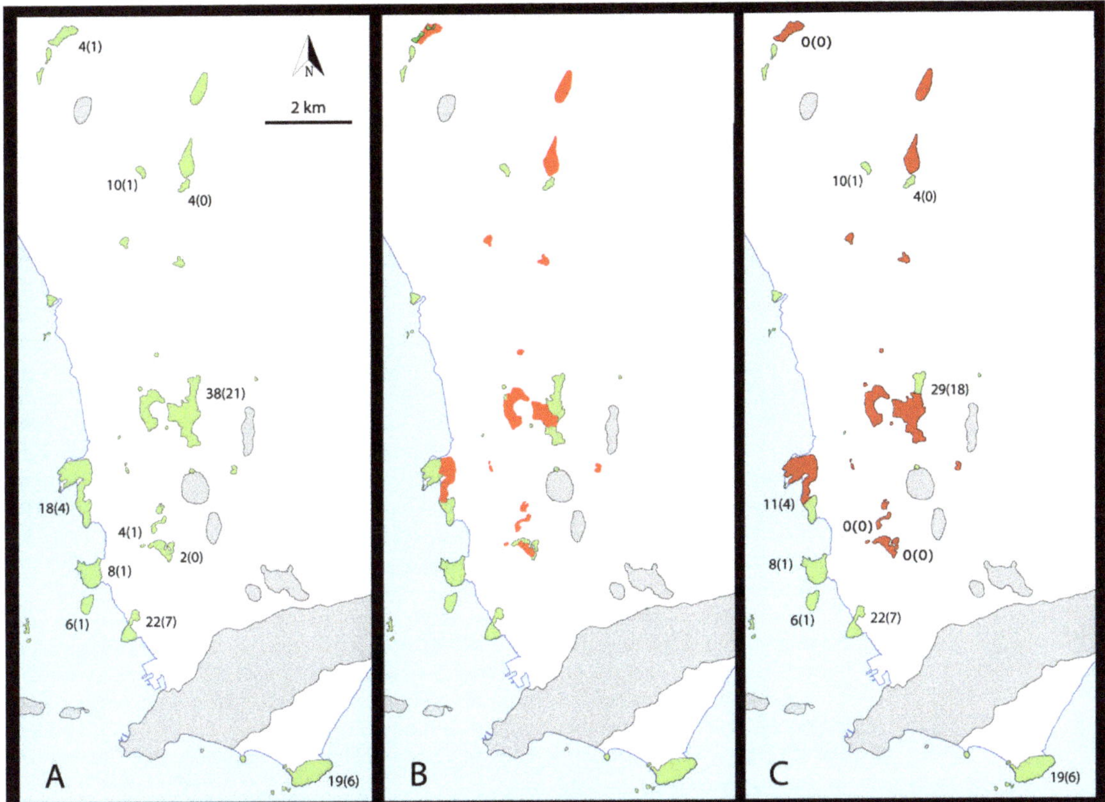

Figure 10. The MDL-HC karst. Blue line, sea coastline; green and red, limestone hills; grey, non-limestone hills; species numbers per hill are extracted from Table 1. (**A**) Limestone hills in the 1980s, before quarrying, in green; numbers are total numbers of troglobionts, with number of single-hill endemics in parentheses, recognized in 2023. (**B**) In 2018: green, limestone hills or parts of hills not quarried; red, limestone hills or parts of hills quarried. (**C**) Scenario for the future: green, limestone hills or parts of hills that will not be quarried; red, limestone hills or parts of hills already quarried or planned to be quarried; numbers are total numbers of troglobionts, with number of single-hill endemics that would remain in parentheses; hills in red without annotations were destroyed before they could be sampled.

Quarrying is also causing severe disturbance to non-karst habitats of MDL-HC, linked to the infrastructure of limestone exploitation (harbour, roads, buildings, mud dispersal, and non-limestone quarrying). The attendant increase in the local population also has adverse effects (water pollution, shrimp ponds, mangrove destruction, and huge pressure on remaining forests of sandstone hills). The ravages caused are obvious on the ground and from satellite imagery (Figure 9).

5.1. How Did We Get Here?

MDL-HC is the only limestone resource in southern Vietnam. In 1993, several hills had already been quarried to elimination by Vietnamese companies, but Nui Bai Voi, including Hang Mo So, was still pristine. The history of limestone exploitation from this time onwards is summarized in a leaflet of the International Finance Corporation (IFC) [110] and various unpublished reports.

In 1993, regional cement supply was falling short of demand in the rapidly growing Vietnamese economy. The companies Ha Tien I and Holderbank approached IFC for a loan to establish a cement plant in the MDL-HC region, called Morning Star Cement (MSC), which would produce 1.4 M tonnes of cement per year. Limestone was to be extracted initially from Nui Bai Voi and Nui Cay Xoai. Later, a large part of Nui Khoe La was added to the concession in order to 'compensate' for the designation of a small part of Nui Bai Voi as a cultural monument.

The tourist potential of the local karst landscape was recognized as significant in the otherwise monotonous Mekong Delta plains, but it did not meet IFC's natural habitat standard [110] and was therefore not taken into consideration as a reason to curtail the MSC project. A first Environmental Impact Assessment (EIA), required by IFC, was carried out in 1995. It focused on technical aspects and neglected karst-dependent biodiversity issues. Just before the board took its decision, IFC received a letter from biologists who happened to be surveying the area at the time, pointing out the risk of serious biodiversity loss. In response, the IFC, following their consultant's advice, considered that these biologists had overstated this risk and failed to "recognize Vietnam's critical need for cement" [110]. It is at least true that no MDL-HC species were on the IUCN Red List at that time. But the second assertion ignored the fact that the large limestone outcrops of central Vietnam could be exploited with much less damage to biodiversity than the small hills of MDL-HC, as was pointed out to IFC in the letter. It was too late to question such a large project and, even though IFC acknowledged the weakness of the EIA, its board approved a $97 M loan for the project [110].

Nevertheless, at the start of the quarrying operations in 1999, after consulting Tony Whitten, Senior Biodiversity Specialist for the East Asia and Pacific Region at the World Bank, IFC commissioned a new EIA from Sinclair Knight Merz (SKM) from Australia and a team of the Vietnamese Sub-Institute of Ecology Resources and Environmental Studies (SIERES) of the Institute of Tropical Biology (ITB) in Ho Chi Minh City. This second EIA [111] provided information on plants and vertebrates, and recognized that the conservation of biodiversity in small, fragmented karst landscapes required a tailored approach.

In 2004, the board of Holcim Vietnam (formerly MSC) agreed to launch a new biodiversity study focusing on Nui Bai Voi. The resulting report highlighted the critical impact that quarrying would have on the local population of langur monkeys, but little else [110].

In 2008, a workshop was organized in Rach Gia by the Center for Biodiversity and Development (CBD) of the Institute of Tropical Biology (ITB) of Ho Chi Minh City, which was attended by Kien Giang Province authorities, Holcim staff, and several biologists. Serious concerns were expressed at this meeting about the increasing risks of species extinction due to the rapid destruction of local karst habitats [112] (Figure 11). In 2009, Holcim and IUCN entered a 4-year partnership to prepare a Biodiversity Action Plan. This partnership was later extended to 2015, with a view to establishing a Nature Reserve at Kien Luong. In 2016–2017, the Hon Chong cement plant was acquired by the Siam City Cement Company (SCCC). In 2018 and 2021, successive 3-year MOUs were signed between IUCN and SCCC, with the same objectives. Despite this long partnership, quarrying has continued unabated and the proposed nature reserve has yet to be implemented.

5.2. Species at Risk

At present, many subterranean and soil invertebrates in the MDL-HC karst, along with several plants, are at great risk of extinction due to quarrying. Figure 10 gives the current number of troglobionts and single-hill endemics per hill, and the number of those anticipated to remain at the end of the planned quarrying operations, under the most favourable scenario (i.e. without enlargement of current concession). The true numbers of threatened or possibly extinct endemics will, however, be even higher. This is because (a) several large hills have been destroyed without ever being surveyed (Nui Com, Nui Trau, Nui Cay Xoai); (b) our survey methods were less effective for certain species-rich groups, such as worms and microcrustaceans; and (c) other species-rich groups collected in large numbers have yet to be studied taxonomically, either entirely or in part.

Figure 11. The MDL-HC karst under quarry: (**A**) artisanal limestone quarrying in Xom Lo Voi; (**B**) Nui Khoe La, in 2005; (**C**) part of Nui Bai Voi, southwest of Hang Mo So, at the beginning of quarrying in 2006; (**D**) Nui Cay Xoai, in 2003. Photos: Louis Deharveng.

Range reduction through site destruction is the most immediate threat for narrow endemics. For instance, the planned quarrying of Nui Bai Voi will reduce the known range of 10 troglobionts to a single locality (Table 1). Of course, these might also occur in other caves, but the extensive sampling and ongoing quarrying make this increasingly improbable. In the MDL-HC karst as a whole, five troglobiotic species are due to lose the only site where they have been found and thus likely become extinct. These numbers only include cave-obligate species, representing a small proportion of the total biodiversity [21].

5.3. Conservation Actions

As faunal data accumulated and the situation became increasingly critical, relevant information was disseminated in concert with Tony Whitten. Thus, international media [26,113], IUCN, and Fauna & Flora International (FFI) were informed of developments in the discovery of remarkable taxa, additions to the IUCN Red Lists, the situation at the MDL-HC karst, and proposed actions for conserving the most biologically significant sites. Moreover, Tony Whitten tirelessly led discussions of the situation with the IFC, official authorities, and mining companies.

Over the course of these efforts, it became evident that IUCN red-listing of endangered species represented the only means to impress the companies and financial institutions involved. A special project was therefore initiated, aimed at red-listing selected MDL-HC species. In 2015 and 2016, 9 troglobiotic and 3 troglophilic arthropods were assessed as CR

(2 species), EN (6), VU (4) and NT (1) (Table 1). We also red-listed non-cave invertebrates and plants from MDL-HC: 4 edaphic arthropods (1 CR, 2 VU, 2 NT), 13 gastropods (3 CR, 4 EN, 5 VU, 1 NT) and 3 plants (1 CR, 2 VU). Several other MDL-HC invertebrates, notably at Hang Mo So, match the IUCN categories of CR or EN and will be assessed in the future.

In this context, Hang Mo So is particularly important because it represents a rich pocket of endemism within the small part of Nui Bai Voi (about 0.11 km^2) that is expected to escape quarrying. It is also the last refuge for other species that previously occurred on parts of Nui Bai Voi that have since been destroyed (Figure 11C). However, Hang Mo So itself is at risk from other pressures. As a major local tourist attraction, it is suffering from growing disturbance. Concrete has already been poured in some passages of the cave, hanging root bundles have been cut to make way for electrical wires, rubbish accumulates in recesses, and new settlements near the cave entrance add to the disturbance. If not properly managed, this human impact will accelerate habitat destruction [10] and could thus lead to further species extinctions. Regulation of tourist visits to Hang Mo So will therefore be a prerequisite for the success of a future nature reserve in the MDL-HC karst.

It should be evident from this brief account that the tiny karst of Hong Chong (MDL-HC) faces a greater and more imminent threat than any other karst in the world. It is home to a diverse cave fauna and its deep-soil fauna is probably the richest in the tropics, hence we risk losing a major hotspot of karstic biodiversity in Asia. The only hope for saving part of this unique fauna is the rapid implementation (as opposed to endless discussion) of an action plan for conserving the small portions of the karst that remain.

Author Contributions: Conceptualization, L.D., J.V. and M.L.; data acquisition: M.L., J.V., C.M.L., M.L.I.J., N.S.L. and Q.T.T.; methodology, A.B., C.M.L., Q.T.T. and T.Q.T.N.; writing—original draft preparation, L.D., A.B. and J.V.; writing—review and editing, L.D., J.V., M.L.I.J., C.K.L., N.S.L., M.L. and H.T.L. All authors have read and agreed to the published version of the manuscript.

Funding: Funding was provided in 2014 by IUCN Vietnam for the project "Baseline survey of biodiversity in Kien Giang Province", which included a cave invertebrate survey in the Hon Chong karst (project number PO1221).

Institutional Review Board Statement: Not applicable.

Data Availability Statement: Not applicable.

Acknowledgments: We thank Katja Anker, Quan Mai, Sibyle Moulin, Jean-Christophe Levet, Pham Van Giau, Nguyen Thanh Hai, Phan Van Hung, and other staff of local authorities for their contributions to and help with fieldwork. We are grateful to Thierry Bourgoin, Henri Dalens, Thierry Deuve, Arnaud Faille, Nestor Fernandez, Julio Ferrer, Sergei Golovatch, Adriano Kury, Elise-Anne Leguin, Wilson Lourenço, Joanna Makol, Jean-Paul Mauriès, Christine Rollard, Wolfgang Schawaller, Clément Schneider, Peter Schwendinger, Stefano Taiti, Lubomir Vidlicka, Douglas Williams and Feng Zhang for identifications and information. We owe a particular debt to the late Tony Whitten, who enthusiastically encouraged our work over many years and fought tirelessly to save the unique invertebrate fauna of the Hon Chong hills.

Conflicts of Interest: The authors declare no conflict of interest.

References

1. Culver, D.C.; Deharveng, L.; Pipan, T.; Bedos, A. An Overview of Subterranean Biodiversity Hotspots. *Diversity* **2021**, *13*, 487. [CrossRef]
2. Niemiller, M.L.; Helf, K.; Toomey, R.S. Mammoth Cave: A Hotspot of Subterranean Biodiversity in the United States. *Diversity* **2021**, *13*, 373. [CrossRef]
3. Zagmajster, M.; Polak, S.; Fišer, C. Postojna-Planina Cave System in Slovenia, a Hotspot of Subterranean Biodiversity and a Cradle of Speleobiology. *Diversity* **2021**, *13*, 271. [CrossRef]
4. Camacho, A.I.; Puch, C. Ojo Guareña: A Hotspot of Subterranean Biodiversity in Spain. *Diversity* **2021**, *13*, 199. [CrossRef]
5. Huang, S.; Zhao, M.; Luo, X.; Bedos, A.; Wang, Y.; Chocat, M.; Tian, M.; Liu, W. Feihu Dong, a New Hotspot Cave of Subterranean Biodiversity from China. *Diversity* **2023**, *15*, 902. [CrossRef]
6. Deharveng, L.; Rahmadi, C.; Suhardjono, Y.R.; Bedos, A. The Towakkalak System, a Hotspot of Subterranean Biodiversity in Sulawesi, Indonesia. *Diversity* **2021**, *13*, 392. [CrossRef]

7. Ferreira, R.L.; Berbert-Born, M.; Souza-Silva, M. The Água Clara Cave System in Northeastern Brazil: The Richest Hotspot of Subterranean Biodiversity in South America. *Diversity* **2023**, *15*, 761. [CrossRef]

8. Laumanns, M. Karsts and Caves of South Vietnam, Part 1: Provinces of Kien Giang, An Giang and Da Nang. *Berl. Höhlenkundliche Berichte* **2011**, *43*, 1–73.

9. Denneborg, M.; Laumanns, M.; Schnadwinkel, M.; Vogt, S. German Speleological Campaign, Cambodia 95/96. *Berl. Höhlenkundliche Berichte* **2002**, *6*, 1–92.

10. Kiernan, K. Human impacts on geodiversity and associated natural values of bedrock hills in the Mekong Delta. *Geoheritage* **2010**, *2*, 101–122. [CrossRef]

11. Pirazzoli, P.A. Marine notches. In *Sea-Level Research: A Manual for the Collection and Evaluation of Data*; van de Plassche, O., Ed.; Geo Books: Norwich, UK, 1986; pp. 361–400.

12. Salomon, J. Les influences climatiques sur la géomorphologie karstique: Exemple des milieux tropicaux et arides. *Quaternaire* **1997**, *8*, 107–117. [CrossRef]

13. Deharveng, L.; Truong, Q.T.; Duong, T.D. Explorations au centre et au sud du Vietnam. *Spelunca* **1995**, *59*, 8–10.

14. Le, C.K. La végétation des collines calcaires de la région de Kien-Luong—Ha Tien. *Nien-San* **1970**, *3*, 121–200.

15. Le, C.K. La végétation des collines calcaires de la région de Kien-Luong—Ha Tien (suite). *Nien-San* **1974**, *4*, 11–90.

16. Truong, Q.T.; Kiew, R.; Vermeulen, J.J. *Begonia bataiensis* Kiew, a new species in section Leprosae (Begoniaceae) from Vietnam. *Gard. Bull. Singap.* **2005**, *57*, 119–123.

17. Middleton, D.J.; Ly, N.S. A new species of *Ornithoboea* (Gesneriaceae) from Vietnam. *Edinb. J. Bot.* **2008**, *65*, 353–357. [CrossRef]

18. Nguyen, V.D.; Luu, H.T.; Nguyen, Q.D.; Hetterscheid, W.L.A. *Amorphophallus kienluongensis* (Araceae), a new species from the Mekong Delta, Southern Vietnam. *Blumea* **2016**, *61*, 1–3. [CrossRef]

19. Deharveng, L.; Le, C.K.; Bedos, A. Vietnam. In *Encyclopaedia Biospeologica Tome III*; Juberthie, C., Decu, V., Eds.; Société Internationale de Biospéologie: Moulis, France; Bucarest, Romania, 2001; pp. 2027–2037.

20. Boutin, C. Cambodge. In *Encyclopaedia Biospeologica Tome III*; Juberthie, C., Decu, V., Eds.; Société Internationale de Biospéologie: Moulis, France; Bucarest, Romania, 2001; pp. 1755–1762.

21. Deharveng, L.; Bedos, A.; Le, C.K.; Le, C.M.; Truong, Q.T. Endemic arthropods of the Hon Chong hills (Kien Giang), an unrivaled biodiversity heritage in Southeast Asia. In *Beleaguered Hills: Managing the Biodiversity of the Remaining Karst Hills of Kien Giang, Vietnam*; Le, C.K., Truong, Q.T., Ly, N.S., Eds.; Nha Xuat Ban Nong Nghiep: Ho Chi Minh, Vietnam, 2009; pp. 31–57.

22. Steiner, H. Karsts and Caves of South Vietnam, Part 1: Provinces of Kien Giang, An Giang and Da Nang. Biospeleology of southern Vietnam. *Berl. Höhlenkundliche Berichte* **2011**, *43*, 53–73.

23. Deharveng, L.; Le, C.K.; Le, C.M.; Bedos, A. Hot issues in karst conservation: The biodiversity of Hon Chong hills (southern Vietnam), with emphasis on invertebrate endemism. In *Trans-KARST 2004, Proceedings of the International Transdisciplinary Conference on Development and Conservation of Karst Regions, Hanoi, Vietnam, 13–18 September 2004*; Batelaan, O., Dusar, M., Masschelein, J., Vu, T.T., Tran, T.V., Nguyen, X.K., Eds.; Research Institute of Geology and Mineral Resources: Hanoi, Vietnam, 2005; p. 40.

24. Humphreys, W.F. Community extinction: The groundwater (stygo-) fauna of Curaçao, Netherlands Antilles. *Hydrobiologia* **2022**, *849*, 4605–4611. [CrossRef]

25. Ng, P.K.L.; Kottelat, M.; Deharveng, L. Tony Whitten and biodiversity: Personal recollections. *Raffles Bull. Zool.* **2020**, *35*, 11–13.

26. Ives, M. *The Big Story. Threatened Vietnam Cave Bugs Draw Little Sympathy*; Associated Press: New York, NY, USA, 2012.

27. Santos, A.J.; Ferreira, R.L.; Buzatto, B.A. Two New Cave-Dwelling Species of the Short-Tailed Whipscorpion Genus *Rowlandius* (Arachnida: Schizomida: Hubbardiidae) from Northeastern Brazil, with Comments on Male Dimorphism. *PLoS ONE* **2013**, *8*, e63616. [CrossRef] [PubMed]

28. Eberhard, S.M.; Howarth, F.G. Undara lava cave fauna in tropical Queensland with an annotated list of Australian subterranean biodiversity hotspots. *Diversity* **2021**, *13*, 326. [CrossRef]

29. Deharveng, L.; Bedos, A. Diversity of Terrestrial Invertebrates in Subterranean Habitats. In *Cave Ecology, Ecological Studies 235*; Moldovan, O.T., Kovác, L., Halse, S., Eds.; Springer Nature: Cham, Switzerland, 2018; pp. 107–172.

30. Faille, A. Beetles. In *Encyclopedia of Caves*; White, W.B., Culver, D.C., Pipan, T., Eds.; Academic Press: New York, NY, USA, 2019; pp. 102–108.

31. Lukić, M. Collembola. In *Encyclopedia of Caves*; White, W.B., Culver, D.C., Pipan, T., Eds.; Academic Press: New York, NY, USA, 2019; pp. 308–319.

32. Fišer, C. Niphargus—A model system for evolution and ecology. In *Encyclopedia of Caves*; White, W.B., Culver, D.C., Pipan, T., Eds.; Academic Press: New York, NY, USA, 2019; pp. 746–755.

33. Porco, D.; Bedos, A.; Greenslade, P.; Janion, C.; Skarzynski, D.; Stevens, M.I.; Jansen van Vuuren, B.; Deharveng, L. Challenging species delimitation in Collembola: Cryptic diversity among common springtails unveiled by DNA barcoding. *Invertebr. Syst.* **2012**, *26*, 470–477. [CrossRef]

34. Lukić, M.; Delic, T.; Pavlek, M.; Deharveng, L.; Zagmajster, M. Distribution pattern and radiation of the European subterranean genus *Verhoeffiella* (Collembola, Entomobryidae). *Zool. Scr.* **2020**, *49*, 86–100. [CrossRef]

35. Jasinska, E.J.; Knott, B.; McComb, A.J. Root mats in ground water: A fauna-rich cave habitat. *J. N. Am. Benthol. Soc.* **1996**, *15*, 508–519. [CrossRef]

36. Howarth, F.G. Ecology of cave Arthropods. *Annu. Rev. Entomol.* **1983**, *28*, 365–389. [CrossRef]

37. Novak, T.; Perc, M.; Lipovšek, S.; Janžekovič, F. Duality of terrestrial subterranean fauna. *Int. J. Speleol.* **2012**, *41*, 181–188. [CrossRef]
38. Oromí, P.; Socorro, S. Biodiversity in the Cueva del Viento lava tube system (Tenerife, Canary Islands). *Diversity* **2021**, *13*, 226. [CrossRef]
39. Vandel, A. *Biospéologie. La Biologie des Animaux Cavernicoles*; Gauthier-Villars: Paris, France, 1964; pp. 1–619.
40. Wynne, J.J.; Howarth, F.G.; Sommer, S.; Dickson, B.G. Fifty years of cave arthropod sampling: Techniques and best practices. *Int. J. Speleol.* **2019**, *48*, 33–48. [CrossRef]
41. Vermeulen, J.J.; Phung, L.C.; Truong, Q.T. New species of terrestrial molluscs (Caenogastropoda, Pupinidae & Pulmonata, Vertiginidae) of the Hon Chong-Ha Tien limestone hills, Southern Vietnam. *Basteria* **2007**, *71*, 81–92.
42. Vermeulen, J.J.; Luu, H.T.; Theary, K.; Anker, K. New species of land snails (Mollusca: Gastropoda: Caenogastropoda and Pulmonata) of the Mekong Delta limestone hills (Cambodia, Vietnam). *Folia Malacol.* **2019**, *27*, 7–41. [CrossRef]
43. Vermeulen, J.; Whitten, T. *Biodiversity and Cultural Property in the Management of Limestone Resources. Lessons from East Asia*; Directions in Development Series; The World Bank: Washington, DC, UAS, 1999; pp. 1–120.
44. Leclerc, P. Considérations paléogéographiques à propos de la découverte en Thaïlande d'Opilioacariens nouveaux (Acari-Notostigmata). *C. R. Soc. Biogéogr.* **1989**, *65*, 162–174.
45. Beron, P. Comparative study of the invertebrate cave faunas of Southeast Asia and New Guinea. *Hist. Nat. Bulg.* **2015**, *21*, 169–210.
46. Fage, L. Scorpions et pédipalpes de l'Indochine française. *Ann. Soc. Entomol. Fr.* **1946**, *113*, 71–81.
47. Miranda, G.S.; Giupponi, A.P.; Prendini, L.; Scharff, N. Systematic revision of the pantropical whip spider family Charinidae Quintero, 1986 (Arachnida, Amblypygi). *Eur. J. Taxon.* **2021**, *772*, 1–409. [CrossRef]
48. Nguyen, T.T.A.; Phung, T.H.L. A new species of the whip spider genus *Weygoldtia* (Arachnida: Amblypygi: Charinidae) from Con Dao National Park, Vietnam. *Acad. J. Biol.* **2022**, *44*, 67–76.
49. Zhu, X.Y.; Wu, S.Y.; Liu, Y.J.; Reardon, C.R.; Roman-Palacios, C.; Li, Z.; He, Z.Q. A new species of whip spider, *Weygoldtia hainanensis* sp. nov., from Hainan, China (Arachnida: Amblypygi: Charinidae). *Zootaxa* **2021**, *5082*, 65–76. [CrossRef]
50. Mammola, S.; Cardoso, P.; Ribera, C.; Pavlek, M.; Isaia, M. A synthesis on cave-dwelling spiders in Europe. *J. Zool. Syst. Evol. Res.* **2018**, *56*, 301–316. [CrossRef]
51. Brignoli, P.M. New or interesting Anapidae (Arachnida, Araneae). *Rev. Suisse Zool.* **1981**, *88*, 109–134. [CrossRef]
52. Decae, A.E. Three new species in the genus *Latouchia* Pocock, 1901 (Araneae, Mygalomorphae, Halonoproctidae) from Vietnam. *Rev. Suisse Zool.* **2019**, *126*, 275–289.
53. Tong, Y.; Li, F.; Song, Y.; Chen, H.; Li, S. Thirty-two new species of the genus *Speocera* Berland, 1914 (Araneae: Ochyroceratidae) from China, Madagascar and Southeast Asia. *Zool. Syst.* **2019**, *44*, 1–75.
54. Lehtinen, P.T. Spiders of the Oriental-Australian region III. Tetrablemmidae, with a world revision. *Acta Zool. Fenn.* **1981**, *162*, 1–151.
55. Lin, Y.; Li, S. New cave-dwelling armored spiders (Araneae, Tetrablemmidae) from Southwest China. *Zookeys* **2014**, *388*, 35–67.
56. Schwendinger, P.; Martens, J. A taxonomic revision of the family Oncopodidae V. *Gnomulus* from Vietnam and China, with the description of five new species (Opiliones, Laniatores). *Rev. Suisse Zool.* **2006**, *113*, 595–615. [CrossRef]
57. Condé, B. Nouveaux palpigrades de Trieste, de Slovénie, de Malte, du Paraguay, de Thaïlande et de Bornéo. *Rev. Suisse Zool.* **1988**, *95*, 723–750. [CrossRef]
58. Condé, B. Palpigrades cavernicoles et endogés de Thaïlande et des Célèbes (première note). *Rev. Suisse Zool.* **1992**, *99*, 655–672. [CrossRef]
59. Condé, B. Palpigrades cavernicoles et endogés de Thaïlande et de Célèbes (2e note). *Rev. Suisse Zool.* **1994**, *101*, 233–263. [CrossRef]
60. Judson, M.L.I. A new and endangered species of the pseudoscorpion genus *Lagynochthonius* from a cave in Vietnam, with notes on chelal morphology and the composition of the Tyrannochthoniini (Arachnida, Chelonethi, Chthoniidae). *Zootaxa* **2007**, *1627*, 53–68. [CrossRef]
61. Judson, M.L.I. A new subfamily of Feaellidae (Arachnida, Chelonethi, Feaelloidea) from Southeast Asia. *Zootaxa* **2017**, *4258*, 1–33. [CrossRef]
62. Harvey, M.S. Two new species of the pseudoscorpion genus *Cybella* (Pseudoscorpiones: Feaellidae) from Malaysian caves. *Zool. Anz.* **2018**, *273*, 124–132. [CrossRef]
63. Harvey, M.S.; Berry, O.; Edward, K.L.; Humphreys, G. Molecular and morphological systematics of hypogean schizomids (Schizomida: Hubbardiidae) in semiarid Australia. *Invertebr. Syst.* **2008**, *22*, 167–194. [CrossRef]
64. Lourenço, W.R.; Duhem, B. Buthid scorpions found in caves; a new species of *Isometrus* Ehrenberg, 1828 (Scorpiones, Buthidae) from southern Vietnam. *C. R. Biol.* **2010**, *333*, 631–636. [CrossRef] [PubMed]
65. Lourenço, W.R.; Leguin, E.A. A new species of *Isometrus* Ehrenberg 1828 (Scorpiones: Buthidae) from Laos. *Acta Arachnol.* **2012**, *61*, 71–76.
66. Golovatch, S.I. Cave Diplopoda of southern China with reference to millipede diversity in Southeast Asia. *Zookeys* **2015**, *510*, 79–94. [CrossRef] [PubMed]
67. Golovatch, S.I.; Liu, W.X. Diversity, distribution patterns, and fauno-genesis of the millipedes (Diplopoda) of mainland China. *Zookeys* **2020**, *930*, 153–198. [CrossRef]

68. Likhitrakarn, N.; Golovatch, S.I.; Thach, P.; Chhuoy, S.; Ngor, P.B.; Srisonchai, R.; Sutcharit, C.; Panha, S. Two new species of the millipede genus *Plusioglyphiulus* Silvestri, 1923 from Cambodia (Diplopoda, Spirostreptida). *Zookeys* **2020**, *938*, 137–151. [CrossRef]

69. Golovatch, S.I.; Geoffroy, J.J.; Mauriès, J.P.; VandenSpiegel, D. New or poorly-known species of the millipede genus *Trachyjulus* Peters, 1864 (Diplopoda: Spirostreptida: Cambalopsidae). *Arthropoda Sel.* **2012**, *21*, 103–129. [CrossRef]

70. Mauriès, J.P.; Golovatch, S.I. *Stemmiulus deharvengi* sp.n., le premier Stemmiulida signalé en Indonésie (Diplopoda: Stemmiulida). *Arthropoda Sel.* **2006**, *15*, 91–98.

71. Srisonchai, R.; Likhitrakarn, N.; Sutcharit, C.; Jeratthitikul, E.; Siriwut, W.; Thrach, P.; Chhuoy, S.; Ngor, P.B.; Panha, S. A new micropolydesmoid millipede of the genus *Eutrichodesmus* Silvestri, 1910 from Cambodia, with a key to species in mainland Southeast Asia (Diplopoda, Polydesmida, Haplodesmidae). *Zookeys* **2020**, *996*, 59–91. [CrossRef]

72. Golovatch, S.I.; Geoffroy, J.J.; VandenSpiegel, D. Review of the millipede family Trichopolydesmidae in the Oriental realm (Diplopoda, Polydesmida), with descriptions of new genera and species. *Zookeys* **2014**, *414*, 19–65. [CrossRef]

73. Schmalfuss, H. World catalog of terrestrial isopods (Isopoda: Oniscidea). *Stuttg. Beitr. Naturk.* **2003**, *(A) 654*, 1–341.

74. Deharveng, L.; Bedos, A. *Rambutsinella*, a new genus of Entomobryidae (Insecta: Collembola) from Southeast Asia. *Raffles Bull. Zool.* **1996**, *44*, 279–285.

75. Deharveng, L. Fauna of Thai caves. II. New Entomobryoidea Collembola from Chiang Dao Cave, Thailand. *Bish. Mus. Occas. Pap.* **1990**, *30*, 279–287.

76. Deharveng, L.; Bedos, A. The cave fauna of southeast Asia. Origin, evolution and ecology. In *Ecosystems of the World 30. Subterranean Ecosystems*; Wilkens, H., Culver, D.C., Humphreys, W.F., Eds.; Elsevier: Amsterdam, The Netherlands, 2000; pp. 603–632.

77. Deharveng, L.; Suhardjono, Y.R. *Pseudosinella maros* sp. n., a troglobitic Entomobryidae (Collembola) from Sulawesi Selatan, Indonesia. *Rev. Suisse Zool.* **2004**, *111*, 979–984. [CrossRef]

78. Yoshii, R. Finding of *Lepidosinella armata* Handschin from East Java. In Proceedings of the 3rd International Seminar on Apterygota, Siena, Italy, 21–26 August 1989; Dallai, R., Ed.; 1989; pp. 89–91.

79. Deharveng, L.; Bedos, A. *Lepidonella lecongkieti* n.sp., premier Collembole cavernicole du Vietnam (Collembola, Insecta). *Bull. Soc. Entomol. Fr.* **1995**, *100*, 21–24. [CrossRef]

80. Deharveng, L.; Jantarit, S.; Bedos, A. Revisiting *Lepidonella* Yosii (Collembola: Paronellidae): Character overview, checklist of world species and reassessment of *Pseudoparonella doveri* Carpenter. *Ann. Soc. Entomol. Fr.* **2018**, *54*, 381–400. [CrossRef]

81. Jantarit, S.; Satasook, C.; Deharveng, L. *Cyphoderus* (Cyphoderidae) as a major component of collembolan cave fauna in Thailand, with description of two new species. *Zookeys* **2014**, *368*, 1–21.

82. Papáč, V.; Palacios-Vargas, J.G. A new genus of Neelidae (Collembola) from Mexican caves. *Zookeys* **2016**, *569*, 37–51.

83. Schneider, C.; Deharveng, L. First record of the genus *Spinaethorax* Papác & Palacios Vargas, 2016 (Collembola, Neelipleona, Neelidae) in Asia, with a new species from a Vietnamese cave. *Eur. J. Taxon.* **2017**, *363*, 1–20.

84. Schneider, C. Morphological review of the order Neelipleona (Collembola) through the redescription of the type species of *Acanthoneelidus*, *Neelides* and *Neelus*. *Zootaxa* **2017**, *4308*, 1–94. [CrossRef]

85. Lucañas, C.C.; Bláha, M.; Rahmadi, C.; Patoka, J. The first *Nocticola* Bolivar 1892 (Blattodea: Nocticolidae) from New Guinea. *Zootaxa* **2021**, *5082*, 294–300. [CrossRef] [PubMed]

86. Moulds, T.; Anderson, J.; Anderson, R.; Nykiel, P. Preliminary Survey of Cave Fauna in the Gunung Mulu World Heritage Area, Sarawak, Malaysia. **2013**, *unpublished*.

87. Hoch, H.; Howarth, F.G. Multiple cave invasions by species of the planthopper genus *Oliarus* in Hawaii (Homoptera: Fulgoroidea: Cixiidae). *Zool. J. Linn. Soc.* **1999**, *127*, 453–475. [CrossRef]

88. Hoch, H.; Mühlethaler, R.; Wachmann, E.; Stelbrink, B.; Wessel, A. *Celebenna thomarosa* gen. n., sp. n. (Hemiptera, Fulgoromorpha, Cixiidae, Bennini) from Indonesia: Sulawesi with notes on its ecology and behaviour. *Dtsch. Entomol. Z.* **2011**, *58*, 241–250. [CrossRef]

89. Casale, A.; Taglianti, A.V.; Juberthie, C.; Decu, V. Coleoptera Carabidae. In *Encyclopaedia Biospeologica Tome II*; Juberthie, C., Decu, V., Eds.; Société Internationale de Biospéologie: Moulis, France; Bucarest, Romania, 1998; pp. 1047–1081.

90. Deuve, T. Description d'un Coléoptère troglobie du genre *Eustra*, découvert dans un karst du Vietnam méridional. *Rev. Fr. Entomol.* **1996**, *18*, 23–26.

91. Ferrer, J. Description d'un nouveau genre de Stenosini troglobie du Vietnam (Coleoptera, Tenebrionidae). *Nouv. Rev. Entomol.* **2004**, *20*, 367–371.

92. Kaszab, Z. Eine neue höhlenbewohnende *Dichillus*-Art aus Thailand (Coleoptera: Tenebrionidae). *Ann. Nat. Hist. Mus. Wien* **1980**, *83*, 585–588.

93. Robinson, G.S. Cave-dwelling Tineid moths: A taxonomic review of the world species (Lepidoptera: Tineidae). *Trans. Br. Cave Res. Assoc.* **1980**, *7*, 83–120.

94. Howarth, F.G.; James, S.A.; Preston, D.J.; Imada, C.T. Identification of roots in lava tube caves using molecular techniques: Implications for conservation of cave arthropod faunas. *J. Insect Conserv.* **2007**, *11*, 251–261. [CrossRef]

95. Trajano, E.; Bichuette, M.E. Diversity of Brazilian subterranean invertebrates, with a list of troglomorphic taxa. *Subterr. Biol.* **2010**, *7*, 1–16.

96. Fernandez, N.; Theron, P.; Rollard, C.; Castillo, E.R. Oribatid mites from deep soils of Hòn Chông limestone hills, Vietnam: The family Lohmanniidae (Acari: Oribatida), with the descriptions of *Bedoslohmannia anneae* n. gen., n. sp., and *Paulianacarus vietnamese* n. sp. *Zoosystema* **2014**, *36*, 771–787. [CrossRef]
97. Meregalli, M.; Osella, G. Studies on Oriental Molytinae. IV. *Anonyxmolytes lilliput* new genus and new species from Vietnam (Coleoptera, Curculionidae). *Ital. J. Zool.* **2007**, *74*, 381–388. [CrossRef]
98. Jaloszynski, P. Subterranean Scydmaeninae of Southeast Asia (Coleoptera: Staphylinidae). *Ann. Zool.* **2017**, *67*, 713–723. [CrossRef]
99. Mauriès, J.P.; Golovatch, S.I.; Geoffroy, J.J. Un nouveau genre et une nouvelle espèce de l'ordre Stemmiulida du Viet-Nam (Diplopoda). *Arthropoda Sel.* **2010**, *19*, 73–80. [CrossRef]
100. Deharveng, L.; Bedos, A. Chapter 18—Diversity Patterns in the Tropics. In *Encyclopedia of Caves*; White, W.B., Culver, D.C., Pipan, T., Eds.; Academic Press: Cambridge, MA, USA, 2019; pp. 146–162.
101. Moseley, M.; Lim, T.W.; Lim, T.T. Fauna reported from Batu caves, Selangor, Malaysia: Annotated checklist and bibliography. *Cave Karst Sci.* **2012**, *39*, 77–92.
102. Chapman, P. Invertebrates in Niah Great Cave, Batu Niah National Park, Sarawak. *Cave Sci.* **1984**, *11*, 89–91.
103. Beck, H.E.; Zimmermann, N.E.; McVicar, T.R.; Vergopolan, N.; Berg, A.; Wood, E.F. Present and future Köppen-Geiger climate classification maps at 1-km resolution. *Sci. Data* **2018**, *5*, 180214. [CrossRef]
104. Isaac, N.J.B.; Pocock, M.J.O. Bias and information in biological records. *Biol. J. Linn. Soc.* **2015**, *115*, 522–531. [CrossRef]
105. Li, F.; Li, S. Paleocene–Eocene and Plio–Pleistocene sea-level changes as "species T pumps" in Southeast Asia: Evidence from *Althepus* spiders. *Mol. Phylogenet. Evol.* **2018**, *127*, 545–555. [CrossRef]
106. Voris, H.K. Maps of Pleistocene sea levels in Southeast Asia: Shorelines, river systems and time durations. *J. Biogeogr.* **2001**, *27*, 1153–1167. [CrossRef]
107. Tamura, T.; Saito, Y.; Sieng, S.; Ben, B.; Kong, M.; Sim, I.; Choup, S.; Akiba, F. Initiation of the Mekong River delta at 8 ka: Evidence from the sedimentary succession in the Cambodian lowland. *Quat. Sci. Rev.* **2009**, *28*, 327–344. [CrossRef]
108. Liu, J.P.; DeMaster, D.J.; Nguyen, T.T.; Saito, Y.; Nguyen, V.L.; Ta, T.K.O.; Li, X. Stratigraphic formation of the Mekong River Delta and its recent shoreline changes. *Oceanography* **2017**, *30*, 72–83. [CrossRef]
109. Mylroie, J.; Tronvig, K. Karst Waters Institute creates Top Ten List of endangered karst ecosystem. *NSS News* **1998**, *56*, 41–43.
110. IFC. *Cement Manufacturing in a Critical Ecosystem. A Guide to Biodiversity for the Private Sector*; Holcim Vietnam Ltd.: Ho Chi Minh City, Vietnam, 2005; pp. 1–3.
111. Sinclair Knight Merz (Australia). *Limestone Biodiversity Study, Hon Chong (EIA Report)*; Sinclair Knight Merz (Australia): St. Leonards, Australia, 2002; (Unpublished report).
112. Le, C.K.; Truong, Q.T.; Ly, N.S. (Eds.) *Beleaguered Hills: Managing the Biodiversity of the Remaining Karst Hills of Kien Giang, Vietnam*, Proceedings of the International Workshop on Biodiversity of Kien Giang Karst, Rach Gia, Vietnam, 5–6 June 2008; Nha Xuat Ban Nong Nghiep: Ho Chi Minh, Vietnam, 2009; pp. 1–199.
113. Lang, C. Vietnam: Unique biodiversity threatened by World Bank-funded cement plant. *World Rainfor. Mov. Bull.* **2003**, *71*, 1–3.

Editorial

Global Subterranean Biodiversity: A Unique Pattern

Louis Deharveng [1,*], Anne Bedos [1], Tanja Pipan [2,3] and David C. Culver [4]

[1] Institut de Systématique, Évolution, Biodiversité (ISYEB), UMR7205, Muséum National d'Histoire Naturelle, Sorbonne Université, EPHE, 45 Rue Buffon, 75005 Paris, France; bedosanne@yahoo.fr

[2] Karst Research Institute ZRC-SAZU, Titov trg 2, SI-6230 Postojna, Slovenia; tanja.pipan@zrc-sazu.si

[3] UNESCO Chair on Karst Education, University of Nova Gorica, Glavni trg 8, SI-5271 Vipava, Slovenia

[4] Department of Environmental Science, American University, 4400 Massachusetts Ave. NW, Washington, DC 20016, USA; dculver@american.edu

* Correspondence: dehar.louis@wanadoo.fr

Citation: Deharveng, L.; Bedos, A.; Pipan, T.; Culver, D.C. Global Subterranean Biodiversity: A Unique Pattern. *Diversity* **2024**, *16*, 157. https://doi.org/10.3390/d16030157

Received: 20 January 2024

Accepted: 5 February 2024

Published: 1 March 2024

1. Introduction

Since the 1980s, with the widespread use of the phrase biodiversity [1], the mapping and analysis of biodiversity has excelled at a rapid pace at a range of scales, from small to global. One example of this is the global maps of biodiversity produced by the World-wide Fund for Wildlife [2]. However, large-scale analyses of subterranean biodiversity—especially the biodiversity of caves and other subterranean habitats, such as soil, epikarst, and the underflow of rivers—are conspicuously absent [3]. There are a number of reasons for this, including difficulties in accessing habitats, incomplete taxonomy, and the dominance of β diversity over α diversity [4,5]. Culver and Sket attempted to circumvent and ignore these problems by concentrating on individual caves and aquifers, as opposed to just regions [6]. Their initial list of 20 caves and aquifers, with 20 or more species limited to subterranean habitats (stygobionts and troglobionts), stimulated interest among researchers in identifying species present in various caves and wells. However, this early attempt and subsequent studies [7,8] remained incomplete, because several large regions remained unexplored and available data of others had not been synthesized. It is why we decided in 2020 to launch a more comprehensive analysis, with the aim to document the richest subterranean biodiversity hotspots at the world scale and to get insight into the understanding of their geographical and ecological pattern. A first special issue of the journal *Diversity* was published in 2021 on the question [9], and a second one was launched in 2022. The present contribution is the closing paper of these special issues, that synthesizes the current state of our knowledge on these hotspots of subterranean biodiversity. With the completion of the second issue, 12 hotspots of subterranean biodiversity are added to the 14 sites analyzed in the former issue. With two recent cases from the literature, we reach a total of 28 hotspots, that represent almost all subterranean biodiversity hotspots documented thus far at the world scale.

2. Goals

We use here this expanded dataset to synthesize the geographic patterns and formulate global maps of subterranean biodiversity. To this end, we summarize the numbers of species found in the different sites, with 25 or more subterranean specialists, stygobionts and troglobionts. For a better understanding of the observed patterns, we give an overview of the distribution of lower biodiversity spots on earth.

Our second goal is to outline some of the challenges encountered in the analysis, and especially in the comparison of subterranean biodiversity. Challenges linked to sampling unevenness are widespread. We use total species number versus species numbers excluding different zoological groups to evaluate the impact of these biases in between-site comparisons. Challenges are also particularly acute in the definition, both theoretical and practical, for what constitutes a cave (or subterranean) limited species—stygobionts and troglobionts.

3. Results

The current map and list of these 28 sites of high subterranean biodiversity (Figure 1, Table 1) represent nearly all of the sites known to contain 25 or more stygobionts and troglobionts. Most of these sites are discussed in individual papers in two Special Issues of *Diversity*, the only exceptions being the Areias Cave System of Brazil [10] and Túnel de la Atlantida of Canary Islands [11]. There are only two sites that are claimed to contain 25 or more species and were not included—Logarček in Slovenia [6] and Sauve Spring in France [12]. The potential occurrence of high-diversity cave faunas in other tropical and temperate regions is briefly discussed further down. We have not included deep soil sites or hyporheic sites unconnected with caves. There are also sites in the hyporheic of rivers with 25 or more stygobionts, the most thoroughly studied being the Danube Flood Plain National Park in Austria [13] and the Rhone River near Lyon [14]. Biodiversity patterns in the hyporheic require a different treatment.

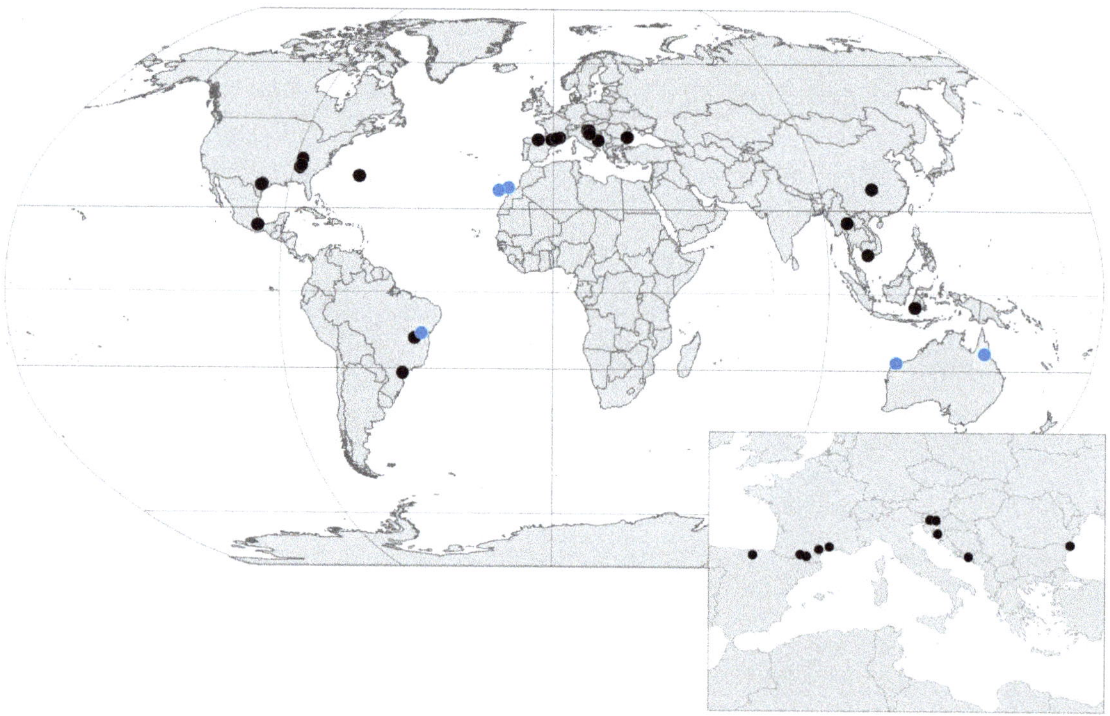

Figure 1. Map of all known subterranean sites with 25 or more stygobionts plus troglobionts. Horizontal lines are the equator, ±23.5° (Tropic of Cancer and Tropic of Capricorn), and the Arctic and Antarctic Circles (±66.5°). Non-karst sites are shown in blue and karst sites are shown in black. Inset map of the Mediterranean region provides a greater resolution. Map courtesy of Magdalena Năpăruş-Aljančič.

Overall, subterranean hotspot sites have been found on all continents except Africa and Antarctica. The latitudinal range of subterranean hotspots is from 25° S to 45° N, including sites in the tropics, thus far only in the seasonal tropics. There are far fewer sites in the Southern Hemisphere, and none of these are farther south than 25° S. There is a concentration of hotspots at 40 to 50° N latitudes (Figure 2). The 40–50° N cluster generally corresponds to the previously described ridge of high cave biodiversity in Europe [15].

Additionally, there is an almost complete absence of hotspots around the equator; only one, the Towakkalak System in Indonesia, occurs within 10° of the equator.

Table 1. Physical characteristics of hotspot caves and wells, arranged by increasing latitude.

Country	Cave	Latitude	Longitude	Features
BRA	Areias Cave System	−24.6	−48.7	Karstic
Tropic of Capricorn		−23.5		
AUS	Robe River Well 2A	−21.6	115.9	Calcrete/ Hypogene
AUS	Undara Lava Tube System	−18.2	144.5	Volcanic
BRA	Água Clara System	−13.8	−44.0	Karstic
BRA	Igatu Cave System	−12.9	−41.4	Silici-clastic
IDN	Towakkalak System	−5.0	119.6	Karstic
Equator		0		
VNM	Hang Mo So	10.2	104.6	Karstic
MEX	Sistema Huautla	18.1	−96.8	Karstic
THA	Tham Chiang Dao	19.4	98.9	Karstic
Tropic of Cancer		23.5		
ESP	Cueva del Viento System	28.4	−16.7	Volcanic
ESP	Túnel de la Atlantida	29.2	−13.5	Volcanic
CHN	Feihu Dong	29.2	109.3	Karstic
USA	Comal Springs	29.7	−98.1	Hypogene
USA	San Marcos Artesian Well	29.9	−97.9	Hypogene
BMU	Walsingham Caves	32.3	−64.8	Hypogene
USA	Fern Cave	34.7	−86.3	Karstic
USA	Crystal-Wonder Cave System	35.3	−85.9	Karstic
USA	Mammoth Cave	37.1	−86.1	Karstic
BIH	Vjetrenica Cave System	42.9	18.0	Karstic
ESP	Ojo Guareña System	43.0	−3.7	Karstic
FRA	Baget System	43.0	1.0	Karstic
FRA	Coume Ouarnède System	43.0	0.9	Karstic
ROU	Movile Cave	43.8	28.6	Hypogene
FRA	Lez Aquifer	43.8	3.8	Karstic
FRA	Cent Fonts	43.8	3.6	Karstic
HRV	Lukina Jama-Trojama Cave System	44.8	15.0	Karstic
SVN	Križna Jama	45.7	14.4	Karstic
SVN	Postojna Planina Cave System	45.8	14.2	Karstic

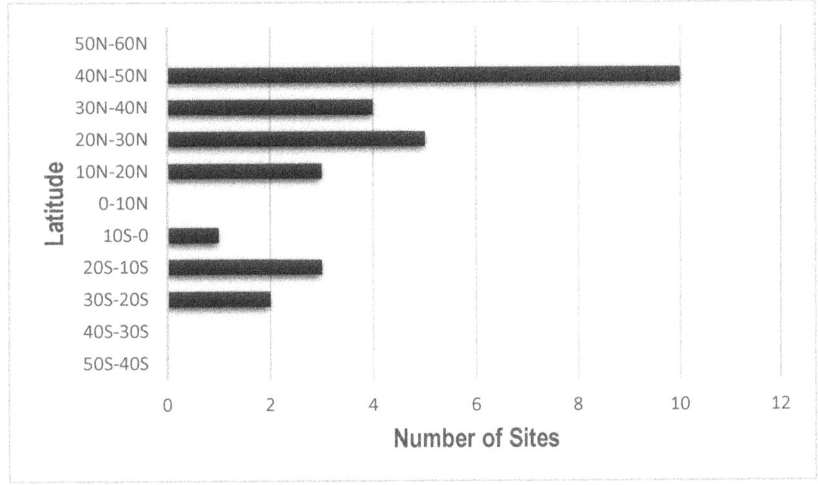

Figure 2. Distribution of subterranean hotspot sites by latitude. See Table 1.

The sites are a mixture of individual caves, hydrologically connected caves in a single karst drainage system, karst aquifers, and non-karstic caves including lava tubes and silicoclastic caves. The caves themselves greatly vary in size and depth, but include the world's longest cave—Mammoth Cave in Kentucky—and one of the deepest—Sistema Huautla in Mexico. On the other hand, many of the caves are less than 1000 m long and only a few meters in depth. Two of the caves are anchialine—tidal caves with a freshwater lens (Geben-Herzberg)—Túnel de la Atlantida (Canary Islands) and Walsingham Caves (Bermuda). The caves and cave systems include both epigenic and hypogenic caves. Epigenic caves (caves formed by falling waters [16]), are organized into subterranean drainage basins [17], which can be delineated by the injection and capture of soluble dyes such as fluorescein [18]. Many of the hotspot caves are epigenic (Table 1), and for some—Água Clara Cave System (Brazil), Vjetrenica Cave System (Bosnia and Hercegovina), Ojo Guareña System (Spain), Coume Ouarnède System (France), Crystal-Wonder Cave System (USA), and Postojna Planina Cave System (Slovenia)—species counts for both the largest caves and the drainage basin are available. For other epigenic caves—Feihu Dong (China), Tham Chiang Dao (Thailand), Towakkalak System (Indonesia), Lukina Jama-Trojama Cave System (Croatia), Sistema Huautla (Mexico), Križna Jama (Slovenia), Fern Cave (USA), and Mammoth Cave (USA)—only data for the caves themselves are available. A few sites are not organized by drainage but instead by their proximity and isolation from other systems, e.g., Undara lava tube system in Australia and Hang Mo So in Vietnam. For other epigenic caves and karst areas—Cent Fonts, Lez aquifer and Baget System (France)— data only for the entire basin are available, with data for individual caves included when available. Baget is of special historical interest as it is the site of extensive studies conducted by R. Rouch, who first suggested that karst basins were the natural units for ecosystem studies [19]. Overall, species counts were used for the entire drainage basin when available.

Some caves are not formed by descending water but ascending water; therefore, they are unconnected to and isolated from surface drainage patterns. Many of these caves are formed by H_2SO_4. The frequency of hypogenic caves is still unclear [16], and many epigenic caves show signs of having hypogenic origins [20]. Two caves—Movile Cave (Romania) and Walsingham Caves (Bermuda)—are hypogenic (Table 1) as are the aquifers associated with the Robe River wells (Australia) and San Marcos Artesian Well (Texas). What was accessible for sampling in these four sites was quite different. Movile Cave is only 240 m long but connected to a much larger, deeper aquifer of between 50 and 100 km^2 [21]; Walsingham Caves comprise a large number of small caves located in a 4 by 0.5 km isolated band of limestone—the Walsingham Tract [22]. The San Marcos Artesian Well samples the 900,000 km^2 Edwards/Trinity Aquifer, and the two Robe River wells (about 1 km apart) sample the iron-rich Robe alluvial aquifer (of unknown size) [23].

The overall global pattern (Figure 1) is a concentration of sites in Europe, centered around latitude 40 N. The only exception are the sites in the Canary Islands; there are no sites in continental Africa. In the Americas, there are two small clusters: one is a small cluster of three sites near the intersection of the borders of Alabama, Tennessee, and Georgia, the other one is a small cluster in Brazil, also with three sites.

There are differences in richness among the hotspots (Figure 3), a point that we address in detail in the section on challenges. Here, we note that only three sites have both 25 or more troglobionts and 25 or more stygobionts. All are in the Dinaric karst, which ranges from northeastern Italy to Montenegro: Vjetrenica Cave System (Bosnia and Hercegovina), and Križna Jama and Postojna Planina Cave System (Slovenia). This supports the longstanding claim from Sket [24] that the Dinaric karst is a global center of subterranean biodiversity. The richest site of terrestrial biodiversity is Vjetrenica Cave System, and the richest site of aquatic biodiversity is Walsingham Caves (Bermuda). Only 7 sites out of 28 have 50 species or more when troglobionts and stygobionts are counted together:

1. Postojna Planina Cave System, Slovenia (105);
2. Vjetrenica Cave System, Bosnia and Hercegovina (93);
3. Walsingham Caves, Bermuda (63);
4. Križna Jama, Slovenia (59);
5. Baget System, France (57);
6. San Marcos Artesian Well, Texas (55);
7. Ojo Guareña System, Spain (54).

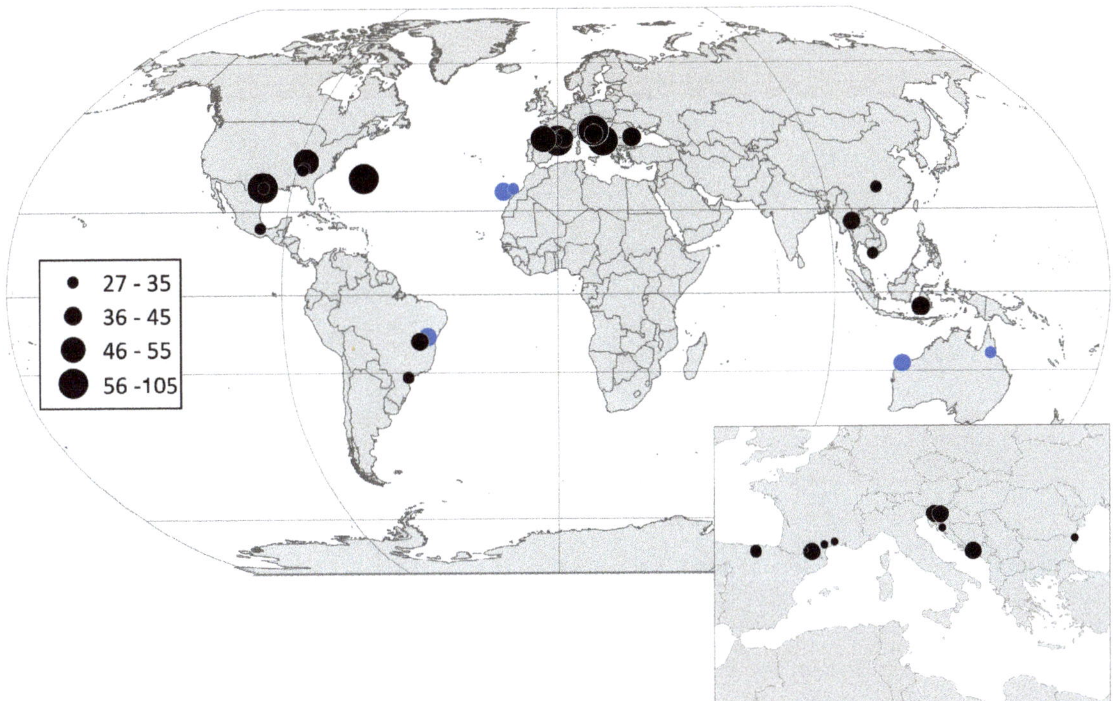

Figure 3. Global map of subterranean hotspots, where the dots are proportional to the number of species (see Table 2, column S+T). Horizontal lines are the Equator, ±23.5° (Tropic of Cancer and Tropic of Capricorn, and Arctic and Antarctic Circles (±66.5°). Non-karst sites are shown in blue and karst sites are shown in black. The inset map of the Mediterranean region provides greater resolution. Map courtesy of Magdalena Năpăruş-Aljančič.

The fauna of many of the sites listed in Table 2 are primarily terrestrial or aquatic; nine sites are exclusively one or the other (Table 2). The richer sites tend to be in Europe and North America, but this may in part be due to more thorough data collection, a point that we address below.

Table 2. Numbers of species per cave under different counting options. S, observed number of stygobionts; T, observed number of troglobionts; AOG, aquatic obscure/undersampled groups (Protista, Rotifera, Nematoda, Nemertea, Oligochaeta, Acari, commensals and parasites); MC, microcrustacea (Copepoda, Ostracoda and Syncarida, not including parasites and commensals already counted in AOG); S*, S minus AOG minus MC; TOG, terrestrial obscure/undersampled groups (Acari, Oligochaeta, commensals and parasites); T*, T minus TOG; un, undescribed species; un%, 100*un/(S+T).

Country	Cave	S	T	S+T	AOG	MC	S*	TOG	T*	S*+T*	un	un%	Source
AUS	Robe River Well 2A (a)	43	0	43	11	21	11	0	0	11	21	49	[23]
AUS	Undara Lava Tube System	1	30	31	0	0	1	0	30	31	25	81	[25]
BIH	Vjetrenica Cave System	48	45	93	8	8	32	1	44	76	6	6	[26]
BMU	Walsingham Caves	63	0	63	8	29	26	0	0	26	0	0	[22]
BRA	Água Clara Cave System	8	33	41	0	0	8	1	32	40	30	73	[27,28]
BRA	Areias Cave System	6	22	28	1	0	5	0	22	27	14	50	[10]
BRA	Igatu Cave System	2	35	37	0	0	2	3	32	34	29	78	[29]
CHN	Feihu Dong	4	23	27	0	1	3	1	22	25	14	52	[30]
ESP	Cueva del Viento System (b)	0	42	42	0	0	0	0	42	42	0	0	[31]
ESP	Ojo Guareña System	46	8	54	14	24	8	2	6	14	23	43	[32]
ESP	Túnel de la Atlantida	34	0	34	0	12	22	0	0	22	4	12	[11]
FRA	Baget System	40	17	57	4	27	9	1	16	25	5	9	[33]
FRA	Cent Fonts (c)	43	1	44	2	19	22	0	1	23	4	9	[12]
FRA	Coume Ouarnède System (d)	17	17	34	1	8	8	2	15	23	1	3	[34]
FRA	Lez Aquifer (e)	39	0	39	2	15	22	0	0	22	7	18	[12,35]
HRV	Lukina Jama-Trojama Cave System	16	25	41	0	0	16	2	23	39	20	49	[36]
IDN	Towakkalak System	10	26	36	0	0	10	1	25	35	18	50	[37]
MEX	Sistema Huautla	0	27	27	0	0	0	0	27	27	10	37	[38]
ROU	Movile Cave (f)	13	25	38	3	3	7	1	24	31	3	8	[21]
SVN	Križna Jama (g)	31	28	59	10	10	11	0	28	39	5	8	[39]
SVN	Postojna Planina Cave System (h)	62	43	105	12	29	21	2	41	62	11	10	[40]
THA	Tham Chiang Dao	4	33	37	2	2	0	1	32	32	17	46	[41]
USA	Comal Springs (i, j)	32	0	32	3	2	27	0	0	27	4	13	[42]
USA	Crystal-Wonder Cave System (k)	8	23	31	0	0	8	1	22	30	3	10	[43]
USA	Fern Cave (l)	8	19	27	2	0	6	1	18	23	7	26	[44]
USA	Mammoth Cave (m)	17	32	49	3	2	12	6	26	38	0	0	[45]
USA	San Marcos Artesian Well (j)	55	0	55	8	15	32	0	0	32	16	29	[42]
VNM	Hang Mo So	0	27	27	0	0	0	1	26	26	20	74	[46]

Notes. (a) one stygophilic species, Tubificidae sp., discarded; two subspecies of humphreysi and their hybrid counted as a single species; (b) the system includes Cueva Felipe Reventón with 38 troglobionts, and Cueva del Viento with 36 troglobionts (erroneously noted 28 in Culver et al. [9]); (c) one additional troglobiont, *Laemostenus (Actenipus) oblongus balmae*; not counted in the abstract of from Prié et al. [12]; (d) four species discarded (strictly hyporheic species found outside cave); (e) a single aquifer, accessed from several wells close each other; (f) species found in nearby springs and wells have been counted; (g) one stygophilic species, *Synurella ambulans*, discarded; (h) eleven species discarded (stygophiles or troglophiles, listed as troglobiotic populations of surface species); (i) Table S2 in [42]; (j) aquifer is greater than 2500 km^2 so single points used; (k) Crystal and Wonder Caves connected by 91 m long surface stream, therefore combined; (l) all taxa considered from the two columns "Historical" and "This Study" of Table 1 in Niemiller et al. [44]; (m) taxa considered from the column "This Study" of Table 1 in Niemiller et al. [45].

4. Challenges

4.1. How to Define Stygobionts, Troglobionts, etc.

Perhaps it is surprising that obligate subterranean dwelling species are difficult to separate from those species that are either transient in caves—in the sense they do not complete their entire life cycle there (sometimes called subtroglophiles [47])—or can survive and reproduce in surface habitats (sometimes called eutroglophiles [47]). Several classifications and terminologies have been proposed [47–51], but for the most part, researchers agree on the core aspects of stygobionts and troglobionts—species that can only survive and reproduce in subterranean habitats, especially caves and aquifers. Some authors [39,40] use the terms stygobionts and troglobionts for species which have subterranean and above-ground populations. These species match the definition of eutroglophiles, and, for practical reasons, particularly with regard to consistency of terminology, were not included in counts. Aside some exceptions [52], a core tenet of speleobiology is that troglobionts and stygobionts often have a convergent morphology known as "troglomorphy" [53], including the reduction of eyes and pigment, increase in size, elongation of the appendages and development of extra-optic sensory structures. A number of troglobionts, however, especially but not necessarily phylogenetically young ones, may display little or no troglomorphy [54]. Subterranean populations of highly variable species show sometimes a reduction or a loss of pigment and eyes, and are considered by some authors as stygobionts or troglobionts. The best known example of this phenomenon is the Mexican cavefish, *Astyanax mexicanus*, with eyed and eyeless populations [55]. Such cases need to be examined carefully, both to determine if they point to genetically separate taxonomic entities, which seems most likely, and if the cave populations are truly troglomorphic (see below). The taxonomic status of *Astyanax mexicanus* is disputed, with some authors claiming that the eyeless cave populations are a separate species [56].

The regressive characters (pigment and eye reduction) shown by some species are often invoked by authors to qualify the species as "troglomorphic" and, by extension, troglobiotic; however, these species may not exhibit the set of characters that define troglomorphy in an adaptive sense. Many of them refer to deep-soil life forms, which share eye and pigment regression with troglomorphic cave-restricted species, but not other traits that define troglomorphy. Such species are, in that case, erroneously classified as troglomorphic. Finally, morphological characters of a number of species or morphospecies, especially in the tropics and subtropics, remain undescribed, making it difficult assignment to life forms.

In general, we have followed various authors in their assessment of ecological status, with some exceptions, especially when they disagree on the status of a species. Approaching troglomorphy as a purely morphological qualification and troglobiotic as a purely ecological one would avoid much confusion. Numerous ambiguities and uncertainties persist in the literature, mostly linked to a loose use of terminology and to disputable assignment of species to ecological categories. There is no completely satisfactory terminology for troglomorphic cave populations of species with non-troglomorphic populations outside of caves, like those of *Astyanax mexicanus* [55]. The statuses of species that have deep-soil species facies, those in Robe River boreholes [23], and of species with deep sea populations, those from Walsingham Caves [22] are disputable because their connection with soil (Robe River) or marine environment (Walsingham Caves) is unknown. These and other problems have led some researchers to reject the terminology or at least reduce its usage [57,58]. On the other hand, redefining troglomorphy and specialization to deep soil (edaphomorphy) based on morphological criteria that are statistically correlated to the occurrence of species in habitats [59] could make such a terminology useful.

Deharveng et al. [46] employed a decision system to determine if a species was troglobiotic in their analysis of Vietnamese caves. These caves are usually short and shallow, with frequent terrestrial and aquatic connections to the exterior, allowing constant inputs of nutrients. This generates cave communities dominated by troglophiles and tramp species, and makes the ecological category assessment of individual species hazardous. They adopted four complementary approaches:

(1) Morphological inference is based on presence or absence of troglomorphic traits. The presence of a set of convergent troglomorphic traits in most arthropods (eye and pigment reduction combined with appendage and size increases compared to surface relatives) points to obligate cave life. Depigmentation and eye reduction are trends shared by many soil and cave arthropods, and Brignoli [60] stressed that the equation "blind = troglobite" has a limited value. When they are combined with appendage shortening and a decrease in size, they qualify a species as euedaphomorphic [59]. It is only when eye and pigment regression are combined with appendage elongation and an increase in size (or other characters recognized as troglomorphic), that they qualify a species as troglomorphic. Statistically, the correlation of troglomorphic and euedaphomorphic life forms with the ecological categories of troglobiont and edaphobiont is one-way and robust. Where the set of troglomorphic traits is not present, such as in many guano-associated and tropical species, we have to rely on other inferences [61].

(2) Parallel sampling inference is based on the absence of species outside subterranean habitats, and allows to assign a status of troglobionts to species that do not exhibit troglomorphy ("obligate troglophiles" of Howarth and Wynne [51]). Statistically meaningful data on the occurrence of species, both inside and outside caves, can be extracted from the literature for well-investigated regions. In lesser known areas such as the tropics, sampling in parallel cave and non-cave habitats may allow us to reasonably assess the ecological status of a species, the strength of such an inference being dependent on sampling efforts and on the rarity of the species.

(3) Taxonomic inference is based on the ecological status of related taxa. Certain groups are known to greatly diversify in subterranean habitats [62–64], while others never colonize such habitats. A species from a group which is not prone to underground diversification and lack troglomorphic traits is less likely to be a troglobiont.

(4) Barcoding inference is based on genetic divergence between populations and species. Within a troglophilic or stygophilic species, molecular analyses may characterize populations that live in caves as different from those that live outside [65], leading to split the original species into cave-restricted and non-cave-restricted lineages or species. Barcoding may conversely lead a species to lose its ecological status of cave-restricted if it is shown to be molecularly inseparable from another species which is not cave-restricted.

4.2. Taxonomic Completeness

As is to be expected with a rare, elusive fauna, sampling for the majority (if not all) of the sites listed in Table 2 is incomplete, and the extent of incompleteness varies among sites. A few sites, most notably Postojna Planina Cave System (Slovenia), Vjetrenica Cave System (Bosnia and Hercegovina), and Mammoth Cave (USA) have a centuries-old history of biological study. Others, especially those in the tropics and subtropics, have a decade long or less history of biological study. Lukić et al. [36] reported that almost all information for the fauna of the Lukina Jama-Trojama Cave System in Croatia dates from the 1990s or later. The same is true for Fern Cave in Alabama [44], Hon Chong in Vietnam [46], and the Água Clara Cave System in Brazil [27,28]. Sampling effort is therefore a major determinant of the richness of species in these caves. We estimated sampling effort by comparing numbers of species with the date of the first listing of the fauna. Other measures, such as total number of publications, are difficult to estimate due to difficulties in defining publication about the site, for example, does it include monographic taxonomic studies where the species in question are a small part of the study? When date of first faunal list and number of species are compared, the regression accounts for approximately one-third of the total variance in species number (Figure 4). This gives pause to more biological interpretations (see below).

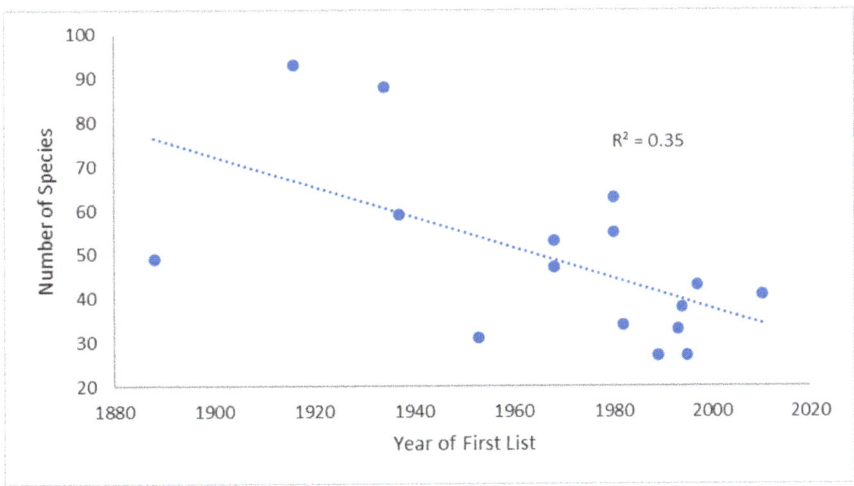

Figure 4. Regression of number of stygobionts and troglobionts against the first complete fauna list when available.

Some groups, e.g., beetles and amphipods, have been searched for and studied in most hotspot sites. For other taxonomic and ecological groups, this is not the case. Among aquatic species, these include parasites and commensals, Protista, Rotifera, Nematoda, Nemertea, Oligochaeta and water mites. For example, subterranean-limited Protista have been reported from Planina Postojna Cave System [40] and Walsingham Caves [22], but have been searched for in few of the other caves (but see [66]). Only three contributions report on subterranean rotifers—Ojo Guareña [32], Robe Valley [23] and Cent Fonts [12,67]—and only for the Robe Valley is it claimed that the rotifers are stygobiotic. The data on obligate subterranean aquatic species summarized in Table 2 are "corrected" for this discrepancy by eliminating species of these groups. A second source of bias in the counting of obligate subterranean aquatic species is microcrustacean fauna—Copepoda, Ostracoda, and Syncarida. For sites with aquatic fauna, at least a few of these species have been reported, but a rich source of these species is the epikarst [68] that has rarely been sampled. In the two caves where it has been sampled, 15 copepod species were discovered in Postojna Planina Cave System [69] and 9 copepod, ostracod, and syncarid species were discovered in Ojo Guareña [32,70] (Table 2). Species numbers with copepods, ostracods, and syncarids deleted are also shown in Table 2.

Although the variability of the level of study of the terrestrial species is similar to that of the aquatic fauna, we were less successful in correcting for this bias. Higher level categories often do not have global distributions [71]; therefore, absence is often the result of this rather than lack of collecting. Acari is one group that has been inconsistently studied among hotspot caves. Mites are rare in caves that lack guano, and this is largely due to the undersampling of their preferred habitats, i.e., cave soils and organic matter. In most cases, they also remain undescribed or unidentified. The only cave with more than two species reported as troglobionts is Mammoth Cave (Table 2), which contains six species. Five of the six Mammoth Cave species were first described in the 19th century, but these data are problematic since several of these species have not been found since [45]. Coleoptera are the best studied, but, while investigated in all hotspot caves, have a biased geographic distribution, being most common in the Holarctic and least common in dry and tropical areas, while the reverse is globally true for Arachnida (Table 3) [59].

Table 3. Relative percentages of troglobionts among arachnids (excluding mites) and beetles in hotspot caves where terrestrial cave fauna includes more than 10 troglobionts. Caves are ranked by decreasing values of the ratio Coleoptera/Arachnida (ratio Co/Ar). N, species number.

Country	Cave System	Arachnida		Coleoptera		Ratio
		N	%	N	%	Co/Ar
FRA	Baget System	2	22.2	7	77.8	3.50
FRA	Coume Ouarnède System	2	25.0	6	75.0	3.00
BIH	Vjetrenica Cave System	10	43.5	13	56.5	1.30
USA	Crystal-Wonder Cave System	5	50.0	5	50.0	1.00
CHN	Feihu Dong	5	50.0	5	50.0	1.00
SVN	Križna Jama	7	50.0	7	50.0	1.00
SVN	Postojna Planina Cave System	10	52.6	9	47.4	0.90
ESP	Cueva del Viento System	16	55.2	13	44.8	0.81
USA	Mammoth Cave	10	55.6	8	44.4	0.80
ROU	Movile Cave	8	61.5	5	38.5	0.63
HRV	Lukina Jama-Trojama Cave System	5	62.5	3	37.5	0.60
USA	Fern Cave	6	66.7	3	33.3	0.50
AUS	Undara Lava Tube System	10	71.4	4	28.6	0.40
BRA	Areias Cave System	6	75.0	2	25.0	0.33
VNM	Hang Mo So	8	80.0	2	20.0	0.25
BRA	Água Clara Cave System	9	81.8	2	18.2	0.22
THA	Tham Chiang Dao	10	83.3	2	16.7	0.20
BRA	Igatu Cave System	13	86.7	2	13.3	0.15
IDN	Towakkalak System	11	91.7	1	8.3	0.09
MEX	Sistema Huautla	18	100.0	0	0.0	0.00

5. Coldspots, Low-Diversity Spots and Undersampled Spots

While patterns of high diversity are emerging at a global level, the distribution of remaining subterranean diversity on Earth has been less thoroughly scrutinized. A better knowledge of the spatial patterns of lower subterranean biodiversity would help our understanding of hotspot patterns, making clearer where coldspots are located, as well as predicting where sites of high diversity may be expected to be found in future investigations. Rough outlines of our knowledge on these lower biodiversity sites are summarized below.

5.1. Where Are the Coldspots?

Large regions where troglobionts and stygobionts are absent or rare have been extensively documented in many parts of the world, especially Canada [72], Germany [73], and Poland [74]. Their biodiversity is assumed to have been largely depleted by Quaternary glaciations. Therefore, troglobionts are restricted to a few refugium massifs [74,75] that remained free of ice during glaciations. Stygobionts are more diversified, having survived under the ice caps [75] or having recolonized from areas unaffected by glaciations.

As expected and confirmed by emerging patterns, hotspots of subterranean diversity are all located within the southern and the northern areas affected by glaciations. However, these are small sites in a very large area comprising diverse environments, where it is usually difficult to know if local low diversity and coldspots are real or the result of undersampling. This uncertainty is reflected in considerable ecological and geographical sampling gaps. Understanding gap distribution would aid our understanding of patterns with high richness.

5.2. Lower Biodiversity Habitats

Some habitats widespread on earth are devoid of hotspots. It may be that they really are low biodiversity habitats or that they are rarely sampled, at least from a biodiversity perspective. Aside from local and special microhabitats, such as hypotelminorheic habitats [76], several of these habitats are extensively distributed—marine caves [77], anchialine caves [78], littoral interstitial habitats [79,80], phreatic and hyporheic of rivers [81], and MSS (milieu souterrain superficiel) and scree in all kinds of rocks [82].

Anchialine. Most hotspots in our survey are filled with or connected to freshwater. Only two are related to anchialine habitats with a marine/freshwater boundary (Túnel de la Atlantida and Walsingham Caves). While much less frequent than freshwater caves, they have a global distribution, and are relatively common in some regions, such as the Yucatan Peninsula and the Mediterranean Sea [83]. The high biodiversity of marine and anchialine caves is often emphasized at a regional level, especially in the Mediterranean Sea [79]. However, strictly marine caves appear to have few stygobionts [77].

Littoral interstitial. There is a dichotomy between freshwater interstitial and littoral interstitial habitats at the level of faunal composition, but not at the level of habitat characteristics, which are similar in most respects except for salinity, nor at the level of habitat continuity, as freshwater and littoral interstitial are largely adjoining [80,84]. Contrary to freshwater interstitial habitats that largely contribute to the biodiversity of several hotspots in our survey, littoral interstitial habitats are absent, despite their global importance. None of the few faunistic datasets of these last habitats currently available in the literature explicitly points to hotspot, but it has been demonstrated that they have a rich and very distinctive fauna, which differs from the freshwater interstitial fauna at high taxonomic level. Delamare Deboutteville [79] reports for instance 95 species in the interstitial of a 300 m long beach of Southern France (Canet-Plage).

Deep Phreatic. Among the hotspots documented in these two Special Issues of *Diversity*, three are clusters of wells that penetrate phreatic waters—Edwards Aquifer (Texas), Robe River (Australia), Lez aquifer (France) (Table 2, Figure 1). These are, to our knowledge, the only deep wells that have been sampled.

Hyporheic. The alluvia of rivers are not included in this study but certainly contain a rich fauna. Danielopol and colleagues [13,81] showed that some European sites (Danube, Rhine and Rhône rivers) hosted a rich interstitial fauna. The Sava River in central Europe [85] and the Flathead River in Montana can be added to this list [86]. For example, the Grand Gravier site along the Rhône River near Lyon, France, yielded 30 species of stygobionts [14]. Although sampling difficulties may be technically challenging, more sampling with Bou Rouch pumps is a promising way to find new hotspots. The challenge here is also to define site units to be compared.

MSS and screes. Although widespread in temperate regions and known to host obligate subterranean fauna [3,87], we do not know of any hotspots of troglobiotic biodiversity in these habitats. Published lists for single sites usually include a few troglobiotic species [82,88–90]. Given serious sampling difficulties and the small number of investigated sites, especially in regions known to harbor rich biodiversity, it cannot be ruled out that future investigations in such sites may provide more troglobionts.

Epikarst. Individual epikarst drips may have up to ten species of stygobiotic copepods, and all the drips in a cave may have up to 15 species [69,91]. On average, about five stygobiotic epikarst copepods are found per cave [91]. Not all species are found in epikarst pools, and thus specialized drip collectors are needed. Only a few epikarst sites have been thoroughly sampled in Slovenia, Italy, Romania, Spain, and the United States [92].

Hypotelminorheic habitats. They harbor even fewer species—four or less in seeps in the upper Potomac basin near Washington, D.C. [93,94]. The spatial extent of these habitats is difficult to assess, but they are locally common and completely unconnected with caves and other deeper subterranean habitats.

5.3. Lower Biodiversity Sites and Regions

While significant progress has been made over the past several decades, especially in the tropics [71,95–98], many large cave areas remain unsampled or undersampled. Even in Europe, sampling is highly uneven [99]. Because of the rarity of species and difficulty of collecting them, multiple trips are needed to obtain a more or less complete list of species [95,97]. As a result, many species, even in well-studied areas, are known only from a handful of specimens, sometimes even one. The few studies on sampling completeness using species accumulation curves, indicate that sampling may require 100 or more caves to contact 70 percent of the fauna [100]. One feature of sampling completeness is that it is not just in low-diversity areas that additional species are expected to be discovered. For example, Zagmajster et al. [100] have shown that more new species of beetles are expected to be found in high-diversity rather than in low-diversity sites. In the same line, new troglobionts among cave beetles of China are mostly described from karsts that were already known to be the richest in cave species [101].

With the possible exception of Mammoth Cave, Vjetrenica Cave System, and Postojna Planina Cave System, authors of articles that focus on individual caves indicated that more species remain to be discovered. Is it possible that the hotspot caves of Table 2 have just been more thoroughly sampled than other caves? While there certainly remain hotspot sites to be discovered, we believe that hotspot species and their mapped distributions are roughly representative of real patterns, even if collections need to be completed for several habitats and regions.

In support to this hypothesis, a number of caves that have been well studied are rich in cave restricted fauna, but clearly not hotspots, and sometimes do not even host any troglobiont. Therefore, their global distribution may help to reinforce and finetune the hotspot patterns. We highlight a few of these below, as well as pointing out that entire regions are unlikely to yield hotspot caves, including the extensive cave system occurring in once glaciated Europe and North America.

5.3.1. Lower-Biodiversity Spots in Africa

Two hotspots have been documented in Africa, both located in the Canary Islands, but none have been recognized in the rest of the continent. The northern fringe of Africa between the Sahara and the Mediterranean Sea, which belongs faunistically to the Mediterranean basin, has no hotspot of subterranean biodiversity, in contrast to the Northern Mediterranean basin where the density of hotspots is the highest in the world [15] (Table 2). The Djurdjura massif in Algeria, that was the subject of intensive investigations conducted 110 years ago by Peyerimhoff (in [102]), is probably the richest of North Africa in cave-restricted fauna, with about 29 species, often troglomorphic, linked to cold caves and snow pits. These karstic features are densely distributed in the massif, but remain generally unconnected, and individual caves have no more than 12 troglobionts.

In sub-Saharan Africa, where karst is limited to some extent, several caves have been significantly sampled for subterranean biodiversity, especially in the Congo Republic, Democratic Republic of Congo, Kenya, Madagascar, South Africa and Tanzania [103]. Phreatic habitats were particularly studied in Somalia [104]. Most of these sites are usually very rich in troglophilic and guano-dependent species, but none has provided more than 20 strictly subterranean species. The Wynberg Cave System in South Africa (with 19 species considered as cave-restricted) is the richest one in sub-Saharan Africa [105], while Kulumuzi cave in Tanzania has a similar number of troglobites, but most with an uncertain ecological status [106].

On the whole, Africa appears to be relatively poor in subterranean biodiversity, with the exception of the Canarian hotspots.

5.3.2. Lower-Biodiversity Spots in Southern Tropical Asia and the Pacific

In lowlands of tropical Asia and the Pacific, about 12 caves have been documented in the literature as having more than 15 cave-restricted species. Three of them were treated as hotspots in the two Special Issues of *Diversity*, as they host more than 25 such species (Towakkalak System in Indonesia, Tham Chiang Dao in Thailand and Hang Mo So in Vietnam). Nine other caves scattered in the lowlands of the same region have between 15 and 24 cave-restricted species: Batu Caves in Malaysia [107] and Saripa Cave in Sulawesi, Indonesia [37] with 24 species, Clearwater Cave in Sarawak [108] with at least 22 species, Ganxiao Dong in Southern China [30] and the Sangki System in Sumatra, Indonesia [71] with 20 species, Ma San Dong in Southern China [71] with 17 species, Batu Lubang in Halmahera, Indonesia [71] and Tham Thon in Laos [71] with 16 species, Tham None in Laos [71] with 15 species. Several other caves which have been well sampled are much less rich than those cited above. With the exception of the Siju caves in Meghalaya, India which have been studied in detail for more than one century and has only 10 cave-restricted species [109], these low-diversity caves seem to be located in oceanic islands. The lava tubes of Hawaii are not particularly diverse, despite having a well-known, highly distinctive fauna [110]. Culver and Pipan [3] reported only 37 troglobionts from all of the Hawaiian Islands, although there are undoubtedly a number of undescribed species [111]. Only eight troglobionts have been recorded in the longest lava tube in the world, Kazamura Cave in Hawaii [111]. By contrast, the Canary Islands, off the coast of Africa, aside from having two lava tube hotspots, are generally rich in cave fauna [112]. The well documented Fapon Cave in the karst of Santo island in Vanuatu has no more than four unambiguous troglobionts [113] and can thus be considered a coldspot for cave-restricted fauna. A clear common feature of these lowland caves is the wide occurrence of guano and the impressive abundance and diversity of its associated fauna, with a number of species difficult to assign to the traditional ecological categories used for temperate cave fauna [114].

Biological data for caves above an elevation of 500 m are extremely limited in Southern Asia and the Pacific, but valuable datasets exist for two cave systems above 2000 m asl., located in the highlands of Papua New Guinea: Atea Kananda, which host 13 potential troglobionts and a few uncertain stygobionts [115], and Selminum Tem, with at least 19 troglobionts and 5 stygobionts [116]. Their fauna differs widely from that lowland caves of the region, being more similar to that of temperate caves, as shown by the near absence of bats and guano-associated species, the absence of some groups of arachnids (amblypygids, schizomids), and the presence of highly troglomorphic species of beetles. Given that several groups of species collected in these caves have not yet been studied [117], their species richness is likely at the currently recognized level of hotspots in tropical Asia.

The pattern of cave-restricted species richness described above for tropical Southern Asia and the Pacific could be driven by the combined effect of the microclimate (which determines the presence of bats, swiftlets and guano) and of the geological history (which accounts for the length of karst isolation from potential sources of colonizers). Whether patterns in American caves of humid tropics match those of the Old World tropics remains to be explored, but the relatively large amount of data available notably for Cuba, and to a lesser extent for Venezuela [61], does not indicate a rich fauna of cave-obligate species.

5.3.3. Low-Biodiversity Spots in the Temperate Zone

Coldspots and Glaciations. Caves of polar areas, roughly north of 50 °C and south of 45 °C in New Zealand and South America, are relatively well studied [72,118–120]. All are very poor in cave-restricted species, especially in troglobionts. Wind Cave in South Dakota, in glaciated North America, and with passages of nearly 250 km, has only two reported troglobionts and no stygobionts [121]. Knight [122], in an extensive review of the aquatic fauna of Swildon's Hole in the Mendip Hills of England, found no stygobionts or troglobionts. In North America, caves even 100 km south of Pleistocene glaciations in the Appalachians have depauperate fauna [123], but those in the Interior Low Plateaus of Indiana have rich fauna [124], both attributed to the effects of the Pleistocene. A similarly low level of cave obligate biodiversity is documented for the Northern Alps, where cave-restricted fauna are assumed to have been extirpated during glaciations. Würm glaciers cover most parts of these mountains, for instance, in the Swiss and Savoy Alps [125], which harbor very few cave-restricted species [126]. As a rule, all caves located in areas of the Northern Hemisphere that were glaciated or affected by permafrost during quaternary glaciations have a low number or are devoid of obligate subterranean taxa [127]. Most terrestrial cave-restricted species have not been able to recolonize following deglaciation, while it has been much easier for a significant number of aquatic interstitial species to do so, which today may be found farther north [127].

Between the Biodiversity Ridge and the Southern Limits of Glaciations in Europe. South of the limits of areas strongly affected by glaciations and the northern limit of the biodiversity ridge described by Culver et al. [15], there are vast territories where cave-restricted species may be present, but where the caves that have been sampled to some extent, are at most moderately rich. For example, the Carpathians, the second largest mountain range in Europe after the Alps, have been investigated for its cave fauna for more than a century. Its richest cave in Romania, aside from Movile [21], according to a recent published inventory [128], has only 16 cave-restricted species (3 troglobionts and 13 stygobionts). In the Jura range, northwest of the Alps, 22 cave-restricted species were reported in Grotte du Pissoir, France, as a result of intensive sampling over several years [129].

Lower Diversity Areas in the European Biodiversity Ridge. All the well-documented sites of the biodiversity ridge did not provide rich cave fauna. For example, the large cave at Predjama Castle in Slovenia is less than 10 km from Postojna Planina Cave System, a hotspot cave (Table 2), and has been the subject of inventories for over 10 years [130]. Only 11 troglobionts have been reported here, compared to 43 from Postojna Planina Cave System. Such moderately rich sites within the ridge of biodiversity are not uncommon. They may be explained by local site characteristics, such as a narrower range of habitats. More interestingly, large regions on the ridge have well investigated biodiversity spots that are only moderately rich. The most obvious is the Alpine range, which spans a large section of the ridge, even in its southwestern part that was weakly affected by glaciations. Alpine caves documented so far have a subterranean biodiversity lower than those of the west of the ridge (the Pyreneo-Cantabric range) and those of its eastern part (Dinarides). The richest subterranean biodiversity spot documented this far in the Alps is the well-studied Arena Cave in the Lessinian Mountains of Italy (24 obligate subterranean species, of which 16 are troglobionts and 8 stygobionts) [131]. None of the other caves investigated in the French Alps contains more than 20 cave-restricted species [132].

Towards the east. The European Biodiversity Ridge was recognized till the eastern Dinarids [15]. Further east, the ridge could be now extended to the Movile Cave hotspot which is located in the latitudinal range of other hotspots. The absence of documented hotspot east of Movile does not allow to extrapolate, but caves hosting 15–20 cave-restricted species are known in the Caucasus, which remains much less known than the regions included in the ridge. This suggests that, in the future, the ridge may be shown to continue much further east.

6. Discussion

6.1. The Emerging Global Pattern and Its Causes

The distribution of hotspot caves shown in Figure 1 is emphatically not one of high tropical diversity with a decline in richness towards the poles. The distribution of hotspot caves is also different in the Nearctic and the Palearctic. With the exception of sites in the Canary Islands, Palearctic hotspots are clustered along 40° N, a ridge of high subterranean biodiversity previously noted [15,133]. They point out that the ridge is the area of highest secondary productivity in Europe. Such a ridge of high biodiversity does not occur in North America, but there is a small area near the combined border of Tennessee, Alabama, and Georgia of similarly high biodiversity and presumed high secondary productivity [15,43,44]. More generally, a difference between Europe and North America is that European mountain ranges are often oriented east–west while North American mountains are north–south in orientation. Mountain range orientation has important implications for the effect of the Pleistocene and other glaciations on faunal migrations. This may partly explain why there is a high-diversity ridge in Europe but not in North America. The relationship between the Pleistocene glaciations and the distribution of cave fauna is that glaciation may be a major driver of the extinction, isolation and subsequent speciation of surface-dwelling terrestrial invertebrates in caves [134,135]. However, it is by no means certain that the Pleistocene is an important driver of either isolation or speciation. Based on molecular clock determinations, many subterranean lineages are considerably older than the Pleistocene (e.g., [136]).

Several types of subterranean sites are more likely to be hotspots than others. They are as follows:

- Phreatic aquifers. Relatively few aquifers have been sampled, usually in wells or springs. Five of these sites are on the hotspot list—San Marcos Artesian Well (Texas), Comal Springs (Texas), Robe River (Australia), Lez aquifer (France), Cent Fonts (France), and Baget System (France). The first three sites are also sites of chemoautotrophy, which acts to increase the resource base of subterranean communities.
- Sites with known chemoautotrophy, including Movile Cave and Walsingham Caves.
- Lava tubes. Canarian lava tubes and Australian lava tubes, which occur very close to the surface, have a high species richness once again possibly due to increased resources, including tree roots [137].

If this pattern proves to be robust, then a major determinant of cave biodiversity is available organic matter. Of course, the availability of organic matter is a complicated issue in itself, and may be dependent on details of topography (e.g., rugosity [138]), temporal distribution of rainfall, vegetation and disturbance.

On the other hand, it is not the entire explanation. The richest sites, those in the Dinaric karst, are, as far as we know, not particularly rich in organic matter relative to the rest of the world. However, the Dinaric karst has several unique features:

- It is next to the Mediterranean Sea, and the marine fauna of the Mediterranean was a source of colonists of subterranean sites, particularly during the Messinian Salinity Crisis.
- It is a region of high annual rainfall, relative to the rest of Europe. Additionally, temperatures are high for that latitude of the Dinarides. Therefore, productivity is higher.

Another exception to this pattern are tropical karsts of the humid tropics, where the best investigated caves (e.g. Niah Cave, Batu Caves and Mulu caves in Asia, Kulumuzi and Shimoni caves in Eastern Africa, or caves in Cuba, Guatemala, Venezuela in central America), particularly rich in guano, are not biodiversity hotspots for troglobionts in the traditional sense, as documented above. It can therefore be hypothesized that the nature of the organic matter available may be as important as its amount.

Some comparison among sites is possible, but first, corrections need to be made to account for differences in taxonomic coverage for different sites and in the proportion of undescribed species. The microcrustacean fauna, especially in epikarst, is often quite rich but it has only been studied in a few sites (e.g., Ojo Guareña); therefore, we eliminated all micro-crustacea (Ostracoda, Copepoda, Syncarida) for further analysis. We did likewise with those few parasites and commensals as well as several aquatic groups that have only been sporadically reported or described, including Protista, Oligochaeta and Nemertina. Finally, we eliminated Acari, which have not been described or studied in most caves, at least in the last hundred years (e.g., Mammoth Cave).

The resulting estimates (Table 2, Figure 5) are clustered into three groups. First, there are low-diversity sites, ones that primarily consist of micro-crustaceans—Ojo Guareña System (Spain) and Robe River Well 2A (Australia). Second, the great majority of caves (23 in all) range in species numbers from 22 to 40, suggesting a wide range of macroscopic stygobiotic and troglobiotic fauna. Among these 23 caves, seven have a rich aquatic and terrestrial fauna (at least 10 species in each ecological category): Coume Ouarnède and Baget Systems, Movile Cave, Towakkalak System, Mammoth Cave, Lukina Jama-Trojama

Cave System, Križna Jama. The third group includes the three richest caves—Cueva del Viento, Postojna Planina Cave System and Vjetrenica Cave System—which have 42 to 76 cave-restricted species. The Cueva del Viento from Canary Islands is exceptional in its exclusively terrestrial fauna, which is clearly as rich in troglobionts as the two world richest hotspot caves—Postojna Planina Cave System and Vjetrenica Cave System—caves that anchor in a sense the two ends of the Dinarides.

6.2. Weighting Species Value in Conservation of Subterranean Sites

Weighting the importance of a species in conservation is common, and there are several types of weighting that are particularly relevant to subterranean site conservation. The first is that obligate subterranean-dwelling species are weighted more than others. In this review, we have ignored non-obligate species for the most part, although some authors of articles focusing on individual caves have included lists of species that maintain permanent populations in subterranean sites—eutroglophiles and stygophiles. As a practical matter, lists of troglophiles and stygophiles are less readily available. For example, no list of troglophiles and stygophiles is available for the well-studied Mammoth Cave since 1968 [139]. A second widely used weighting takes into account the number of occurrences of species. For example, if each species is to have an equal weight overall, and it occurs in n sites, each occurrence is given a weight of $1/n$. This is especially important for subterranean fauna for its high levels of single-site endemism. In the eastern U.S., 211 of 467 troglobionts were single-cave endemics [140]. A related weighting is that of extent of the range. A species may be common within a very narrow range, such as many Cambalopsidae in tropical Asia [141], or widespread, such as the copepod *Acanthocyclops hispanicus* Kiefer, 1937 in Southern Europe. A fourth weighting is that of abundance. Some species are common in the sites where they are found, and others are extremely rare. In some cases, a species is known from only one or two specimens, such as the milliped *Euzkadiulus sarensis* (Mauriès, 1970) from Grotte de Sare in Pyrenees, and troglobionts of many tropical caves, such as *Eostemmiulus coecus* Mauriès, Golovatch & Geoffroy, 2010 from Hang Mo So in Vietnam. Distribution disjunction may be another weighting option, with species of disjunct geographical distribution being given a greater weight [96]. Related to this is the possibility of providing extra weighting to type localities. A fifth weighting is that of phylogenetic distinctness. The subterranean fauna is replete with examples of monotypic genera and even supra-generic taxa. Among these are Glacicavicolini from the USA, with its unique species, *Glacicavicola bathyscioides* Westcott, 1968 (Coleoptera), or the Collembola *Bessoniella procera* Deharveng & Thibaud, 1989, the only species of the subfamily Bessoniellinae, known from a few caves of a small Pyrenean massif. More generally, a proxy of phylogenetic distinctness is the number of supra-specific taxa. A sixth possible weighting is to give species with extreme morphological modifications (troglomorphy of Christiansen [53]) more weighting than a less modified species, as suggested by Gallão & Bichuette [97].

With the above weighting schemes, there is an implicit weighting of species richness for the simple reason that more species will result in higher weights for richer sites; however, there are exceptions. If only single-site endemics are given any weight, then the result is that more isolated sites are weighted higher, and richer caves in larger areas, but with low endemism, are given lower weights. The same may happen if phyletically isolated species, which are often geographically isolated, are given more weight. Valuing geographic or phyletic isolation makes sense biologically, but it might be more effective to consider species richness and this kind of weighting separately than in combination.

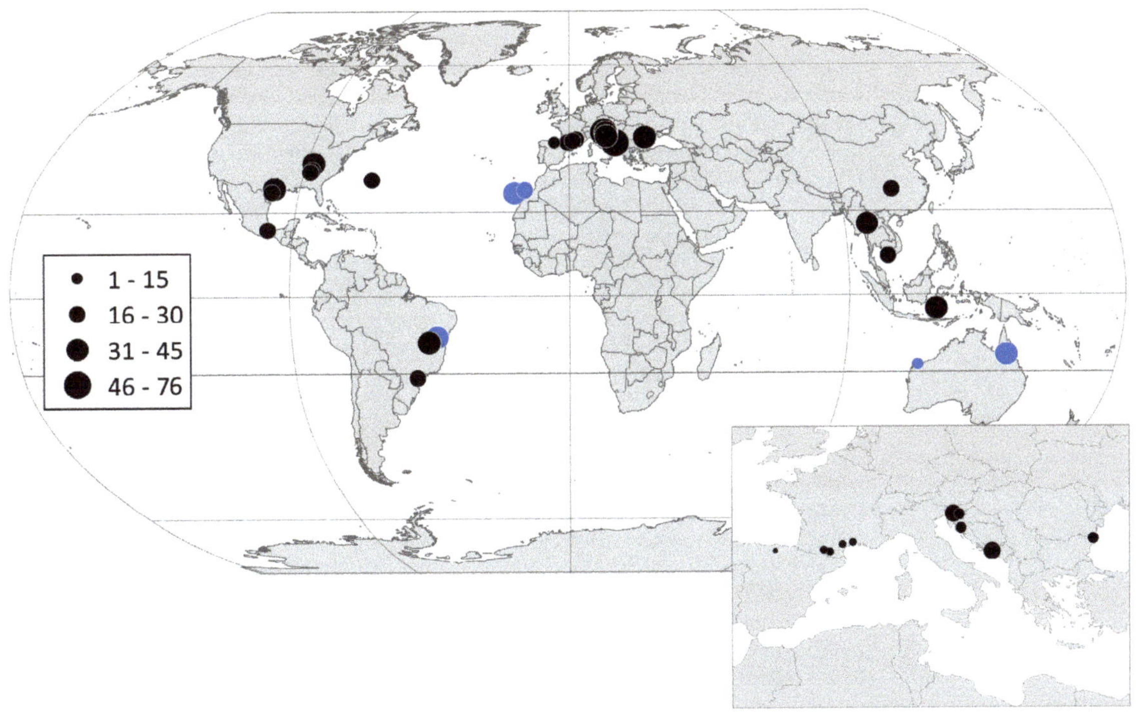

Figure 5. Global map of subterranean hotspots, where the dots are proportional to the number of species (see Table 2, column S*+T*). Horizontal lines are the Equator, ±23.5° (Tropic of Cancer and Tropic of Capricorn, and Arctic and Antarctic Circles (±66.5°). Non-karst sites are shown in blue and karst sites are shown in black. Inset map of the Mediterranean region provides greater resolution. Map courtesy of Magdalena Năpăruş-Aljančič.

Simple species count is the approach that was adopted in the different papers of the Special Issues of *Diversity*. It is classically used to evaluate site biological diversity, especially for conservation purpose, as done for instance at a large scale in Europe [142] or Brazil [143]. The different options of species weighting mentioned above would probably rank subterranean hotspots differently than the basic approach based on species count. It is obviously an important issue for site selection in a conservation perspective, that would deserve deeper investigations.

6.3. Vulnerabilities and Threats

It has been argued that subterranean organisms, as a consequence of less variable environmental conditions in the subterranean realm, are more susceptible to many types of environmental perturbation, including global warming [144], and the tone of many papers is that subterranean habitats and their fauna are delicate and vulnerable [145].

Factual evidence on the vulnerability of aquatic fauna is ambiguous and generally lacking [146], but Mammola et al. [147] provide at least a scenario for considering the effect of climate change on subterranean fauna. Subterranean fauna is certainly vulnerable to climate change, but this threat is not as immediate as some others, with regard to causing extinction in the short term. There is a growing body of evidence for short-term variations in subterranean microclimate, including daily and annual temperature cycles in caves [148], yet it is remarkable how little is known about long-term fluctuations in temperature in subterranean habitats, including changes over the last few decades [149,150].

Several authors have listed the threats and vulnerabilities of subterranean fauna [51,151]. They vary from site to site and from species to species. For example, the complete destruction of limestone hills for cement production in Vietnam is an immediate and irreversible threat to all subterranean communities, that is leading several microendemic species to extinction [45]. Hotspot sites that have been documented in these two Special Issues of *Diversity* are likewise vulnerable due their particularly rich endemic fauna, and call for vigilance. Their vulnerability is dependent on the size and isolation of karst or hydrogeological units available for the cave-restricted fauna. In this regard, tropical tower karsts with hills scattered on non-limestone terrain are especially vulnerable to local limestone exploitation [46,151]. Additionally, cave-restricted species are vulnerable to a variety of environmental perturbations, at an extent that is unknown in most cases. Danielopol & Marmonier [81] demonstrated however that some groundwater crustacean species, such as *Proasellus slavus* (Remy, 1948), are "regulators", able to maintain more or less unaltered activity independently of variable environmental conditions. More generally, cave-restricted species are clearly sensitive to desiccation, but contrary to what is sometimes stated in the literature, many seem to be able to cope with large ranges of temperatures [84].

Cave animals may be more fragile at the individual level than most surface organisms in the face of certain chemical or climatic disturbances [152]. However, they are usually protected in cave habitats from the most severe disturbances that affect surface habitats [153], including, to some extent, from climate warming. It may be the reason why cave communities subsist almost unaltered in regions where surface environment and fauna have been severely degraded by deforestation or other land-use changes. This certainly also occurred at the geological scale, and may explain why the proportion of relictual taxa of various ages is much higher in subterranean than in surface habitats, as can be easily inferred from biodiversity inventories involving cave and non-cave species [154].

6.4. Protection Strategies

There are a number of protection strategies available for subterranean habitats, the most prominent of which are site protection, acquisition by government agencies (e.g., Mammoth Cave National Park), and legislation (e.g., European Habitat Directive). Site protection of course depends on whether subsurface sites are explicitly protected or, in fact, inadvertently protected. For example, many caves in the mountains of Montana are protected since they are located in the Bob Marshall Wilderness Complex of Flathead National Forest. The Lez aquifer is protected because the aquifer is used for the water supply of the municipality of Montpelier. The overall efficacy of protection varies from site to site and country to country, and is dependent on resources committed to education and to protection enforcement. It is worth noting that several sites on the hotspot list are on government-owned land. They include the following:

- Fern Cave (National Wildlife Refuge);
- Igatu Cave System (Chapada Diamantina National Park);
- Lukina Jama–Trojama Cave System (Velebit National Park);
- Mammoth Cave (National Park);
- Ojo Guareña System (National Monument);
- Movile Cave (owned by the municipality of Mangalia);
- Tham Chiang Dao (Chiang Dao Wildlife Sanctuary);
- Towakkalak System (Bantimurung-Bulu Saraung National Park);
- Vjetrenica Cave System (owned by the municipality of Ravno).

Other forms of cave protection involve conservation through private ownership and show caves. In the United States and elsewhere, non-governmental organizations have been created for the purpose of protecting caves by acquiring cave entrances and surrounding properties. The largest such organization in the U.S., the Southeastern Cave Conservancy, owns 32 cave preserves with more than 170 caves. Show caves, such as Križna Jama, Tham Chiang Dao, and Cueva del Viento, are protected because of the commercial value of the intact cave.

Few if any of these protections are complete. Portions of the aquifer and cave may be outside the protected area, or devoted to touristic visit, and protection strategies themselves are often inadequate. Some of the most recent examples come from the strategies to protect cave-dwelling bats. A common technique for bat protection is the creation of a gate at the cave entrance to prevent human access. However, some bat species, such as *Myotis grisescens* (Howell, 1909), the gray bat, are sensitive to gates and have difficulty passing through, making them more vulnerable to predators such as snakes and owls [155]. The species *Miniopterus schreibersi* Kuhl, 1817, emblematic of conservation efforts in Europe, is also gate-sensitive. The effectiveness of gating caves for bat protection was determined to be inconclusive based on a meta-analysis of 21 case studies [156]. Gates placed externally to the cave entrance are in any case less of a deterrent to bats. Some gates also restrict the access of other small mammals, impeding the flow of organic matter into caves, and thus negatively impacting the terrestrial cave community. Rigid general rules of protection seem to be ineffective; instead, a demonstration of real risks and solutions is required.

Acknowledgments: We thank the many contributors of papers to both editions of Subterranean Biodiversity Hotspots, as well as the many reviewers. The Managing Editor and Assistant Editors coordinated efficiently editorial comments and other features related to the publication. Magdalena Năpăruş-Aljančič prepared the maps.

Conflicts of Interest: The authors declare no conflict of interest.

References

1. Wilson, E.O. (Ed.) *Biodiversity*; National Academy Press: Washington, DC, USA, 1988.
2. Olson, D.M.; Dinerstein, E.; Wikramanayake, E.D.; Burgess, N.D.; Powell, G.V.N.; Underwood, J.A.; D'Amico, J.; Itoua, I.; Strand, H.E.; Morrison, J.C.; et al. Terrestrial ecoregions of the world: A new map of life on earth. *Bioscience* **1988**, *51*, 933–938. [CrossRef]
3. Culver, D.C.; Pipan, T. *Shallow Subterranean Habitats. Ecology, Evolution, and Conservation*; Oxford University Press: Oxford, UK, 2014.
4. Gibert, J.; Deharveng, L. Subterranean ecosystems: A truncated functional diversity. *Bioscience* **2002**, *52*, 473–481. [CrossRef]
5. Malard, F.; Boutin, C.; Camacho, A.I.; Ferreira, D.; Michel, G.; Sket, B.; Stoch, F. Diversity patterns of stygobiotic crustaceans across multiple spatial scales in western Europe. *Freshw. Biol.* **2009**, *54*, 756–776. [CrossRef]
6. Culver, D.C.; Sket, B. Hotspots of subterranean biodiversity in caves and wells. *J. Cave Karst Stud.* **2000**, *62*, 11–17.
7. Culver, D.C.; Pipan, T. Subterranean ecosystems. In *Encyclopedia of Biodiversity*, 7th ed.; Levin, S.A., Ed.; Academic Press: Waltham, MA, USA, 2013; Volume 7, pp. 49–62.
8. Culver, D.C.; Pipan, T. *Biology of Caves and Other Subterranean Habitats*, 2nd ed.; Oxford University Press: Oxford, UK, 2019.
9. Culver, D.C.; Deharveng, L.; Pipan, T.; Bedos, A. An overview of subterranean biodiversity hotspots. *Diversity* **2021**, *13*, 487. [CrossRef]
10. Souza-Silva, M.; Cerqueira, R.F.V.; Pellegrini, T.G.; Ferreira, R.L. Habitat selection of cave-restricted fauna in a new hotspot of subterranean biodiversity in Neotropics. *Biodiv. Conserv.* **2021**, *30*, 4223–4250. [CrossRef]
11. Martínez, A.; Gonzalez, B.C. Volcanic Anchialine Habitats of Lanzarote. In *Cave Ecology*; Moldovan, O.T., Kováč, L., Halse, S., Eds.; Springer Nature: Cham, Switzerland, 2018; pp. 399–414.
12. Prié, V.; Alonzo, C.; Bou, C.; Galassi, D.M.P.; Marmonier, P.; Dole-Olivier, M.-J. The Cent Fonts aquifer: An overlooked subterranean biodiversity hotspot in a stygobiont-rich region. *Diversity* **2024**, *16*, 50. [CrossRef]
13. Danielopol, D.L.; Pospisil, P. Hidden biodiversity in the groundwater of the Danube Flood Plain National Park, Austria. *Biodiv. Conserv.* **2001**, *10*, 1711–1721. [CrossRef]
14. Dole, M.-J. La domaine aquatique souterrain de la plaine alluviale du Rhône à l'est de Lyon. I Diversité hydrologique et biocénotique de trois stations representatives de la dynamique fluviale. *Vie Et Milieu* **1984**, *33*, 219–229.
15. Culver, D.C.; Deharveng, L.; Bedos, A.; Lewis, J.J.; Madden, M.; Reddell, J.R.; Sket, B.; Trontelj, P.; White, D. The mid-latitude biodiversity ridge in terrestrial cave fauna. *Ecography* **2006**, *29*, 120–128. [CrossRef]
16. Klimchouk, A. Types and settings of hypogene karst. In *Hypogene Karst Regions and Caves of the World*; Klimchouk, A., Palmer, A.N., De Waele, J., Auler, A.S., Audra, P., Eds.; Springer Nature: Cham, Switzerland, 2017; pp. 1–42.
17. De Waele, J.; Gutiérrez, F. *Karst Hydrogeology, Geomorphology, and Caves*; Wiley Blackwell: Hoboken, NJ, USA, 2022.
18. Jones, W.K. Water tracing in karst aquifers. In *Encyclopedia of Caves*, 3rd ed.; White, W.B., Culver, D.C., Pipan, T., Eds.; Elsevier: London, UK, 2019; pp. 1144–1155.
19. Rouch, R. Considérations sur l'écosystème karstique. *Compte Rendu Acad. Sci.* **1977**, *284*, 1101–1103.
20. Doctor, D.H.; Orndorff, W. Hypogene caves of the central Appalachian Shenandoah Valley. In *Hypogene Karst Regions and Caves of the World*; Klimchouk, A., Palmer, A.N., De Waele, J., Auler, A.S., Audra, P., Eds.; Springer Nature: Cham, Switzerland, 2017; pp. 691–708.

21. Brad, T.; Iepure, S.; Sarbu, S.M. The Chemoautotrophically Based Movile Cave Groundwater Ecosystem, a Hotspot of Subterranean Biodiversity. *Diversity* **2021**, *13*, 128, Erratum in *Diversity* **2021**, *13*, 461. [CrossRef]

22. Iliffe, T.M.; Calderón-Gutiérrez, F. Bermuda's Walsingham Caves: A global hotspot for anchialine stygobionts. *Diversity* **2021**, *13*, 352. [CrossRef]

23. Clark, H.L.; Buzatto, B.A.; Halse, S.A. A hotspot of arid zone subterranean biodiversity: The Robe Valley in Western Australia. *Diversity* **2021**, *13*, 482. [CrossRef]

24. Sket, B. The nature of biodiversity in subterranean waters and how it is endangered. *Biodiv. Conserv.* **1999**, *8*, 1319–1338. [CrossRef]

25. Eberhard, S.M.; Howarth, F.G. Undara Lava Cave Fauna in Tropical Queensland with an Annotated List of Australian Subterranean Biodiversity Hotspots. *Diversity* **2021**, *13*, 326. [CrossRef]

26. Delić, T.; Pipan, T.; Ozimec, R.; Culver, D.C.; Zagmajster, M. The subterranean species of the Vjetrenica Cave System in Bosnia and Hercegovina. *Diversity* **2023**, *15*, 912. [CrossRef]

27. Ferreira, R.L.; Berbert-Born, M.; Souza-Silva, M. The Água Clara Cave System in northeastern Brazil: The richest hotspot of subterranean biodiversity in South America. *Diversity* **2023**, *15*, 761. [CrossRef]

28. Ferreira, R.L.; Souza-Silva, M. Beyond expectations: Recent discovery of new cave-restricted species elevates Água Clara Cave System to the richest hotspot of subterranean biodiversity in the Neotropics. *Diversity* **2023**, *15*, 1215. [CrossRef]

29. Gallão, J.E.; Ribeiro, D.B.; Bichuette, M.E. There and back again, the siliciclastic Igatu hotspot caves: Expanding the data for the subterranean fauna in Brazil, Chapada Diamantina region. *Diversity* **2023**, *15*, 991. [CrossRef]

30. Huang, S.; Zhao, M.; Luo, X.; Bedos, A.; Wang, Y.; Chocat, M.; Tian, M.; Liu, W. Feihu Dong, a new hotspot cave of subterranean biodiversity from China. *Diversity* **2023**, *15*, 902. [CrossRef]

31. Oromí, P.; Socorro, S. Biodiversity in the Cueva del Viento lava tube system (Tenerife, Canary Islands). *Diversity* **2021**, *13*, 226. [CrossRef]

32. Camacho, A.I.; Puch, C. Ojo Guareña, a hotspot of subterranean biodiversity in Spain. *Diversity* **2021**, *13*, 199. [CrossRef]

33. Bréhier, F.; Defave, D.; Faille, A.; Bedos, A. The Baget-karstic system and the interstitial environment of Lachein, a hotspot of subterranean biodiversity in the Pyrenees (France). *Diversity* **2024**, *16*, 62. [CrossRef]

34. Faille, A.; Deharveng, L. The Coume Ouarnède system, a hotspot of subterranean biodiversity in Pyrenees (France). *Diversity* **2021**, *13*, 419. [CrossRef]

35. Jourde, H.; Dörfliger, N.; Maréchal, J.-C.; Batiot-Guilhe, C.; Bouvier, C.; Courrioux, G.; Desprats, J.-F.; Fullgraf, T.; Ladouche, B.; Léonardi, V.; et al. *Project Gestion Multi-Usages de l'Hydrosystème Karstique du Lez—Synthèse des Connaissances Récentes et Passées—Rapport Final*; BRGM, Hydrosciences Montpellier, G-eau, Agence de l'eau Rhône-Méditerranée-Corse: Montpellier, France, 2011.

36. Lukić, M.; Fišer, C.; Delić, T.; Bilandžija, H.; Pavlek, M.; Komerički, A.; Dražina, T.; Jalžić, B.; Ozimec, R.; Slapnik, R.; et al. Subterranean fauna of the Lukina Jama-Trojama cave system in Croatia: The deepest cave of the Dinaric karst. *Diversity* **2023**, *15*, 726. [CrossRef]

37. Deharveng, L.; Rahmadi, C.; Suhardjono, Y.R.; Bedos, A. The Towakkalak System, a hotspot of subterranean biodiversity in Sulawesi, Indonesia. *Diversity* **2021**, *13*, 199. [CrossRef]

38. Francke, O.F.; Monjaraz-Ruedas, R.; Cruz-López, J. Biodiversity of the Huautla Cave System, Oaxaca, Mexico. *Diversity* **2021**, *13*, 429. [CrossRef]

39. Polak, S.; Pipan, T. The subterranean fauna of Križna jama, Slovenia. *Diversity* **2021**, *13*, 210. [CrossRef]

40. Zagmajster, M.; Polak, S.; Fišer, C. Postojna Planina Cave System in Slovenia, a hotspot of subterranean biodiversity and a cradle of speleobiology. *Diversity* **2021**, *13*, 271. [CrossRef]

41. Deharveng, L.; Ellis, M.; Bedos, A.; Jantarit, S. Tham Chiang Dao: A hotspot of subterranean biodiversity in Northern Thailand. *Diversity* **2023**, *15*, 1076. [CrossRef]

42. Hutchins, B.T.; Gibson, J.R.; Diaz, P.H.; Schwartz, B.F. Stygobiont diversity in the San Marcos Artesian Well and Edwards Aquifer groundwater ecosystem, Texas, USA. *Diversity* **2021**, *13*, 234. [CrossRef]

43. Niemiller, M.L.; Zigler, K.S.; Hinkle, A.; Stephen, C.D.R.; Cramphorn, B.; Higgs, J.; Mann, N.; Miller, B.T.; Kendall Niemiller, K.D.; Smallwood, K.; et al. The Crystal-Wonder Cave System: A new hotspot o In subterranean biodiversity in the southern Cumberland Plateau of south-central Tennessee. *Diversity* **2023**, *15*, 801. [CrossRef]

44. Niemiller, M.L.; Slay, M.E.; Inebnit, T.; Miller, B.; Tobin, B.; Cramphorn, B.; Hinkle, A.; Jones, B.D.; Mann, N.; Kendall Niemiller, K.D.; et al. Fern Cave: A hotspot of subterranean biodiversity in the Interior Low Plateau karst region of Alabama in the southeastern United States. *Diversity* **2023**, *15*, 633. [CrossRef]

45. Niemiller, M.L.; Helf, K.; Toomey, R.S. Mammoth Cave: A hotspot of subterranean biodiversity in the United States. *Diversity* **2021**, *13*, 199. [CrossRef]

46. Deharveng, L.; Le, C.K.; Bedos, A.; Judson, M.L.I.; Le, C.M.; Lukić, M.; Luu, H.T.; Ly, N.S.; Nguyen, T.Q.T.; Truong, Q.T.; et al. A hotspot of subterranean biodiversity on the brink: Mo So cave and the Hon Chong karst of Vietnam. *Diversity* **2023**, *15*, 1058. [CrossRef]

47. Sket, B. Can we agree on an ecological classification of subterranean animals? *J. Nat. Hist.* **2008**, *42*, 1549–1563. [CrossRef]

48. Racovitza, E.G. Essai sur les problèmes biospéologiques. *Arch. Zool. Expér. Gen.* **1907**, *4*, 371–488.

49. Barr, T.C. Cave ecology and the evolution of troglobites. *Evol. Biol.* **1968**, *2*, 35–102.

50. Trajano, E.; de Carvalho, M.R. Towards a biologically meaningful classification of subterranean organisms: A critical analysis of the Schiner-Racovitza system from a historical perspective, difficulties of its application and implications for conservation. *Subterr. Biol.* **2017**, *22*, 1–26. [CrossRef]
51. Howarth, F.G.; Wynne, J.J. Influence of the physical environment on terrestrial cave diversity. In *Cave Biodiversity. Speciation and Diversity of Subterranean Fauna of Caves*; Wynne, J.J., Ed.; Johns Hopkins University Press: Baltimore, MD, USA; Elsevier: London, UK, 2022; pp. 1–56.
52. Romero, A. *Cave Biology: Life in Darkness*; Cambridge University Press: Cambridge, UK, 2009.
53. Christiansen, K.A. Proposition pour la classification des animaux cavernicoles. *Spelunca Mémoires* **1962**, *2*, 75–78.
54. Culver, D.C.; Pipan, T. Shifting paradigms of the evolution of cave life. *Acta Carsologica* **2015**, *44*, 415–425. [CrossRef]
55. Ormelas-García, C.P.; Pedraza-Lara, C. Phylogeny and evolutionary history of *Astyanax mexicanus*. In *Biology and Evolution of the Mexican Cavefish*; Keene, A., Yoshizawa, M., McGaugh, S.E., Eds.; Elsevier; Academic Press: Amsterdam, The Netherland, 2016; pp. 77–92.
56. Wilkens, H.; Strecker, U. *Evolution in the Dark*; Springer: Berlin/Heidelberg, Germany, 2017.
57. Culver, D.C.; Pipan, T.; Fišer, Ž. Ecological and evolutionary jargon in subterranean biology. In *Groundwater Ecology and Evolution*, 2nd ed.; Malard, F., Griebler, C., Rétaux, S., Eds.; Elsevier; Academic Press: London, UK, 2019; pp. 308–319.
58. Martinez, A.; Mammola, S. Specialized terminology reduces the number of citations of scientific papers. *Proc. Royal Soc. Ser. B* **2021**, *288*, 20202581. [CrossRef]
59. Deharveng, L.; Bedos, A. Diversity of terrestrial invertebrates in subterranean habitats. In *Cave Ecology*; Moldovan, O.T., Kováč, L., Halse, S., Eds.; Springer Nature: Cham, Switzerland, 2018; pp. 107–172.
60. Brignoli, P.M. On some cave spiders from Papua, New Guinea. In Proceedings of the Eight International Congress of Speleology, Bowling Green, KY, USA, 18–24 July 1981; Volume 1, pp. 110–112.
61. Chapman, P. The invertebrate fauna of caves of the Serrania de San Luis, Edo. Falcon, Venezuela. *Trans. Br. Cave Res. Assoc.* **1980**, *7*, 179–199.
62. Faille, A. Beetles. In *Encyclopedia of Caves*, 3rd ed.; White, W.B., Culver, D.C., Pipan, T., Eds.; Elsevier: London, UK, 2019; pp. 102–108.
63. Lukić, M. Collembola. In *Encyclopedia of Caves*, 3rd ed.; White, W.B., Culver, D.C., Pipan, T., Eds.; Elsevier: London, UK, 2019; pp. 308–319.
64. Fišer, Z. *Niphargus*—A model system for evolution and ecology. In *Encyclopedia of Caves*, 3rd ed.; White, W.B., Culver, D.C., Pipan, T., Eds.; Elsevier: London, UK, 2019; pp. 746–755.
65. Lukić, M.; Delić, T.; Pavlek, M.; Deharveng, L.; Zagmajster, M. Distribution pattern and radiation of the European subterranean genus *Verhoeffiella* (Collembola, Entomobryidae). *Zool Scr.* **2019**, *49*, 86–100. [CrossRef]
66. Gittleson, S.M.; Hoover, R.L. Cavernicolous protozoa: Review of the literature and new studies in Mammoth Cave, Kentucky. *Ann. Speleol.* **1969**, *24*, 737–776.
67. Olivier, M.J.; Martin, D.; Bou, C.; Prié, V. Interprétation du Suivi Hydrobiologique de la Faune Stygobie, Réalisé sur le Système Karstique des Cent Fonts Lors du Pompage d'essai. In *Système karstique des Cent Fonts. Simulation de scénarios de Gestion de la Ressource*; BRGM/RP+54865-FR: Montpellier, France, 2006; pp. 235–275.
68. Pipan, T. *Epikarst—A Promising Habitat*; Založba ZRC: Ljubljana, Slovenia, 2005.
69. Pipan, T.; Culver, D.C. Regional species richness in an obligate subterranean dwelling fauna—Epikarst copepods. *J. Biogeoggraph.* **2007**, *34*, 854–861. [CrossRef]
70. Camacho, A.I.; Valdecasas, A.G.; Rodriguez, J.; Cuezva, S.; Lario, J.; Sánchez-Moral, S. Habitats constraints in epikarstic waters of an Iberian Peninsula cave system. *Int. J. Limnol.* **2006**, *42*, 127–140. [CrossRef]
71. Deharveng, L.; Bedos, A. Biodiversity in the tropics. In *Encyclopedia of Caves*, 3rd ed.; White, W.B., Culver, D.C., Pipan, T., Eds.; Elsevier: London, UK, 2019; pp. 146–162.
72. Peck, S.B. A review of the cave fauna of Canada, and the composition and ecology of the invertebrate fauna of caves and mines in Ontario. *Canad. J. Zool.* **1988**, *66*, 1197–1213. [CrossRef]
73. Weber, D.; Binder, H.; Pust, J. Germany. In *Encyclopaedia Biospeologica Tome 3*; Juberthie, C., Decu, V., Eds.; Société de Biospéologie: Moulis, France; Bucarest, Romania, 2001; pp. 693–702.
74. Dumnicka, E.; Galas, J.; Najberek, K.; Urban, J. The influence of Pleistocene glaciations on the distribution of obligate subterranean aquatic invertebrate fauna in Poland. *Zool. Anz.* **2020**, *286*, 90–99. [CrossRef]
75. Christian, E.; Spötl, C. Karst geology and cave fauna of Austria: A concise review. *Int. J. Speleol.* **2010**, *39*, 3. [CrossRef]
76. Culver, D.C.; Pipan, T.; Gottstein, S. Hypotelminorheic—A unique freshwater habitat. *Subterr. Biol.* **2006**, *4*, 1–8.
77. Gerovasileiou, V.; Nike Bianchi, C. Mediterranean marine caves: A synthesis of current knowledge. *Oceanogr. Mar. Biol. Ann. Rev.* **2021**, *59*, 1–88.
78. Iliffe, T.M.; Alvarez, F. Research in anchialine caves. In *Cave Ecology*; Moldovan, O.T., Kováč, V., Halse, S., Eds.; Springer: Cham, Switzerland, 2018; pp. 383–398.
79. Delamare Deboutteville, C. *Biologie des Eaux Souterraines Littorales et Continentales*; Hermann: Paris, France, 1960.
80. Svedmark, B. The interstitial fauna of marine sand. *Biol. Rev.* **1964**, *39*, 1–42. [CrossRef]
81. Danielopol, D.L.; Marmonier, P. Aspects of research on groundwater along the Rhône, Rhine, and Danube. *Reg. Rivers Res. Manag.* **1992**, *7*, 5–16. [CrossRef]

82. Růžička, V.; Klimeš, L. Spider (Araneae) communities of scree slopes in the Czech Republic. *J. Arachnol.* **2005**, *33*, 280–289. [CrossRef]

83. Jaume, D.; Boxshall, G.A. Life in extreme ocean environments: Anchialine caves. In *Marine Ecology*; UNESCO-EOLSS: Oxford, UK, 2009; pp. 230–251.

84. Vandel, A. *Biospeleology: The Biology of Cavernicolous Animals*; Freeman, B.F., Translator; Pergamon Press: New York, NY, USA, 1965.

85. Prevorčnik, S.; Remškar, A.; Fišer, C.; Sket, B.; Bračko, G.; Delić, T.; Mori, N.; Brancelj, N.; Zagmajster, M. Interstitial fauna of the Sava River in eastern Slovenia. *Natura Sloveniae* **2019**, *21*, 13–23. [CrossRef]

86. Ward, J.V.; Stanford, J.A.; Voelz, N.J. Spatial distribution patterns of Crustacea in the floodplain aquifer of an alluvial river. *Hydrobiologia* **1994**, *287*, 11–17. [CrossRef]

87. Juberthie, C.; Delay, B.; Bouillon, M. Extension du milieu souterrain en zone non calcaire: Description d'un nouveau milieu et de son peuplement par les Coléoptères troglobies. *Mém. Biospéol.* **1980**, *7*, 19–52.

88. Jiménez-Valverde, A.; Gilgado, J.D.; Sendra, A.; Pérez-Suárez, G.; Herrero-Borgoñón, J.J.; Ortuño, V.M. Exceptional invertebrate diversity in a scree slope in Eastern Spain. *J. Insect Conserv.* **2015**, *19*, 713–728. [CrossRef]

89. Jureková, N.; Raschmanová, N.; Miklisová, D.; Kováč, L. Mesofauna at the soil-scree interface in a deep karst environment. *Diversity* **2021**, *13*, 242. [CrossRef]

90. Gouze, A. Données thermiques sur le milieu souterrain superficiel et les horizons du sol sus-jacents. *Mém. Biospéol.* **1988**, *15*, 175–187.

91. Pipan, T.; Culver, D.C.; Papi, F.; Kozel, P. Partitioning diversity in subterranean invertebrates: The epikarst fauna of Slovenia. *PLoS ONE* **2018**, *13*, e0185991. [CrossRef]

92. Pipan, T.; Culver, D.C. Epikarst: An important aquatic and terrestrial habitat. In *Encyclopedia of Inland Waters*, 2nd ed.; Mehner, T., Tockner, K., Eds.; Elsevier: Amsterdam, The Netherland, 2022; pp. 437–438.

93. Culver, D.C.; Holsinger, J.R.; Feller, D.J. The fauna of seepage springs and other shallow subterranean habitats in the mid-Atlantic Piedmont and Coastal Plain, U.S.A. *Northeast. Nat.* **2012**, *19*, 1–42. [CrossRef]

94. Burch, E.; Culver, D.C.; Alonzo, M.; Malloy, E.J. Landscape features and forest maturity promote the occurrence of macroinvertebrates specialized for seepage springs in urban forests in Washington, DC. *Aquat. Conserv. Mar. Freshw. Ecosyst.* **2022**, *32*, 922–929. [CrossRef]

95. Howarth, F.G. Why the delay in recognizing terrestrial obligate cave species in the tropics? *Int. J. Speleol.* **2023**, *52*, 23–43. [CrossRef]

96. Souza-Silva, M.; Ferreira, R.L. The first two hotspots of subterranean biodiversity in South America. *Subterr. Biol.* **2016**, *19*, 1–21. [CrossRef]

97. Gallão, J.E.; Bichuette, M.E. Taxonomic distinctness and conservation of a new high biodiversity subterranean area in Brazil. *An. Acad. Bras. Cienc.* **2015**, *87*, 209–217. [CrossRef] [PubMed]

98. Beron, P. Comparative study of the invertebrate cave faunas of Southeast Asia and New Guinea. *Hist. Nat. Bulg.* **2015**, *21*, 169–210.

99. Deharveng, L.; Stoch, F.; Gibert, J.; Bedos, A.; Galassi, D.; Zagmajster, M.; Brancelj, A.; Camacho, A.; Fiers, F.; Martin, P.; et al. Groundwater biodiversity in Europe. *Freshw. Biol.* **2009**, *54*, 709–728. [CrossRef]

100. Zagmajster, M.; Culver, D.C.; Christman, M.C.; Sket, B. Evaluating the sampling bias in pattern of subterranean species richness—Combining approaches. *Biodiv. Conserv.* **2010**, *19*, 3035–3048. [CrossRef]

101. Tian, M.; Huang, S.; Wang, X.; Tang, M. Contributions to the knowledge of subterranean trechine beetles in southern China's karsts: Five new genera (Insecta, Coleoptera, Carabidae, Trechinae). *ZooKeys* **2016**, *564*, 121. [CrossRef]

102. Jeannel, R.; Racovitza, E. Biospeologica XXXIX—Enumérations des grottes visitées 1913-1917 (sixième série). *Arch. Zool. Exp. Gén.* **1918**, *57*, 203–470.

103. Juberthie, C.; Decu, V. Africa, généralités. In *Encyclopaedia Biospeologica Tome 3*; Juberthie, C., Decu, V., Eds.; Société de Biospéologie: Moulis, France; Bucarest, Romania, 2001; pp. 1461–1462.

104. Messana, G.; Chelazzi, L.; Baccetti, N. Biospeleology of Somalia. Mugdile and Showli Berdi Caves. *Monitore Zoologico Italiano Supplemento* **1985**, *20*, 325–340. [CrossRef]

105. Ferreira, R.L.; Giribet, G.; Du Preez, G.; Ventouras, O.; Janion, C.; Silva, M.S. The Wynberg cave system, the most important site for cave fauna in South Africa at risk. *Subterr. Biol.* **2010**, *36*, 73–81. [CrossRef]

106. Juberthie, C.; Decu, V. Tanzanie. In *Encyclopaedia Biospeologica Tome 3*; Juberthie, C., Decu, V., Eds.; Société de Biospéologie: Moulis, France; Bucarest, Romania, 2001; pp. 1703–1712.

107. Moseley, M.; Lim, T.W.; Lim, T.T. Fauna reported from Batu Caves, Selangor, Malaysia: Annotated checklist and bibliography. *Cave Karst Sci.* **2012**, *39*, 77–92.

108. Chapman, P. The invertebrate fauna of the caves of Gunung Mulu National Park. *Sarawak Mus. J.* **1984**, *30*, 1–8.

109. Harries, D.B.; Kharkongor, I.J.; Saikia, U. The biota of Siju Cave, Meghalaya, India: A comparison of biological records from 1922 and from 2019. *Cave Karst Sci.* **2020**, *47*, 119–130.

110. Howarth, F.G. Cavernicoles in lava tubes in the island of Hawaii. *Science* **1972**, *175*, 325–326. [CrossRef] [PubMed]

111. Wessel, A.; Hoch, H.; Asche, M.; von Rintelen, T.; Stelbrink, B.; Heck, U.; Stone, F.D.; Howarth, F.G. Founder effects initiated rapid species radiation in Hawaiian cave planthoppers. *Proc. Nat. Acad. Sci. USA* **2013**, *110*, 9391–9396. [CrossRef] [PubMed]

112. Oromí, P. Researches in lava tubes. In *Cave Ecology*; Moldovan, O.T., Kováč, Ľ., Halse, S., Eds.; Springer Nature: Cham, Switzerland, 2018; pp. 369–382.
113. Deharveng, L.; Lips, J.; Rahmadi, C. Focus on guano. In *The Natural History of Santo: Caves and Soils*; Bouchet, P., Le Guyader, H., Pascal, O., Eds.; Patrimoines Naturels; Muséum National d'Histoire Naturelle: Paris, France; IRD: Marseille, France; Pro-Natura International: Paris, France, 2011; Volume 70, pp. 300–305.
114. Decu, V. Bat guano ecosystem a new classifications and some considerations with special references to Neotropical data. *Res. Exp. Biol.* **1981**, *1981*, 9–15.
115. Smith, G.B. Biospeleology. In *Report of the 1978 Speleological Expedition to the Atea Kananda, Southern Highlands, Papua New Guinea*; James, J.M., Dyson, H.J., Eds.; Speleological Research Council: Sydney, Australia, 1980; pp. 121–129.
116. Chapman, P. Speleobiology. *Trans. Brit. Cave Res. Assoc.* **1975**, *3*, 192–203.
117. Beron, P. Interim zoological results from the British Speleological Expedition to Papua New Guinea, 1975. *Cave Karst Sci.* **2018**, *45*, 77–82.
118. Martin, P.; De Broyer, C.; Fiers, F.; Michel, G.; Sablon, R.; Wouters, K. Biodiversity of Belgian groundwater fauna in relation to environmental conditions. *Freshw. Biol.* **2009**, *54*, 814–829. [CrossRef]
119. Kocot-Zalewska, J.; Domagała, P.J. Terrestrial invertebrate fauna of Polish caves–A summary of 100 years of research. *Subterr. Biol.* **2020**, *33*, 45–69. [CrossRef]
120. May, B.M.; Decu, V.; Juberthie, C. New Zealand. In *Encyclopaedia Biospeologica Tome 3*; Juberthie, C., Decu, V., Eds.; Société de Biospéologie: Moulis, France; Bucarest, Romania, 2001; pp. 2079–2091.
121. Culver, D.C.; Christman, M.C.; Elliott, W.R.; Hobbs, H.H., III; Reddell, J.R. The North American obligate cave fauna: Regional patterns. *Biodiv. Conserv.* **2003**, *12*, 441–468. [CrossRef]
122. Knight, L.R.F.D. The aquatic macro-invertebrate fauna of Swildon's Hole, Mendip Hills, Somerset, UK. *Cave Karst Sci.* **2011**, *38*, 81–92.
123. Holsinger, J.R. The cave fauna of Pennsylvania. In *Geology and Biology of Pennsylvania Caves*; White, W.B., Ed.; Pennsylvania Geological Survey: Harrisburg, PA, USA, 1976; pp. 72–87.
124. Barr, T.C. A synopsis of cave beetles of the genus *Pseudanophthalmus* of the Mitchell Plain in southern Indiana (Coleoptera, Carabidae). *Ame. Midland Nat.* **1960**, *63*, 307–320. [CrossRef]
125. Coutterand, S.; Schoeneich, P.; Nicoud, G. Le Lobe Glaciaire Lyonnais au Maximum Würmien: Glacier du Rhône ou/et Glaciers Savoyard? HAL, 2011. pp. 11–22. Available online: https://shs.hal.science/halshs-00389085 (accessed on 15 December 2023).
126. Strinati, P. Faune cavernicole de la Suisse. *Ann. Spéléol.* **1966**, *21*, 5–268 and 357–371.
127. Assmann, T.; Casale, A.; Drees, C.; Habel, J.C.; Matern, A.; Schuldt, A. The dark side of relict species biology: Cave animals as ancient lineages. In *Relict Species: Phylogeography and Conservation Biology*; Springer: Berlin/Heidelberg, Germany, 2010; pp. 91–103.
128. Moldovan, O.T.; Iepure, S.; Brad, T.; Kenesz, M.; Mirea, I.C.; Năstase-Bucur, R. Database of Romanian cave invertebrates with a Red List of cave species and a list of hotspot/coldspot caves. *Biodivers. Data J.* **2020**, *8*, e53571. [CrossRef]
129. Gibert, J. Ecologie d'un système karstique jurassien. Hydrogéologie, dérive animale, transits de matières, dynamique de la population de *Niphargus* (Crustacé Amphipode). *Mém. Biospéologie* **1986**, *13*, 1–379.
130. Šebela, S.; Kogovšek, B.; Kozel, P.; Mulec, J.; Petrič, M.; Pipan, T. *Report of ZRK ZRC SAZU on the Performance of the Concession in the Postojna and Predjama Cave Systems in 2022: Climate Monitoring, Biological Monitoring, Expert Supervision, Water Monitoring and Inventory of Research*; Znanstvenoraziskovalni Center SAZU, Inštitut za Raziskovanje Krasa: Postojna, Slovenia, 2023; p. 85.
131. Latella, L. Does size matter? Two subterranean biodiversity hotspots in the Lessini Mountains in the Veneto Prealps in Northern Italy. *Diversity* **2024**, *16*, 25. [CrossRef]
132. Juberthie, C.; Ginet, R. France. In *Encyclopaedia Biospeologica Tome 1*; Juberthie, C., Decu, V., Eds.; Société de Biospéologie: Moulis, France; Bucarest, Romania, 1994; pp. 665–692.
133. Zagmajster, M.; Eme, D.; Fišer, C.; Galassi, D.; Marmonier, P.; Stoch, F.; Cornu, J.F.; Malard, F. Geographic variation in range size and beta diversity of groundwater crustaceans: Insights from habitats with low thermal seasonality. *Glob. Ecol. Biogeog.* **2014**, *10*, 1135–1145. [CrossRef]
134. Jeannel, R. *Les Fossiles Vivants des Cavernes*; Gallimard: Paris, France, 1943.
135. Barr, T.C. Observations on the ecology of caves. *Amer. Natur.* **1967**, *101*, 475–491. [CrossRef]
136. Saclier, N.; Duchemin, L.; Konecny-Dupré, L.; Grison, P.; Eme, D.; Martin, C.; Callou, C.; Malard, F. A collaborative backbone resource for comparative studies of subterranean evolution: The World Asellidae database. *Mol. Ecol. Resour.* **2024**, *24*, e13882. [CrossRef] [PubMed]
137. Howarth, F.G.; James, S.A.; McDowell, W.; Preston, D.J.; Imada, C.T. Identification of roots in lava tube caves using molecular techniques: Implications for conservation of cave arthropod faunas. *J. Insect Conserv.* **2007**, *11*, 251–261. [CrossRef]
138. Christman, M.C.; Doctor, D.H.; Niemiller, M.L.; Weary, D.J.; Young, J.A.; Zigler, K.S.; Culver, D.C. Predicting the occurrence of cave-inhabiting fauna based on features of the Earth surface environment. *PLoS ONE* **2016**, *11*, e0160408. [CrossRef] [PubMed]
139. Barr, T.C. Ecological studies in the Mammoth Cave system of Kentucky. I. The biota. *Int. J. Speleol.* **1968**, *3*, 147–204. [CrossRef]
140. Christman, M.C.; Culver, D.C.; Madden, M.; White, D. Patterns of endemism of the eastern North American cave fauna. *J. Biogeogr.* **2005**, *32*, 1441–1452. [CrossRef]

141. Golovatch, S.I.; Geoffroy, J.J.; Mauriès, J.P.; Vanden Spiegel, D. New species of the millipede genus *Glyphiulus* Gervais, 1847 from the *granulatus*-group (Diplopoda: Spirostreptida: Cambalopsidae). *Arthropoda Selecta* **2011**, *20*, 65–114. [CrossRef]
142. Michel, G.; Malard, F.; Deharveng, L.; Di Lorenzo, T.; Sket, B.; De Broyer, C. Reserve selection for conserving groundwater biodiversity. *Freshw. Biol.* **2009**, *54*, 861–876. [CrossRef]
143. Souza Silva, M.; Martins, R.P.; Ferreira, R.L. Cave conservation priority index to adopt a rapid protection strategy: A case study in Brazilian Atlantic rain forest. *Environ. Manag.* **2015**, *55*, 279–295. [CrossRef] [PubMed]
144. Mammola, S.; Piano, E.; Cardoso, P.; Vernon, P.; Domìnguez-Viller, D.; Culver, D.C.; Pipan, T.; Isaia, M. Climate change going deep: The effects of global climatic alterations on cave ecosystems. *Anthr. Rev.* **2019**, *6*, 98–116. [CrossRef]
145. Fong, D.W. Management of subterranean fauna in karst. In *Karst Management*; van Beynen, P.E., Ed.; Springer: Cham, Switzerland, 2011; pp. 201–224.
146. Di Lorenzo, T.; Avramov, M.; Galassi, D.M.P.; Iepure, S.; Mammola, S.; Reboleira, A.S.P.S.; Hervant, F. Physiological tolerance and ecotoxicological constraints on groundwater fauna. In *Groundwater Ecology and Evolution*, 2nd ed.; Malard, F., Griebler, D., Retaux, S., Eds.; Elsevier; Academic Press: London, UK, 2023; pp. 457–482.
147. Mammola, S.; Goodacre, S.L.; Isaia, M. Climate change may drive cave spiders to extinction. *Ecography* **2018**, *41*, 233–243. [CrossRef]
148. Medina, M.J.; Antić, D.; Borges, P.A.; Borko, Š.; Fišer, C.; Lauritzen, S.E.; Martín, J.L.; Oromí, P.; Pavlek, M.; Premate, E.; et al. Temperature variation in caves and its significance for subterranean ecosystems. *Sci. Rep.* **2023**, *13*, 20735. [CrossRef] [PubMed]
149. Šebela, S.; Turk, J.; Pipan, T. Cave micro-climate and tourism: Towards 200 years (1819–2015) at Postojnska jama (Slovenia). *Cave Karst Sci.* **2015**, *42*, 78–85.
150. Šebela, S. *Natural and Anthropogenic Impacts on Cave Climates: Postojna and Predjama Show Caves (Slovenia)*; Elsevier: London, UK, 2021.
151. Vermeulen, J.; Whitten, T. *Biodiversity and Cultural Property in the Management of Limestone Resources*; World Bank: Washington, DC, USA, 1999.
152. Thibaud, J.M. *Biologie et Écologie des Collemboles Hypogastruridae Édaphiques et Cavernicoles*; Éditions du Muséum: Paris, France, 1970.
153. Wynne, J.J.; Bernard, E.C.; Howarth, F.G.; Sommer, S.; Soto-Adames, F.N.; Taiti, S.; Mockford, E.L.; Horrocks, M.; Pakarati, L.; Pakarti-Hotus, V. Disturbance relicts in a rapidly changing world: The Rapa Nui (Easter Island) factor. *Bioscience* **2014**, *64*, 711–718. [CrossRef]
154. Deharveng, L.; Bedos, A. The cave fauna of southeast Asia. Origin, evolution, and ecology. In *Subterranean Ecosystems*; Wilkens, H., Culver, D.C., Humphreys, W.F., Eds.; Elsevier: Amsterdam, The Netherland, 2000; pp. 603–632.
155. Spanjer, G.R.; Fenton, M.B. Behavioral responses of bats to gates at caves and mines. *Wildl. Soc. Bull.* **2005**, *33*, 1101–1112. [CrossRef]
156. Meierhofer, M.B.; Johnson, J.S.; Perez-Jimenez, J.; Ito, F.; Webela, P.W.; Wiantoro, S.; Bernard, E.; Tanalgo, K.C.; Hughes, A.; Cardoso, P.; et al. Effective conservation of subterranean-roosting bats. *Conserv. Biol.* **2023**, *38*, e14157. [CrossRef]

MDPI

St. Alban-Anlage 66

4052 Basel

Switzerland

www.mdpi.com

Diversity Editorial Office

E-mail: diversity@mdpi.com

www.mdpi.com/journal/diversity